Advance Praise for *Head First Statistics*

"Head First Statistics is by far the most entertaining, attention-catching study guide on the market. By presenting the material in an engaging manner, it provides students with a comfortab[illegible] to learn an otherwise cumbersome subject. The explanation of the topics is presented in a mar[illegible] to students of all levels."

— Ariana Anderson, Teaching Fellow/PhD candidate in Statis[illegible]

"Head First is an intuitive way to understand statistics using simple, real-life examples that make [illegible] fun and natural."

— Michael Prerau, computational neuroscientist and statistics instructor, Boston University

"Thought Head First was just for computer nerds? Try the brain-friendly way with statistics and you'll change your mind. It really works."

— Andy Parker

"This book is a great way for students to learn statistics—it is entertaining, comprehensive, and easy to understand. A perfect solution!"

— Danielle Levitt

"Down with dull statistics books! Even my cat liked this one."

— Cary Collett

Praise for other *Head First* books

"Kathy and Bert's *Head First Java* transforms the printed page into the closest thing to a GUI you've ever seen. In a wry, hip manner, the authors make learning Java an engaging 'what're they gonna do next?' experience."

—Warren Keuffel, Software Development Magazine

"Beyond the engaging style that drags you forward from know-nothing into exalted Java warrior status, Head First Java covers a huge amount of practical matters that other texts leave as the dreaded "exercise for the reader..." It's clever, wry, hip and practical—there aren't a lot of textbooks that can make that claim and live up to it while also teaching you about object serialization and network launch protocols. "

—Dr. Dan Russell, Director of User Sciences and Experience Research
IBM Almaden Research Center (and teaches Artificial Intelligence at Stanford University)

"It's fast, irreverent, fun, and engaging. Be careful—you might actually learn something!"

—Ken Arnold, former Senior Engineer at Sun Microsystems
Co-author (with James Gosling, creator of Java), *The Java Programming Language*

"I feel like a thousand pounds of books have just been lifted off of my head."

—Ward Cunningham, inventor of the Wiki and founder of the Hillside Group

"Just the right tone for the geeked-out, casual-cool guru coder in all of us. The right reference for practical development strategies—gets my brain going without having to slog through a bunch of tired stale professor-speak."

—Travis Kalanick, Founder of Scour and Red Swoosh
Member of the MIT TR100

"There are books you buy, books you keep, books you keep on your desk, and thanks to O'Reilly and the Head First crew, there is the penultimate category, Head First books. They're the ones that are dog-eared, mangled, and carried everywhere. Head First SQL is at the top of my stack. Heck, even the PDF I have for review is tattered and torn."

— Bill Sawyer, ATG Curriculum Manager, Oracle

"This book's admirable clarity, humor and substantial doses of clever make it the sort of book that helps even non-programmers think well about problem-solving."

— Cory Doctorow, co-editor of Boing Boing
Author, *Down and Out in the Magic Kingdom*
and *Someone Comes to Town, Someone Leaves Town*

Praise for other *Head First* books

"I received the book yesterday and started to read it...and I couldn't stop. This is definitely très 'cool.' It is fun, but they cover a lot of ground and they are right to the point. I'm really impressed."

— Erich Gamma, IBM Distinguished Engineer, and co-author of *Design Patterns*

"One of the funniest and smartest books on software design I've ever read."

— Aaron LaBerge, VP Technology, ESPN.com

"What used to be a long trial and error learning process has now been reduced neatly into an engaging paperback."

— Mike Davidson, CEO, Newsvine, Inc.

"Elegant design is at the core of every chapter here, each concept conveyed with equal doses of pragmatism and wit."

— Ken Goldstein, Executive Vice President, Disney Online

"I ♥ Head First HTML with CSS & XHTML—it teaches you everything you need to learn in a 'fun coated' format."

— Sally Applin, UI Designer and Artist

"Usually when reading through a book or article on design patterns, I'd have to occasionally stick myself in the eye with something just to make sure I was paying attention. Not with this book. Odd as it may sound, this book makes learning about design patterns fun.

"While other books on design patterns are saying 'Buehler… Buehler… Buehler…' this book is on the float belting out 'Shake it up, baby!'"

— Eric Wuehler

"I literally love this book. In fact, I kissed this book in front of my wife."

— Satish Kumar

Other related books from O'Reilly

Statistics Hacks™

Statistics in a Nutshell

Mind Hacks™

Mind Performance Hacks™

Your Brain: The Missing Manual

Other books in O'Reilly's *Head First* series

Head First Java™

Head First Object-Oriented Analysis and Design (OOA&D)

Head First HTML with CSS and XHTML

Head First Design Patterns

Head First Servlets and JSP

Head First EJB

Head First PMP

Head First SQL

Head First Software Development

Head First JavaScript

Head First Ajax (2008)

Head First Physics (2008)

Head First Rails (2008)

Head First Web Design (2008)

Head First Programming (2008)

Head First PHP & MySQL (2008)

Head First Algebra (2008)

Head First Statistics

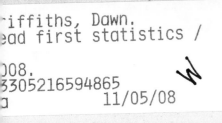

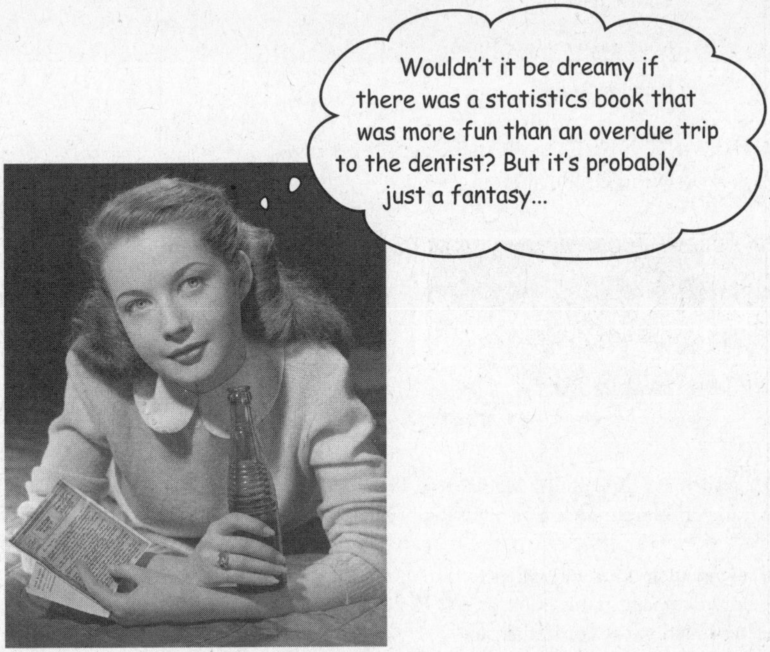

Wouldn't it be dreamy if there was a statistics book that was more fun than an overdue trip to the dentist? But it's probably just a fantasy...

Dawn Griffiths

O'REILLY®

Beijing • Cambridge • Köln • Sebastopol • Taipei • Tokyo

Head First Statistics

by Dawn Griffiths

Printed in the United States of America.

Published by O'Reilly Media, Inc., 1005 Gravenstein Highway North, Sebastopol, CA 95472.

O'Reilly Media books may be purchased for educational, business, or sales promotional use. Online editions are also available for most titles (*safari.oreilly.com*). For more information, contact our corporate/institutional sales department: (800) 998-9938 or *corporate@oreilly.com*.

Series Creators: Kathy Sierra, Bert Bates

Series Editor: Brett D. McLaughlin

Editor: Sanders Kleinfeld

Design Editor: Louise Barr

Cover Designers: Louise Barr, Steve Fehler

Production Editor: Brittany Smith

Indexer: Julie Hawks

Page Viewers: David Griffiths, Mum and Dad

Printing History:

August 2008: First Edition

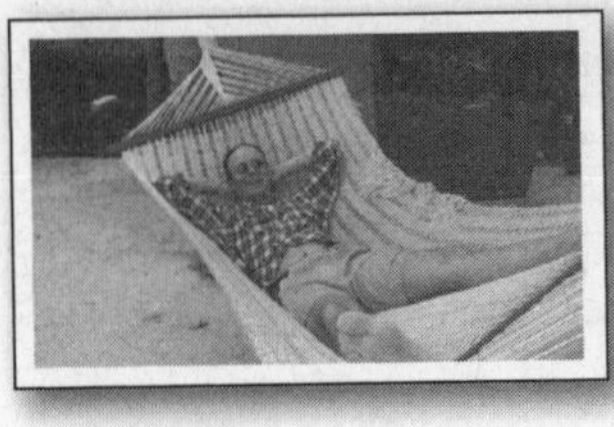

David

Mum and Dad

No snorers were harmed in the making of this book, although a horse lost its toupee at one point and suffered a minor indignity in front of the other horses. Also a snowboarder picked up a few bruises along the way, but nothing serious.

ISBN: 978-0-596-52758-7

[M]

To David, Mum, Dad, and Carl. Thanks for the support and believing I could do it. But you'll have to wait a while for the car.

Author of Head First Statistics

Dawn Griffiths started life as a mathematician at a top UK university. She was awarded a First-Class Honours degree in Mathematics, but she turned down a PhD scholarship studying particularly rare breeds of differential equations when she realized people would stop talking to her at parties. Instead she pursued a career in software development, and she currently combines IT consultancy with writing and mathematics.

When Dawn's not working on Head First books, you'll find her honing her Tai Chi skills, making bobbin lace or cooking nice meals. She hasn't yet mastered the art of doing all three at the same time. She also enjoys traveling, and spending time with her lovely husband, David.

Dawn has a theory that **Head First Bobbin Lacemaking** might prove to a be a big cult hit, but she suspects that Brett and Laurie might disagree.

Table of Contents (Summary)

Table of Contents (the real thing)

Intro

Your brain on statistics. Here *you* are trying to *learn* something, while here your *brain* is doing you a favor by making sure the learning doesn't *stick*. Your brain's thinking, "Better leave room for more important things, like which wild animals to avoid and whether naked snowboarding is a bad idea." So how *do* you trick your brain into thinking that your life depends on knowing statistics?

1

visualizing information

First Impressions

Can't tell your facts from your figures?

Statistics help you make sense of confusing sets of data. They **make the complex simple**. And when you've found out what's really going on, you need a way of **visualizing** it and **telling everyone else**. So if you want to pick the best chart for the job, grab your coat, pack your best slide rule, and join us on a ride to Statsville.

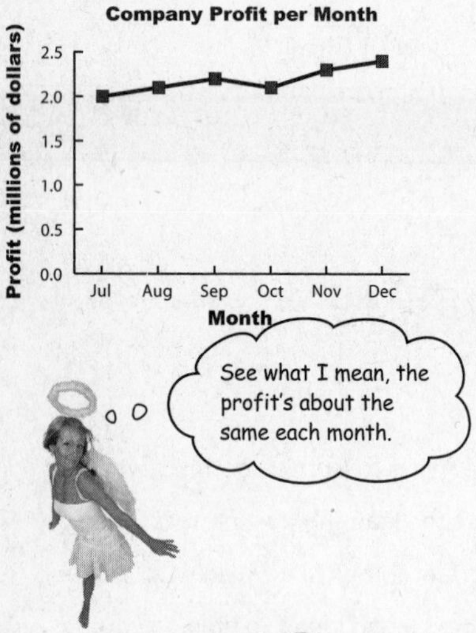

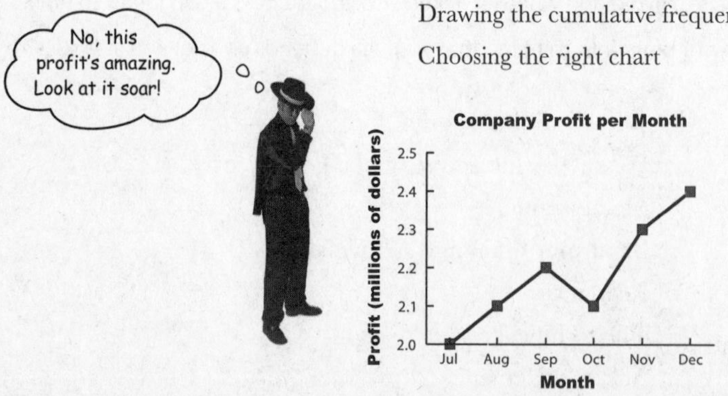

2 measuring central tendency

The Middle Way

Sometimes you just need to get to the heart of the matter.

It can be difficult to see patterns and trends in a big pile of figures, and finding the **average** is often the first step towards seeing the bigger picture. With averages at your disposal, you'll be able to quickly find the most representative values in your data and draw important conclusions. In this chapter, we'll look at several ways to calculate one of the most important statistics in town—mean, median, and mode—and you'll start to see how to effectively **summarize data** as concisely and usefully as possible.

The Health Club
Statsville's Premier Spa

Age 20

Age 20

Age 21

Age 20

Age 19

measuring variability and spread

3 Power Ranges

Not everything's reliable, but how can you tell?

Averages do a great job of giving you a typical value in your data set, but they **don't tell you the full story**. OK, so you know where the center of your data is, but often the mean, median, and mode alone aren't enough information to go on when you're summarizing a data set. In this chapter, we'll show you how to take your data skills to the next level as we begin to analyze **ranges and variation**.

05

All three players have the same average score for shooting, but I need some way of choosing between them. Think you can help?

calculating probabilities

4 Taking Chances

Life is full of uncertainty.

Sometimes it can be impossible to say what will happen from one minute to the next. But certain events are more likely to occur than others, and that's where **probability theory** comes into play. Probability lets you **predict the future** by assessing how likely outcomes are, and knowing what could happen helps you make **informed decisions**. In this chapter, you'll find out more about probability and learn how to take control of the future!

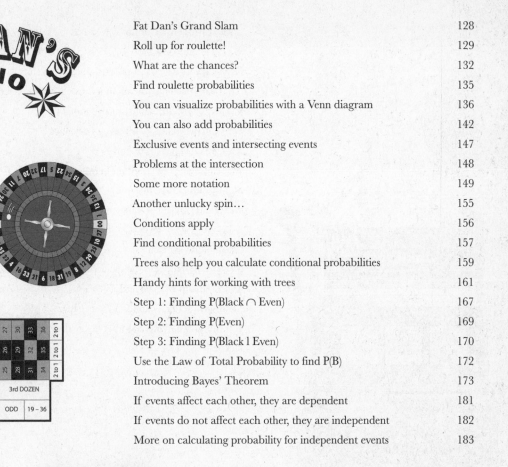

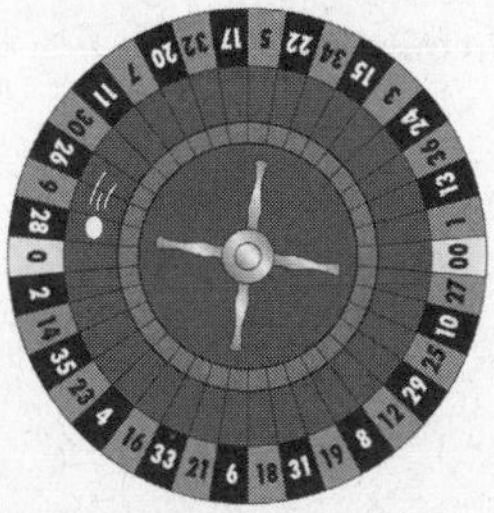

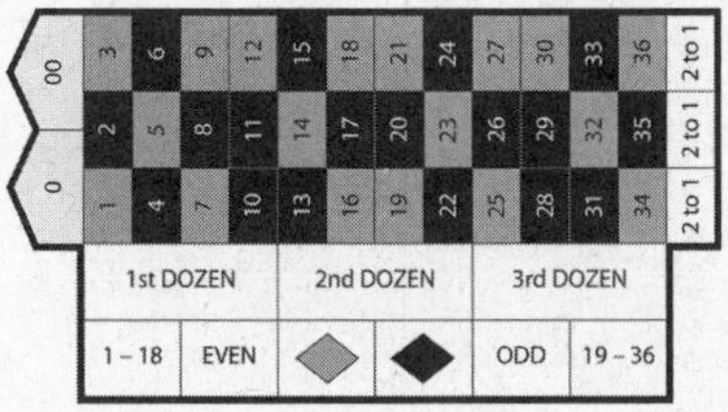

5

using discrete probability distributions

Manage Your Expectations

Unlikely events happen, but what are the consequences?

So far we've looked at how probabilities tell you how likely certain events are. What probability *doesn't* tell you is the **overall impact** of these events, and what it means to you. Sure, you'll sometimes make it big on the roulette table, but is it really worth it with all the money you lose in the meantime? In this chapter, we'll show you how you can use probability to **predict long-term outcomes**, and also **measure the certainty** of these predictions.

6 permutations and combinations

Making Arrangements

Sometimes, order is important.

Counting **all the possible ways** in which you can order things is time consuming, but the trouble is, this sort of information is **crucial** for calculating some probabilities. In this chapter, we'll show you a **quick way** of deriving this sort of information without you having to figure out what all of the possible outcomes are. Come with us and we'll show you how to **count the possibilities**.

7 geometric, binomial, and poisson distributions

Keeping Things Discrete

Calculating probability distributions takes time.

So far we've looked at how to calculate and use probability distributions, but wouldn't it be nice to have something **easier to work with**, or just **quicker to calculate**? In this chapter, we'll show you some **special probability distributions** that follow very definite patterns. Once you know these patterns, you'll be able to use them to **calculate probabilities, expectations, and variances in record time**. Read on, and we'll introduce you to the geometric, binomial and Poisson distributions.

8 using the normal distribution

Being Normal

Discrete probability distributions can't handle every situation.

So far we've looked at probability distributions where we've been able to specify exact values, but this isn't the case for every set of data. Some types of data just **don't fit** the probability distributions we've encountered so far. In this chapter, we'll take a look at how **continuous probability distributions** work, and introduce you to one of the most important probability distributions in town—the **normal distribution**.

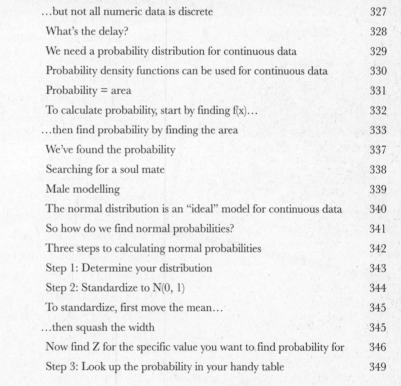

9 using the normal distribution ii

Beyond Normal

If only all probability distributions were normal.

Life can be so much ***simpler*** with the normal distribution. Why spend all your time working out individual probabilities when you can look up entire ranges in one swoop, and still leave time for game play? In this chapter, you'll see how to **solve more complex problems** in the blink of an eye, and you'll also find out how to bring some of that normal goodness to **other probability distributions**.

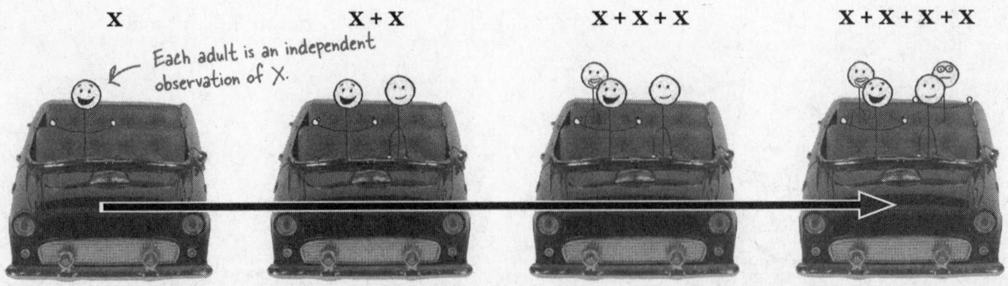

10 using statistical sampling

Taking Samples

Statistics deal with data, but where does it come from?

Some of the time, data's easy to collect, such as the ages of people attending a health club or the sales figures for a games company. But what about the times when data isn't so easy to collect? Sometimes the number of things we want to collect data about are so huge that it's difficult to know where to start. In this chapter, we'll take a look at how you can **effectively gather data** in the real world, in a way that's efficient, accurate, and can also save you time and money to boot. Welcome to the world of sampling.

estimating your population

11 Making Predictions

Wouldn't it be great if you could tell what a population was like, just by taking one sample?

Before you can claim **full sample mastery**, you need to know how to use your samples to best effect once you've collected them. This means using them to **accurately predict** what the population will be like and coming up with a way of saying how **reliable** your predictions are. In this chapter, we'll show you how knowing your sample helps you **get to know your population**, and vice versa.

This is awesome! We have a lot of impressive statistics we can use in our advertising.
10/40 people prefer pink!

constructing confidence intervals

12 Guessing with Confidence

Sometimes samples don't give quite the right result.

You've seen how you can use point estimators to estimate the **precise value** of the population mean, variance, or proportion, but the trouble is, how can you be certain that your estimate is completely accurate? After all, your assumptions about the population rely on just one sample, and what if your sample's off? In this chapter, you'll see **another way of estimating population statistics**, one that **allows for uncertainty**. Pick up your probability tables, and we'll show you the ins and outs of **confidence intervals**.

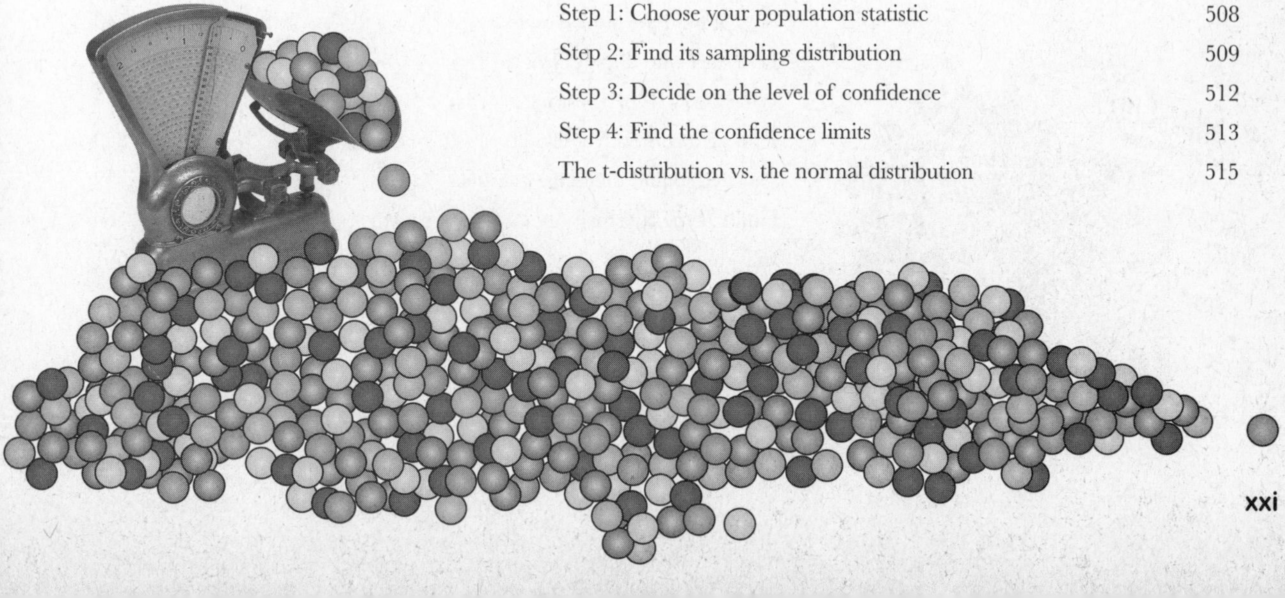

13

using hypothesis tests

Look at the Evidence

Not everything you're told is absolutely certain.

The trouble is, how do you know when what you're being told isn't right? ***Hypothesis tests*** give you a way of using samples to test whether or not statistical claims are likely to be true. They give you a way of ***weighing the evidence*** and testing whether extreme results can be explained by ***mere coincidence***, or whether there are darker forces at work. Come with us on a ride through this chapter, and we'll show you how you can use hypothesis tests to confirm or allay your deepest suspicions.

SnoreCull
SnoreCull Tablets: The Ultimate Remedy for Snoring
90% Success Rate!

the χ^2 distribution

14 There's Something Going On...

Sometimes things don't turn out quite the way you expect.

When you model a situation using a particular probability distribution, you have a good idea of how things are likely to turn out long-term. But what happens if there are differences between **what you expect and what you get?** How can you tell whether your discrepancies come down to normal fluctuations, or whether they're a sign of an underlying problem with your probability model instead? In this chapter, we'll show you how you can use the χ^2 distribution to **analyze your results** and sniff out **suspicious results**.

15

correlation and regression

What's My Line?

Have you ever wondered how two things are connected?

So far we've looked at statistics that tell you about just one variable—like men's height, points scored by basketball players, or how long gumball flavor lasts—but there are other statistics that tell you about the **connection between variables**. Seeing how things are connected can give you a lot of information about the real world, information that you can use to your advantage. Stay with us while we show you the **key to spotting connections**: correlation and regression.

Feel that funky rhythm, baby.
Sweet! But is that a rain cloud I see up there?

leftovers

i

The Top Ten Things (we didn't cover)

Even after all that, there's a bit more. There are just a few more things we think you need to know. We wouldn't feel right about ignoring them, even though they only need a brief mention. So before you put the book down, take a read through these **short but important statistics tidbits**.

statistics tables

ii

Looking Things up

Where would you be without your trusty probability tables? Understanding your probability distributions isn't quite enough. For some of them, you need to be able to **look up your probabilities** in standard **probability tables**. In this appendix you'll find tables for the **normal, t and X^2 distributions** so you can look up probabilities to your heart's content.

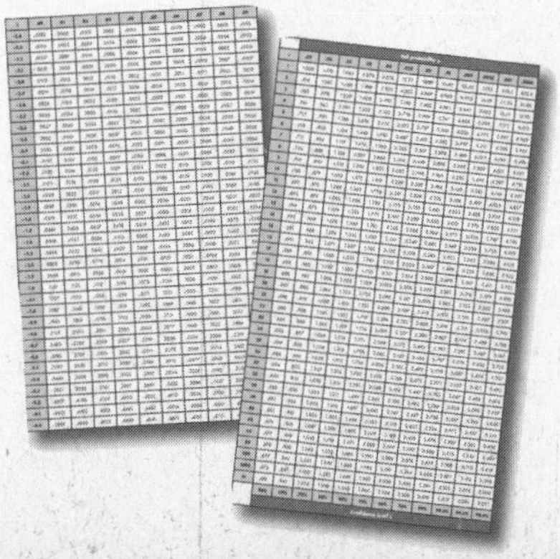

Intro

In this section we answer the burning question: "So why DID they put that in a Statistics book?"

Who is this book for?

If you can answer "yes" to all of these:

1. Do you need to understand statistics for a **course**, for your **line of work**, or just because **you think it's about time you learned what standard deviation means or how to find the probability of winning at roulette?**

2. Do you want to **learn**, **understand**, and **remember** how to **use probability and statistics to get the right results, every time**?

3. Do you prefer **stimulating dinner party conversation** to **dry**, **dull**, **academic lectures**?

this book is for you.

Who should probably back away from this book?

If you can answer "yes" to any of these:

1. **Are you someone who's never studied basic algebra?**

   (You don't need to be advanced, but you should understand basic addition and subtraction, multiplication and division.)

2. Are you a kick-butt statistician looking for a ***reference* book**?

3. Are you **afraid to try something different**? Would you rather have a root canal than mix stripes with plaid? Do you believe that a statistics book can't be serious if Venn diagrams are anthropomorphized?

this book is not for you.

[Note from marketing: this book is for anyone with a credit card.]

We know what you're thinking

"How can *this* be a serious book on statistics?"

"What's with all the graphics?"

"Can I actually *learn* it this way?"

We know what your *brain* is thinking

Your brain thinks THIS is important.

Your brain craves novelty. It's always searching, scanning, *waiting* for something unusual. It was built that way, and it helps you stay alive.

So what does your brain do with all the routine, ordinary, normal things you encounter? Everything it *can* to stop them from interfering with the brain's *real* job—recording things that *matter*. It doesn't bother saving the boring things; they never make it past the "this is obviously not important" filter.

How does your brain *know* what's important? Suppose you're out for a day hike and a tiger jumps in front of you, what happens inside your head and body?

Neurons fire. Emotions crank up. *Chemicals surge.*

And that's how your brain knows...

This must be important! Don't forget it!

But imagine you're at home, or in a library. It's a safe, warm, tiger-free zone. You're studying. Getting ready for an exam. Or trying to learn some tough technical topic your boss thinks will take a week, ten days at the most.

Great. Only 700 more dull, dry, boring pages.
Your brain thinks THIS isn't worth saving.

Just one problem. Your brain's trying to do you a big favor. It's trying to make sure that this *obviously* non-important content doesn't clutter up scarce resources. Resources that are better spent storing the really *big* things. Like tigers. Like the danger of fire. Like how you should never have posted those "party" photos on your Facebook page.

And there's no simple way to tell your brain, "Hey brain, thank you very much, but no matter how dull this book is, and how little I'm registering on the emotional Richter scale right now, I really *do* want you to keep this stuff around."

We think of a "Head First" reader as a learner.

So what does it take to *learn* something? First, you have to *get* it, then make sure you don't *forget* it. It's not about pushing facts into your head. Based on the latest research in cognitive science, neurobiology, and educational psychology, *learning* takes a lot more than text on a page. We know what turns your brain on.

Some of the Head First learning principles:

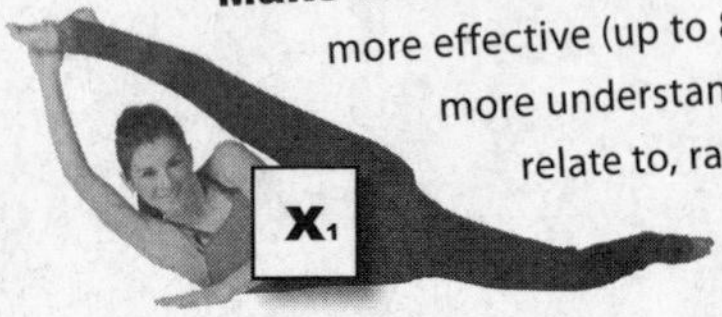

Make it visual. Images are far more memorable than words alone, and make learning much more effective (up to 89% improvement in recall and transfer studies). It also makes things more understandable. **Put the words within or near the graphics** they relate to, rather than on the bottom or on another page, and learners will be up to *twice* as likely to solve problems related to the content.

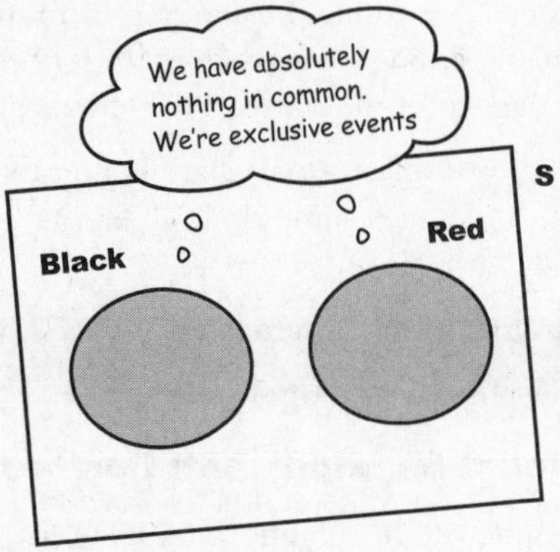

Use a conversational and personalized style. In recent studies, students performed up to 40% better on post-learning tests if the content spoke directly to the reader, using a first-person, conversational style rather than taking a formal tone. Tell stories instead of lecturing. Use casual language. Don't take yourself too seriously. Which would *you* pay more attention to: a stimulating dinner party companion, or a lecture?

Get the learner to think more deeply. In other words, unless you actively flex your neurons, nothing much happens in your head. A reader has to be motivated, engaged, curious, and inspired to solve problems, draw conclusions, and generate new knowledge. And for that, you need challenges, exercises, and thought-provoking questions, and activities that involve both sides of the brain and multiple senses.

Get—and keep—the reader's attention. We've all had the "I really want to learn this but I can't stay awake past page one" experience. Your brain pays attention to things that are out of the ordinary, interesting, strange, eye-catching, unexpected. Learning a new, tough, technical topic doesn't have to be boring. Your brain will learn much more quickly if it's not.

Touch their emotions. We now know that your ability to remember something is largely dependent on its emotional content. You remember what you care about. You remember when you *feel* something. No, we're not talking heart-wrenching stories about a boy and his dog. We're talking emotions like surprise, curiosity, fun, "what the...?", and the feeling of "I Rule!" that comes when you solve a puzzle, learn something everybody else thinks is hard, or realize you know something that "I'm more mathematically inclined than thou" Bob from class *doesn't*.

Metacognition: thinking about thinking

If you really want to learn, and you want to learn more quickly and more deeply, pay attention to how you pay attention. Think about how you think. Learn how you learn.

Most of us did not take courses on metacognition or learning theory when we were growing up. We were *expected* to learn, but rarely *taught* to learn.

But we assume that if you're holding this book, you really want to learn statistics. And you probably don't want to spend a lot of time. If you want to use what you read in this book, you need to *remember* what you read. And for that, you've got to *understand* it. To get the most from this book, or *any* book or learning experience, take responsibility for your brain. Your brain on *this* content.

The trick is to get your brain to see the new material you're learning as Really Important. Crucial to your well-being. As important as a tiger. Otherwise, you're in for a constant battle, with your brain doing its best to keep the new content from sticking.

So just how *DO* you get your brain to treat statistics like it was a hungry tiger?

There's the slow, tedious way, or the faster, more effective way. The slow way is about sheer repetition. You obviously know that you *are* able to learn and remember even the dullest of topics if you keep pounding the same thing into your brain. With enough repetition, your brain says, "This doesn't *feel* important to him, but he keeps looking at the same thing *over* and *over* and *over*, so I suppose it must be."

The faster way is to do ***anything that increases brain activity,*** especially different *types* of brain activity. The things on the previous page are a big part of the solution, and they're all things that have been proven to help your brain work in your favor. For example, studies show that putting words *within* the pictures they describe (as opposed to somewhere else in the page, like a caption or in the body text) causes your brain to try to makes sense of how the words and picture relate, and this causes more neurons to fire. More neurons firing = more chances for your brain to *get* that this is something worth paying attention to, and possibly recording.

A conversational style helps because people tend to pay more attention when they perceive that they're in a conversation, since they're expected to follow along and hold up their end. The amazing thing is, your brain doesn't necessarily *care* that the "conversation" is between you and a book! On the other hand, if the writing style is formal and dry, your brain perceives it the same way you experience being lectured to while sitting in a roomful of passive attendees. No need to stay awake.

But pictures and conversational style are just the beginning…

Here's what WE did:

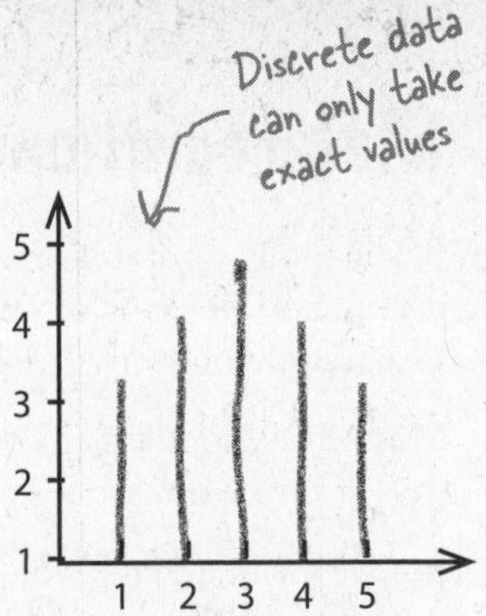

We used ***pictures***, because your brain is tuned for visuals, not text. As far as your brain's concerned, a picture really *is* worth a thousand words. And when text and pictures work together, we embedded the text *in* the pictures because your brain works more effectively when the text is *within* the thing the text refers to, as opposed to in a caption or buried in the text somewhere.

We used ***redundancy***, saying the same thing in *different* ways and with different media types, and *multiple senses*, to increase the chance that the content gets coded into more than one area of your brain.

We used concepts and pictures in ***unexpected*** ways because your brain is tuned for novelty, and we used pictures and ideas with at least *some* ***emotional*** *content*, because your brain is tuned to pay attention to the biochemistry of emotions. That which causes you to *feel* something is more likely to be remembered, even if that feeling is nothing more than a little ***humor***, ***surprise***, or ***interest.***

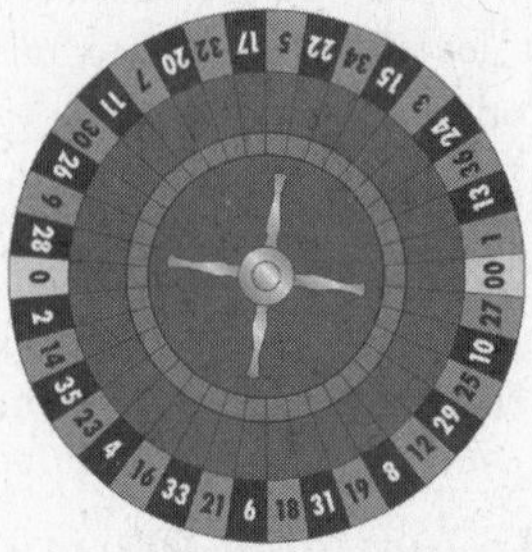

We used a personalized, ***conversational style***, because your brain is tuned to pay more attention when it believes you're in a conversation than if it thinks you're passively listening to a presentation. Your brain does this even when you're *reading.*

We included more than 80 ***activities***, because your brain is tuned to learn and remember more when you ***do*** things than when you *read* about things. And we made the exercises challenging-yet-do-able, because that's what most people prefer.

We used ***multiple learning styles***, because *you* might prefer step-by-step procedures, while someone else wants to understand the big picture first, and someone else just wants to see an example. But regardless of your own learning preference, *everyone* benefits from seeing the same content represented in multiple ways.

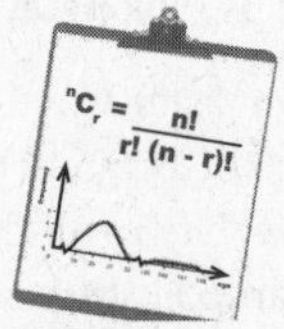

We include content for ***both sides of your brain***, because the more of your brain you engage, the more likely you are to learn and remember, and the longer you can stay focused. Since working one side of the brain often means giving the other side a chance to rest, you can be more productive at learning for a longer period of time.

And we included ***stories*** and exercises that present ***more than one point of view,*** because your brain is tuned to learn more deeply when it's forced to make evaluations and judgments.

We included ***challenges***, with exercises, and by asking ***questions*** that don't always have a straight answer, because your brain is tuned to learn and remember when it has to *work* at something. Think about it—you can't get your *body* in shape just by *watching* people at the gym. But we did our best to make sure that when you're working hard, it's on the *right* things. That ***you're not spending one extra dendrite*** processing a hard-to-understand example, or parsing difficult, jargon-laden, or overly terse text.

We used ***people***. In stories, examples, pictures, etc., because, well, because *you're* a person. And your brain pays more attention to *people* than it does to *things*.

Here's what YOU can do to bend your brain into submission

So, we did our part. The rest is up to you. These tips are a starting point; listen to your brain and figure out what works for you and what doesn't. Try new things.

Cut this out and stick it on your refrigerator.

1 Slow down. The more you understand, the less you have to memorize.

Don't just *read*. Stop and think. When the book asks you a question, don't just skip to the answer. Imagine that someone really *is* asking the question. The more deeply you force your brain to think, the better chance you have of learning and remembering.

2 Do the exercises. Write your own notes.

We put them in, but if we did them for you, that would be like having someone else do your workouts for you. And don't just *look* at the exercises. **Use a pencil.** There's plenty of evidence that physical activity *while* learning can increase the learning.

3 Read the "There are No Dumb Questions"

That means all of them. They're not optional sidebars—***they're part of the core content!*** Don't skip them.

4 Make this the last thing you read before bed. Or at least the last challenging thing.

Part of the learning (especially the transfer to long-term memory) happens *after* you put the book down. Your brain needs time on its own, to do more processing. If you put in something new during that processing time, some of what you just learned will be lost.

5 Drink water. Lots of it.

Your brain works best in a nice bath of fluid. Dehydration (which can happen before you ever feel thirsty) decreases cognitive function.

6 Talk about it. Out loud.

Speaking activates a different part of the brain. If you're trying to understand something, or increase your chance of remembering it later, say it out loud. Better still, try to explain it out loud to someone else. You'll learn more quickly, and you might uncover ideas you hadn't known were there when you were reading about it.

7 Listen to your brain.

Pay attention to whether your brain is getting overloaded. If you find yourself starting to skim the surface or forget what you just read, it's time for a break. Once you go past a certain point, you won't learn faster by trying to shove more in, and you might even hurt the process.

8 Feel something.

Your brain needs to know that this *matters*. Get involved with the stories. Make up your own captions for the photos. Groaning over a bad joke is *still* better than feeling nothing at all.

9 Practice solving problems!

There's only one way to truly master statistics: **practice answering questions**. And that's what you're going to do throughout this book. Using statistics is a skill, and the only way to get good at it is to practice. We're going to give you a lot of practice: every chapter has exercises that pose problems for you to solve. Don't just skip over them—a lot of the learning happens when you solve the exercises. We included a solution to each exercise—don't be afraid to **peek at the solution** if you get stuck! (It's easy to get snagged on something small.) But try to solve the problem before you look at the solution. And definitely make sure you understand what's going on before you move on to the next part of the book.

Read Me

This is a learning experience, not a reference book. We deliberately stripped out everything that might get in the way of learning whatever it is we're working on at that point in the book. And the first time through, you need to begin at the beginning, because the book makes assumptions about what you've already seen and learned.

We begin by teaching basic ways of representing and summarizing data, then move on to probability distributions, and then more advanced techniques such as hypothesis testing.

While later topics are important, the first thing you need to tackle is fundamental building blocks such as charting, averages, and measures of variability. So we begin by showing you basic statistical problems that you actually solve yourself. That way you can immediately do something with statistics, and you will begin to get excited about it. Then, a bit later in the book, we show you how to use probability and probability distributions. By then you'll have a solid grasp of statistics fundamentals, and can focus on *learning the concepts*. After that, we show you how to apply your knowledge in more powerful ways, such as how to conduct hypothesis tests. We teach you what you need to know at the point you need to know it because that's when it has the most value.

We cover the same general set of topics that are on the AP and A Level curriculum.

While we focus on the overall learning experience rather than exam preparation, we provide good coverage of the AP and A Level curriculum. This means that while you work your way through the topics, you'll gain the deep understanding you need to get a good grade in whatever exam it is you're taking. This is a far more effective way of learning statistics than learning formulae by rote, as you'll feel confident about what you need when, and how to use it.

We help you out with online resources.

Our readers tell us that sometimes you need a bit of extra help, so we provide online resources, right at your fingertips. We give you an online forum where you can go to seek help, online papers, and other resources too. The starting point is **http://www.headfirstlabs.com/books/hfstats/**

The activities are NOT optional.

The exercises and activities are not add-ons; they're part of the core content of the book. Some of them are to help with memory, some are for understanding, and some will help you apply what you've learned. ***Don't skip the exercises.*** The crossword puzzles are the only thing you don't *have* to do, but they're good for giving your brain a chance to think about the words and terms you've been learning in a different context.

The redundancy is intentional and important.

One distinct difference in a Head First book is that we want you to *really* get it. And we want you to finish the book remembering what you've learned. Most reference books don't have retention and recall as a goal, but this book is about *learning*, so you'll see some of the same concepts come up more than once.

The Brain Power and Brain Barbell exercises don't have answers.

For some of them, there is no right answer, and for others, part of the learning experience of the activities is for you to decide if and when your answers are right. In some of the Brain Power and Brain Barbell exercises, you will find hints to point you in the right direction.

The technical review team

Ariana Anderson

Cary Collett

Dru Kleinfeld

Danielle Levitt

Andy Parker

Michael Prerau

Not pictured (but just as awesome):
Jeffrey Maddelena and Matt Vadeboncoeur

Thanks also to Keith Fahlgren, Bruce Frey, and Leanne Lockhart for technical feedback.

Technical Reviewers:

Ariana Anderson is teaching assistant and PhD Candidate in the Department of Statistics at UCLA. Her research interests include data-mining and pattern recognition.

Dru Kleinfeld is a graduate of Cornell University, class of 2007, with a BA in Economics. Dru currently lives in New York City and works in the Human Resources Department of Morgan Stanley.

Danielle Levitt is currently a first year medical student in Tel Aviv, Israel. In her free time she enjoys listening to music, swimming in the ocean, and spending time with my friends and family.

Having started his career as a research physicist, **Andy Parker** thought he knew a thing or two about statistics. Sadly, having read this book, that turned out not to be the case. Andy spends most of his time now, worrying about what other important things he may have forgotten.

Michael J. Prerau is a researcher in Computational Neuroscience creating new statistical methods to analyze how the neurons encode information in the brain. He is a Ph.D. student in the Program in Neuroscience at Boston University, as well as a research associate in the Neuroscience Statistics Research Laboratory in the Department of Anesthesia and Critical Care at Massachusetts General Hospital.

Matthew Vadeboncoeur is a graduate student studying ecosystem ecology at the Complex Systems Research Center at the University of New Hampshire.

Acknowledgments

My editor:

Heartfelt thanks go to my editor, **Sanders Kleinfeld**. Sanders has been a delight to work with, and over the course of phone calls, emails and chat clients has become a good friend. He's amazingly dedicated and hard-working, and the advice and support I've received from him have been first-rate. Thanks Sanders! You're awesome and I couldn't have done this without you.

Sanders Kleinfeld

The O'Reilly team:

A big thank you goes to **Brett McLaughlin** for flying me over to Boston for the Head First boot camp, and giving me the opportunity of a lifetime. Brett's instincts for Head First are phenomenal, and I've truly appreciated all the guidance he's given me over the course of the book. Thanks, Brett.

Lou Barr

It's hard to imagine what this book would have been like without **Lou Barr**. Lou is an amazing graphics designer who has sprinkled magic through all the pages of this book. Not only that, she's added so much to the overall learning experience. No challenge has been too great for her, and I'm so grateful to her.

Catherine Nolan

The rest of the Head First team also deserve a great big thank you. **Catherine Nolan** helped me through the early stages of the book and made me feel at home in Head First land, **Brittany Smith** kept production running smoothly, **Laurie Petrycki** trusted me enough to let me write this book (and borrow her office earlier in the year—thanks Laurie!), **Keith McNamara** did a great job organizing the tech review team, and **Caitrin McCullough** managed everything on the website. Thanks guys!

A special mention goes to **Kathy Sierra** and **Bert Bates** for being the original masterminds behind this wonderful series of books. It's an honor to be part of it.

Brett McLaughlin

My family and friends:

I wish there was space to mention everyone who has helped me along the way, but special thanks must go to David, Mum, Dad, Steve Harvey, Gill Chester, Paul Burgess, Andy Tatler, and Peter Walker. You guys have kept me going, and I can't tell you how much I've appreciated your support and encouragement. Thank you.

Safari® Books Online

When you see a Safari® icon on the cover of your favorite technology book that means the book is available online through the O'Reilly Network Safari Bookshelf.

Safari offers a solution that's better than e-books. It's a virtual library that lets you easily search thousands of top tech books, cut and paste code samples, download chapters, and find quick answers when you need the most accurate, current information. Try it for free at `http://safari.oreilly.com`.

1 visualizing information

First Impressions

I want to look clean and pretty, so I give the right impression.

Can't tell your facts from your figures?

Statistics help you make sense of confusing sets of data. They **make the complex simple**. And when you've found out what's really going on, you need a way of **visualizing** it and **telling everyone else**. So if you want to pick the best chart for the job, grab your coat, pack your best slide rule, and join us on a ride to Statsville.

Statistics are everywhere

Everywhere you look you can find statistics, whether you're browsing the Internet, playing sports, or looking through the top scores of your favorite video game. But what actually *is* a statistic?

Statistics are numbers that summarize raw facts and figures in some meaningful way. They present key ideas that may not be immediately apparent by just looking at the raw data, and by data, we mean facts or figures from which we can draw conclusions. As an example, you don't have to wade through lots of football scores when all you want to know is the league position of your favorite team. You need a statistic to quickly give you the information you need.

The *study* of statistics covers where statistics come from, how to calculate them, and how you can use them effectively.

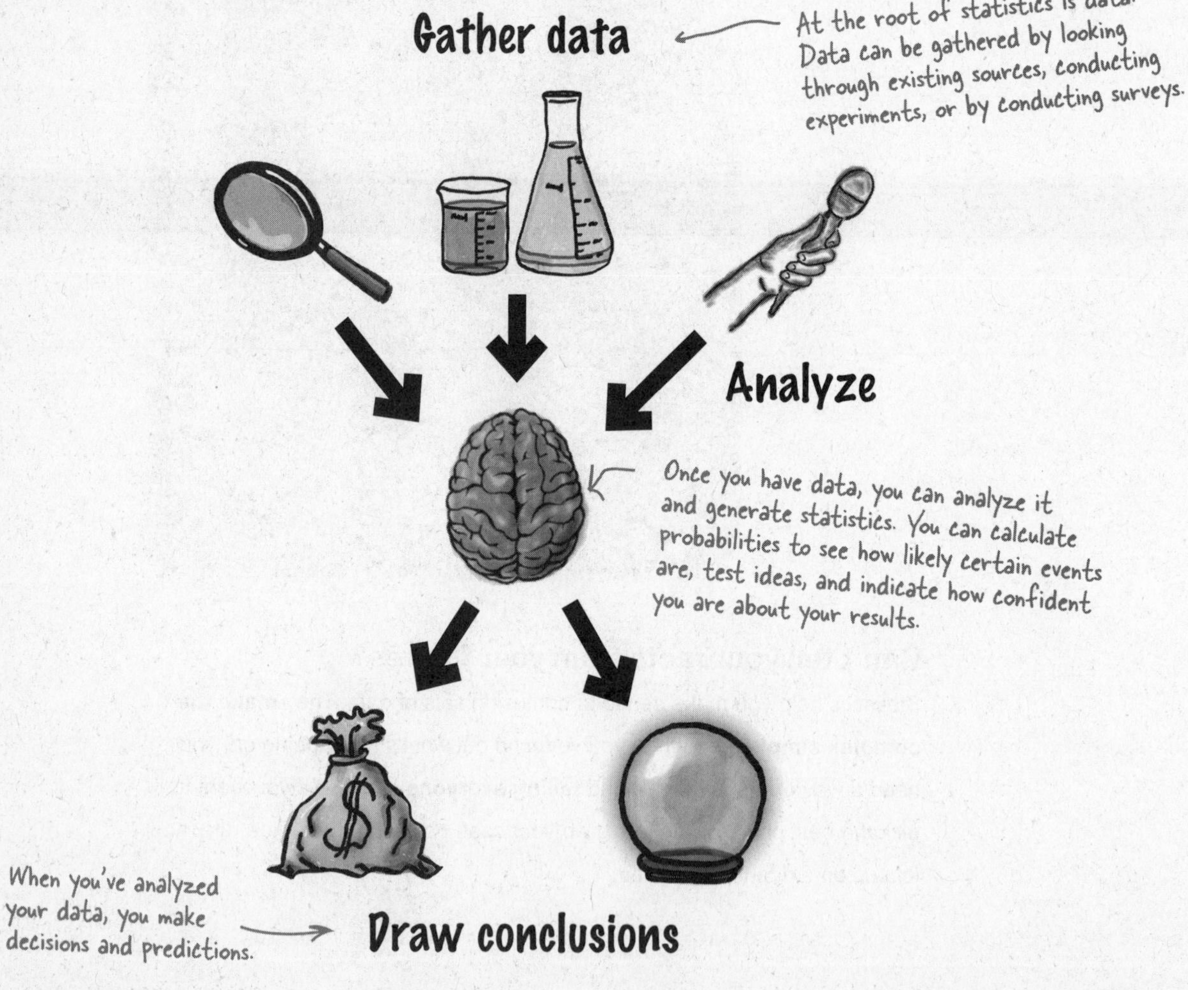

Gather data
At the root of statistics is data. Data can be gathered by looking through existing sources, conducting experiments, or by conducting surveys.
Analyze
Once you have data, you can analyze it and generate statistics. You can calculate probabilities to see how likely certain events are, test ideas, and indicate how confident you are about your results.
When you've analyzed your data, you make decisions and predictions.
Draw conclusions

But why learn statistics?

Understanding what's really going on with statistics empowers you. If you really *get* statistics, you'll be able to make objective decisions, make accurate predictions that seem inspired, and convey the message you want in the most effective way possible.

Statistics can be a convenient way of summarizing key truths about data, but there's a dark side too.

You can use statistics to help explain things about the world.
You can say what you want with statistics, even lie.

Statistics are based on facts, but even so, they can sometimes be misleading. They can be used to tell the truth—or to lie. The problem is how do you know when you're being told the truth, and when you're being told lies?

Having a good understanding of statistics puts you in a strong position. You're much better equipped to tell when statistics are inaccurate or misleading. In other words, studying statistics is a good way of making sure you don't get fooled.

As an example, take a look at the profits made by a company in the latter half of last year.

Month	Jul	Aug	Sep	Oct	Nov	Dec
Profit (millions)	2.0	2.1	2.2	2.1	2.3	2.4

The profit's holding steady, but it's nothing special.
This stock's so hot it's smokin'

How can there be two interpretations of the same set of data? Let's take a closer look.

A tale of two charts

So how can we explore these two different interpretations of the same data? What we need is some way of visualizing them. If you need to visualize information, there's no better way than using a chart or graph. They can be a quick way of summarizing raw information and can help you get an impression of what's going on at a glance. But you need to be careful because even the simplest chart can be used to subtly mislead and misdirect you.

Here are two time graphs showing a companies profits for six months. They're both based on the same information, so why do they look so different? They give drastically different versions of the same information.

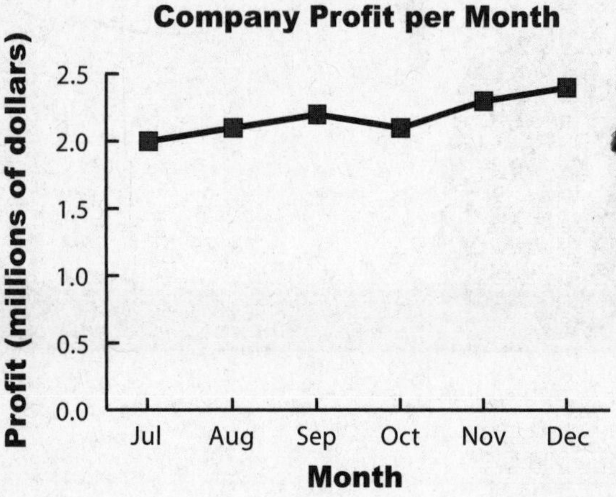

Both of these charts are based on the same information, but they look wildly different. What's going on?

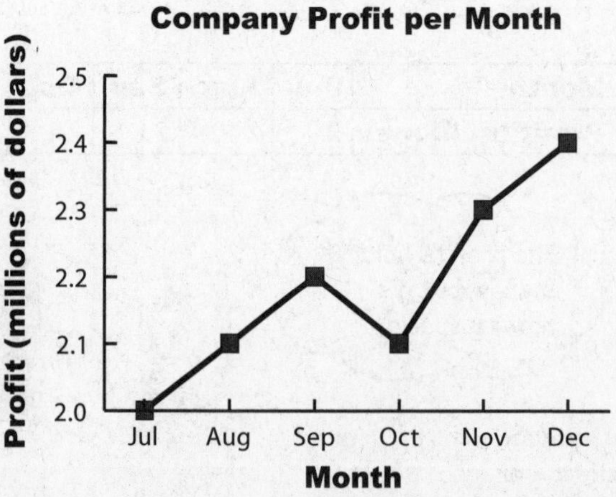

Sharpen your pencil

Take a look at the two charts on the facing page. What would you say are the key differences? How do they give such different first impressions of the data?

there are no Dumb Questions

Q: Why not just go on the data? Why chart it?

A: Sometimes it's difficult to see what's really going on just by looking at the raw data. There can be patterns and trends in the data, but these can be very hard to spot if you're just looking at a heap of numbers. Charts give you a way of literally seeing patterns in your data. They allow you to visualize your data and see what's really going on in a quick glance.

Q: What's the difference between information and data?

A: *Data* refers to raw facts and figures that have been collected. *Information* is data that has some sort of added meaning.

As an example, take the numbers 5, 6, and 7. By themselves, these are just numbers. You don't know what they mean or represent. They're *data*. If you're then told that these are the ages of three children, you have *information* as the numbers are now meaningful.

Sharpen your pencil Solution

Take a look at the two charts. What would you say are the key differences? How do they give such different first impressions of the data?

Both charts are based on the same underlying data, but they each send a different message.

The first chart shows that the profit is relatively steady. It achieves this by having the vertical axis start at 0, and then plotting the profit for each month against this.

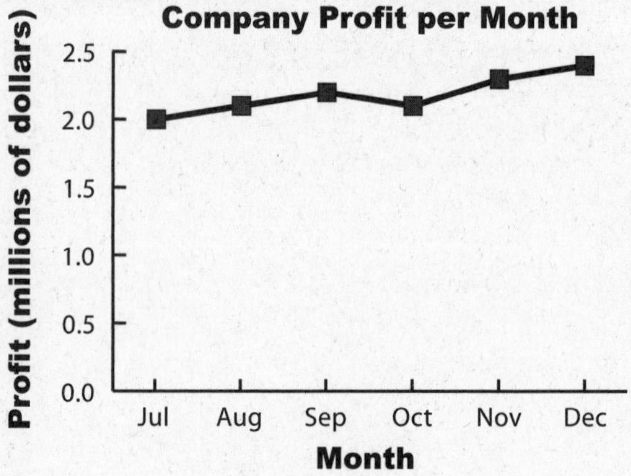

Look, the vertical axes are different on each chart.

The second chart gives a different impression by making the vertical axis start at a different place and adjusting the scale accordingly. At a glance, the profits appear to be rising dramatically each month. It's only when you look closer that you see what's really going on.

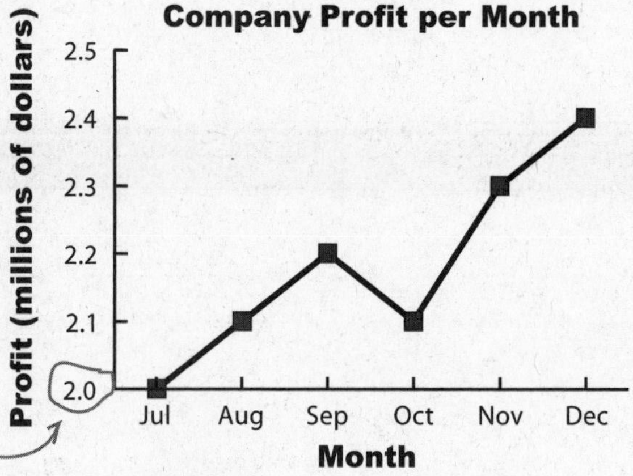

The axis for this chart starts at 2.0, not 0. No wonder the profit looks so awesome.

Why should I care about charts? Chart software can handle everything for you, that's what it's there for.

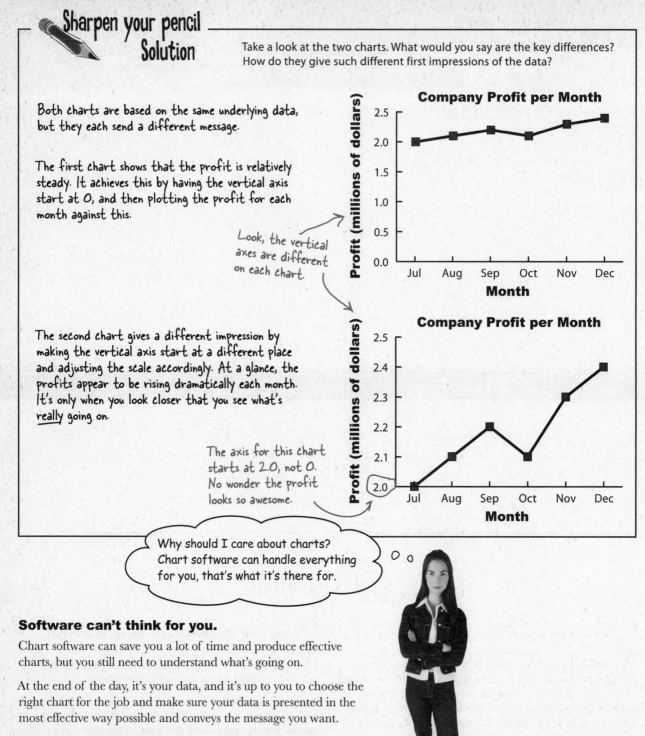

Software can't think for you.

Chart software can save you a lot of time and produce effective charts, but you still need to understand what's going on.

At the end of the day, it's your data, and it's up to you to choose the right chart for the job and make sure your data is presented in the most effective way possible and conveys the message you want.

Software can translate data into charts, but it's up to **you** to make sure the chart is right.

Manic Mango needs some charts

One company that needs some charting expertise is Manic Mango, an innovative games company that is taking the world by storm. The CEO has been invited to deliver a keynote presentation at the next worldwide games expo. He needs some quick, slick ways of presenting data, and he's asked you to come up with the goods. There's a lot riding on this. If the keynote goes well, Manic Mango will get extra sponsorship revenue, and you're bound to get a hefty bonus for your efforts.

Manic Mango

The first thing the CEO wants to be able to do is compare the percentage of satisfied players for each game genre. He's started off by plugging the data he has through some charting software, and here are the results:

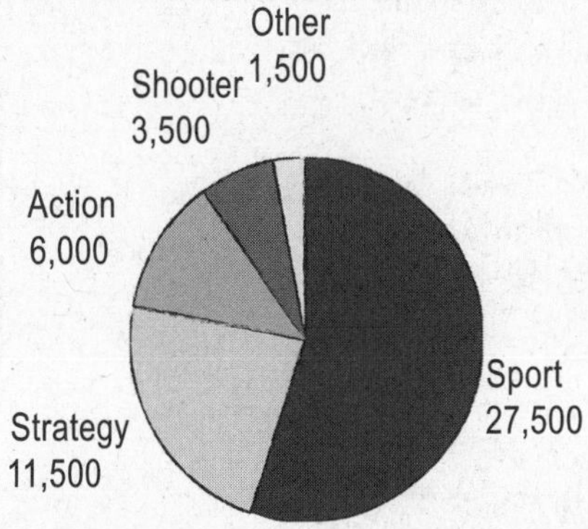

Units Sold per Genre

Brain Power

Take a good look at the pie chart that the CEO has produced. What does each slice represent? What can you infer about the relative popularity of different video game genres?

The humble pie chart

Pie charts work by splitting your data into distinct groups or categories. The chart consists of a circle split into wedge-shaped slices, and each slice represents a group. The size of each slice is proportional to how many are in each group compared with the others. The larger the slice, the greater the relative popularity of that group. The number in a particular group is called the ***frequency***.

Pie charts divide your entire data set into distinct groups. This means that if you add together the frequency of each slice, you should get 100%.

Let's take a closer look at our pie chart showing the number of units sold per genre:

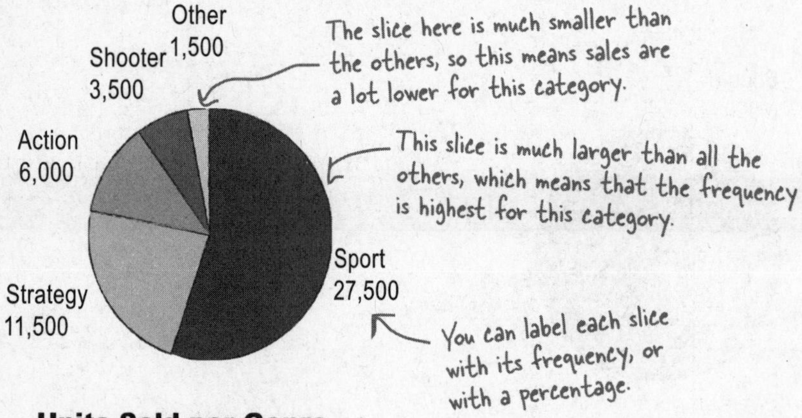

Other
1,500
Shooter
3,500
Action
6,000
Strategy
11,500
Sport
27,500
The slice here is much smaller than the others, so this means sales are a lot lower for this category.
This slice is much larger than all the others, which means that the frequency is highest for this category.
You can label each slice with its frequency, or with a percentage.

Units Sold per Genre

Genre	Units sold
Sports	27,500
Strategy	11,500
Action	6,000
Shooter	3,500
Other	1,500

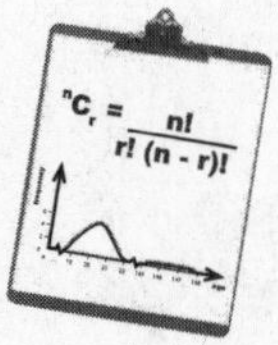

Vital Statistics

Frequency

Frequency describes how many items there are in a particular group or interval. It's like a count of how many there are.

So when are pie charts useful?

We've seen that the size of each slice represents the relative frequency of each group of data you're showingg. Because of this, pie charts can be useful if you want to compare basic proportions. It's usually easy to tell at a glance which groups have a high frequency compared with the others. Pie charts are less useful if all the slices have similar sizes, as it's difficult to pick up on subtle differences between the slice sizes.

So what about the pie chart that the Manic Mango CEO has created?

Chart failure

Creating a pie chart worked out so great for displaying the units sold per genre that the CEO's decided to create another to chart consumer satisfaction with Manic Mango's game. The CEO needs a chart that will allow him to compare the percentage of satisfied players for each game genre. He's run the data through the charting software again, but this time he's not as impressed.

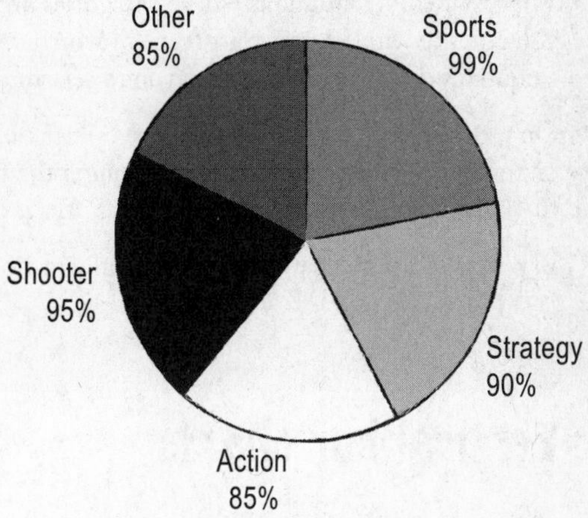

% Players Satisfied per Genre

What happened here? All the slices are the same size, but the percentages are all different and are much larger than the slices. Can you help me fix this chart? Now?

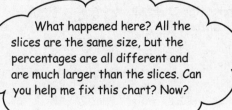

Manic Mango's CEO

Pie charts are used to compare the proportions of different groups or categories, but in this case there's little variation between each group.

It's difficult to take in at a glance which category has the highest level of player satisfaction.

It's also generally confusing to label pie charts with percentages that don't relate to the overall proportion of the slice. As an example, the Sports slice is labelled 99%, but it only fills about 20% of the chart. Another problem is that we don't know whether there's an equal number of responses for each genre, so we don't know whether it's fair to compare genre satisfaction in this way.

Pie charts show proportions

BRAIN POWER

Take a look at the data, and think about the problems there are with this chart. What would be a better sort of chart for this kind of information?

Bar charts can allow for more accuracy

A better way of showing this kind of data is with a ***bar chart***. Just like pie charts, bar charts allow you to compare relative sizes, but the advantage of using a bar chart is that they allow for a greater degree of precision. They're ideal in situations where categories are roughly the same size, as you can tell with far greater precision which category has the highest frequency. It makes it easier for you to see small differences.

On a bar chart, each bar represents a particular category, and the length of the bar indicates the value. The longer the bar, the greater the value. All the bars have the same width, which makes it easier to compare them.

Bar charts can be drawn either vertically or horizontally.

Vertical bar charts

Vertical bar charts show categories on the horizontal axis, and either frequency or percentage on the vertical axis. The height of each bar indicates the value of its category. Here's an example showing the sales figures in units for five regions, A, B, C, D, and E:

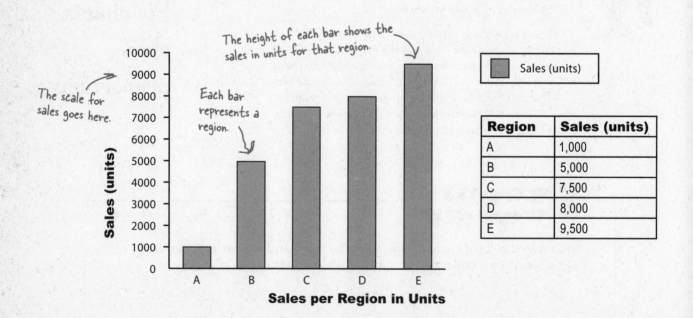

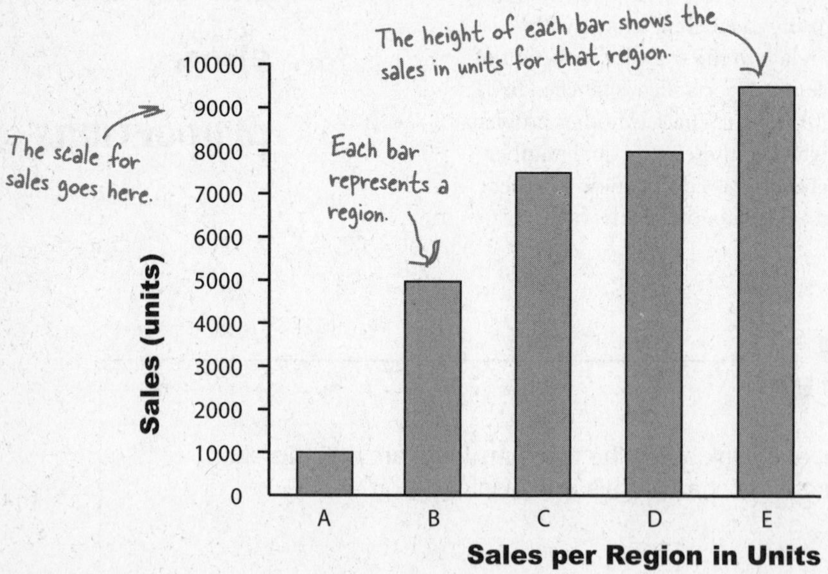

Region	Sales (units)
A	1,000
B	5,000
C	7,500
D	8,000
E	9,500

Horizontal bar charts

Horizontal bar charts are just like vertical bar charts except that the axes are flipped round. With horizontal bar charts, you show the categories on the vertical axis and the frequency or percentage on the horizontal axis.

Here's a horizontal bar chart for the CEO's genre data from page 9. As you can see, it's much easier to quickly gauge which category has the highest value, and which the lowest.

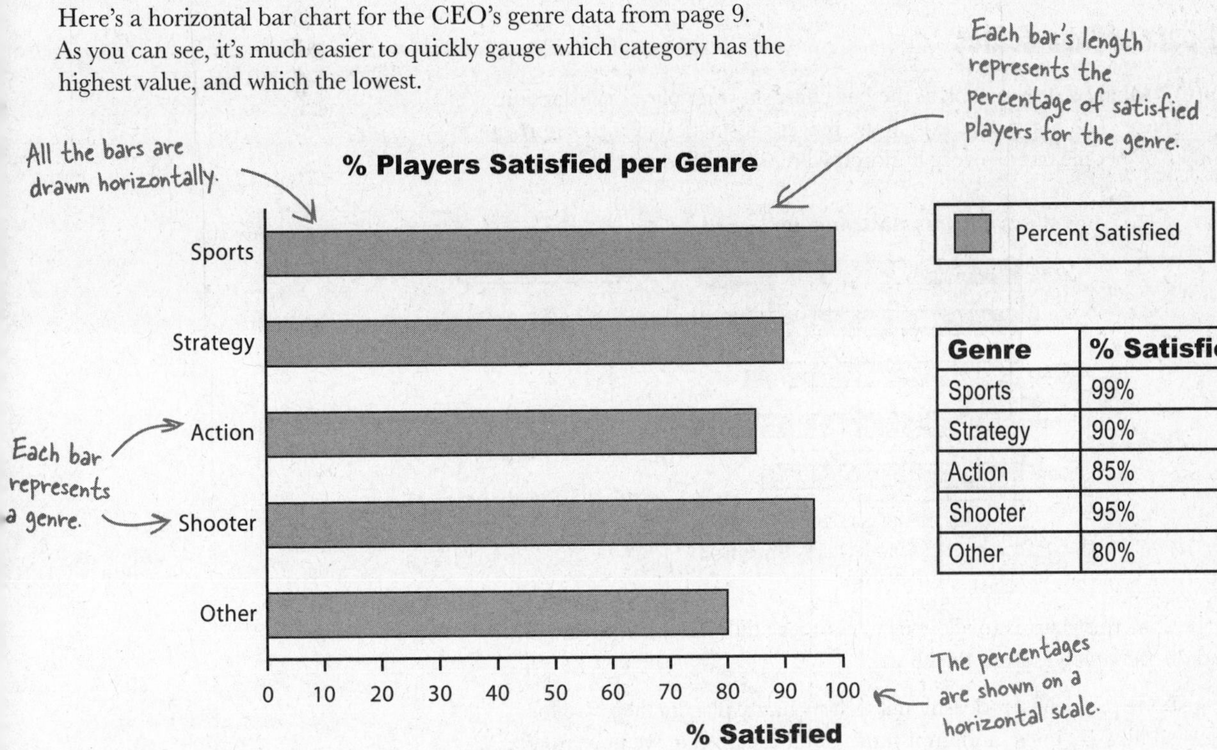

Genre	% Satisfied
Sports	99%
Strategy	90%
Action	85%
Shooter	95%
Other	80%

Vertical bar charts tend to be more common, but horizontal bar charts are useful if the names of your categories are long. They give you lots of space for showing the name of each category without having to turn the bar labels sideways.

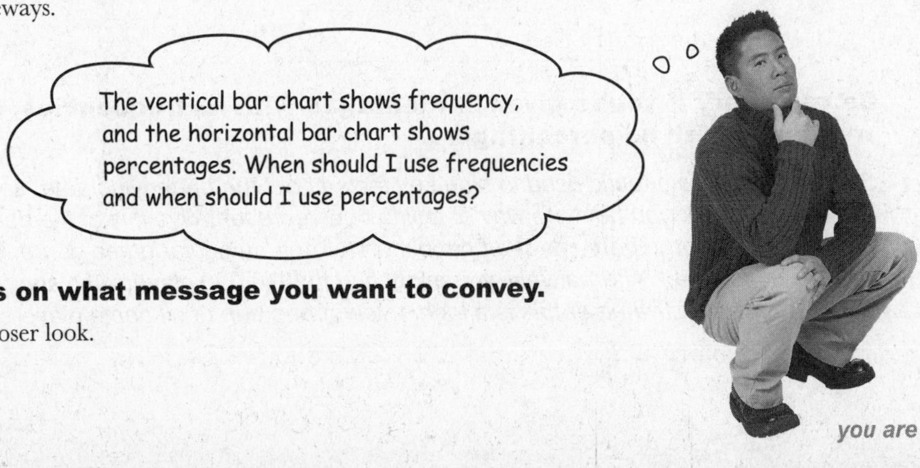

The vertical bar chart shows frequency, and the horizontal bar chart shows percentages. When should I use frequencies and when should I use percentages?

It depends on what message you want to convey.

Let's take a closer look.

It's a matter of scale

Understanding scale allows you to create powerful bar charts that pick out the key facts you want to draw attention to. But be careful—scale can also conceal vital facts about your data. Let's see how.

Using percentage scales

Let's start by taking a deeper look at the bar chart showing player satisfaction per game genre. The horizontal axis shows player satisfaction as a ***percentage***, the number of people out of every hundred who are satisfied with this genre.

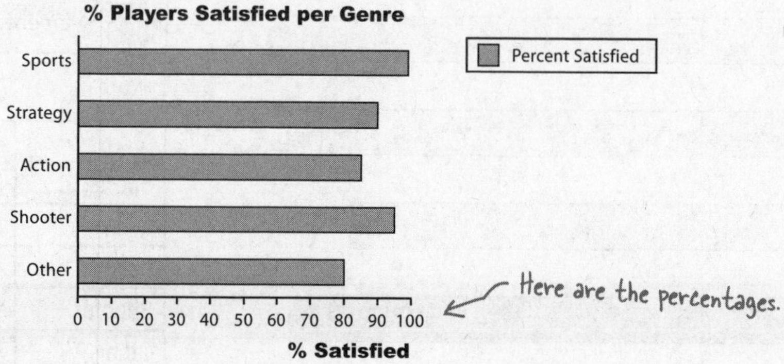

The purpose of this chart is to allow us to compare different percentages and also read off percentages from the chart.

There's just one problem—it doesn't tell us how many players there are for each genre. This may not sound important, but it means that we have no idea whether this reflects the views of all players, some of them, or even just a handful. In other words, we don't know how representative this is of players as a whole. The golden rule for designing charts that show percentages is to try and indicate the frequencies, either on the chart or just next to it.

Watch it!

Be very wary if you're given percentages with no frequencies, or a frequency with no percentage.

Sometimes this is a tactic used to hide key facts about the underlying data, as just based on a chart, you have no way of telling how representative it is of the data. You may find that a large percentage of people prefer one particular game genre, but that only 10 people were questioned. Alternatively, you might find that 10,000 players like sports games most, but by itself, you can't tell whether this is a high or low proportion of all game players.

Using frequency scales

You can show frequencies on your scale instead of percentages. This makes it easy for people to see exactly what the frequencies are and compare values.

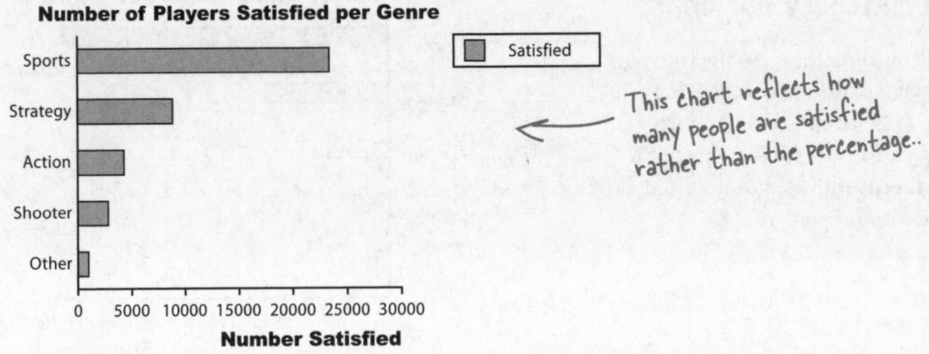

This chart reflects how many people are satisfied rather than the percentage..

Normally your scale should start at 0, but watch out! Not every chart does this, and as you saw earlier on page 6, using a scale that *doesn't* start at 0 can give a different first impression of your data. This is something to watch out for on other people's charts, as it's very easy to miss and can give you the wrong impression of the data.

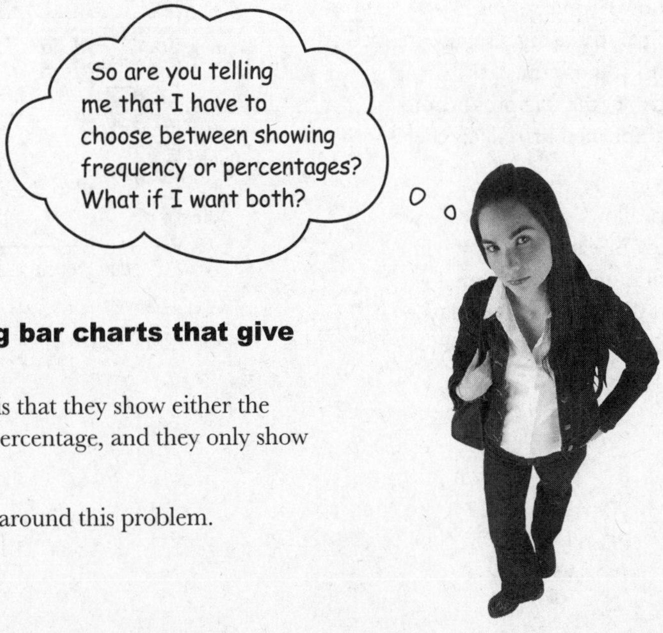

So are you telling me that I have to choose between showing frequency or percentages? What if I want both?

There are ways of drawing bar charts that give you more flexibility.

The problem with these bar charts is that they show either the number of satisfied players or the percentage, and they only show satisfied players.

Let's take a look at how we can get around this problem.

Dealing with multiple sets of data

With bar charts, it's actually really easy to show more than one set of data on the same chart. As an example, we can show both the percentage of satisfied players and the percentage of dissatisfied players on the same chart.

The split-category bar chart

One way of tackling this is to use one bar for the frequency of satisfied players and another for those dissatisfied, for each genre. This sort of chart is useful if you want to **compare frequencies**, but it's difficult to see proportions and percentages.

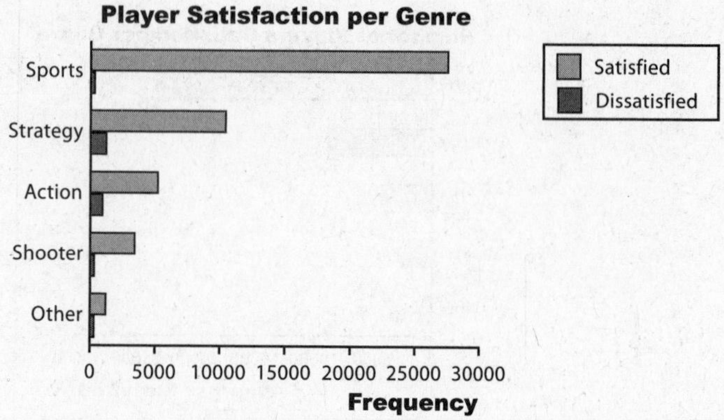

The segmented bar chart

If you want to show frequencies *and* percentages, you can try using a segmented bar chart. For this, you use one bar for each category, but you split the bar proportionally. The overall length of the bar reflects the total frequency.

This sort of chart allows you to quickly see the total frequency of each category—in this case, the total number of players for each genre—and the frequency of player satisfaction. You can see proportions at a glance, too.

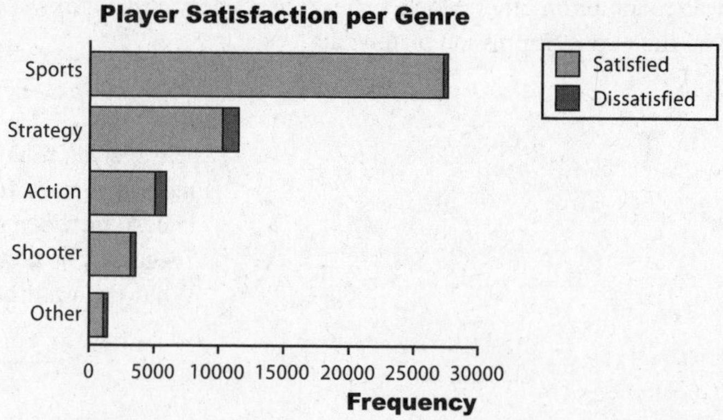

Exercise

The CEO needs another chart for the keynote presentation. Here's the data; see if you can sketch the bar chart.

Continent	**Sales (units)**
North America	1,500
South America	500
Europe	1,500
Asia	2,000
Oceania	1,000
Africa	500
Antarctica	1

Sharpen your pencil

Here's another chart generated by the software. Which genre sold the most in 2007? How did this genre fare in 2006?

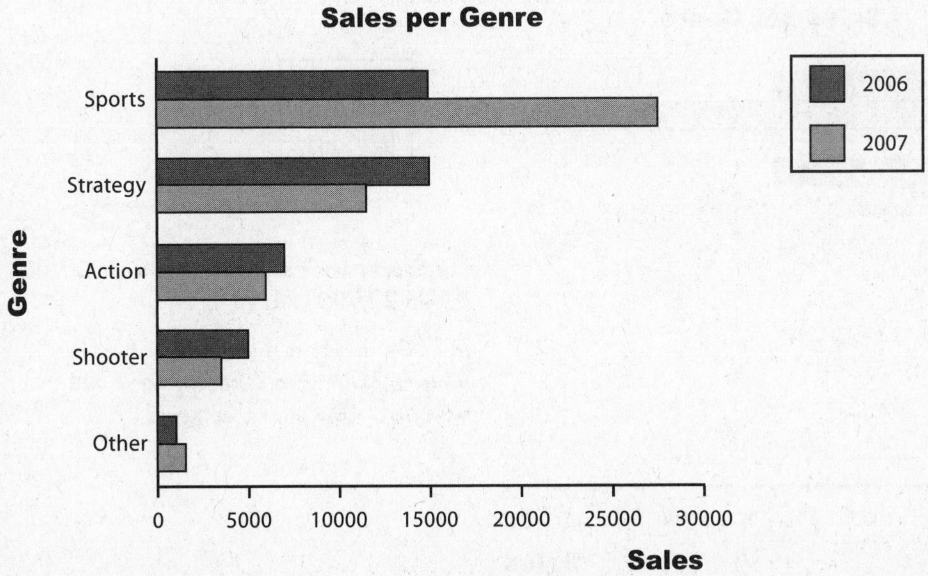

Exercise Solution

The CEO needs another chart for the keynote presentation. Here's the data, see if you can create the chart.

Continent	Sales (units)
North America	1,500
South America	500
Europe	1,500
Asia	2,000
Oceania	1,000
Africa	500
Antarctica	1

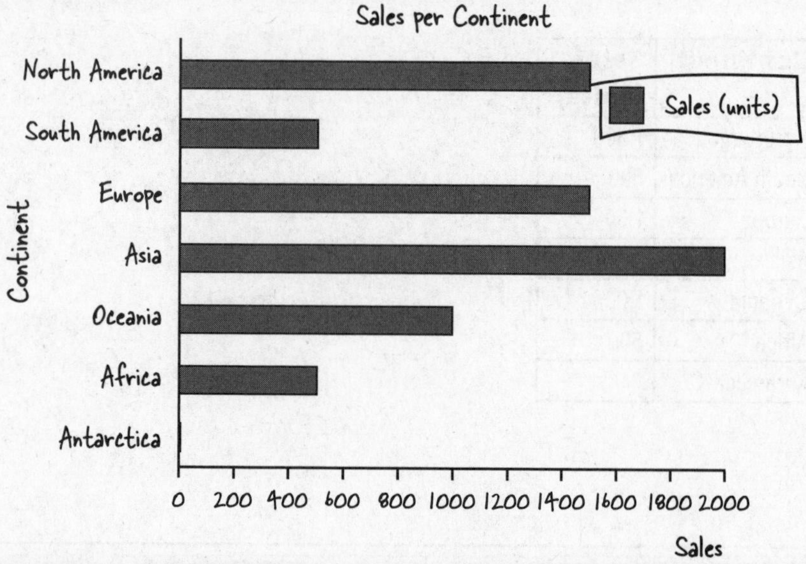

Sharpen your pencil Solution

Here's another chart generated by the software. Which genre sold the most in 2007? How did this genre fare in 2006?

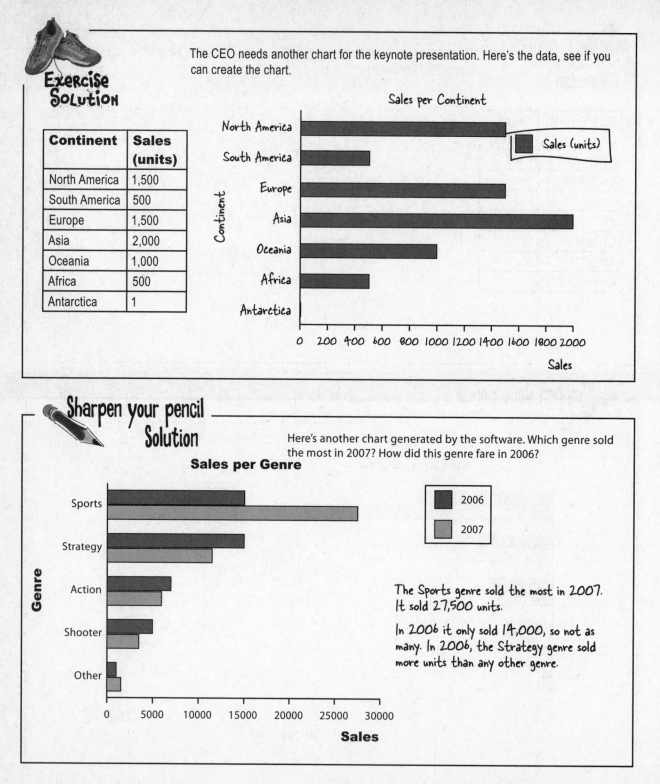

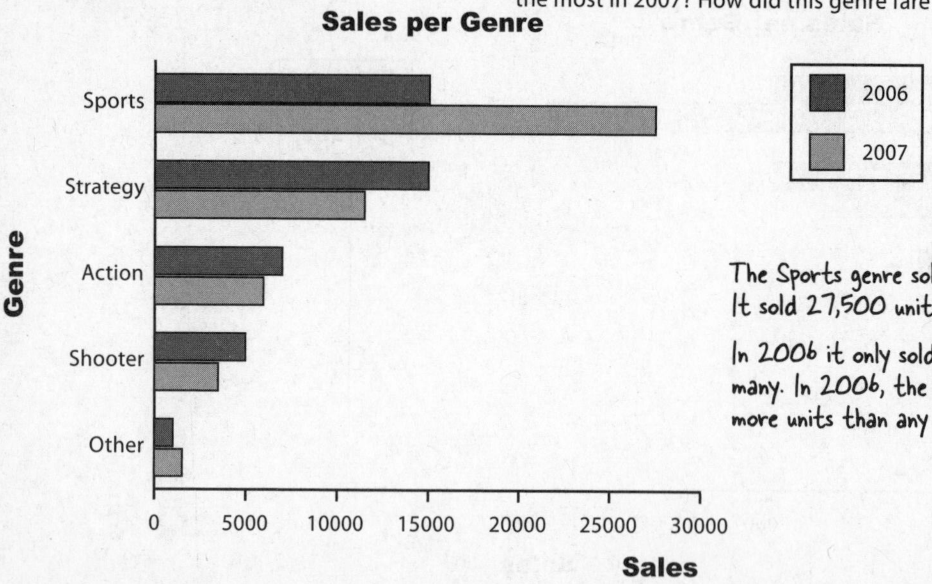

The Sports genre sold the most in 2007. It sold 27,500 units.

In 2006 it only sold 14,000, so not as many. In 2006, the Strategy genre sold more units than any other genre.

Your bar charts rock

The CEO is thrilled with the bar charts you've produced, but there's more data he needs to present at the keynote.

Nice work! Those charts are going to be a big hit at the expo. I've got another assignment for you. We've been testing a new game with a group of volunteers, and we need a chart to show the breakdown of scores per game. Here's the data:

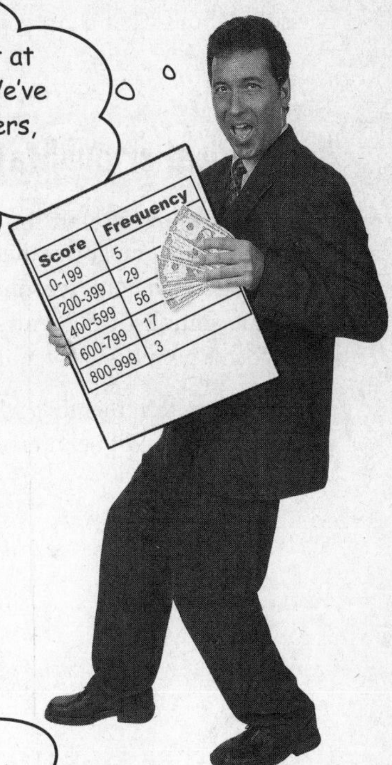

People can score between 0 and 999, and the data is broken into groups. As an example, players scored between 0 and 199 on 5 occasions.

Score	Frequency
0-199	5
200-399	29
400-599	56
600-799	17
800-999	3

The frequency is the number of times a score within each range was achieved.

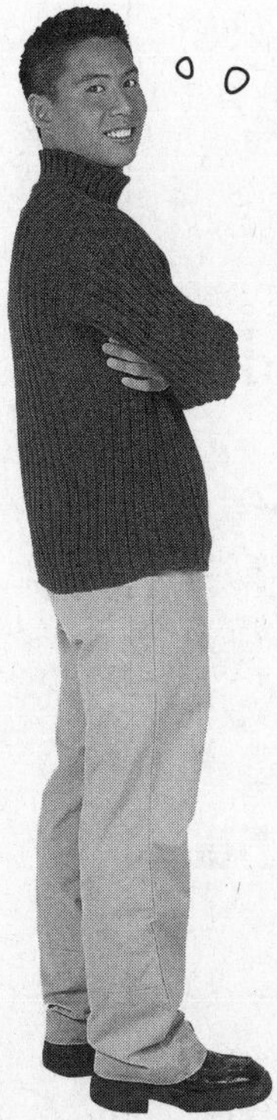

This data looks different from the other types of data we've seen so far. I wonder if that means we treat it differently?

BRAIN POWER

Look back through the chapter. How do you think this type of data is different? What impact do you think this could have on charts?

Categories vs. numbers

When you're working with charts, one of the key things you need to figure out is what sort of data you're dealing with. Once you've figured that out, you'll find it easier to make key decisions about what chart you need to best represent your data.

Categorical or qualitative data

Most of the data we've seen so far is ***categorical***. The data is split into categories that describe qualities or characteristics. For this reason, it's also called ***qualitative*** data. An example of qualitative data is game genre; each genre forms a separate category.

The key thing to remember with qualitative data is that the data values can't be interpreted as numbers.

Numerical or quantitative data

Numerical data, on the other hand, deals with **numbers**. It's data where the values have meaning as numbers, and that involves measurements or counts. Numerical data is also called ***quantitative*** data because it describes quantities.

So what impact does this have on the chart for Manic Mango?

Dealing with grouped data

The latest set of data from the Manic Mango CEO is numeric and, what's more, the scores are grouped into intervals. So what's the best way of charting data like this?

The scores are numeric and grouped into intervals

Score	Frequency
0-199	5
200-399	29
400-599	56
600-799	17
800-999	3

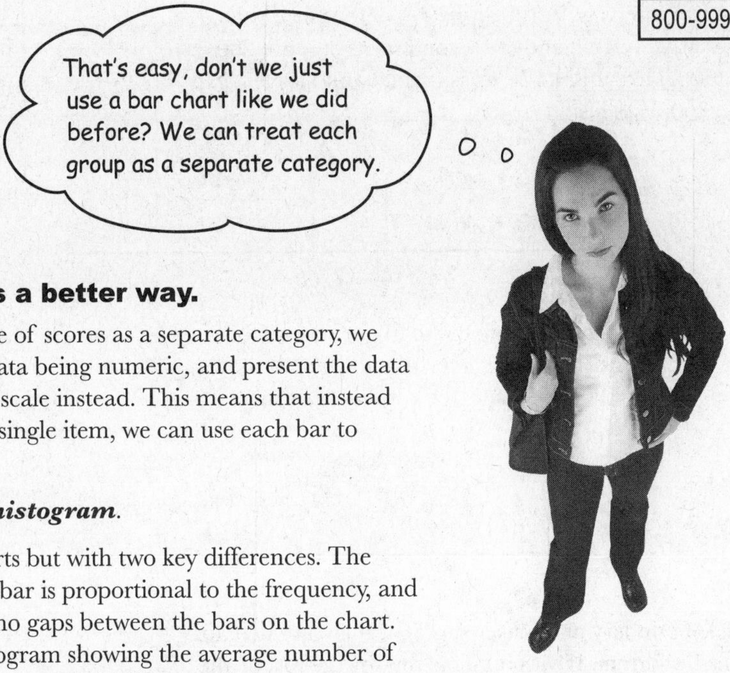

That's easy, don't we just use a bar chart like we did before? We can treat each group as a separate category.

We could, but there's a better way.

Rather than treat each range of scores as a separate category, we can take advantage of the data being numeric, and present the data using a continuous numeric scale instead. This means that instead of using bars to represent a single item, we can use each bar to represent a *range* of scores.

To do this, we can create a ***histogram***.

Histograms are like bar charts but with two key differences. The first is that the area of each bar is proportional to the frequency, and the second is that there are no gaps between the bars on the chart. Here's an example of a histogram showing the average number of games bought per month by households in Statsville:

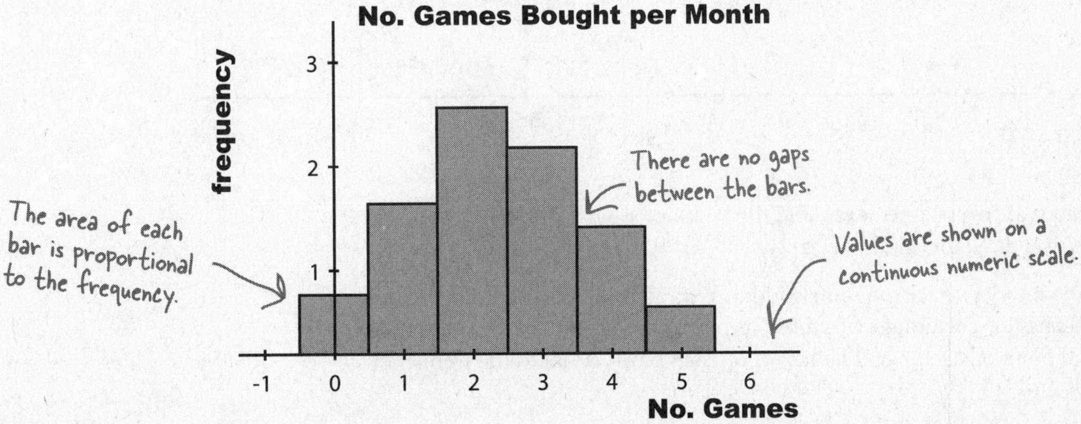

To make a histogram, start by finding bar widths

Score	Frequency
0–199	5
200–399	29
400–599	56
600–799	17
800–999	3

The first step to creating a histogram is to look at each of the intervals and work out how wide each of them needs to be, and what range of values each one needs to cover. While doing this, we need to make sure that there will be no gaps between the bars on the histogram.

Let's start with the first two intervals, 0–199 and 200–399. At face value, the first interval finishes at score 199, and the second starts at score 200. The problem with plotting it like this, however, is that it would leave a gap between score 199 and 200, like this:

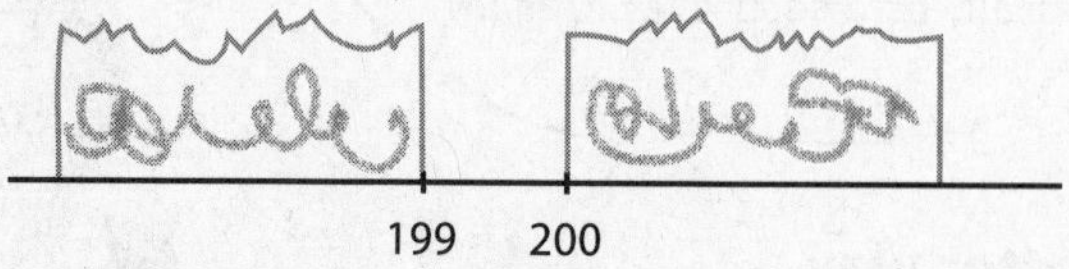

Histograms shouldn't have gaps between the bars, so to get around this, we extend their ranges slightly. Instead of one interval ending at score 199 and the next starting at score 200, we make the two intervals meet at 199.5, like this:

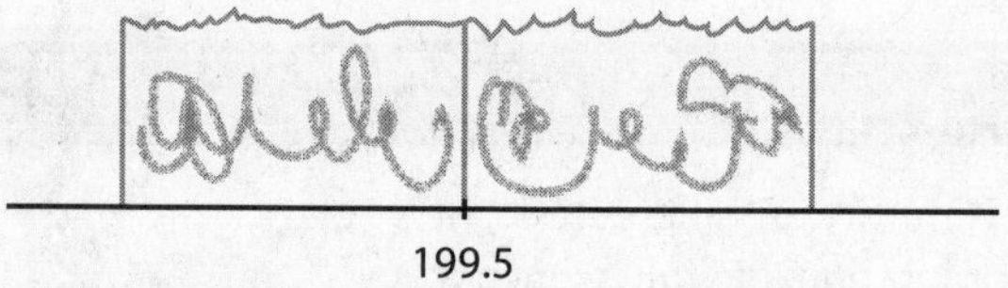

Doing this forms a single boundary and makes sure that there are no gaps between the bars on the histogram. If we complete this for the rest of the intervals, we get the following boundaries:

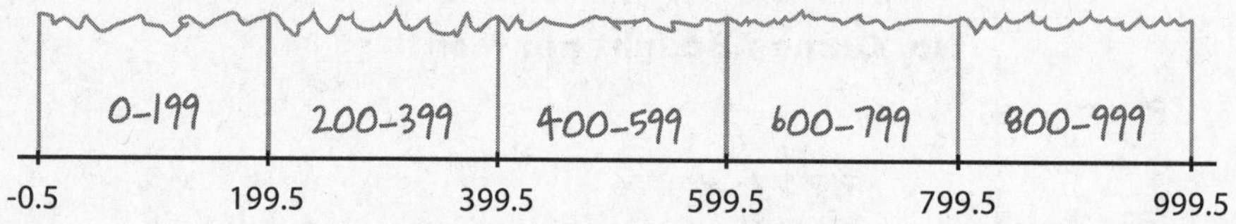

Each interval covers 200 scores, and the width of each interval is 200. Each interval has the same width.

As all the intervals have the same width, we create the histogram by drawing vertical bars for each range of scores, using the boundaries to form the start and end point of each bar. The height of each bar is equal to the frequency.

Here's a reminder of the data for Manic Mango.

Score	Frequency
0–199	5
200–399	29
400–599	56
600–799	17
800–999	3

See if you can use the class boundaries to create a histogram for this data. Remember, the frequency goes on the vertical axis.

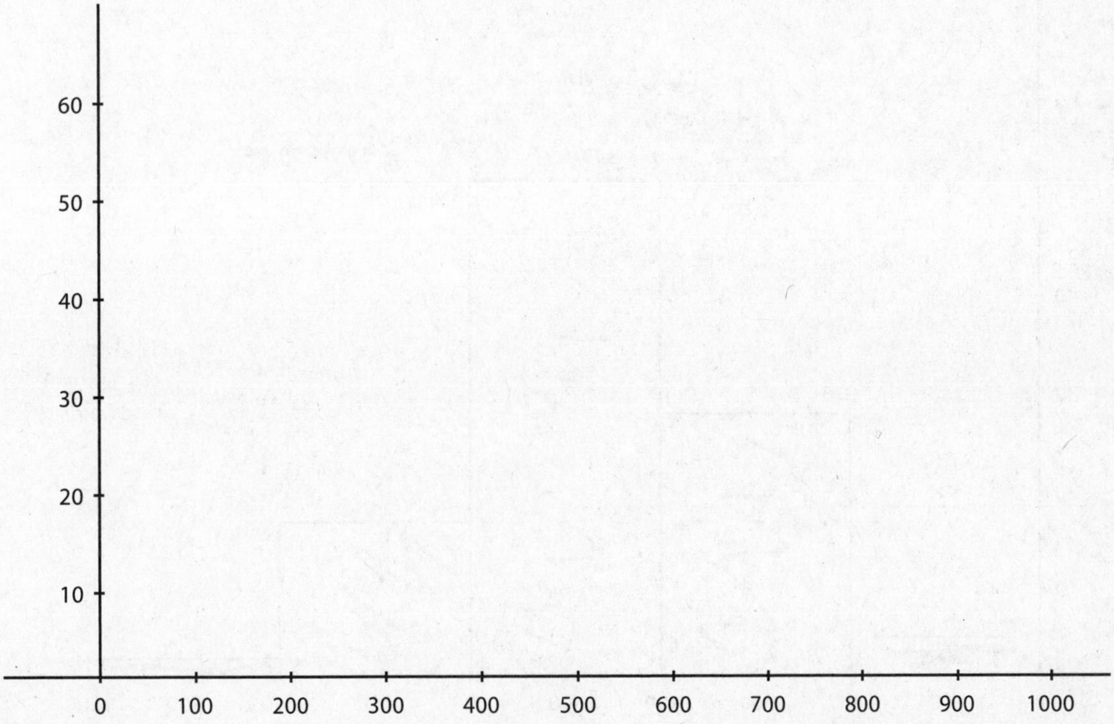

Exercise Solution

Here's a reminder of the data for Manic Mango.

Score	Frequency
0–199	5
200–399	29
400–599	56
600–799	17
800–999	3

See if you can use the class boundaries to create a histogram for this data. Remember, the frequency goes on the vertical axis.

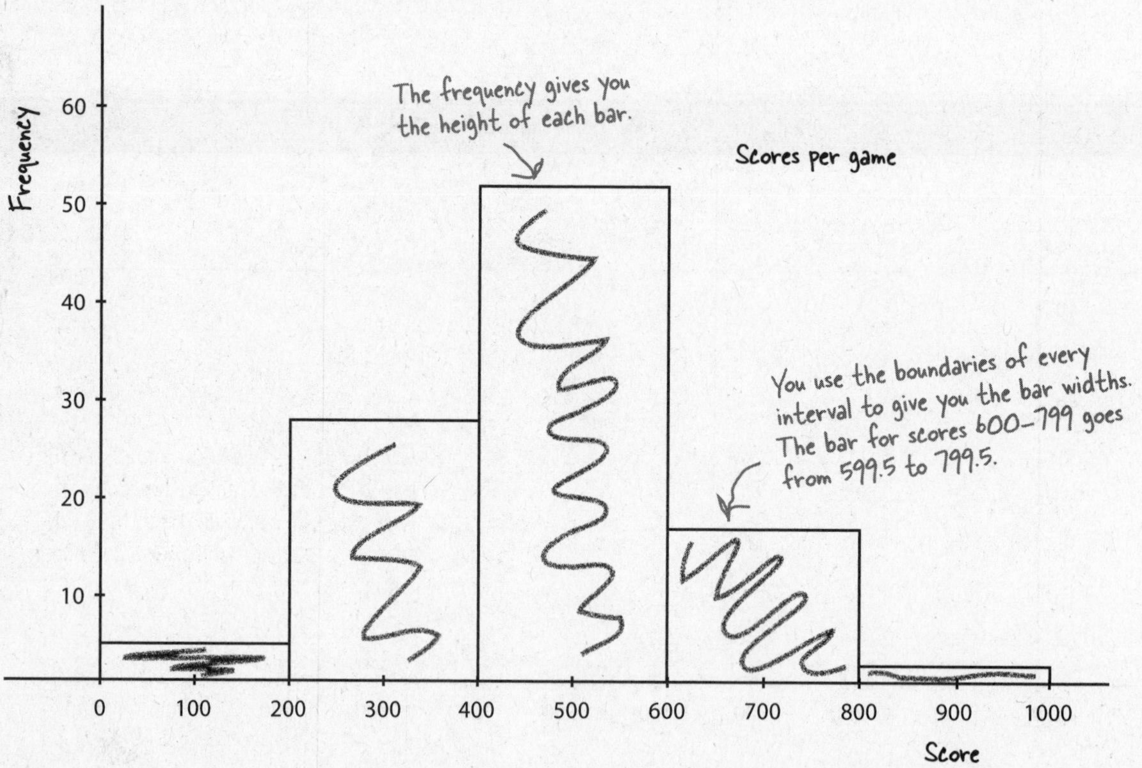
Scores per game
Frequency
Score
0
10
20
30
40
50
60
100
200
300
400
500
600
700
800
900
1000
The frequency gives you the height of each bar.
You use the boundaries of every interval to give you the bar widths. The bar for scores 600–799 goes from 599.5 to 799.5.

there are no Dumb Questions

Q: So is a histogram basically for grouped numeric data?

A: Yes it is. The advantage of a histogram is that because its numeric, you can use it to show the width of each interval as well as the frequency.

Q: What about if the intervals are different widths? Can you still use a histogram?

A: Absolutely. It's more common for the interval widths to be equal size, but with a histogram they don't have to be. There are a couple more steps you need to go through to create a histogram with unequal sized intervals, but we'll show you that very soon.

Q: Why shouldn't histograms have gaps between the bars?

A: There are at least two good reasons. The first is to show that there are no gaps in the values, and that every value is covered. The second is so that the width of the interval reflects the range of the values you're covering. As an example, if we drew the interval 0–199 as extending from value 0 to value 199, the width on the chart would only be 199 – 0 = 199.

Q: So why do we make the bars meet midway between the two?

A: The bars have to meet, and it's usually at the midway point, but it all comes down to how you round your values. When you round values, you normally round them to the nearest whole number. This means that the range of values from -0.5 to 0.5 all round to 0, and so when we show 0 on a histogram, we show it using the range of values from -0.5 to 0.5.

Q: Are there any exceptions to this?

A: Yes, age is one exception. If you have to represent the age range 18–19 on a histogram, you would normally represent this using an interval that goes from 18 to 20. The reason for this is that we typically classify someone as being 19, for example, up until their 20th birthday. In effect, we round ages down.

BULLET POINTS

- The **frequency** is a statistical way of saying how many items there are in a category.
- **Pie charts** are good for showing basic proportions.
- **Bar charts** give you more flexibility and precision.
- **Numerical data** deals with numbers and quantities; categorical data deals with words and qualities.
- **Horizontal bar charts are used for categorical data**, particularly where the category names are long.
- **Vertical bar charts are used for numerical data**, or categorical data if the category names are short.
- You can show **multiple sets of data on a bar chart**, and you have a choice of how to do this. You can compare frequencies by showing related bars side-by-side on a split-category bar chart. You can show proportions and total frequencies by stacking the bars on top of each other on a segmented bar chart.
- **Bar chart scales** can show either percentages or frequencies.
- Each chart comes in a number of different varieties.

Manic Mango needs another chart

The CEO is very pleased with the histogram you've created for him—so much so, that he wants you to create another histogram for him. This time, he wants a chart showing for how long Manic Mango players tend to play online games over a 24-hour period. Here's the data:

This is the number of hours people play for →

Hours	Frequency
0–1	4,300
1–3	6,900
3–5	4,900
5–10	2,000
10–24	2,100

← This is frequency with which people play for this lengh of time

There's something funny about that data. It's grouped like last time, but the intervals aren't all the same width.

He's right, the interval widths aren't all equal.

If you take a look at the intervals, you can see that they're different widths. As an example, the 10–24 range covers far more hours than the 0–1 range.

If we had access to the raw data, we could look at how we could construct equal width intervals, but unfortunately this is all the data we have. We need a way of drawing a histogram that makes allowances for the data having different widths.

BRAIN POWER

For histograms, the frequency is proportional to the area of each bar. How would you use this to create a histogram for this data? What do you need to be aware of?

I think we can just create this in the same way we did before—it's no big deal. We draw bars on a numeric scale; it's just that this time the bars are different widths.

Do you think she's right?

Here's a sketch of the chart, using frequency on the vertical scale and drawing bar widths proportional to their interval size. Do you see any problems?

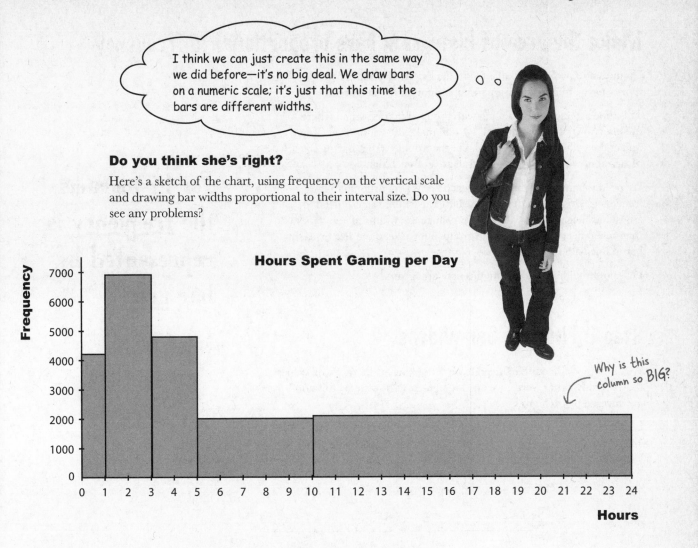

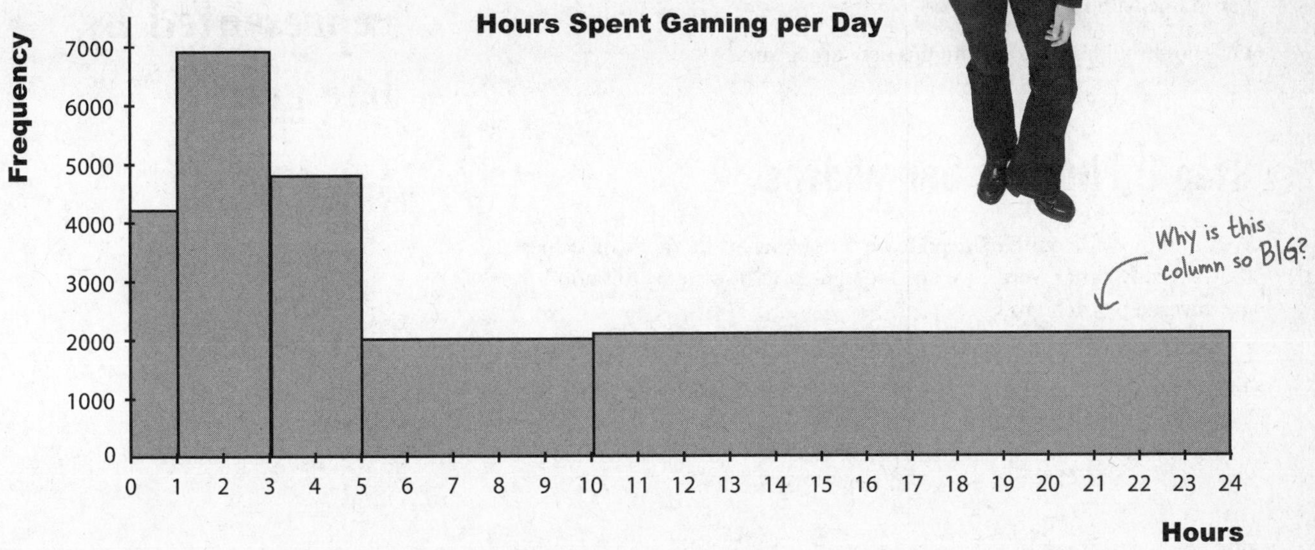

A histogram's bar area must be proportional to frequency

The problem with this chart is that making the width of each bar reflect the width of each interval has made some of the bars look disproportionately large. Just glancing at the chart, you might be left with a misleading impression about how many hours per day people really play games for. As an example, the bar that takes up the largest area is the bar showing game play of 10–24 hours, even though most people don't play for this long.

As this is a histogram, we need to make the bar area proportional to the frequency it represents. As the bars have unequal widths, what should we do to the bar height?

Make the area of histogram bars proportional to frequency

Up until now, we've been able to use the height of each bar to represent the frequency of a particular number or category.

This time around, we're dealing with grouped numeric data where the interval widths are unequal. We can make the width of each bar reflect the width of each interval, but the trouble is that having bars of different widths affects the overall area of each bar.

We need to make sure the area of each bar is proportional to its frequency. This means that if we adjust bar width, we also need to adjust bar height. That way, we can change the widths of the bars so that they reflect the width of the group, but we keep the size of each bar in line with its frequency.

Let's go through how to create this new histogram.

For histograms, the frequency is represented by bar area

Step 1: Find the bar widths

We find how wide our bars need to be by looking at the range of values they cover. In other words, we need to figure out how many full hours are covered by each group.

Let's take the 1–3 group. This group covers 2 full hours, 1–2 and 2–3. This means that the width of the bar needs to be 2, with boundaries of 1 and 3.

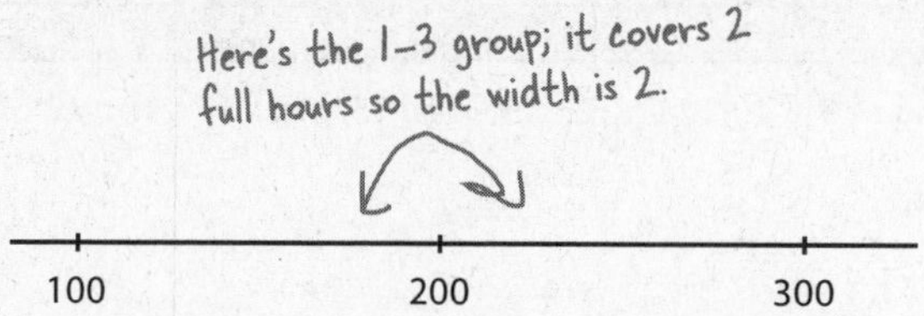
Here's the 1–3 group; it covers 2 full hours so the width is 2.
100
200
300

If we calculate the rest of the widths, we get:

Hours	Frequency	Width
0–1	4,300	1
1–3	6,900	2
3–5	4,900	2
5–10	2,000	5
10–24	2,100	14

Now that we've figured out the bar widths, we can move onto working out the heights.

Step 2: Find the bar heights

Now that we have the widths of all the groups, we can use these to find the heights the bars need to be. Remember, we need to adjust the bar heights so that the overall area of each bar is proportional to the group's frequency.

First of all, let's take the area of each bar. We've said that frequency and area are equivalent. As we already know what the frequency of each group is, we know what the areas should be too:

Area of bar = Frequency of group

We were given these right at the start, so we know what area we're aiming for.

Now each bar is basically just a rectangle, which means that the area of each bar is equal to the width multiplied by the height. As the area gives us the frequency, this means:

Frequency = Width of bar × Height of bar

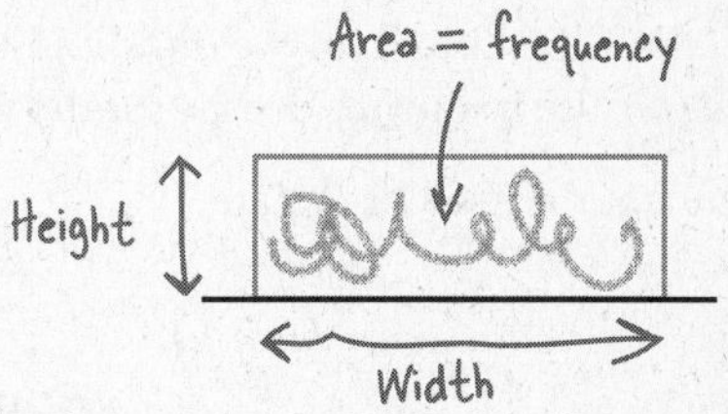

We found the widths of the bars in the last step, which means that we can use these to find what height each bar should be. In other words,

$$\textbf{Height of bar} = \frac{\textbf{Frequency}}{\textbf{Width of bar}}$$

The height of the bar is used to measure how **concentrated** the frequency is for a particular group. It's a way of measuring how densely packed the frequency is, a way of saying how thick or thin on the ground the numbers are. The height of the bar is called the ***frequency density***.

Sharpen your pencil

What should the height of each bar be? Complete the table.

Hours	Frequency	Width	Height (Frequency Density)
0–1	4,300	1	4,300 ÷ 1 = 4,300
1–3	6,900	2	
3–5	4,900	2	
5–10	2,000	5	
10–24	2,100	14	

Sharpen your pencil Solution

What should the height of each bar be? Complete the table.

Hours	Frequency	Width	Height (Frequency Density)
0–1	4,300	1	4,300 ÷ 1 = 4,300
1–3	6,900	2	6,900 ÷ 2 = 3,450
3–5	4,900	2	4,900 ÷ 2 = 2,450
5–10	2,000	5	2,000 ÷ 5 = 400
10–24	2100	14	2,100 ÷ 14 = 150

Step 3: Draw your chart—a histogram

Now that we've worked out the widths and heights of each bar, we can draw the histogram. We draw it just like before, except that this time, we use frequency density for the vertical axis and not frequency.

Here's our revised histogram.

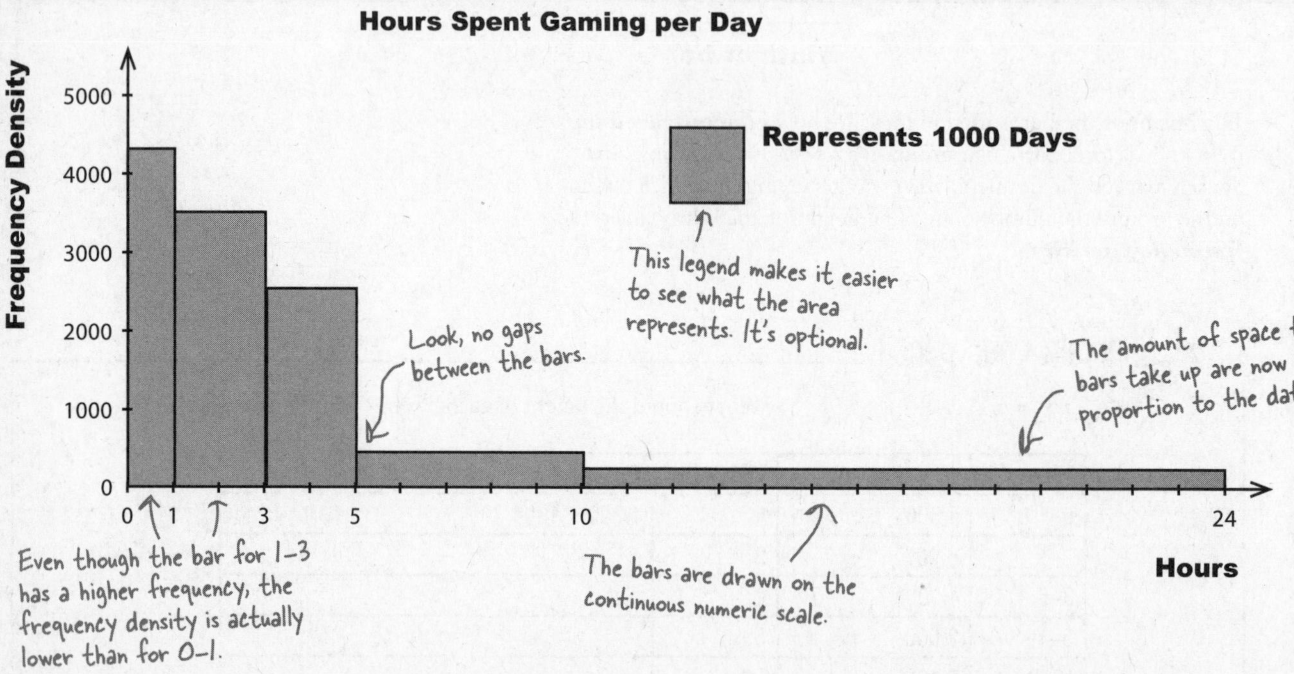

Hours Spent Gaming per Day
Frequency Density
5000
4000
3000
2000
1000
0
Represents 1000 Days
This legend makes it easier to see what the area represents. It's optional.
Look, no gaps between the bars.
The amount of space th
bars take up are now in
proportion to the data
0
1
3
5
10
24
Hours
Even though the bar for 1–3 has a higher frequency, the frequency density is actually lower than for 0–1.
The bars are drawn on the continuous numeric scale.

Frequency Density Up Close

Frequency density refers to the concentration of values in data. It's related to frequency, but it's not the same thing. Here's an analogy to demonstrate the relationship between the two.

Imagine you have a quantity of juice that you've poured into a glass like this:

Here's all your juice in the glass. It comes up to this level.

What if you then pour the *same quantity* of juice into a *different sized glass*, say one that's wider? What happens to the level of the juice? This time the glass is wider, so the level the juice comes up to is lower.

The level of the juice varies in line with the width of the glass; the wider the glass, the lower the level. The converse is true too; the narrower the glass, the higher the level of juice.

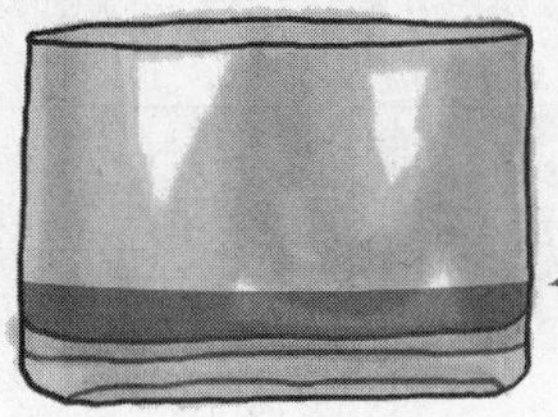
The glass is wider, so the level isn't as high.

So what does juice have to do with frequency density?

Juice = Frequency

Imagine that instead of pouring juice into glasses, you're "pouring" frequency into the bars on your chart. Just as you know the width of the glass, you know what width your bars are. And just like the space the juice occupies in the glass (width x height) tells you the quantity of juice in the glass, the area of the bar on the graph is equivalent to its frequency.

The frequency density is then equal to the height of the bar. Keeping with our analogy, it's equivalent to the level your juice comes to in each glass. Just as a wider glass means the juice comes to a lower level, a wider bar means a lower frequency density.

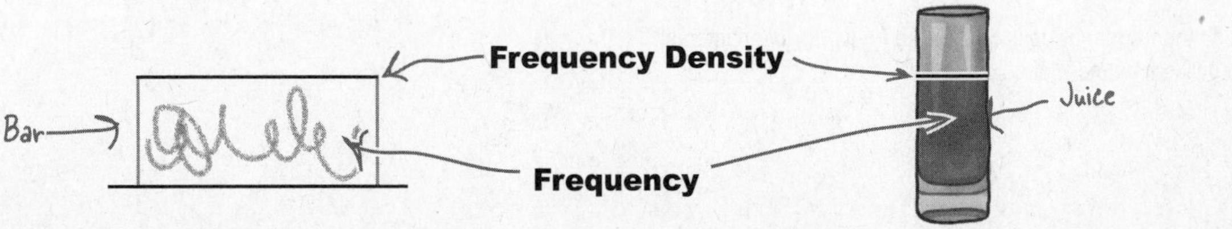
Frequency Density
Bar
Juice
Frequency

BULLET POINTS

- **Frequency density** relates to how concentrated the frequencies are for grouped data. It's calculated using

$$\text{Frequency density} = \frac{\text{Frequency}}{\text{Group width}}$$

- A **histogram** is a chart that specializes in grouped data. It looks like a bar chart, but the height of each bar equates to frequency density rather than frequency.
- When drawing histograms, the width of each bar is proportional to the width of its group. The bars are shown on a continuous numeric scale.
- In a histogram, the frequency of a group is given by the area of its bar.
- A histogram has no gaps between its bars.

there are no Dumb Questions

Q: Why do we use area to represent frequency when we're graphing histograms?

A: It's a way of making sure the relative sizes of each group stay in proportion to the data, and stay honest. With grouped data, we need a visual way of expressing the width of each group and also its frequency. Changing the width of the bars is an intuitive way of reflecting the group range, but it has the side effect of making some of the bar sizes look disproportionate.

Adjusting the bar height and using the area to represent frequency is a way around this. This way, no group is misrepresented by taking up too much or too little space.

Q: What's frequency density again?

A: Frequency density is a way of indicating how concentrated values are in a particular interval. It gives you a way of comparing different intervals that may be different widths. It makes the frequency proportional to the area of a bar, rather than height.

To find the frequency density, take the frequency of an interval, and divide it by the width.

Q: If I have grouped numeric data, but all the intervals are the same width, can I use a normal bar chart?

A: Using a histogram will better represent your data, as you're still dealing with grouped data. You really want your frequency to be proportional to its area, not height.

Q: Do histograms *have* to show grouped data? Can you use them for individual numbers as well as groups of numbers?

A: Yes, you can. The key thing to remember is to make sure there are no gaps between the bars and that you make each bar 1 wide. Normally you do this by positioning your number in the center of the bar.

As an example, if you wanted to draw a bar representing the individual number 1, then you'd draw a bar ranging from 0.5 to 1.5, with 1 in the center.

Exercise

Here's a histogram representing the number of levels completed in each game of Cows Gone Wild. How many games have been played in total? Assume each level is a whole number.

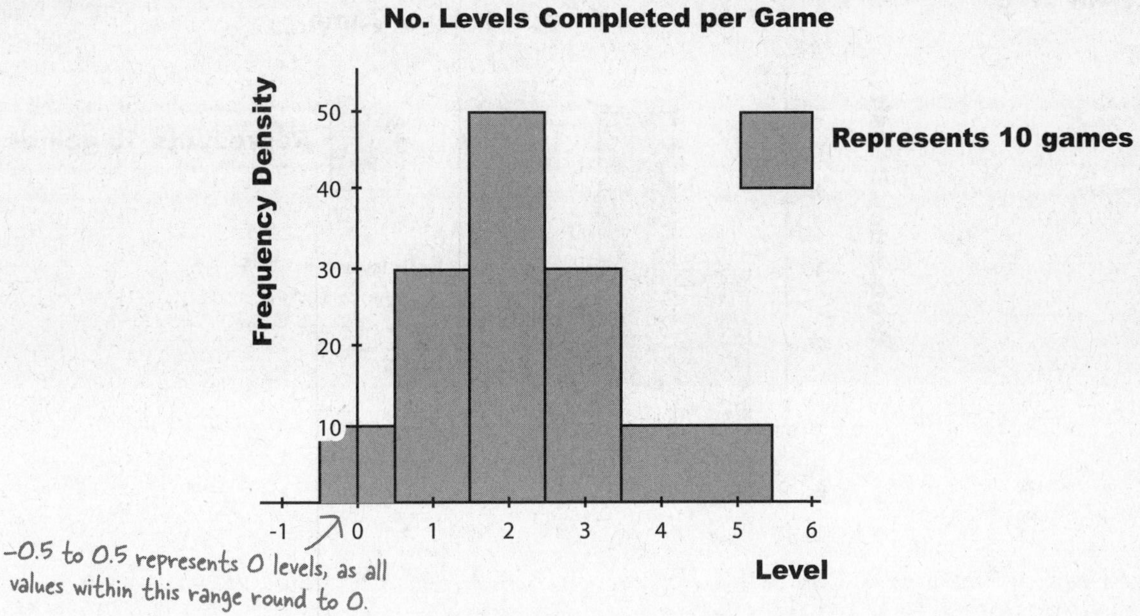

Here's a histogram representing the number of levels completed in each game of Cows Gone Wild. How many games have been played in total? Assume each level is a whole number.

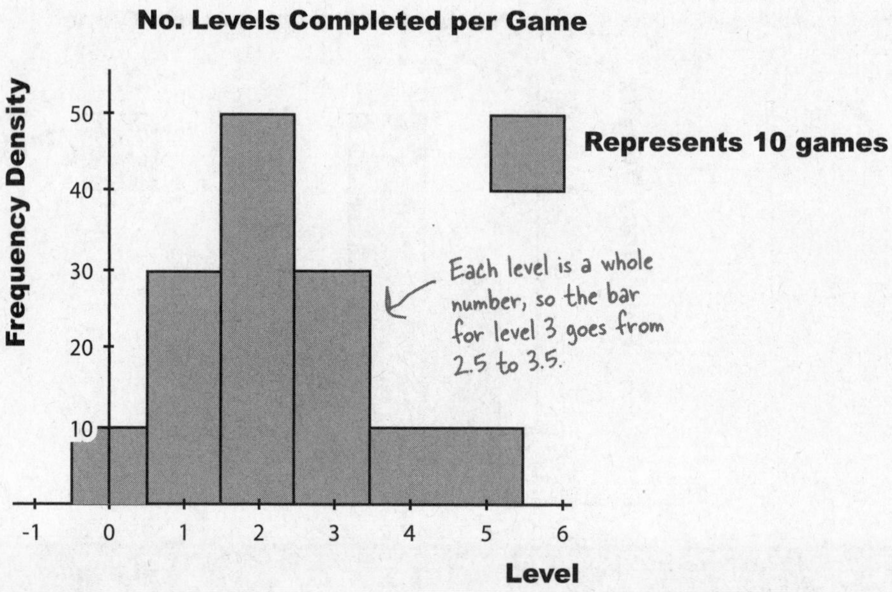

We need to find the total number of games played, which means we need to find the total frequency.

The total frequency is equal to the area of each bar added together. In other words, we multiply the width of each bar by its frequency density to get the frequency, and then add the whole lot up together.

Level	Width	Frequency Density	Frequency
0	1	10	1x10 = 10
1	1	30	1x30 = 30
2	1	50	1x50 = 50
3	1	30	1x30 = 30
4-5	2	10	2x10 = 20

Total Frequency = 10 + 30 + 50 + 30 + 20

= 140

Histograms can't do everything

While histograms are an excellent way to display grouped numeric data, there are still some kinds of this data they're not ideally suited for presenting—like running totals…

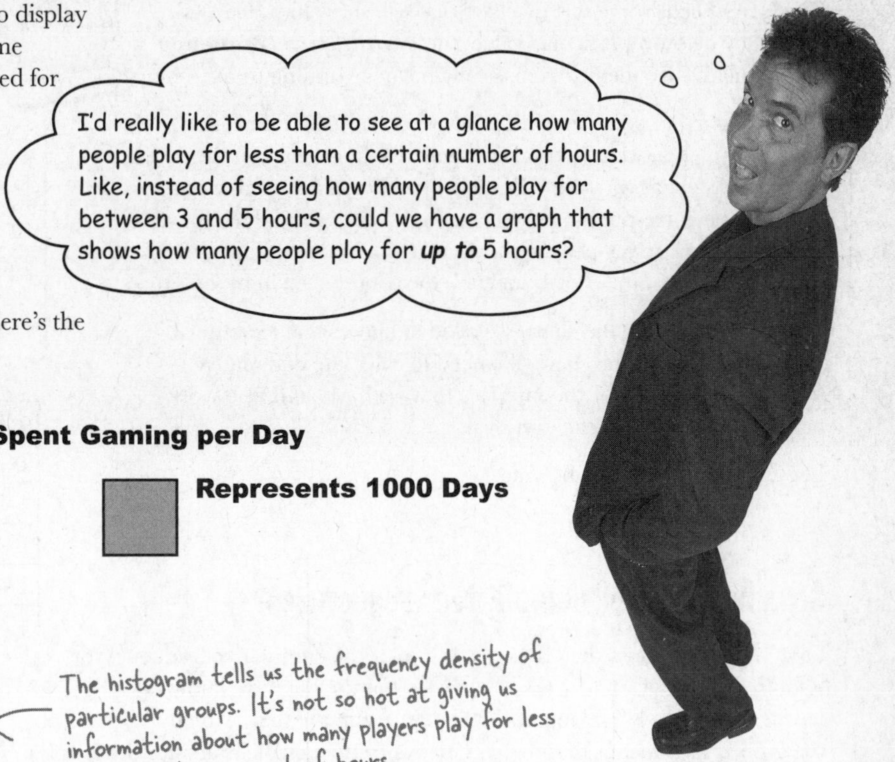
I'd really like to be able to see at a glance how many people play for less than a certain number of hours. Like, instead of seeing how many people play for between 3 and 5 hours, could we have a graph that shows how many people play for ***up to*** 5 hours?

Let's see if we can help the CEO out. Here's the histogram we had before.

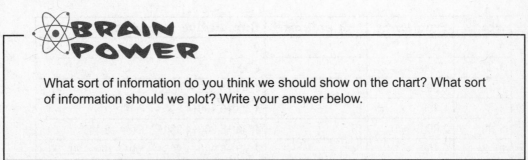

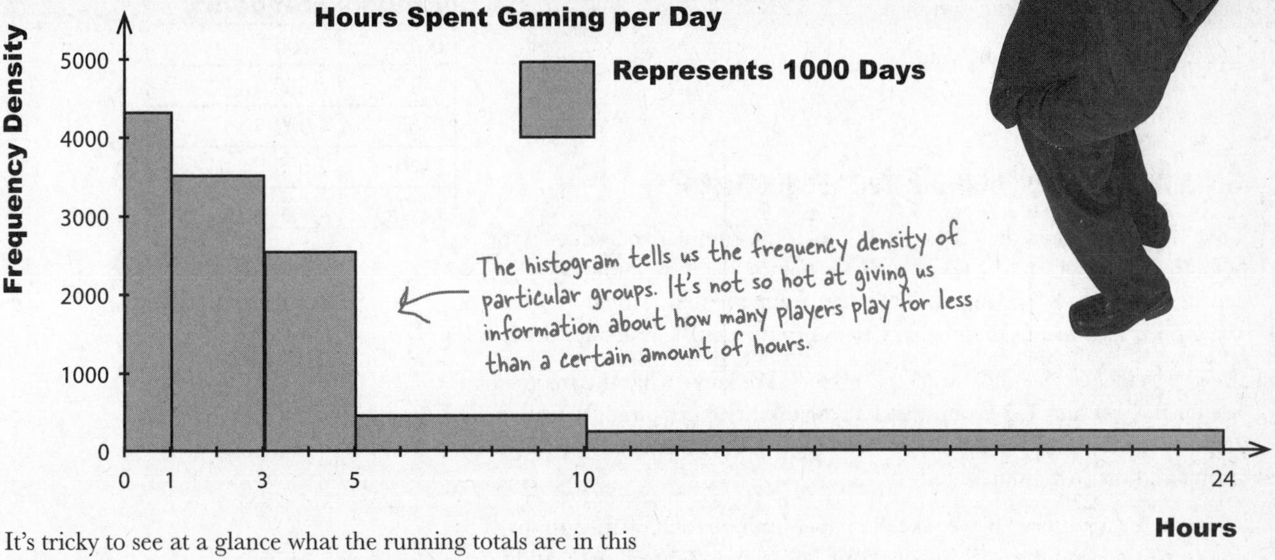
Hours Spent Gaming per Day
Represents 1000 Days
Frequency Density
5000
4000
3000
2000
1000
0
0
1
3
5
10
24
Hours
The histogram tells us the frequency density of particular groups. It's not so hot at giving us information about how many players play for less than a certain amount of hours.

It's tricky to see at a glance what the running totals are in this chart. In order to find the frequency of players playing for up to 5 hours, we need to add different frequencies together. We need another sort of chart…but what?

BRAIN POWER

What sort of information do you think we should show on the chart? What sort of information should we plot? Write your answer below.

Introducing cumulative frequency

The CEO needs some sort of chart that will show him the total frequency below a particular value: the ***cumulative frequency***. By cumulative frequency, we basically mean a running total.

What we need to come up with is some sort of graph that shows hours on the horizontal axis and cumulative frequency on the vertical axis. That way, the CEO will be able to take a value and read off the corresponding frequency up to that point. He'll be able to find out how many people play for up to 5 hours, 6 hours, or whatever other number of hours he's most interested in at the time.

Before we can draw the chart, we need to know what exactly we need to plot on the chart. We need to calculate cumulative frequencies for each of the intervals that we have, and also work out the upper limit of each interval.

Let's start by looking at the data.

Vital Statistics

Cumulative Frequency

The total frequency up to certain value. It's basically a running total of the frequencies.

Hours	Frequency
0–1	4,300
1–3	6,900
3–5	4,900
5–10	2,000
10–24	2,100

Here's the data.

So what are the cumulative frequencies?

First off, let's suppose the CEO needs to plot the cumulative frequency, or total frequency, of up to 1 hour. If we look at the data, we know that the frequency of the 0–1 group is 4300, and we can see that is the upper limit of the group. This means that the cumulative frequency of hours up to 1 is 4300.

Next, let's look at the total frequency up to 3. We know what the frequencies are for the 0–1 and 1–3 groups, and 3 is again the upper limit. To find the total frequency of hours up to 3, we add together the frequency of the 0–1 group and the 1–3 group.

Can you see a pattern? If we take the upper limit of each of the groups of hours, we can find the total frequency of hours up to that value by adding together the frequencies. Applying this to all the groups gives us

Hours	Frequency	Upper limit	Cumulative frequency
0	0	0	0
0–1	4,300	1	4,300
1–3	6,900	3	4,300+6,900 = 11,200
3–5	4,900	5	4,300+6,900+4,900 = 16,100
5–10	2,000	10	4,300+6,900+4,900+2,000 = 18,100
10–24	2,100	24	4,300+6,900+4,900+2,000+2,100 = 20,200

We've added in 0, as you can't play games for LESS than 0 hours a week.

Drawing the cumulative frequency graph

Now that we have the upper limits and cumulative frequencies, we can plot them on a chart. Draw two axes, with the vertical one for the cumulative frequency and the horizontal one for the hours. Once you've done that, plot each of the upper limits against its cumulative frequency, and then join the points together with a line like this:

Watch it!

Cumulative frequencies can never decrease.

If your cumulative frequency decreases at any point, check your calculations.

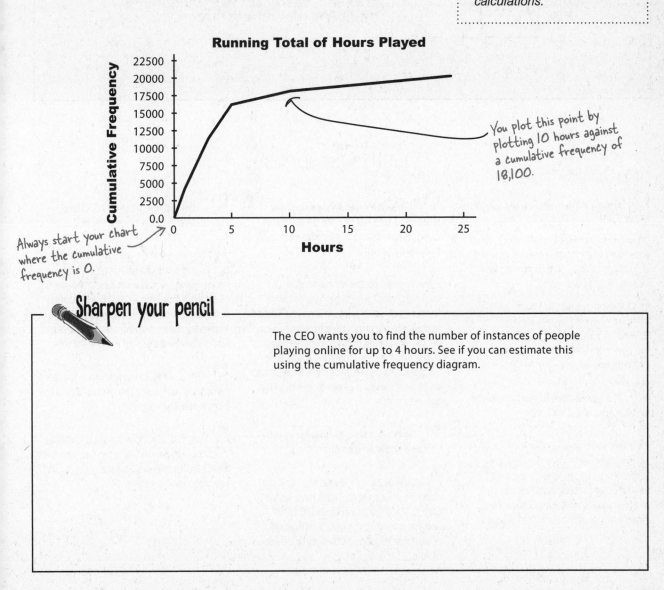

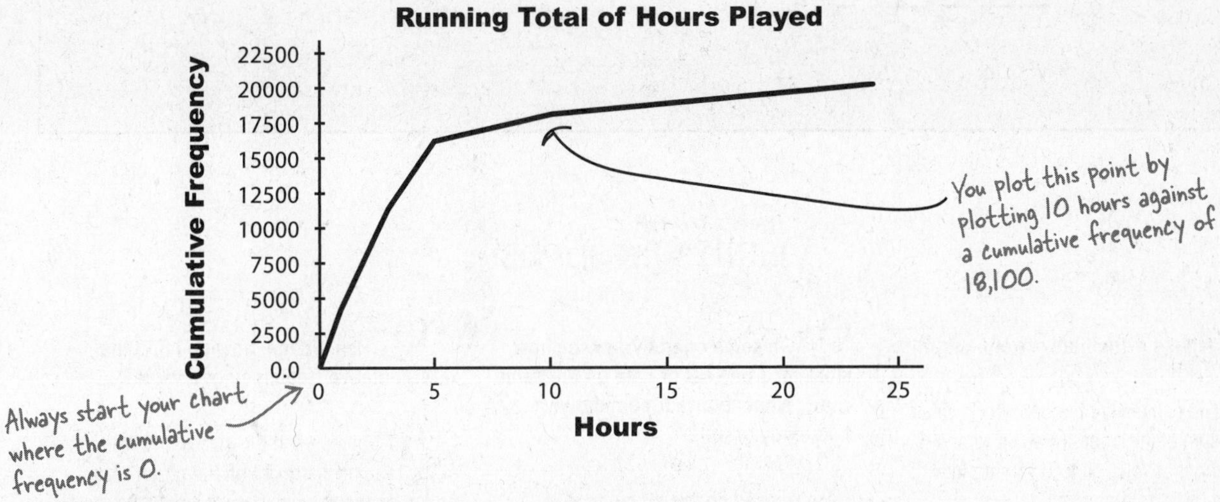

Sharpen your pencil

The CEO wants you to find the number of instances of people playing online for up to 4 hours. See if you can estimate this using the cumulative frequency diagram.

Sharpen your pencil Solution

The CEO wants you to find the number of instances of people playing online for less than 4 hours. See if you can estimate this using the cumulative frequency diagram.

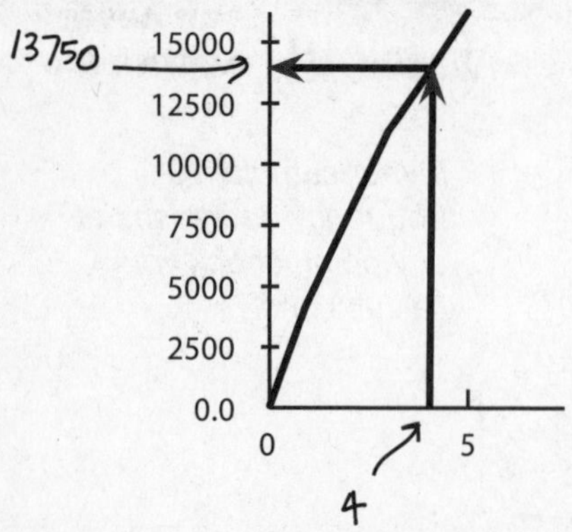

To do this, we find 4 on the horizontal axis, find where this value meets the line of the graph, and read off the corresponding cumulative frequency on the vertical axis.

This gives us an answer of approximately 13,750. In other words, there are approximately 13,750 instances of people playing online for under 4 hours.

there are no Dumb Questions

Q: What's a cumulative frequency?

A: The cumulative frequency of a value is the sum of the frequencies up to and including that value. It tells you the total frequency up to that point.

As an example, suppose you have data telling you how old people are. The cumulative frequency for value 27 tells you how many people there are up to and including age 27.

Q: Are cumulative frequency graphs just for grouped data?

A: Not at all; you can use them for any sort of numeric data. The key thing is whether you want to know the total frequency up to a particular value, or whether you're more interested in the frequencies of particular values instead.

Q: On some charts you can show more than one set of data on the same chart. What about for cumulative frequency graphs?

A: You can do this for cumulative frequency graphs by drawing a separate line for each set of data. If, say, you wanted to compare the cumulative frequencies by gender, you could draw one line showing males and the other females. It would be far more effective to show both lines on one chart, as it makes it easier to compare the two sets of data.

Q: Is there a limit to how many lines you can show on one chart?

A: There's no specific limit, as it all depends on your data. Don't have so many lines that the graph becomes cluttered and you can no longer use it to read off cumulative frequencies and compare sets of data.

Q: Remind me, how do I find the cumulative frequency of a value?

A: You can find the cumulative frequency by reading it straight off the graph. You locate the value you want to find the cumulative frequency for on the horizontal axis, find where this meets the cumulative frequency curve, and then read the value of cumulative frequency off the vertical axis.

Q: If I already know the cumulative frequency, can I use the graph to find the corresponding value?

A: Yes you can. Look for the cumulative frequency on the vertical axis, find where it meets the cumulative frequency curve, and then read off the value.

During the Manic Mango keynote, the CEO wants to explain how he wants to target particular age groups. He has a cumulative frequency graph showing the cumulative frequency of the ages, but he needs the frequencies too, and the dog ate the piece of paper they were written on. See if you can use the cumulative frequency graph to estimate what the frequencies of each group are.

The upper limit is 18 because someone is classed as being 17 from the point of their 17th birthday up until the point they turn 18. Ages are generally rounded down.

Age group	Upper limit	Cumulative frequency	Frequency
<0	0	0	0
0–17	18		
18–24			
25–39			
40–54			
55–79			
80–99			

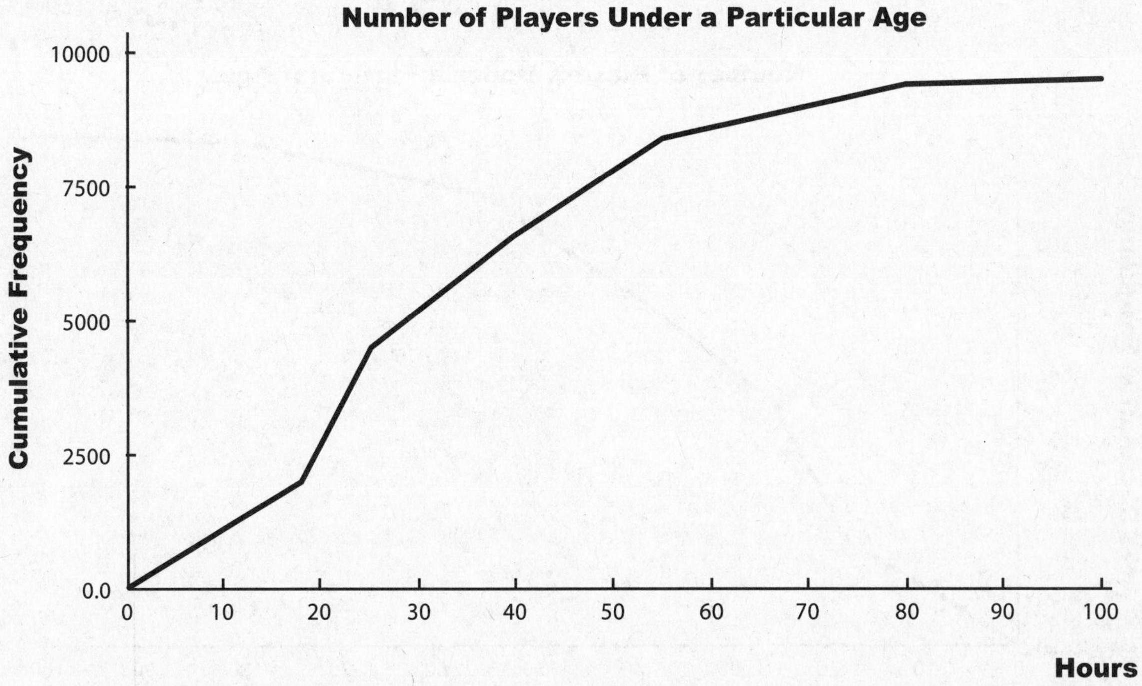

Exercise Solution

During the Manic Mango keynote, the CEO wants to explain how he wants to target particular age groups. He has a cumulative frequency graph showing the cumulative frequency of the ages, but he needs the frequencies too, and the dog ate the piece of paper they were written on. See if you can use the cumulative frequency graph to piece together what the frequencies of each group are.

Age group	Upper limit	Cumulative frequency	Frequency
<0	0	0	0
0–17	18	2,000	2,000
18–24	25	4,500	4,500 - 2,000 = 2,500
25–39	40	6,500	6,500 - 4,500 = 2,000
40–54	55	8,500	8,500 - 6,500 = 2,000
55–79	80	9,400	9,400 - 8,500 = 900
80–99	100	9,500	9,500 - 9,400 = 100

Use the chart to find the cumulative frequencies.

Don't worry if you get slightly different results—they're just estimates.

You can find the frequencies by taking the current cumulative frequency, and subtracting the previous one.

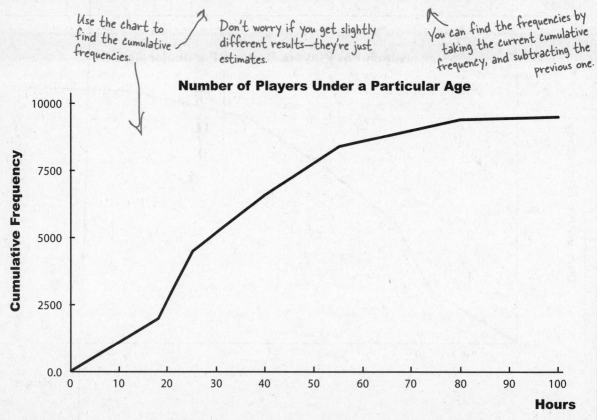

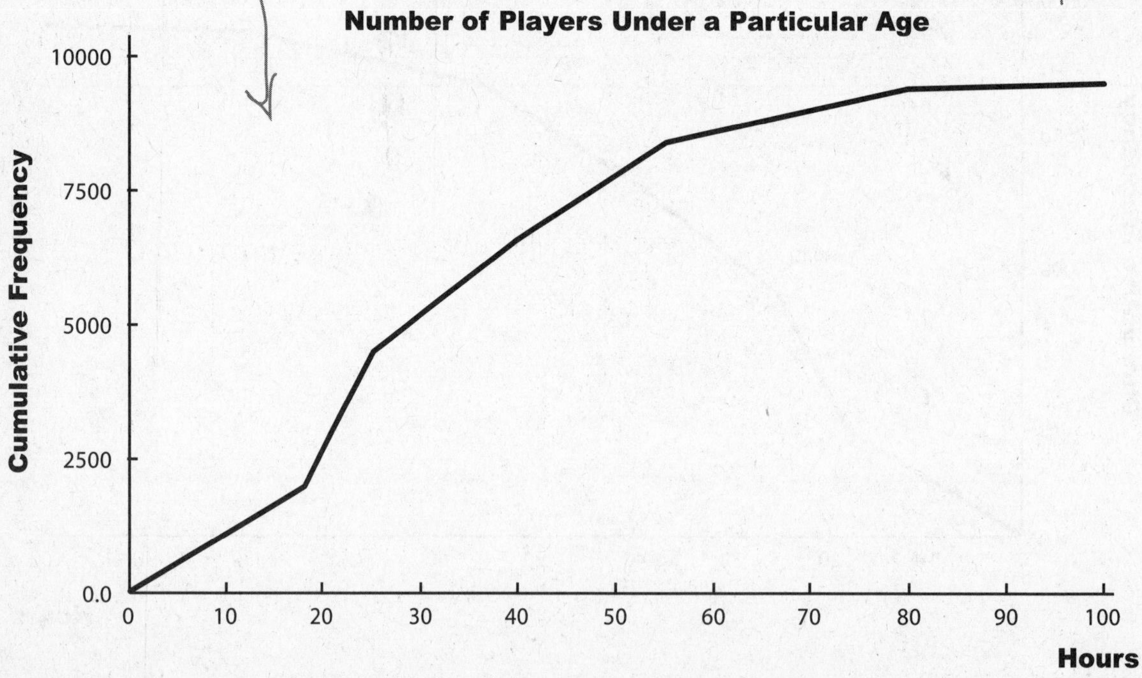

Choosing the right chart

The CEO is really happy with your work on cumulative frequency graphs, and your bonus is nearly in the bag. He's nearly finished preparing for the keynote, but there's just one more thing he needs: a chart showing Manic Mango profits compared with the profits of their main rivals. Which chart should he use?

Exercise

Here are two possible charts that the CEO could use in his keynote. Your task is to annotate each one, and say what you think the strengths and weaknesses are of each one relative to the other. Which would *you* pick?

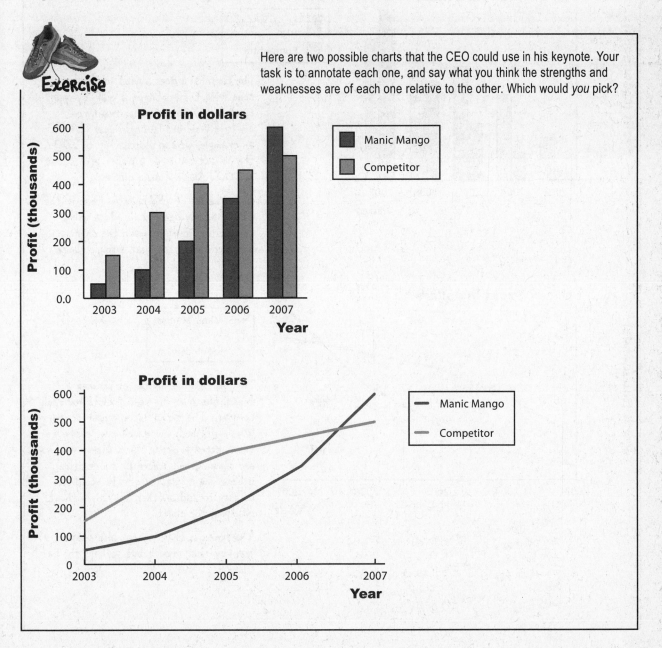

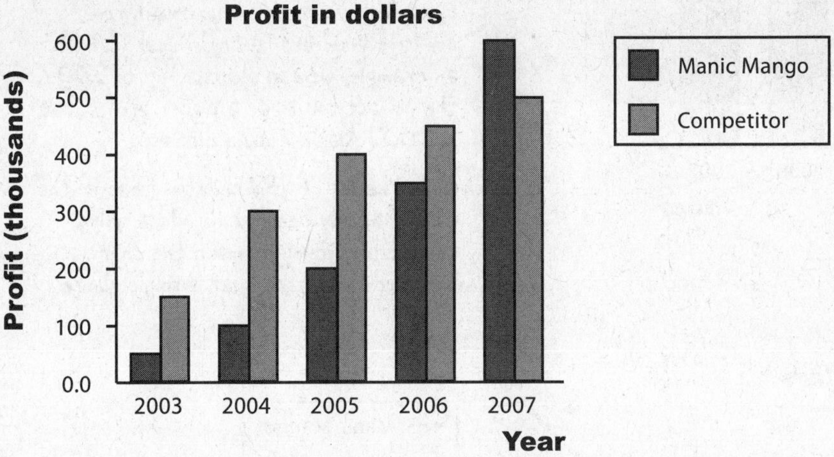

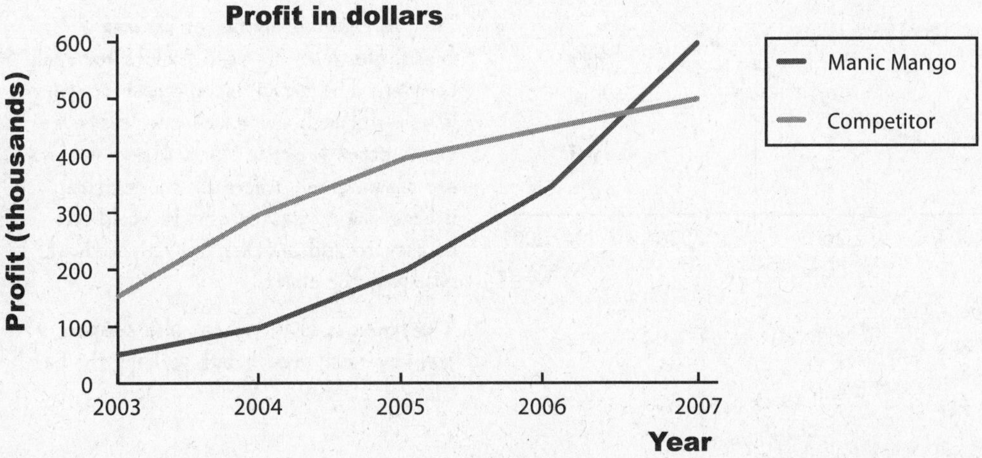

Here are two possible charts that the CEO could use in his keynote. Your task is to annotate each one, and say what you think the strengths and weaknesses are of each one relative to the other. Which would you pick?

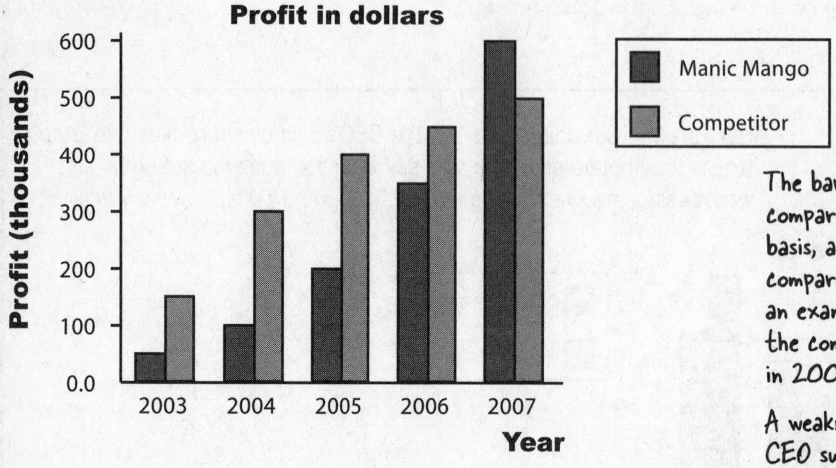

The bar chart does a good job of comparing the profit on a year-by-year basis, and it's great if you want to compare profits in an individual year. As an example, we can see that up to 2007, the competitor made a bigger profit, but in 2007 Manic Mango did.

A weaknesses of this chart is that if the CEO suddenly decided to add a third competitor, it might make the chart a bit harder to take in at a single glance.

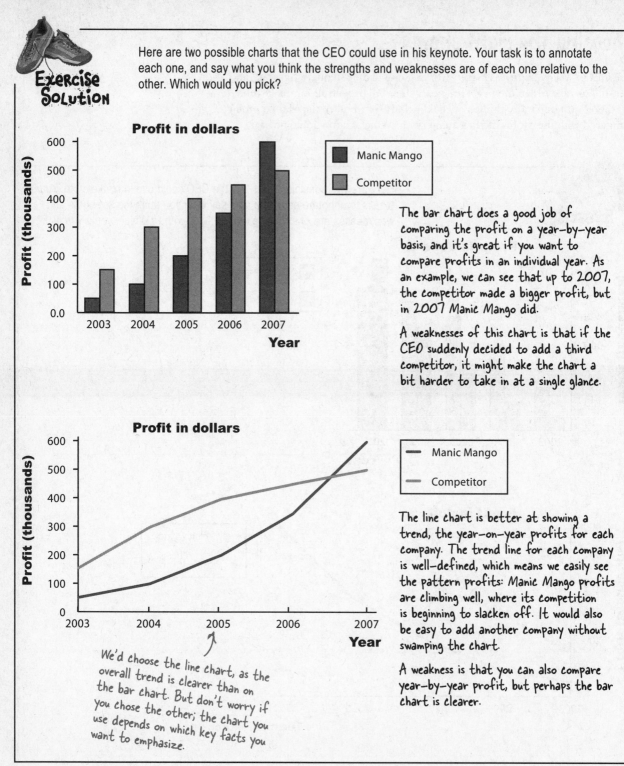

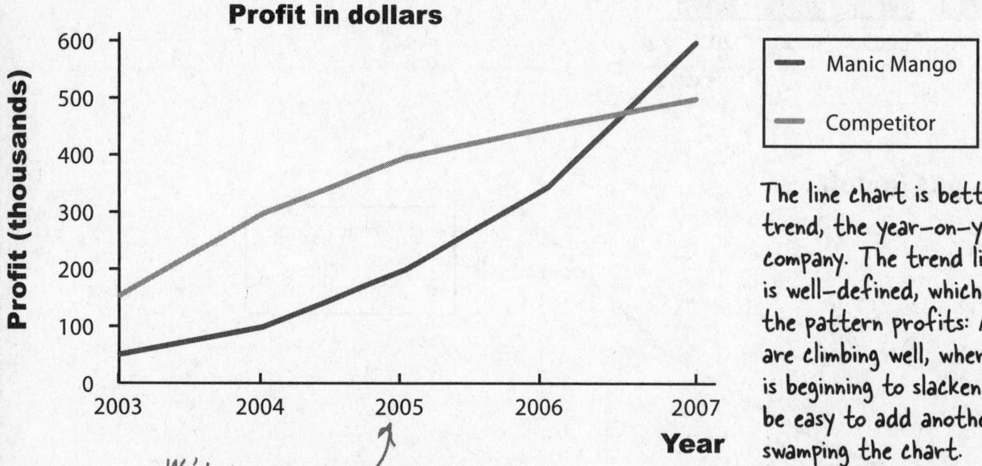

The line chart is better at showing a trend, the year-on-year profits for each company. The trend line for each company is well-defined, which means we easily see the pattern profits: Manic Mango profits are climbing well, where its competition is beginning to slacken off. It would also be easy to add another company without swamping the chart.

A weakness is that you can also compare year-by-year profit, but perhaps the bar chart is clearer.

We'd choose the line chart, as the overall trend is clearer than on the bar chart. But don't worry if you chose the other; the chart you use depends on which key facts you want to emphasize.

Line Charts Up Close

Line charts are good at showing trends in your data. For each set of data, you plot your points and then join them together with lines. You can easily show multiple sets of data on the same chart without it getting too cluttered. Just make sure it's clear which line is which.

As with other sorts of charts, you have a choice of showing frequency or percentages on the vertical axis. The scale you use all depends on what key facts you want to draw out.

Line charts are often used to show time measurements. Time always goes on the horizontal axis, and frequency on the vertical. You can read off the frequency for any period of time by choosing the time value on the horizontal axis, and reading off the corresponding frequency for that point on the line.

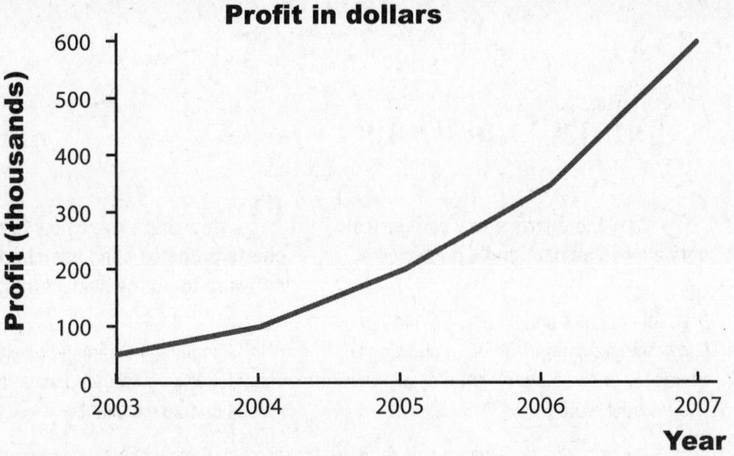

Line charts should be used for numerical data only, and not categorical. This is because it makes sense to compare different categories, but not to draw a trend line. Only use a line chart if you're comparing categories over some numerical unit such as time, and in that case you'd use a separate line for each category.

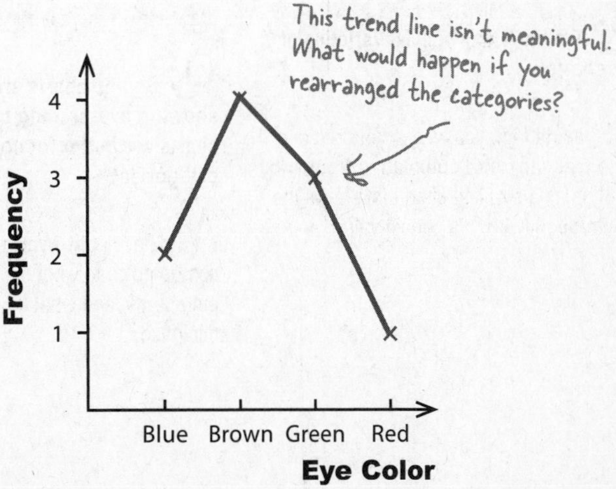

BULLET POINTS

- **Cumulative frequency** is the total frequency up to a particular value. It's a running total of the frequencies.
- Use a cumulative frequency graph to plot the upper limit of each group of data against cumulative frequency.
- Use a line chart if you want to show trends, for example over time.
- You can show more than one set of data on a line chart. Use one line for each set of data, and make sure it's clear which line is which.
- You can use line charts to make basic predictions as it's easy to see the shape of the trend. Just extend the trend line, trying to keep the same basic shape.
- **Don't use line charts to show categorical data** unless you're showing trends for each category, for example over time. If you do this, draw one line per category.

there are no Dumb Questions

Q: Are line charts the same thing as time series charts? I think I've heard that name used before.

A: A time series chart is really a line chart that focuses on time intervals, just like the examples we used. A line chart doesn't have to focus on just time, though.

Q: Are there any special varieties of line charts?

A: Yes. In fact, you've encountered one of them already. The **cumulative frequency graph** is a type of line chart that shows the total frequency up to a certain value

Q: Can line charts show categorical data as well as data that's numeric?

A: Line charts should only be used to show categorical data if you're showing trends for each category, and use a separate line for each category.

What you *shouldn't* do is use a line chart to draw lines from category to category.

Q: So line charts are better for showing overarching trends, and bar charts are better for comparing values or categories?

A: That's right. Which chart you use really comes down to what message you want to put across, and what key facts you want to minimize.

Q: Now that I know how to create charts properly, can I use charting software to do the heavy lifting?

A: Absolutely! Charting software can save you a lot of time and hard work, and the results can be excellent.

The key thing with using software to produce your charts is to remember that the software can't think for you. You still have to decide which chart best represents your key facts, and you have to check that the software produces exactly what you expect it to.

Manic Mango conquered the games market!

You've helped produce some killer charts for Manic Mango, and thanks to you, the keynote was a huge success. Manic Mango has gained tons of extra publicity for their games, and money from sponsorship and advertising is rolling in. The only thing left for you to do is think about all the things you could do and the places you could go with your well-earned bonus.

You've had your first taste of how statistics can help you and what you can achieve by understanding what's really going on. Keep reading and we'll show you more things you can do with statistics, and really start to flex those statistics muscles.

Nice work with those charts! We've got investors lining up outside the office. Take a long vacation, on me!

2 measuring central tendency

The Middle Way

People say I'm just an average golfer, but I'll show them I'm really mean.

Sometimes you just need to get to the heart of the matter.

It can be difficult to see patterns and trends in a big pile of figures, and finding the **average** is often the first step towards seeing the bigger picture. With averages at your disposal, you'll be able to quickly find the most representative values in your data and draw important conclusions. In this chapter, we'll look at several ways to calculate one of the most important statistics in town—mean, median, and mode—and you'll start to see how to effectively **summarize data** as concisely and usefully as possible.

Welcome to the Health Club

The Statsville Health Club prides itself on its ability to find the perfect class for everyone. Whether you want to learn how to swim, practice martial arts, or get your body into shape, they have just the right class for you.

The Health Club
Statsville's Premier Spa

The staff at the health club have noticed that their customers seem happiest when they're in a class with people their own age, and happy customers always come back for more. It seems that the key to success for the health club is to work out what a typical age is for each of their classes, and one way of doing this is to **calculate the average**. The average gives a representative age for each class, which the health club can use to help their customers pick the right class.

Here are the current attendees of the Power Workout class:

How do we work out the average age of the Power Workout class?

A common measure of average is the mean

It's likely that you've been asked to work out averages before. One way to find the average of a bunch of numbers is to add all the numbers together, and then divide by how many numbers there are.

In statistics, this is called the ***mean***.

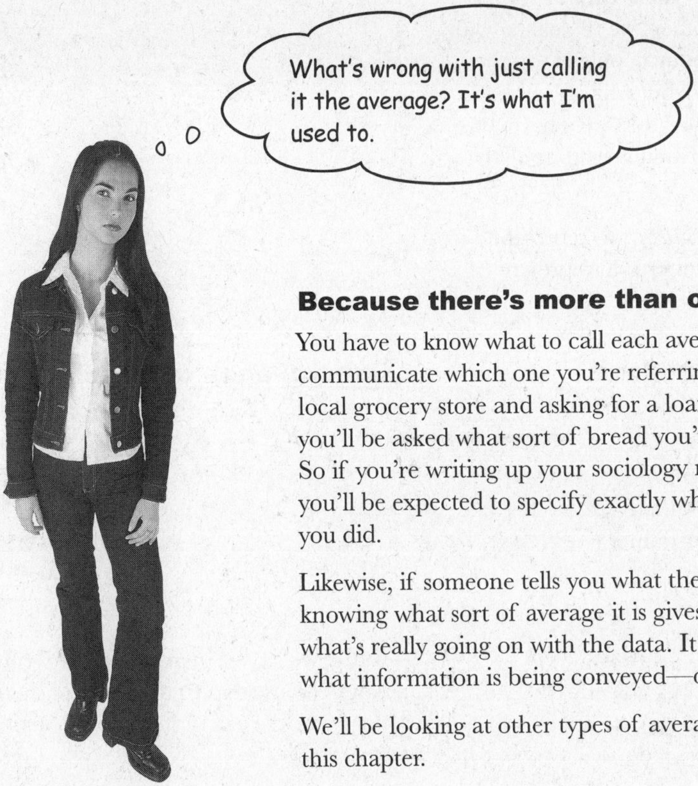

What's wrong with just calling it the average? It's what I'm used to.

Because there's more than one sort of average.

You have to know what to call each average, so you can easily communicate which one you're referring to. It's a bit like going to your local grocery store and asking for a loaf of bread. The chances are you'll be asked what sort of bread you're after: white, whole-grain, etc. So if you're writing up your sociology research findings, for example, you'll be expected to specify exactly what kinds of average calculations you did.

Likewise, if someone tells you what the average of a set of data is, knowing what sort of average it is gives you a better understanding of what's really going on with the data. It can give you vital clues about what information is being conveyed—or, in some cases, concealed.

We'll be looking at other types of averages, besides the mean, later in this chapter.

Mean math

If you want to really excel with statistics, you'll need to become comfortable with some common stats notation. It may look a little strange at first, but you'll soon get used to it.

Letters and numbers

Almost every statistical calculation involves adding a bunch of numbers together. As an example, if we want to find the mean of the Power Workout class, we first have to add the ages of all the class attendees together.

The problem statisticians have is how to generalize this. We don't necessarily know in advance how many numbers we're dealing with, or what they are. We currently know how many people are in the Power Workout class and what their ages are, but what if someone else joins the class? If we could only generalize this, we'd have a way of showing the calculation without rewriting it every time the class changes.

Statisticians get around this problem by using *letters* to represent *numbers*. As an example, they might use the letter **x** to represent ages in the Power Workout class like this:

Specific ages of class attendees → **General ages of class attendees**

19 20 20 20 21 → x_1 x_2 x_3 x_4 x_5

Each x represents one of the class ages.

Each x represents the age of a separate person in the class. It's a bit like labeling each person with a particular number *x*.

We use x_1 as a general way of representing this particular girl's age. She's 19 at the moment, but when she becomes 20, we'll still know her age as x_1. We won't have to rewrite any of our calculations.

Now that we have a general way of writing ages, we can use our *x*'s to represent them in calculations. We can write the sum of the 5 ages in the class as

$$\text{Sum} = x_1 + x_2 + x_3 + x_4 + x_5$$

But what if we don't know how many numbers we have to sum? What if we don't know how many people are in the class?

Dealing with unknowns

Statisticians use letters to represent unknown numbers. But what if we don't know how many numbers we might have to add together? Not a problem—we'll just call the number of values *n*. If we didn't know how many people were in the Power Workout class, we'd just say that there were *n* of them, and write the sum of all the ages as:

$$\text{Sum} = x_1 + x_2 + x_3 + x_4 + x_5 + \ldots + x_n$$

The "..." is a quick way of saying "and so on." In other words, just keep on adding x's.

In this case, x_n represents the age of the *n*th person in the class. If there were 18 people in the class, this would be x_{18}, the age of the 18th person.

Writing out all those x's looks like it could get arduous...

We can take another shortcut.

Writing $x_1 + x_2 + x_3 + x_4 + \ldots + x_n$ is a bit like saying "add age 1 to age 2, then add age 3, then add age 4, and keep on adding ages up to age *n*." In day-to-day conversation it's unlikely we'd phrase it like this. We're far more likely to say "add together all of the ages." It's quicker, simpler, and to the point.

We can do something similar in math notation by using the summation symbol Σ, which is the Greek letter Sigma. We can use Σx (pronounced "sigma x") as a quick way of saying "add together the values of all the *x*'s."

It all adds up now...

$$x_1 + x_2 + x_3 + x_4 + x_5 + \ldots + x_n = \Sigma x$$

Do you see how much quicker and simpler this is? It's just a mathematical way of saying "add your values together" without having to explicitly say what each value is.

Now that we know some handy math shortcuts, let's see how we can apply this to the mean.

Back to the mean

We can use math notation to represent the mean.

To find the mean of a group of numbers, we add them all together, and then divide by how many there are. We've already seen how to write summations, and we've also seen how statisticians refer to the total count of a set of numbers as n.

If we put these together, we can write the mean as:

$$\frac{\Sigma x}{n}$$

Add all the numbers together...

...then divide by however many there are.

In other words, this is just a math shorthand way of saying "add together all of the numbers, and then divide by how many numbers there are."

The mean has its own symbol

The mean is one of the most commonly used statistics around, and statisticians use it so frequently that they've given it a symbol all of its own: μ. This is the Greek letter mu (pronounced "mew"). Remember, it's just a quick way of representing the mean.

The mean is one of the most frequently used statistics. It can be represented with the symbol μ.

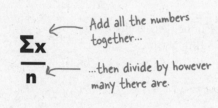

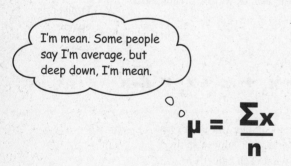

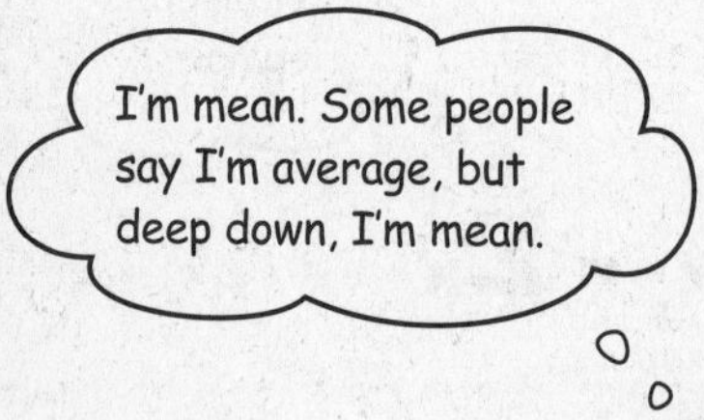

I'm mean. Some people say I'm average, but deep down, I'm mean.

$$\mu = \frac{\Sigma x}{n}$$

Sharpen your pencil

Have a go at calculating the mean age of the Power Workout class? Here are their ages.

How many people there are of each age →

Age	19	20	21
Frequency	1	3	1

Five Minute Mystery

The Case of the Ambiguous Average

The staff at a local company are feeling mutinous about perceived unfair pay. Most of them are paid $500 per week, a few managers are on a higher salary, and the CEO takes home $49,000 per week.

"The average salary here is $2,500 per week, and we're only paid $500," say the workers. "This is unfair, and we demand more money."

One of the managers overhears this and joins in with the demands. "The average salary here is $10,000 per week, and I'm only paid $4,000. I want a raise."

The CEO looks at them all. "You're all wrong; the average salary is $500 per week. Nobody is underpaid. Now get back to work."

What's going on with the average? Who do you think is right?

Sharpen your pencil Solution

Have a go at calculating the mean age of the Power Workout class? Here are their ages.

Age	19	20	21
Frequency	1	3	1

To find μ, we need to add all the people's ages, and divide by how many there are.

This gives us

$$\mu = \frac{19 + 20 + 20 + 20 + 21}{5}$$

Remember that there are 3 people of age 20.

$$= \frac{100}{5}$$

$$= 20$$

The mean age of the class is 20.

Handling frequencies

When you calculate the mean of a set of numbers, you'll often find that some of the numbers are repeated. If you look at the ages of the Power Workout class, you'll see we actually have 3 people of age 20.

It's really important to make sure that you include the *frequency* of each number when you're working out the mean. To make sure we don't overlook it, we can include it in our formula.

If we use the letter **f** to represent frequency, we can rewrite the mean as

$$\mu = \frac{\Sigma fx}{\Sigma f}$$

Multiply each number by its frequency, then add the results together.

Sum of the frequencies

This is just another way of writing the mean, but this time explicitly referring to the frequency. Using this for the Power Workout class gives us

$$\mu = \frac{1 \times 19 + 3 \times 20 + 1 \times 21}{5}$$

$$= 20$$

It's the same calculation written slightly differently.

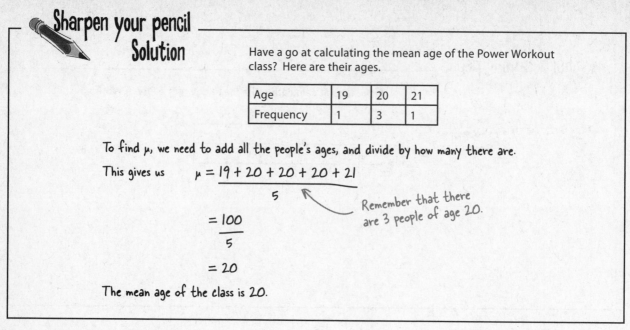

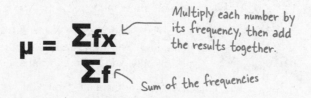

Back to the Health Club

Here's another hopeful customer looking for the perfect class. Can you help him find one?

I want a nice quiet class on a Tuesday evening where I can meet people my age. Do you think you can help me?

This sounds easy enough to sort out. According to the brochure, the Health Club has places available in three of its Tuesday evening classes. The first class has a mean age of 17, the second has a mean of 25, and the mean age of the third one is 38. Clive needs to find the class with an average student age that's closest to his own.

Meet Clive, a man in his late fifties who wants an exercise class composed of other middle-aged folks.

BRAIN BARBELL

Look at the mean ages for each class. Which class should Clive attend?

Everybody was Kung Fu fighting

Clive went along to the class with the mean age of 38. He was expecting a gentle class where he could get some nonstrenuous exercise and meet other people his own age. Unfortunately...

I ended up in the Kung Fu class with lots of young 'uns and a few ancient masters. My back will never be the same again.

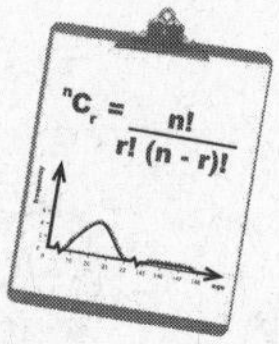

Vital Statistics

Mean

$$\mu = \frac{\sum x}{n}$$

$$\mu = \frac{\sum fx}{\sum f}$$

What could have gone wrong?

The last thing Clive expected (or wanted) was a class that was primarily made up of teenagers. Why do you think this happened?

We need to examine the data to find out. Let's see if sketching the data helps us see what the problem is.

Exercise

Sketch the histograms for the Kung Fu and Power Workout classes. (If you need a refresher on histograms, flip back to Chapter 1.) How do the shapes of the distributions compare? Why was Clive sent to the wrong class?

Power Workout Classmate Ages

Age	19	20	21
Frequency	1	3	1

Kung Fu Classmate Ages

Age	19	20	21	145	147
Frequency	3	6	3	1	1

Sketch the histograms for the Kung Fu and Power Workout classes. (If you need a refresher on histograms, flip back to Chapter 1.) How do the shapes of the distributions compare? Why was Clive sent to the wrong class?

Power Workout Classmate Ages

Age	19	20	21
Frequency	1	3	1

Kung Fu Classmate Ages

Age	19	20	21	145	147
Frequency	3	6	3	1	1

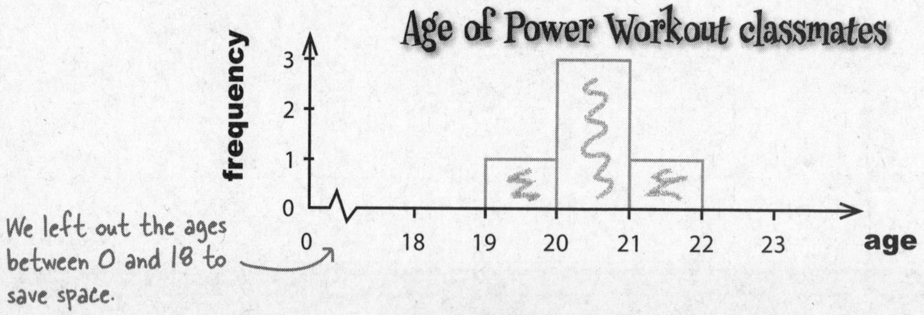

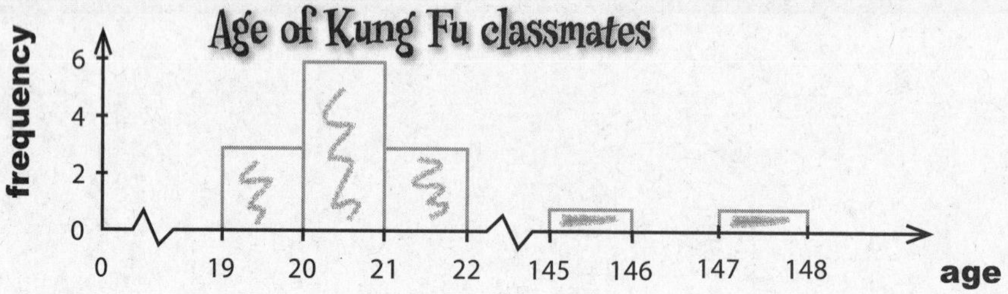

Sharpen your pencil

Do you think the mean can ever be the highest value in a set of numbers? Under what circumstances?

Our data has outliers

Did you see the difference in the shape of the charts for the Power Workout and Kung Fu classes? The ages of the Power Workout class form a smooth, symmetrical shape. It's easy to see what a typical age is for people in the class.

The shape of the chart for the Kung Fu class isn't as straightforward. Most of the ages are around 20, but there are two masters whose ages are much greater than this. Extreme values such as these are called ***outliers***.

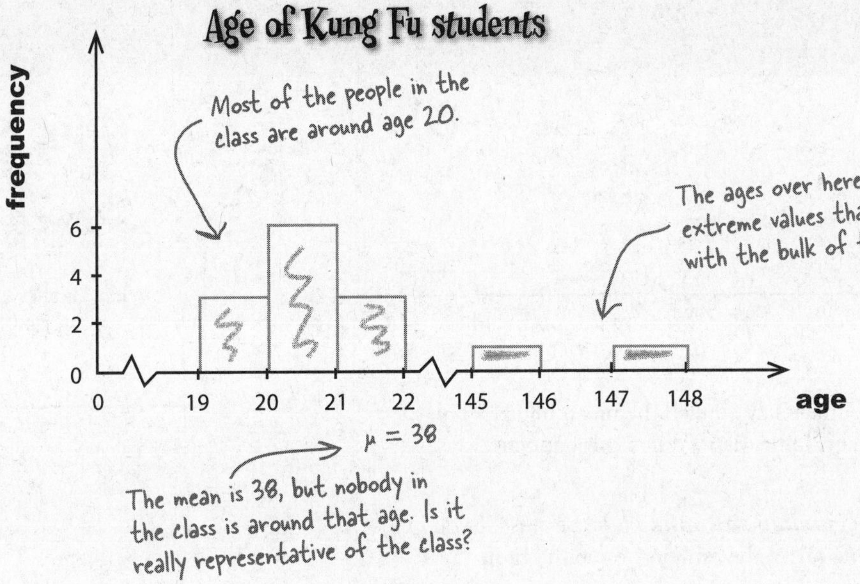
Age of Kung Fu students
frequency
6
4
2
0
0
19
20
21
22
145
146
147
148
age
Most of the people in the class are around age 20.
The ages over here are outliers. They're extreme values that don't really fit in with the bulk of the data.
μ = 38
The mean is 38, but nobody in the class is around that age. Is it really representative of the class?

BRAIN POWER

What would the mean have been if the ancient masters weren't part of the class? Compare this with the actual mean. What does this tell you about the effect of the outliers?

The ~~butler~~ outliers did it

If you look at the data and chart of the Kung Fu class, it's easy to see that most of the people in the class are around 20 years old. In fact, this would be the mean if the ancient masters weren't in the class.

We can't just ignore the ancient masters, though; they're still part of the class. Unfortunately, the presence of people who are way above the "typical" age of the class distorts the mean, pulling it upwards.

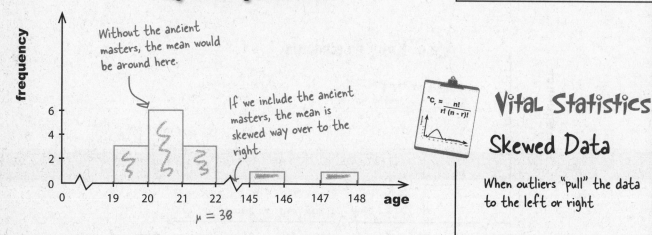

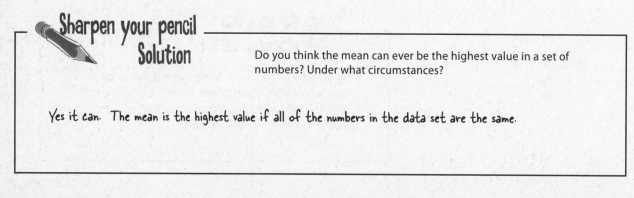

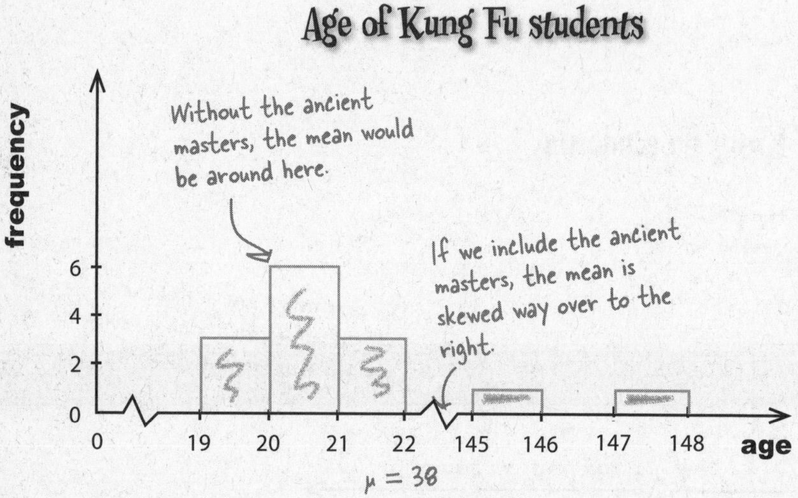

Age of Kung Fu students
frequency
Without the ancient masters, the mean would be around here.
If we include the ancient masters, the mean is skewed way over to the right.
0
2
4
6
0
19
20
21
22
145
146
147
148
age
μ = 38

Vital Statistics

Outlier

An extreme high or low value that stands out from the rest of the data

Vital Statistics

Skewed Data

When outliers "pull" the data to the left or right

Can you see how the outliers have pulled the mean higher? This effect is caused by outliers in the data. When this happens, we say the data is ***skewed***.

The Kung Fu class data is ***skewed to the right*** because if you line the data up in ascending order, the outliers are on the right.

Let's take a closer look at this.

Sharpen your pencil Solution

Do you think the mean can ever be the highest value in a set of numbers? Under what circumstances?

Yes it can. The mean is the highest value if all of the numbers in the data set are the same.

Skewed Data Up Close

Skewed to the right

Data that is skewed to the right has a “tail” of high outliers that trail off to the right. If you look at a right-skewed chart, you can see this tail. The high outliers in the Kung Fu class data distort the mean, pulling it higher—that is, to the right.

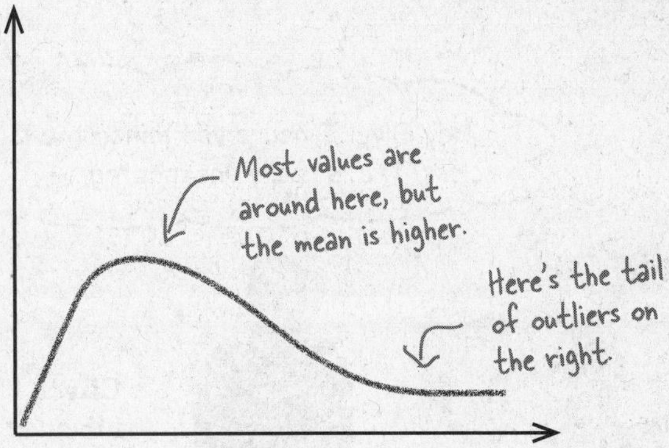

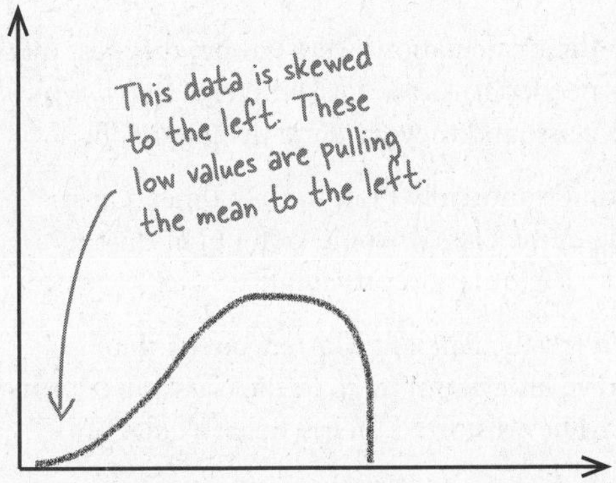

Most values are around here, but the mean is higher.

Skewed to the left

Here’s a chart showing data that is skewed to the left. Can you see the tail of outliers on the left? This time the outliers are low, and they pull the mean over to the left. In this situation, the mean is lower than the majority of values.

Symmetric data

In an ideal world, you’d expect data to be symmetric. If the data is symmetric, the mean is in the middle. There are no outliers pulling the mean in either direction, and the data has about the same shape on either side of the center.

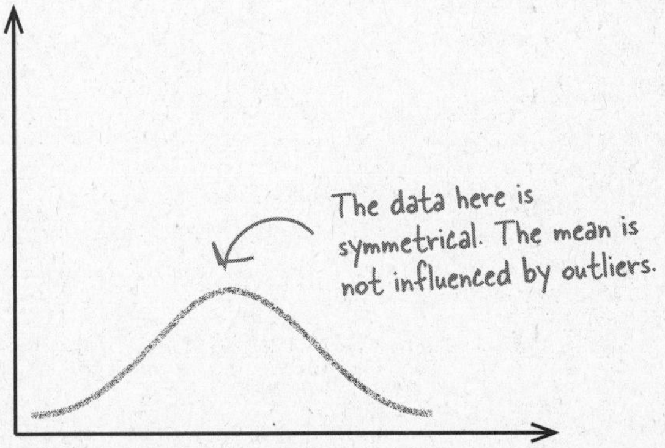

The data here is symmetrical. The mean is not influenced by outliers.

Watercooler conversation

Hey Clive! I heard you joined the Kung Fu class. That's really unexpected...

Clive: They told me the average age for the class is about 38, so I thought I'd fit in alright. I had to sit down after 5 minutes before my legs gave out.

Bendy Girl: But I didn't see anyone that age in the class, so there must have been some sort of mistake in their calculations. Why would they tell you that?

Clive: I don't think their calculations were wrong; they just didn't tell me what I really needed to know. I asked them what a typical sort of age is for the class, and they gave me the mean, 38.

Bendy Girl: And that's not really typical, is it? I mean, just looking at the people in the class, I would've thought that a younger age would be a bit more representative.

Clive: If only they'd left the Ancient Masters out of their calculations, I would've known not to go to the class. That's what did it; I'm sure of it. They distorted their whole calculation.

Bendy Girl: Well, if the Ancient Masters are such a big problem, why can't they just ignore them? Maybe that way they could come up with a more typical age for the class...

Finding the median

If the mean becomes misleading because of skewed data and outliers, then we need some other way of saying what a typical value is. We can do this by, quite literally, taking the middle value. This is a different sort of average, and it's called the ***median***.

To find the median of the Kung Fu class, line up all the ages in ascending order, and then pick the middle value, like this:

19 19 20 20 20 21 21 100 102

Here's the number in the middle. This is the median, 20.

If you line all the ages up in ascending order, the value 20 is exactly halfway along. Therefore, the median of the Kung Fu class is 20.

What if there had been an even number of people in the class?

19 20 20 20 21 21 100 102

If there's an even number of people in the class, there will be no single middle number.

If you have an even set of numbers, just take the **mean of the two middle numbers** (add them together, and divide by 2), and that's your median. In this case, the median is 20.5.

The median is always in the middle. It's the middle value.

BRAIN POWER

We've seen that if you have 9 numbers, the median is the number at position 5. If you have 8 numbers, it's the number at position 4.5 (halfway between the numbers at position 4 and 5). What about if you have *n* numbers?

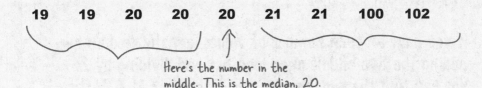

How to find the median in three steps:

1. Line your numbers up in order, from smallest to largest.

2. If you have an odd number of values, the median is the one in the middle. If you have *n* numbers, the middle number is at position $(n + 1) / 2$.

3. If you have an even number of values, get the median by adding the two middle ones together and dividing by 2. You can find the midpoint by calculating $(n + 1) / 2$. The two middle numbers are on either side of this point.

there are no Dumb Questions

Q: Is it still OK to use the mean with skewed data if I really want to?

A: You can, and people often do. However, in this situation the mean won't give you the best representation of what a typical value is. You need the median.

Q: You say that, but surely the whole point of the mean is that it gives a typical value. It's the average.

A: The big danger is that the mean will give a value that doesn't exist in the data set. Take the Kung Fu class as an example. If you were to go into the class and pick a person at random, the chances are that person would be around 20 years old because most people in the class are that sort of age. Just going with the mean doesn't give you that impression. Finding the median can give you a more accurate perspective on the data.

But sometimes even the median will give a value that's not in the data set, like our example on the previous page. That's precisely why there's more than one sort of average; sometimes you need to use different methods in order to accurately say what a typical value is.

Q: So is the median better than the mean?

A: Sometimes the median is more appropriate than the mean, but that doesn't make it better. Most of the time you'll need to use the mean because it usually offers significant advantages over the median. The mean is more stable when you are sampling data. We'll come back to this later in the book.

Q: How do I use the mean or median with categorical data? What about examples like the data on page 9 of Chapter 1?

A: You can only find the mean and median of numerical data. Don't worry, though, there's another sort of average that deals with just this problem that we'll explore later on.

Q: I always get right- and left-skewed data mixed up. How do I remember which is which?

A: Skewed data has a "tail" of outliers. To see which direction the data is skewed in, find the direction the tail is pointing in. For example, right-skewed data has a tail that points to the right.

BE the data

Your job is to play like you're the data, and say what the median is for each set, whether the data is skewed, and whether the mean is higher or lower than the median. Give reasons why.

Values	1	2	3	4	5	6	7	8
Frequency	4	6	4	4	3	2	1	1

Values	1	4	6	8	9	10	11	12
Frequency	1	1	2	3	4	4	5	5

BE the data Solution

Your job is to play like you're the data, and say what the median is for each set, whether the data is skewed, and whether the mean is higher or lower than the median. Give reasons why.

Values	1	2	3	4	5	6	7	8
Frequency	4	6	4	4	3	2	1	1

There are 25 numbers, and if you line them all up, the median is half way along, i.e., 13 numbers along. The median is 3. The data is skewed to the right, which pulls the mean higher. Therefore, the mean is higher than the median.

Values	1	4	6	8	9	10	11	12
Frequency	1	1	2	3	4	4	5	5

The median here is 10. The data is skewed to the left, so the mean is pulled to the left. Therefore, the mean is lower than the median.

If the data is skewed to the right, the mean is to the right of the median (higher).

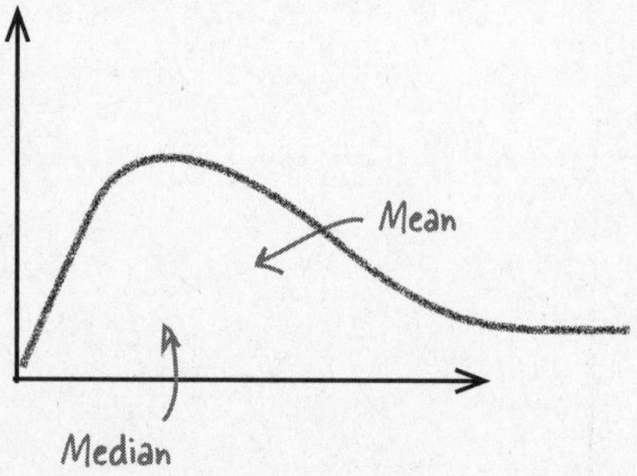

If the data is skewed to the left, the mean is to the left of the median (lower).

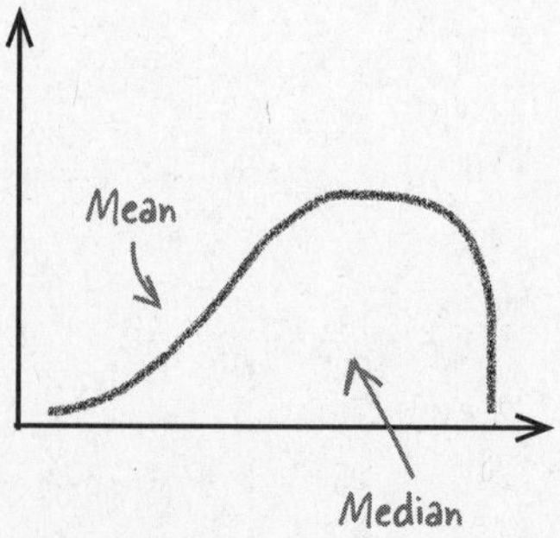

Business is booming

Your work on averages is really paying off. More and more people are turning up for classes at the Health Club, and the staff is finding it much easier to find classes to suit the customers.

This teenager is after a swimming class where he can make new friends his own age.

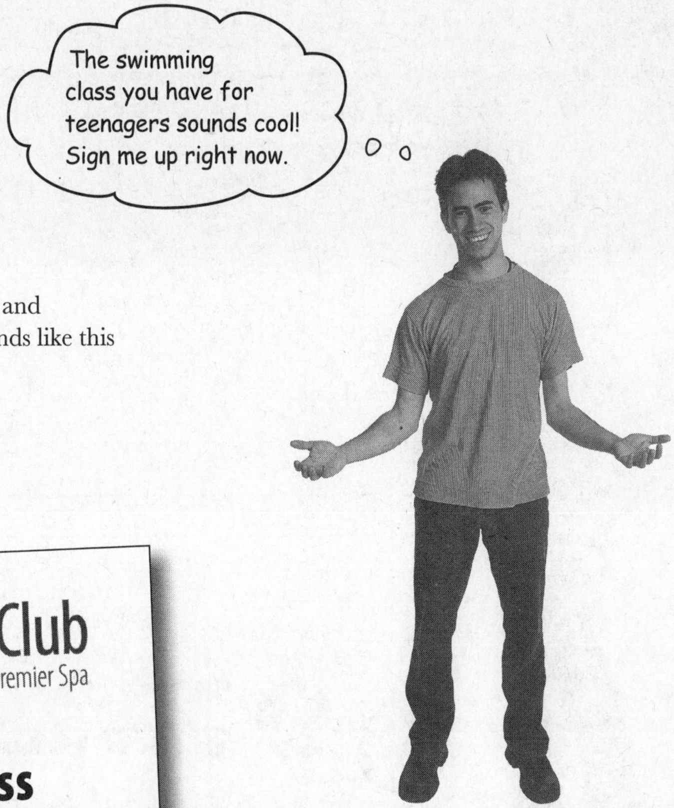

The swimming class you have for teenagers sounds cool! Sign me up right now.

The swimming class has a mean age of 17, and coincidentally, that's the median too. It sounds like this class will be perfect for him.

Let's see what happens...

The Little Ducklings swimming class

The Little Ducklings class meets at the swimming pool twice a week. In this class, parents teach their very young children how to swim, and they all have lots of fun splashing about in the water.

Look who turned up for lessons...

Stop! Where's your baby?

What the heck?

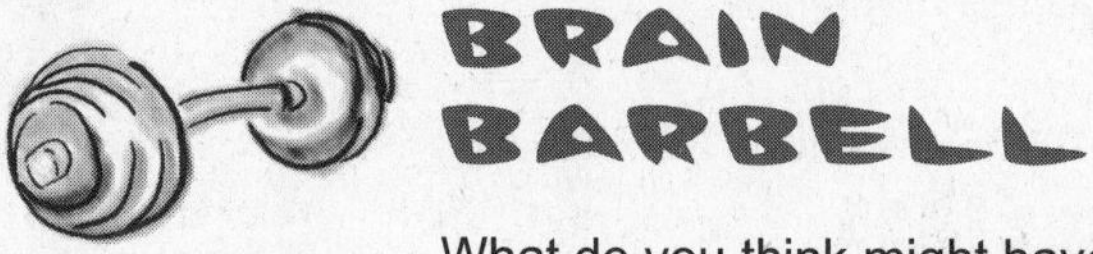

BRAIN BARBELL

What do you think might have gone wrong this time?

Frequency Magnets

Here are the ages of people who go to the Little Ducklings class, but some of the frequencies have fallen off. Your task is to put them in the right slot in the frequency table. Nine children and their parents go to the class, and the mean and median are both 17.

Age	1	2	3	31	32	33
Frequency	3		2	2		

3 4 4

Sharpen your pencil

When you've figured out the frequencies for the Little Ducklings class, sketch the histogram. What do you notice?

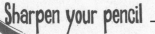

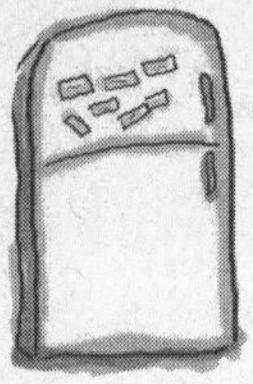

Frequency Magnets

Here are the ages of people who go to the Little Ducklings class, but some of the frequencies have fallen off. Your task is to put them in the right slot in the frequency table. Nine children and their parents go to the class, and the mean and median are both 17.

Age	1	2	3	31	32	33
Frequency	3	4	2	2	4	3

We're told there are 9 children, so the frequencies of the children must add up to 9. There must be 4 children of age 2.

The mean is 17. If we substitute in a and b for the unknown frequencies, we get

$$\frac{1 \times 3 + 2 \times 4 + 3 \times 2 + 31 \times 2 + 32a + 33b}{18} = 17$$

Multiply both sides by 18.

$$3 + 8 + 6 + 62 + 32a + 33b = 17 \times 18 = 306$$

$$32a + 33b = 306 - (3 + 8 + 6 + 62) = 306 - 79$$

$$32a + 33b = 227$$

As $32a + 33b$ is odd, this means that b must be 3, and a must be 4.

Sharpen your pencil Solution

When you've figured out the frequencies for the Little Ducklings class, sketch the histogram. What do you notice?

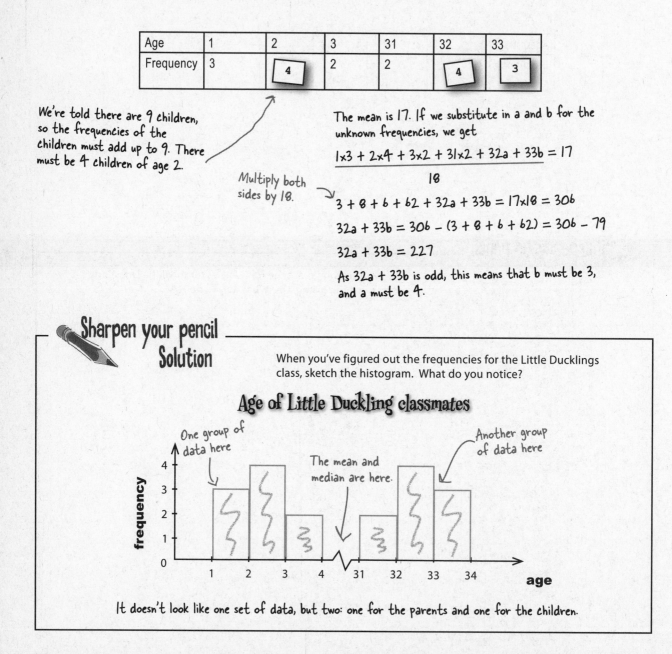

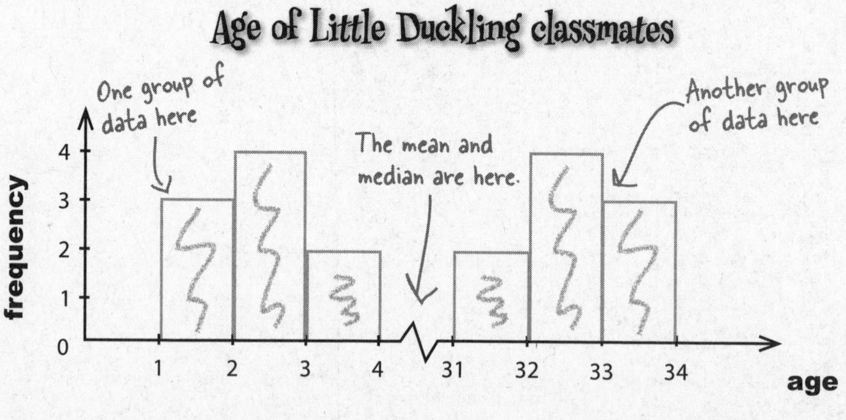

It doesn't look like one set of data, but two: one for the parents and one for the children.

What went wrong with the mean and median?

Let's take a closer look at what's going on.

Here are the ages of people who go to the Little Ducklings class.

1 1 1 2 2 2 2 3 3 | 31 31 32 32 32 32 33 33 33

There's an even number of values, so the median is halfway between 3 and 31. Take the mean of these two numbers—(3 +31)/2—and you get 17.

The mean and median for the class are both 17, even though there are no 17-year-olds in the class!

But what if there had been an odd number of people in the class. Both the mean and median would still have been misleading. Take a look:

1 1 1 2 2 2 2 2 3 (3) 31 31 32 32 32 32 33 33 33

If we add another 2-year-old to the class, the median becomes 3. But what about the adults?

If another two-year-old were to join the class, like we see above, the median would still be 3. This reflects the age of the children, but doesn't take the adults into account.

1 1 1 2 2 2 2 2 3 (31) 31 31 32 32 32 32 33 33 33

If we add another 31-year-old to the class, the median instead becomes 31. This time, we ignore the kids!

If another 33-year-old were added to the class instead, the median would be 31. But that fails to reflect all the kids in the class.

Whichever value we choose for the average age, it seems misleading.

What should we do for data like this?

Sharpen your pencil

Here's where you have to really think about how you can best give a representative age (or ages) for the Little Ducklings class. Here's a reminder of the data:

Age	1	2	3	31	32	33
Frequency	3	4	2	2	4	3

1. Why do you think the mean and median both failed for this data? Why are they misleading?

2. If you had to pick one age to represent this class, what would it be? Why?

3. What if you could pick **two** ages instead? Which two ages would you pick, and why?

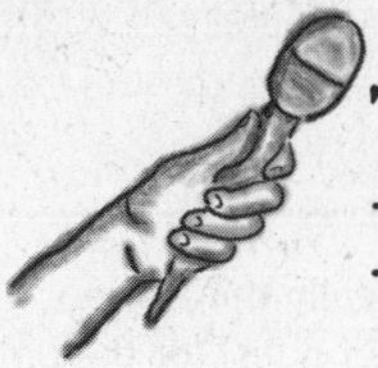

The Mean Exposed

This week's interview: The many types of average

Head First: Hey, Average, great to have you on the show...

Mean: Please, call me Mean.

Head First: Mean? But I thought you were Average. Did we mix up the guest list?

Mean: Not at all. You see, there's more than one type of Average in Statsville, and I'm one of them, the Mean.

Head First: There's more than one Average? That sounds kinda complicated.

Mean: Not really, not once you get used to it. You see, we all say what a typical value is for a set of numbers, but we have different opinions about how to say what that is.

Head First: So which one of you is the real Average? You know, the one where you add all the numbers together, and then divide by however many numbers there are?

Mean: That's me, but please don't call me the "real" Average; the other guys might get offended. The truth is that a lot of people new to Statsville see me as being Mr. Average. I have the same calculation that students see when they first encounter Averages in basic arithmetic. It's just that in Statsville, I'm called Mean to differentiate between the other sorts of Average.

Head First: So do you have any other names?

Mean: Well, I do have a symbol, μ. All the rock stars have them. Well, some of them do. I do anyway. It's Greek, so that makes me exotic.

Head First: So why are any of the other sorts of Average needed?

Mean: I hate to say it, but I have weaknesses. I lose my head a bit when I deal with data that has outliers. Without the outliers I'm fine, but then when I see outliers, I get kinda mesmerized and move towards them. It's led to a few problems. I can sometimes end up well away from where most of the values are. That's where Median comes in.

Head First: Median?

Mean: He's so level-headed when it comes to outliers. No matter what you throw at him, he always stays right in the middle of the data. Of course, the downside of the Median is that you can't calculate him as such; you can only work out what position he should be in. It makes him a bit less useful further down the line.

Head First: Do the two of you ever have the same value?

Mean: We do if the data's symmetric; otherwise, there tends to be differences between us. As a general rule, if there are outliers, then I tend to wander towards them, while Median stays where he is.

Head First: We're running out of time, so here's one final question. Are there any situations where both you and Median have problems saying what a typical value is?

Mean: I'm afraid there is. Sometimes we need a little helping hand from another sort of Average. He doesn't get out all that much, but he's a useful guy to know. Stick around, and I'll show you some of the things he's up to.

Head First: Sounds great!

Sharpen your pencil Solution

Here's where you have to really think about how you can best give a representative age (or ages) for the Little Ducklings class. Here's a reminder of the data:

Age	1	2	3	31	32	33
Frequency	3	4	2	2	4	3

1. Why do you think the mean and median both failed for this data? Why are they misleading?

Both the mean and median are misleading for this set of data because neither fully represents the typical ages of people in the class. The mean suggests that teenagers go to the class, when in fact there are none. The median also has this problem, but it can fluctuate wildly if other people join the class.

2. If you had to pick one age to represent this class, what would it be? Why?

It's not really possible to pick a single age that fully represents the ages in the class. The class is really made up of two sets of ages, those of the children and those of the parents. You can't really represent both of these groups with a single number.

3. What if you could pick **two** ages instead? Which two ages would you pick, and why?

As it looks like there are two sets of data, it makes sense to pick two ages to represent the class, one for the children and one for the parents. We'd choose 2 and 32, as these are the two age groups with the most people in them

Introducing the mode

In addition to the mean and median, there's a third type of average called the ***mode***. The mode of a set of data is the most popular value, the value with the highest frequency. Unlike the mean and median, the mode absolutely *has* to be a value in the data set, and it's the most frequent value.

Sometimes data can have more than one mode. If there is more than one value with the highest frequency, then each one of these values is a mode. If the data looks as though it's representing more than one trend or set of data, then we can give a mode for each set. If a set of data has two modes, then we call the data ***bimodal***.

This is exactly the situation we have with the Little Ducklings class. There are really two sets of ages we're looking at, one for parents and one for children, so there isn't a single age that's totally representative of the entire class. Instead, we can say what the mode is for each set of ages. In the Little Ducklings class, ages 2 and 32 have the highest frequency, so these ages are both modes. On a chart, the modes are the ones with the highest frequencies.

Age	1	2	3	31	32	33
Frequency	3	4	2	2	4	3

These two values are the most popular, so they are both modes.

Age of Little Duckling classmates

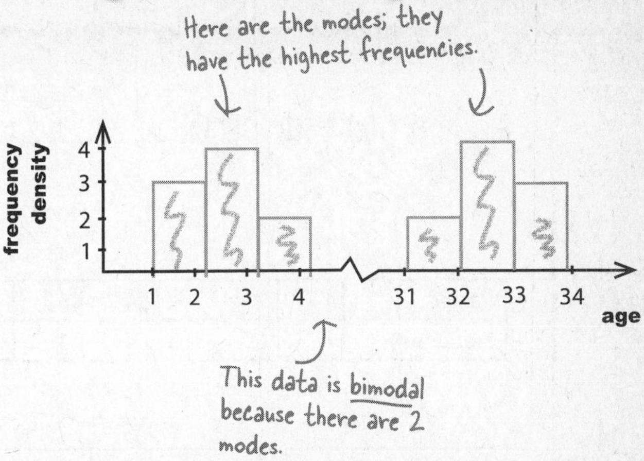

It even works with categorical data

The mode doesn't just work with numeric data; it works with categorical data, too. In fact, it's the *only* sort of average that works with categorical data. When you're dealing with categorical data, the mode is the most frequently occurring category.

You can also use it to specify the highest frequency *group* of values. The category or group with the highest frequency is called the ***modal class***.

Number of sessions by class type

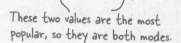

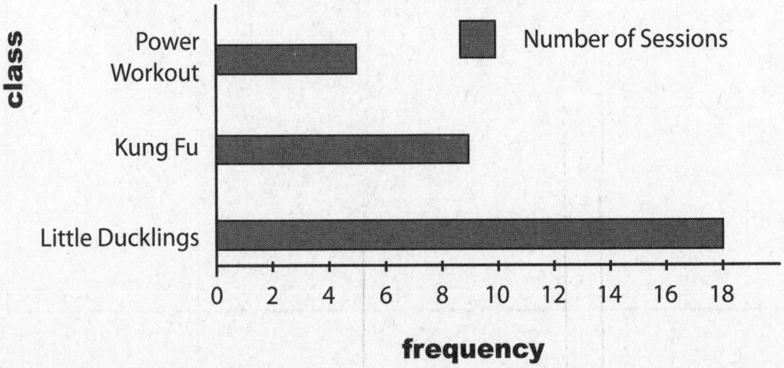

Three steps for finding the mode:

1. **Find all the distinct categories or values in your set of data.**

2. **Write down the frequency of each value or category.**

3. **Pick out the one(s) with the highest frequency to get the mode.**

Sharpen your pencil

Find the mode for the following sets of data.

Values	1	2	3	4	5	6	7	8
Frequency	4	6	4	4	3	2	1	1

Category	Blue	Red	Green	Pink	Yellow
Frequency	4	5	8	1	3

Values	1	2	3	4	5
Frequency	2	3	3	3	3

When do you think the mode is most useful?

When is the mode least useful?

Congratulations!

Your efforts at the Health Club are proving to be a huge success, and demand for classes is high.

Three cheers for M-O-D-E! Most of the class is the same age as me!
An experienced tennis coach like me earns a median salary of $33/hour.
My mean golf score is two under par. But don't tell the ladies my median score is two **over** par.
I can run a mile in a mean of 25 minutes, but that includes a stop at Starbuzz Coffee on the way.
I kick butt at soccer and statistics.
I lose a mean of 7 teeth per hockey match.
Median time spent underwater each day: 24 minutes

Sharpen your pencil Solution

Find the mode for the following sets of data.

Values	1	2	3	4	5	6	7	8
Frequency	4	6	4	4	3	2	1	1

The mode here is 2, as it has the highest frequency.

Category	Blue	Red	Green	Pink	Yellow
Frequency	4	5	8	1	3

This time the mode is Green.

Values	1	2	3	4	5
Frequency	2	3	3	3	3

This set of data has several modes: 2, 3, 4, and 5.

When do you think the mode is most useful?

When the data set has a low number of modes, or when the data is categorical instead of numerical. Neither the mean nor the median can be used with categorical data.

When is the mode least useful?

When there are many modes

Vital Statistics

Mode

The mode has to be in the data set. It's the only average that works with categorical data.

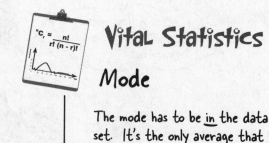

Exercise

Complete the table below. For each type of average we've encountered in the chapter, write down how to calculate it, and then give the circumstances in which you should use each one. Try your hardest to fill this out without looking back through the chapter.

Average	How to calculate	When to use it
Mean (μ)		When the data is fairly symmetric and shows just the one trend.
Median		
Mode		

Exercise Solution

Complete the table below. For each type of average we've encountered in the chapter, write down how to calculate it, and then give the circumstances in which you should use each one. Try your hardest to fill this out without looking back through the chapter.

Average	How to calculate	When to use it
Mean (μ)	Use either $\frac{\sum x}{n}$ (x is each value. n is the number of values.) or $\frac{\sum fx}{\sum f}$ (f is the frequency of each x.)	When the data is fairly symmetric and shows just the one trend.
Median	Line up all the values in ascending order. If there are an odd number of values, the median is the one in the middle. If there are an even number of values, add the two middle ones together, and divide by two.	When the data is skewed because of outliers.
Mode	Choose the value(s) with the highest frequency. If the data is showing two clusters of data, report a mode for each group.	When you're working with categorical data. When the data shows two or more clusters of data. (The only type of average you can calculate for categorical data is the mode.)

Sharpen your pencil

The generous CEO of Starbuzz Coffee wants to give all his employees a pay raise. He's not sure whether to give everyone a straight $2,000 raise, or whether to increase salaries by 10%. The mean salary is $50,000, the median is $20,000, and the mode is $10,000.

a) What happens to the mean, median, and mode if everyone at Starbuzz is given a $2,000 pay raise?

b) What happens to the mean, median, and mode if everyone at Starbuzz is given a 10% pay raise instead?

c) Which sort of pay raise would you prefer if you were earning the mean wage? What about if you were on the same wage as the mode?

Sharpen your pencil Solution

The generous CEO of Starbuzz Coffee wants to give all his employees a pay rise. He's not sure whether to give everyone a straight $2,000 raise, or whether to increase salaries by 10%. The mean salary is $50,000, the median is $20,000, and the mode is $10,000.

a) What happens to the mean, median, and mode if everyone at Starbuzz is given a $2,000 pay raise?

Mean: If x represents the original wages, and n the number of employees,

$$\mu = \frac{\sum(x + 2000)}{n}$$

$$= \frac{\sum x}{n} + \frac{\sum 2000}{n}$$

The original mean

There are n lots of 2000.

$$= 50{,}000 + \frac{2000n}{n}$$

$$= \$52{,}000$$

Adding $2,000 to everyone's salary increases the mean, median, and mode by $2,000.

Median: Every wage has $2,000 added to it, and this includes the middle value—the median. The new median is $20,000 + $2,000 = $22,000.

Mode: The most common wage or mode is $10,000, and with the $2,000 pay raise, this becomes $10,000 + $2000 = $12,000.

b) What happens to the mean, median, and mode if everyone at Starbuzz is given a 10% pay raise instead?

This time, all of the wages are multiplied by 1.1 (which is 100% + 10%).

Mean: $\mu = \dfrac{\sum(1.1x)}{n}$

$$= \frac{1.1\sum x}{n}$$

$$= 1.1 \times 50{,}000$$

$$= \$55{,}000$$

Increasing everyone's salary by 10% increases the mean, median, and mode by 10%.

Median: Every wage is multiplied by 1.1, and this includes the middle value—the median. The new median is $20,000 × 1.1 = $22,000.

Mode: The most common wage or mode is $10,000, and if we multiply this by 1.1, it becomes $10,000 × 1.1 = $12,000.

c) Which sort of pay raise would you prefer if you were earning the mean wage? What about if you were on the same wage as the mode?

If you earn the mean wage, you'll get a larger pay increase if you get a 10% pay raise. If you earn the mode wage, you'll get more money if you ask for the straight $2,000 pay increase.

The Case of the Ambiguous Average: Solved

What's going on with the average? Who do you think is right?

The workers, the managers, and the CEO are each using a different sort of average.

The workers are using the median, which minimizes the effect of the CEO's salary.

The managers are using the mean. The large salary of the CEO is skewing the data to the right, which is making the mean artificially high.

The CEO is using the mode. Most workers are paid $500 per week, and so this is the mode of the salaries.

So who's right? In a sense, they all are, although it has to be said that each group of people are using the average that best supports what they want. Remember, statistics can be informative, but they can also be misleading. For balance, we think that the most appropriate average to use in this situation is the median because of the outliers in the data.

3 measuring variability and spread

Power Ranges

Don't worry about the dinner, Mother. When you have an oven with a lower standard deviation, you'll never burn anything again.

Not everything's reliable, but how can you tell?

Averages do a great job of giving you a typical value in your data set, but they **don't tell you the full story**. OK, so you know where the center of your data is, but often the mean, median, and mode alone aren't enough information to go on when you're summarizing a data set. In this chapter, we'll show you how to take your data skills to the next level as we begin to analyze **ranges and variation**.

Wanted: one player

The Statsville All Stars are the hottest basketball team in the neighborhood, and they're the favorite to win this year's league. There's only one problem—due to a freak accident, they're a player down. They need a new team member, and fast.

The new recruit must be good all-round, but what the coach really needs is a reliable shooter. If he can trust the player's ability to get the ball in the basket, they're on the team.

The coach has been conducting trials all week, and he's down to three players. The question is, which one should he choose?

All three players have the same average score for shooting, but I need some way of choosing between them. Think you can help?
All three players had the same average score in the trials, so how should the coach decide which to pick?
The Statsville All Stars coach
05

We need to compare player scores

Here are the scores of the three players:

Points scored per game	7	8	9	10	11	12	13
Frequency	1	1	2	2	2	1	1

Here, frequency tells us the number of games where the player got each score. This player scored 9 points in 2 games, and 12 points in 1 game.

Points scored per game	7	9	10	11	13
Frequency	1	2	4	2	1

Points scored per game	3	6	7	10	11	13	30
Frequency	2	1	2	3	1	1	1

Each player has a mean, median, and mode score of 10 points, but if you look at their scores, you'll see they've all achieved it in different ways. There's a difference in how consistently the players have performed, which the average can't measure.

What we need is a way of differentiating between the three sets of scores so that we can pick the most suitable player for the team. We need some way of comparing the sets of data in addition to the average—but what?

BRAIN POWER

What information in addition to the average would help the coach make his decision?

Use the range to differentiate between data sets

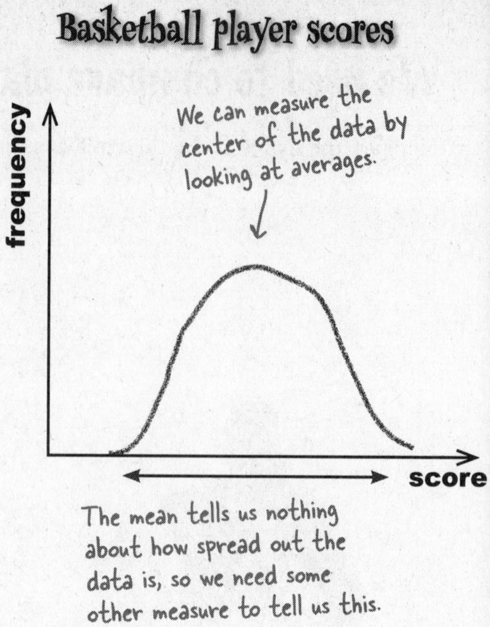
Basketball player scores
frequency
We can measure the center of the data by looking at averages.
score
The mean tells us nothing about how spread out the data is, so we need some other measure to tell us this.

So far we've looked at calculating averages for sets of data, but quite often, the average only gives part of the picture. Averages give us a way of determining where the center of a set of data is, but they don't tell us how the data varies. Each player has the same average score, but there are clear differences between each data set. We need some other way of measuring these differences.

We can differentiate between each set of data by looking at the way in which the scores spread out from the average. Each player's scores are distributed differently, and if we can measure how the scores are dispersed, the coach will be able to make a more informed decision.

Measuring the range

We can easily do this by calculating the range. The ***range*** tells us over how many numbers the data extends, a bit like measuring its width. To find the range, we take the largest number in the data set, and then subtract the smallest.

The smallest value is called the ***lower bound***, and the largest value is the ***upper bound***.

Let's take a look at the set of scores for one of the players and see how this works. Here are the scores:

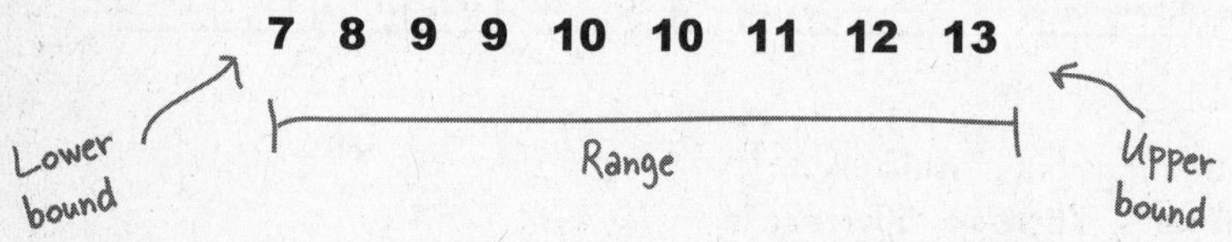
7 8 9 9 10 10 11 12 13
Lower bound
Range
Upper bound

To calculate the range, we subtract the lower bound from the upper bound. Looking at the data, the smallest value is 7, which means that this is the lower bound. Similarly, the upper bound is the largest value, or 13. Subtracting the lower bound from the upper bound gives us:

Range = upper bound - lower bound

= 13 - 7

= 6

so the range of this set of data is 6.

The range is a simple and easy way of measuring how spread out values are, and it gives us another way of comparing sets of data.

Vital Statistics

Range

The range is a way of measuring how spread out a set of values are. It's given by

Upper bound – Lower bound

where the upper bound is the highest value, and the lower bound the lowest.

Exercise

Work out the mean, lower bound, upper bound, and range for the following sets of data, and sketch the charts. Are values dispersed in the same way? Does the range help us describe these differences?

Score	8	9	10	11	12
Frequency	1	2	3	2	1

Score	8	9	10	11	12
Frequency	1	0	8	0	1

Exercise Solution

Work out the mean, lower bound, upper bound, and range for the following sets of basketball scores, and sketch the charts. Are values dispersed in the same way? Does the range help us describe these differences?

Score	8	9	10	11	12
Frequency	1	2	3	2	1

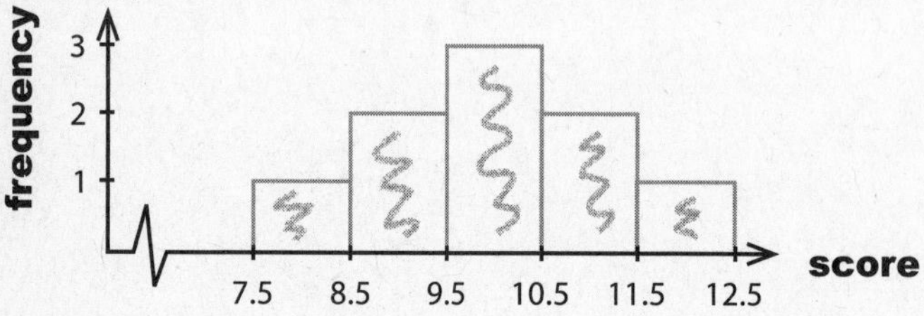

frequency
3
2
1
score
7.5
8.5
9.5
10.5
11.5
12.5

$\mu = 10$

Lower bound = 8

Upper bound = 12

Range = 12 − 8

= 4

Look, these results are the same even though the data's different.

Score	8	9	10	11	12
Frequency	1	0	8	0	1

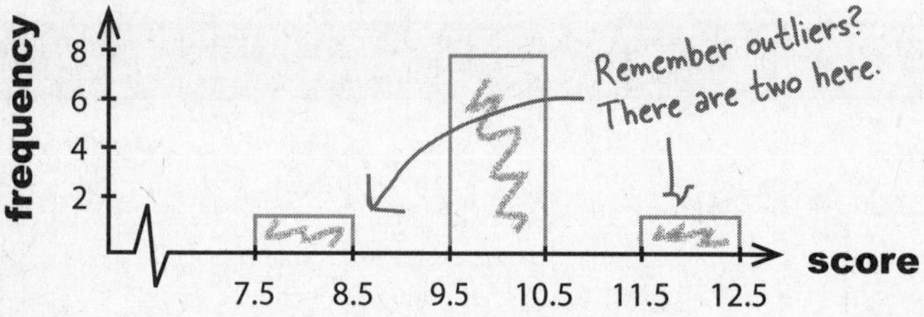

frequency
8
6
4
2
Remember outliers? There are two here.
score
7.5
8.5
9.5
10.5
11.5
12.5

$\mu = 10$

Lower bound = 8

Upper bound = 12

Range = 12 − 8

= 4

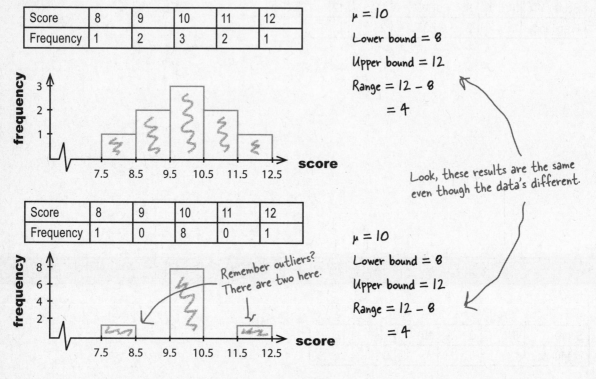

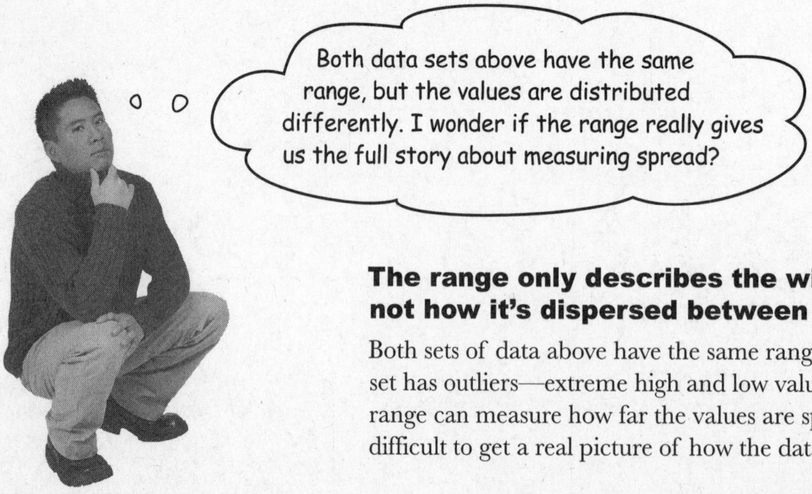

Both data sets above have the same range, but the values are distributed differently. I wonder if the range really gives us the full story about measuring spread?

The range only describes the width of the data, not how it's dispersed between the bounds.

Both sets of data above have the same range, but the second set has outliers—extreme high and low values. It looks like the range can measure how far the values are spread out, but it's difficult to get a real picture of how the data is distributed.

The problem with outliers

The range is a simple way of saying what the spread of a set of data is, but it's often not the best way of measuring how the data is distributed within that range. If your data has outliers, using the range to describe how your values are dispersed can be very misleading because of its sensitivity to outliers. Let's see how.

Imagine you have a set of numbers as follows:

Lower bound of 1

Upper bound of 5

1 1 1 2 2 2 2 3 3 3 3 3 4 4 4 4 5 5 5

Here's the data on a vertical line chart (a type of bar chart that uses lines instead of bars). Each line represents the frequency of each number in the data set.

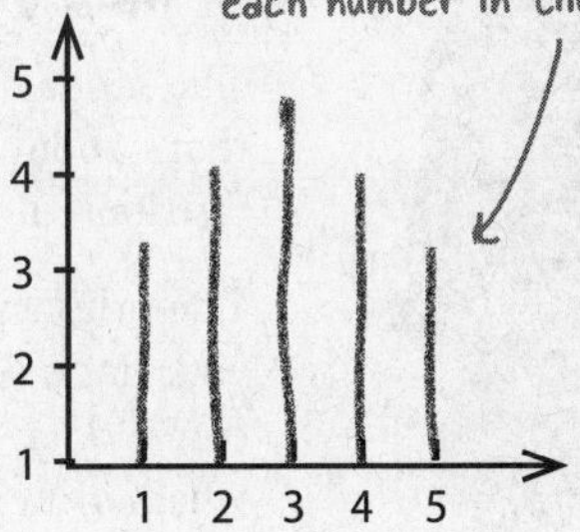

Here, numbers are fairly evenly distributed between the lower bound and upper bound, and there are no outliers for us to worry about. The range of this set of numbers is 4.

But what happens if we introduce an outlier, like the number 10?

The lower bound is still 1.

But the upper bound has increased to 10.

1 1 1 2 2 2 2 3 3 3 3 3 4 4 4 4 5 5 5 10

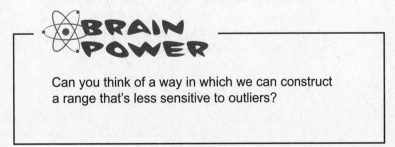

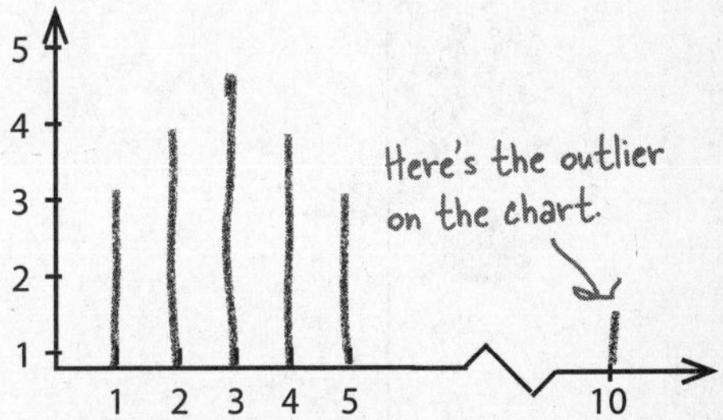

Our lower bound is the same, but the upper bound has gone up to 10, giving us a new range of 9. The range has increased by 5 just because we added one extra number, an outlier.

Without the outlier, the two sets of data would be identical, so why should there be such a big difference in how we describe how the values are distributed?

BRAIN POWER

Can you think of a way in which we can construct a range that's less sensitive to outliers?

So is it a bad idea to use the range then?

The range is a great quick-and-dirty way to get an idea of how values are distributed, but it's a bit limited.

The range tells you how far apart your highest and lowest values are, but that's about it. It only provides a very basic idea of how the values are distributed.

The primary problem with the range is that it only describes the width of your data. Because the range is calculated using the most extreme values of the data, it's impossible to tell what that data actually looks like—and whether it contains outliers. There are many different ways of constructing the same range, and sometimes this additional information is important.

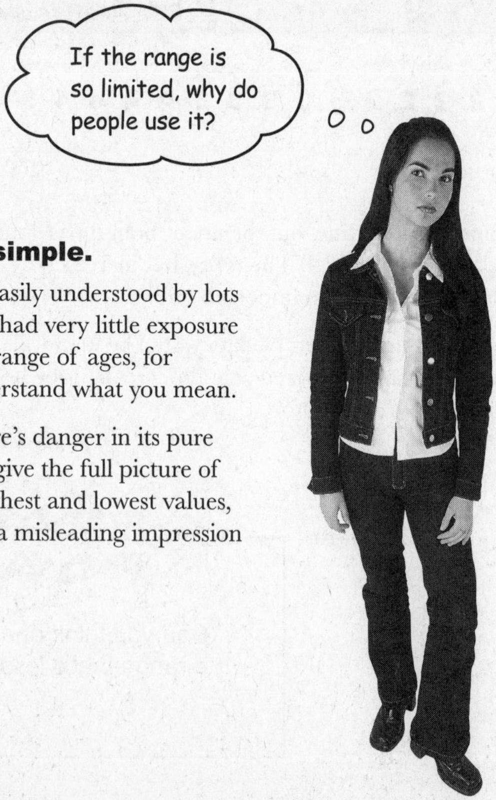

If the range is so limited, why do people use it?

Mainly because it's so simple.

The range is so simple that it's easily understood by lots of people, even those who have had very little exposure to statistics. If you talk about a range of ages, for example, people will easily understand what you mean.

Be careful, though, because there's danger in its pure simplicity. As the range doesn't give the full picture of what's going on between the highest and lowest values, it's easy for it to be used to give a misleading impression of the underlying data.

We need to get away from outliers

The main problem with the range is that, by definition, it includes outliers. If data has outliers, the range will include them, even though there may be only one or two extreme values. What we need is a way of negating the impact of these outliers so that we can best describe how values are dispersed.

One way out of this problem is to look at a kind of *mini* range, one that ignores the outliers. Instead of measuring the range of the whole set of data, we can find the range of part of it, the part that doesn't contain outliers.

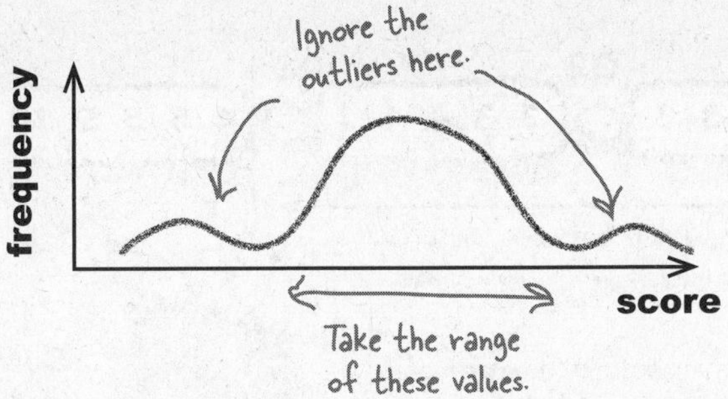

We need a consistent way of doing this.

One of the problems with ignoring outliers on an ad hoc basis is that it's difficult to compare sets of data. How do we know that all sets of data are omitting outliers in exactly the same way?

We need to make sure that we use the same mini range definition for all the sets of data we're comparing. But how?

Quartiles come to the rescue

One way of constructing a mini range is to just use values around the center of the data. We can construct a range in this way by first lining up the values in ascending order, and then splitting the data into four equally sized chunks, with each chunk containing one quarter of the data.

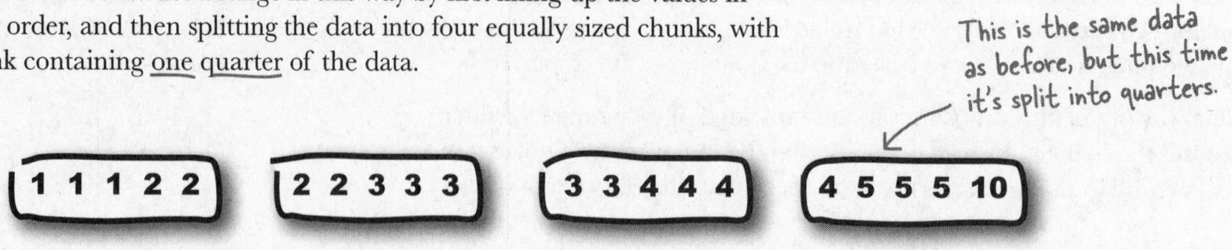

We can then construct a range using the values that fall between the two outer splits:

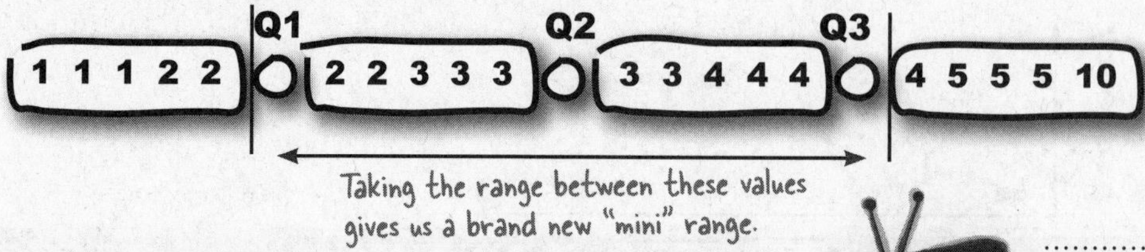

The values that split the data into equal chunks are known as ***quartiles***, as they split the data into quarters. Finding quartiles is a bit like finding the median. Instead of finding the value that splits the data in half, we're finding the values that split the data into quarters.

The lowest quartile is known as the ***lower quartile***, or first quartile (Q1), and the highest quartile is known as the ***upper quartile***, or third quartile (Q3). The quartile in the middle (Q2) is the median, as it splits the data in half. The range of the values in these two quartiles is called the ***interquartile range*** (IQR).

Watch it!

Some textbooks refer to quartiles as the set of values within each quarter of the data.

We're not. We're using the term quartile to specifically refer to the values that split the data into quarters.

Interquartile range =
Upper quartile – Lower quartile

The interquartile range gives us a standard, repeatable way of measuring how values are dispersed. It's another way in which we can compare different sets of data. But what about outliers? Does the interquartile range help us deal with these too? Let's take a look.

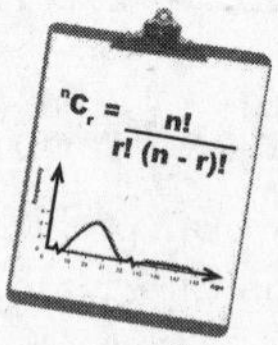

Vital Statistics

Quartiles

Quartiles are values that split your data into quarters. The lowest quartile is called the lower quartile, and the highest quartile is called the upper quartile.

The middle quartile is the median.

The interquartile range excludes outliers

The good thing about the interquartile range is that it's a lot less sensitive to outliers than the range is.

The upper and lower quartiles are positioned so that the lower quartile has 25% of the data below it, and the upper quartile has 25% of the data above it. This means that the interquartile range only uses the central 50% of the data, so outliers are disregarded. As we've said before, outliers are extreme high or low values in the data, so by only considering values around the center of the data, we automatically exclude any outliers.

Here's our data again. Can you see how the interquartile range effectively ignores any outliers?

The interquartile range includes the middle part of the data...

25% of the data in the set

Q1   Q2   Q3

| 1 1 1 2 2 | ○ | 2 2 3 3 3 | ○ | 3 3 4 4 4 | ○ | 4 5 5 5 10 |

25% of the data in the set

...but excludes the two outer quarters where any outliers live.

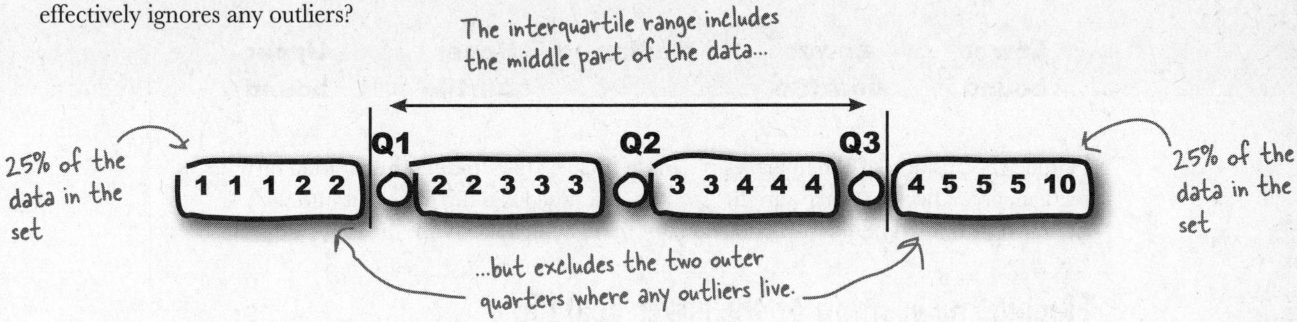

As the interquartile range only uses the central 50% of the data, outliers are excluded irrespective of whether they are extremely high or extremely low. They *can't* be in the middle. This means that any outliers in the data are effectively cut out.

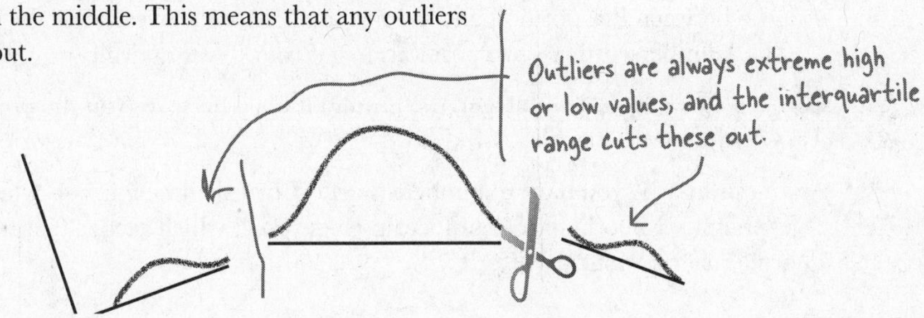

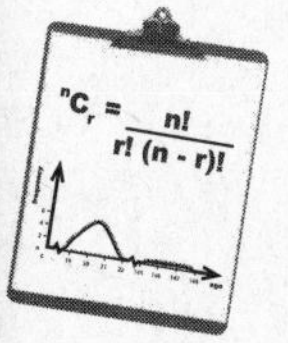

Outliers are always extreme high or low values, and the interquartile range cuts these out.

Vital Statistics

$$^nC_r = \frac{n!}{r!(n-r)!}$$

Interquartile Range

A "mini range" that's less sensitive to outliers. You find it by calculating

Upper quartile – Lower quartile

Excluding the outliers with the interquartile range means that we now have a way of comparing different sets of data without our results being distorted by outliers. Before we can figure out the interquartile range, though, we have to work out what the quartiles are. Flip the page, and we'll show you how it's done.

Quartile anatomy

Finding the quartiles of a set of data is a very similar process to finding the median. If you line all of your values up in ascending order, the median is the value right in the very center. If you have *n* numbers, it's the number that's at position $(n + 1) \div 2$, and if this falls halfway between two numbers, you take their average.

If we then further split the data into quarters, the quartiles are the values at each of these splits. The lowest is the lower quartile, and the highest is the upper quartile:

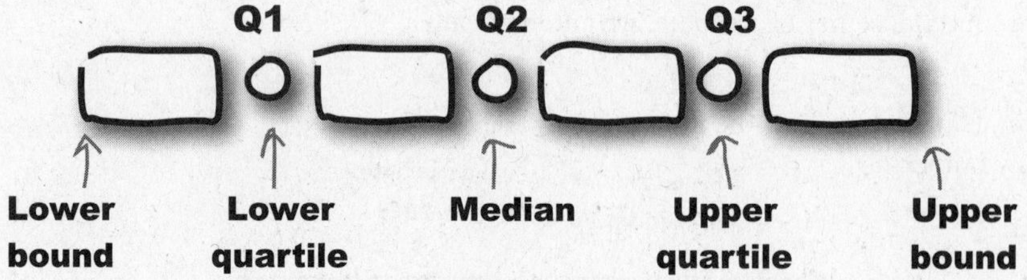

Finding the position of the quartiles is slightly trickier than finding the position of the median, as we need to make sure the values we choose keep the data split into the right proportions. There is a way of doing it though; let's start with the lower quartile.

Finding the position of the lower quartile

1. First, start off by calculating $n \div 4$.
2. If this gives you an integer, then the lower quartile is positioned halfway between this position and the next one. Take the average of the numbers at these two positions to get your lower quartile.
3. If $n \div 4$ is *not* an integer, then round it up. This gives you the position of the lower quartile.

As an example, if you have 6 numbers, start off by calculating $6 \div 4$, which gives you 1.5. Rounding this number up gives you 2, which means that the lower quartile is at position 2.

Finding the position of the upper quartile

1. Start off by calculating $3n \div 4$.
2. If it's an integer, then the upper quartile is positioned halfway between this position and the next. Add the two numbers at these positions together and divide by 2.
3. If $3n \div 4$ is *not* an integer, then round it up. This new number gives you the position of the upper quartile.

Exercise

It's time to put your quartile skills into practice. Here are the scores for one of the players:

Points scored per game	3	6	7	10	11	13	30
Frequency	2	1	2	3	1	1	1

1. What's the range of this set of data?

2. What are the lower and upper quartiles?

3. What's the interquartile range?

Exercise Solution

Here are the scores for one of the players:

Points scored per game	3	6	7	10	11	13	30
Frequency	2	1	2	3	1	1	1

1. What's the range of this set of data?

The lower bound of this set of data is 3, as that's the lowest number of points scored. The upper bound is 30, as that's the highest. This gives us

Range = upper bound – lower bound

= 30 – 3

= 27

2. What are the lower and upper quartiles?

Let's start with the lower quartile. There are 11 numbers, and calculating 11 ÷ 4 gives us 2.75. Rounding this number up gives us the position of the lower quartile, so the lower quartile is at position 3. This means that the lower quartile is 6.

Now let's find the upper quartile. 3 x 11 ÷ 4 gives us 8.25, and rounding this up gives us 9 – the upper quartile is at position 9. This means that the upper quartile is 11.

3 3 [6] 7 7 [10] 10 10 [11] 13 30

Lower quartile (6) — Median (10) — Upper quartile (11)

3. What's the interquartile range?

The interquartile range is the lower bound subtracted from the upper bound.

Interquartile range = upper bound – lower bound

= 11 – 6

= 5

This is MUCH lower than the range, as it excludes outliers.

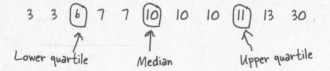

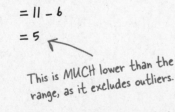

there are no Dumb Questions

Q: I get why mean, median, and mode are useful, but why do I need to know how the data is spread out?

A: Averages offer you only a one-dimensional view of your data. They tell you what the center of your data is, but that's it. While this can be useful, it's often not enough. You need some other way of summarizing your data in addition to the average.

Q: So is the median the same as the interquartile range?

A: No. The median is the middle value of the data, and the interquartile range is the range of the middle 50% of the values.

Q: What's the point of all this quartiles stuff? It seems like a really tedious way to calculate ranges.

A: The problem with using the range to measure how your data is dispersed is that it's very sensitive to outliers. It gives you the difference between the lower and upper bounds of your data, but just one outlier can make a huge difference to the result.

We can get around this by focusing only on the central 50% of the data, as this excludes outliers. This means finding quartiles, and using the interquartile range. So even though finding quartiles is trickier than finding the lower and upper bounds, there are definite advantages.

Q: Should I always use the interquartile range to measure the spread of data?

A: In a lot of cases, the interquartile range is more meaningful than the range, but it all depends on what information you really need. There are other ways of measuring how values are dispersed that you might want to consider too; we'll come to these later.

Q: Would I ever want to look at just one quartile of my data instead of the range or the interquartile range?

A: It's possible. For example, you might be interested in what the high values look like, so you'd just look at what values are in the upper quarter of your data set, using the upper quartile as a cut-off point.

Q: Would I ever want to break my data into smaller pieces than quarters? How about breaking my data into, say, 10 pieces instead of 4?

A: Yes, there are times when you might want to do this. Turn the page, and we'll show you more...

BULLET POINTS

- The **upper** and **lower bounds** of the data are the highest and lowest values in the data set.
- The **range** is a simple way of measuring how values are dispersed. It's given by:

  range = upper bound - lower bound
- The range is very sensitive to outliers.
- The interquartile range is less sensitive to outliers than the range.
- **Quartiles** are values that split your data into quarters. The highest quartile is called the upper quartile, and the lowest quartile is called the lower quartile. The middle quartile is the median.
- The **interquartile range** is the range of the central 50% of the data. It's given by calculating

  upper quartile - lower quartile

We're not just limited to quartiles

So far we've looked at how the range and interquartile range give us ways of measuring how values are dispersed in a set of data. The range is the difference between the highest number and the lowest, while the interquartile range focuses on the middle 50% of the data.

So are they the only sorts of ranges I can use? Do I get any other options?

There are other sorts of ranges we can use in addition to the range and interquartile range.

Our original problem with the range was that it's extremely sensitive to outliers. To get around this, we divided the data into quarters, and we used the interquartile range to provide us with a cut-down range of the data.

While the interquartile range is quite common, it's not the only way of constructing a mini range. Instead of splitting the data into quarters, we could have split it into some other sort of percentage and used that for our range instead.

As an example, suppose we'd divided our set of data into tenths instead of quarters so that each segment contains 10% of the data. We'd have something like this:

This is the same set of data, but it's now split into 10 equally sized chunks. Each chunk contains 10% of the data.
1 1
1 2
2 2
2 3
3 3
3 3
4 4
4 4
5 5
5 10
We can use these divisions to create a brand new mini range.

If you break up a set of data into percentages, the values that split the data are called ***percentiles***. In the case above, our data is split into tenths, so the values are called ***deciles***.

We can use percentiles to construct a new range called the ***interpercentile range***.

So what are percentiles?

Percentiles are values that split your data into percentages in the same way that quartiles split data into quarters. Each percentile is referred to by the percentage with which it splits the data, so the 10th percentile is the value that is 10% of the way through the data. In general, the *x*th percentile is the value that is *k*% of the way through the data. It's usually denoted by P_k.

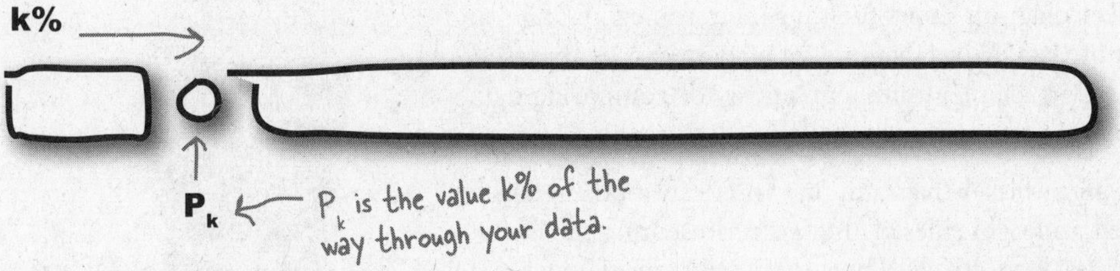
k%
P_k
P_k is the value k% of the way through your data.

Quartiles are actually a type of percentile. The lower quartile is P_{25}, and the upper quartile is P_{75}. The median is P_{50}.

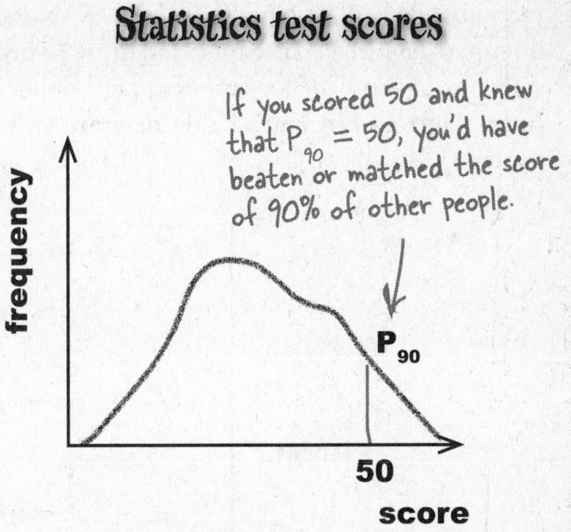
Statistics test scores
If you scored 50 and knew that P_{90} = 50, you'd have beaten or matched the score of 90% of other people.
frequency
P_{90}
50
score

Percentile uses

Even though the interpercentile range isn't that commonly used, the percentiles themselves are useful for benchmarking and determining rank or position. They enable you to determine how high a particular value is relative to all the others. As an example, suppose you heard you scored 50 on your statistics test. With just that number by itself, you'd have no idea how well you'd done relative to anyone else. But if you were told that the 90th percentile for the exam was 50, you'd know that you scored the same as or better than 90% of the other people.

Finding percentiles

You can find percentiles in a similar way to how you find quartiles.

1. First of all, line all your values up in ascending order.
2. To find the position of the *k*th percentile out of *n* numbers, start off by calculating $k\left(\frac{n}{100}\right)$.
3. If this gives you an integer, then your percentile is halfway between the value at position $k\left(\frac{n}{100}\right)$ and the next number along. Take the average of the numbers at these two positions to give you your percentile.
4. If $k\left(\frac{n}{100}\right)$ is *not* an integer, then round it up. This then gives you the position of the percentile.

As an example, if you have 125 numbers and want to find the 10th percentile, start off by calculating 10 × 125 ÷ 100. This gives you a value of 12.5. Rounding this number up gives you 13, which means that the 10th percentile is the number at position 13.

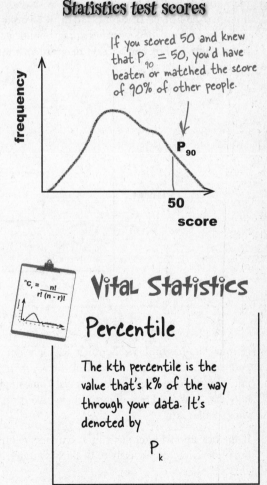

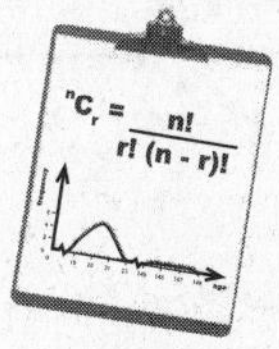
$^nC_r = \frac{n!}{r!\,(n-r)!}$

Vital Statistics

Percentile

The kth percentile is the value that's k% of the way through your data. It's denoted by

P_k

Box and whisker plots let you visualize ranges

We've talked a lot about different sorts of ranges, and it would be useful to be able to compare the ranges of different sets of data in a visual way. There's a chart that specializes in showing different types of ranges: the ***box and whisker*** diagram, or ***box plot***.

A box and whisker diagram shows the range, interquartile range, and median of a set of data. More that one set of data can be represented on the same chart, which means it's a great way of comparing data sets.

3 3 6 7 7 10 10 10 11 13 30

To create a box and whisker diagram, first you draw a box against a scale with the left and right sides of the box representing the lower and upper quartiles, respectively. Then, draw a line inside the box to mark the value of the median. This box shows you the extent of the interquartile range. After that, you draw "whiskers" to either side of your box to show the lower and upper bounds and the extent of the range. Here's a box and whisker diagram for the scores of our player from page 95:

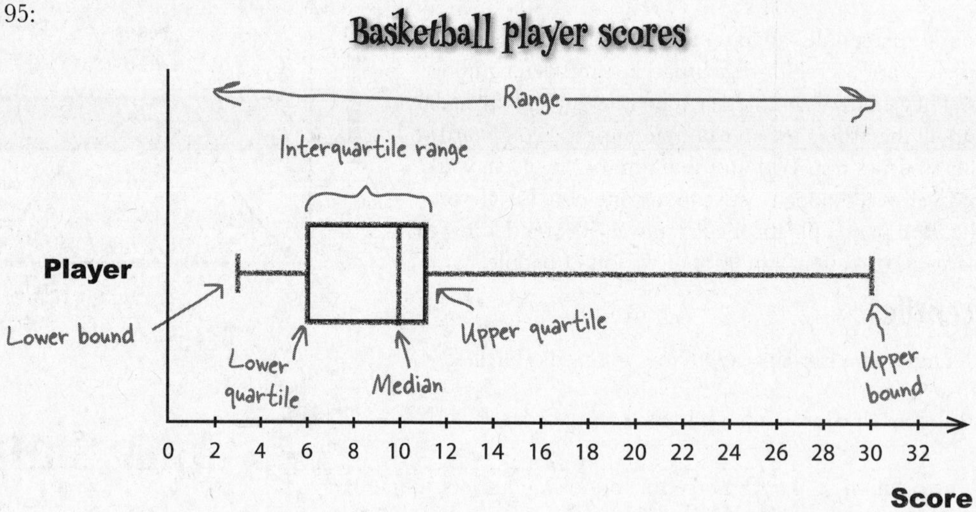

If your data has outliers, the range will be wider. On a box and whisker diagram, the length of the whiskers increases in line with the upper and lower bounds. You can get an idea of how data is skewed by looking at the whiskers on the box and whisker diagram.

If the box and whisker diagram is symmetric, this means that the underlying data is likely to be fairly symmetric, too.

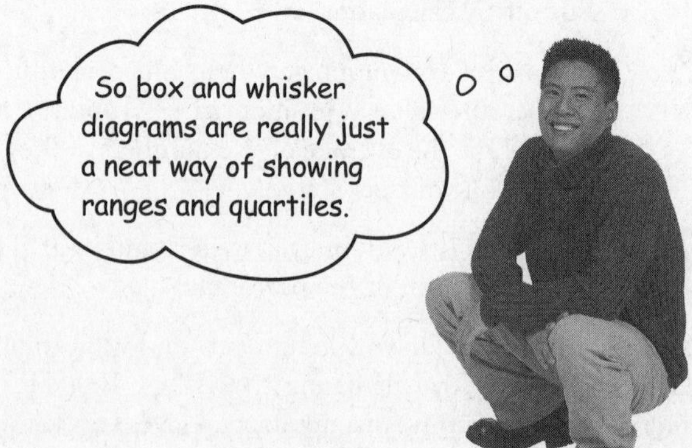

Exercise

Here are box and whisker diagrams for two more basketball players. Compare the ranges of their scores. If you had to choose between having player A or player B on the team, which would you pick? Why?

Basketball Player A and B scores

Player A

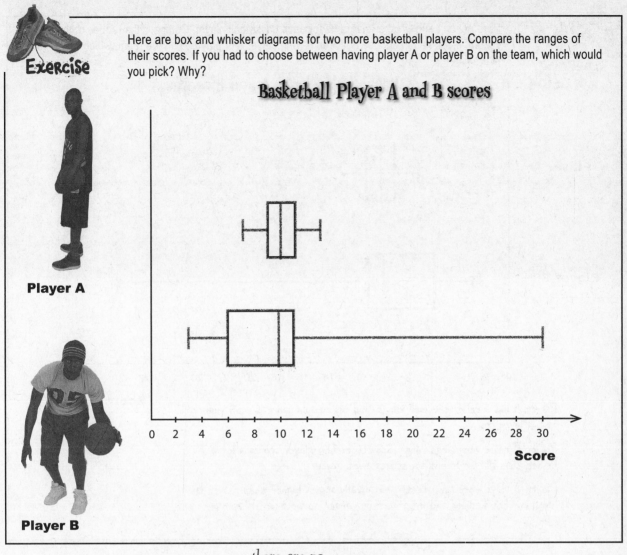

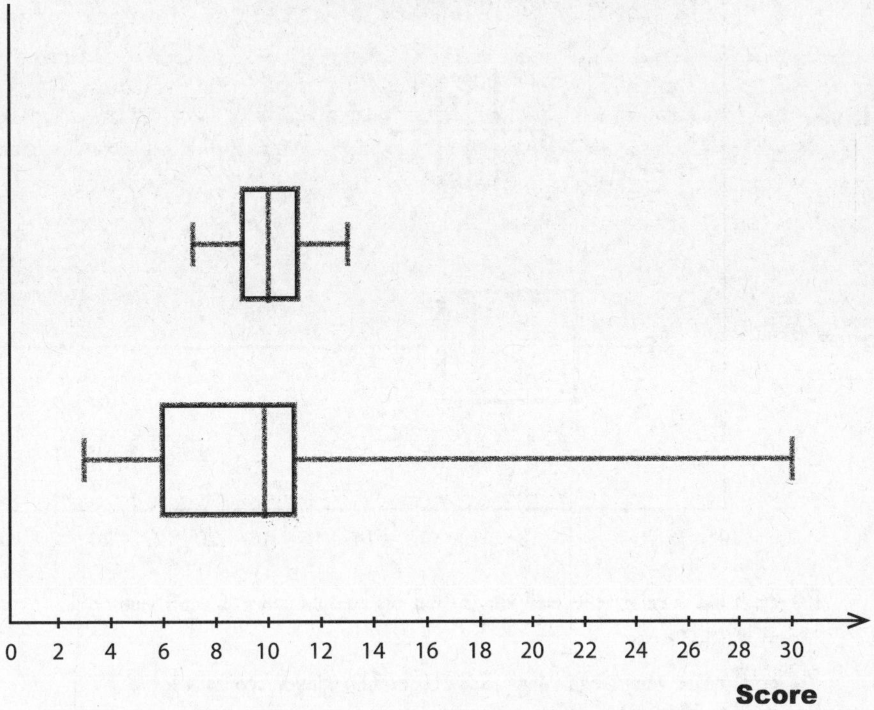

Player B

there are no Dumb Questions

Q: I'm sure I've seen box and whisker diagrams that look a bit different than this.

A: There are actually several versions of box and whisker diagrams. Some have deliberately shorter whiskers and explicitly show outliers as dots or stars extending beyond the whiskers. This makes it easier to see how many outliers there are and how extreme they really are. Other diagrams show the mean as a dot, so you can see where it's positioned in relation to the median. If you're taking a statistics course, it would be a good idea to check which version of the box and whisker diagram is likely to be used.

Q: So if you show the mean as a dot, is it to the left or right of the median?

A: If the data is skewed to the right, then the mean will be to the right of the median, and the whisker on the right will be longer than that of the left. If the data is skewed to the left, the mean will be to the left of the median, and the whisker on the left will be the longest.

Exercise Solution

Here are box and whisker diagrams for each basketball player. Compare the ranges of their scores. If you had to choose between having player A or player B on the team, which would you pick? Why?

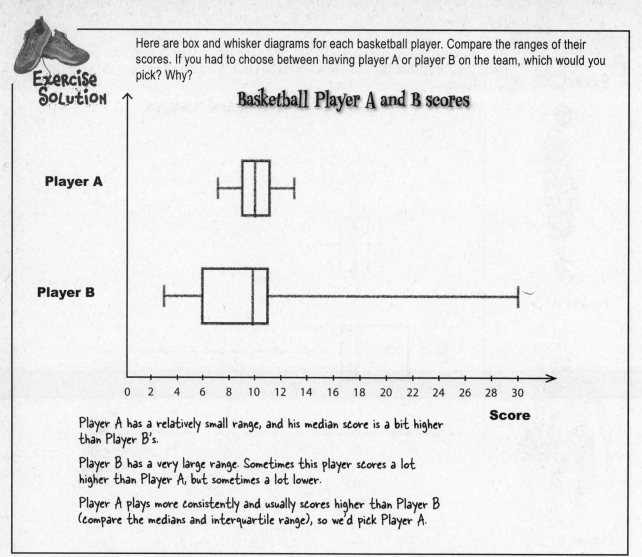

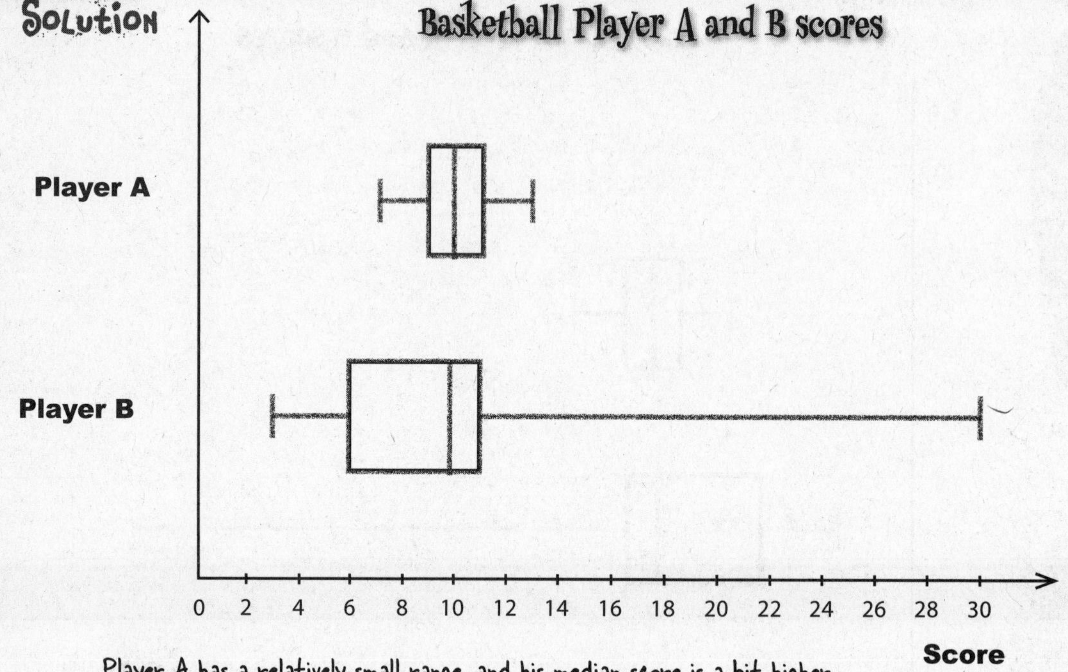

Player A has a relatively small range, and his median score is a bit higher than Player B's.

Player B has a very large range. Sometimes this player scores a lot higher than Player A, but sometimes a lot lower.

Player A plays more consistently and usually scores higher than Player B (compare the medians and interquartile range), so we'd pick Player A.

BULLET POINTS

- **Percentiles** split your data into percentages. They're useful for benchmarking.
- The *k*th percentile is *k*% of the way through your data. It's denoted by P_k.
- An **interpercentile range** is like the interquartile range but, this time, between two percentiles.
- **Box and whisker diagrams**, or box plots, are a useful way of showing ranges and quartiles on a chart. A box shows where the quartiles and interquartile range are, and the whiskers give the upper and lower bounds. More than one set of data can be shown on the same chart, so they're useful for comparisons.

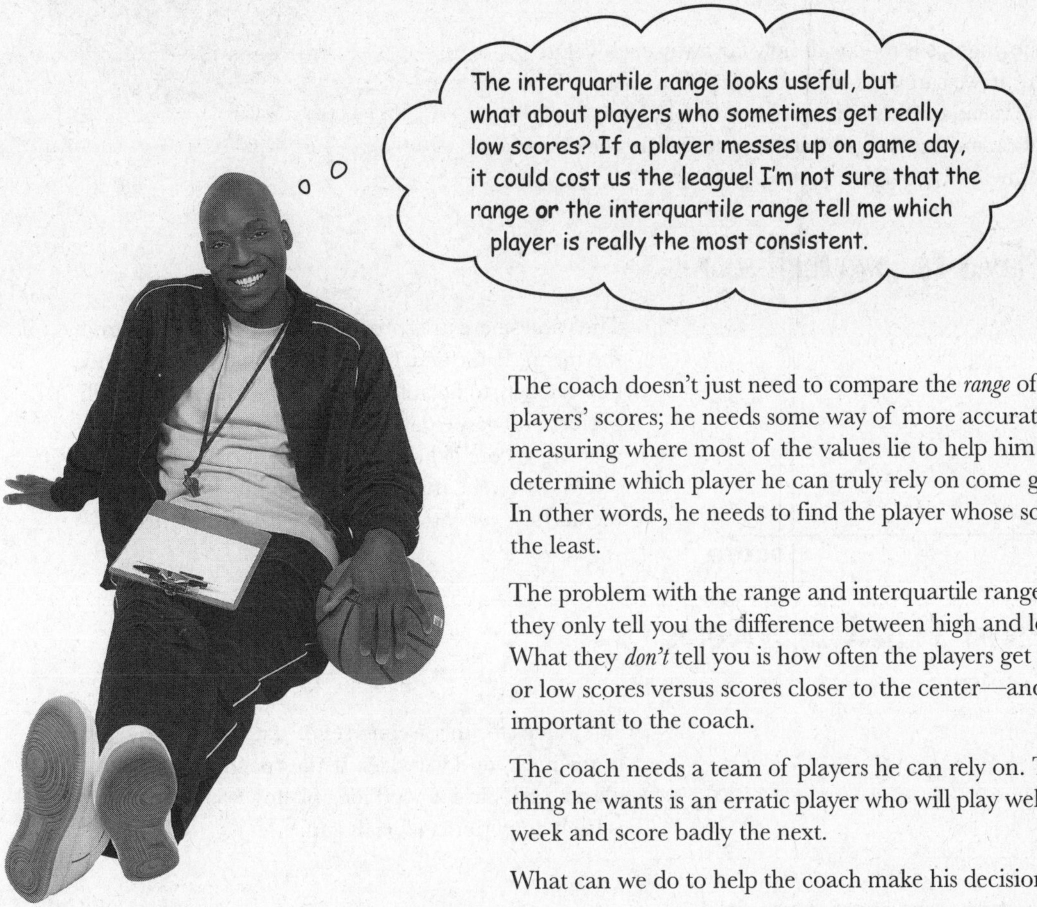

The interquartile range looks useful, but what about players who sometimes get really low scores? If a player messes up on game day, it could cost us the league! I'm not sure that the range **or** the interquartile range tell me which player is really the most consistent.

The coach doesn't just need to compare the *range* of the players' scores; he needs some way of more accurately measuring where most of the values lie to help him determine which player he can truly rely on come game day. In other words, he needs to find the player whose scores vary the least.

The problem with the range and interquartile range is that they only tell you the difference between high and low values. What they *don't* tell you is how often the players get these high or low scores versus scores closer to the center—and that's important to the coach.

The coach needs a team of players he can rely on. The last thing he wants is an erratic player who will play well one week and score badly the next.

What can we do to help the coach make his decision?

How can we more accurately measure variability?

Variability is more than just spread

We don't just want to measure the spread of each set of scores; we want some way of using this to see how reliable the player is. In other words, we want to be able to measure the variability of the players' scores.

One way of achieving this is to look at how far away each value is from the mean. If we can work out some sort of average distance from the mean for the values, we have a way of measuring variation and spread. The smaller the result, the closer values are to the mean. Let's take a look at this.

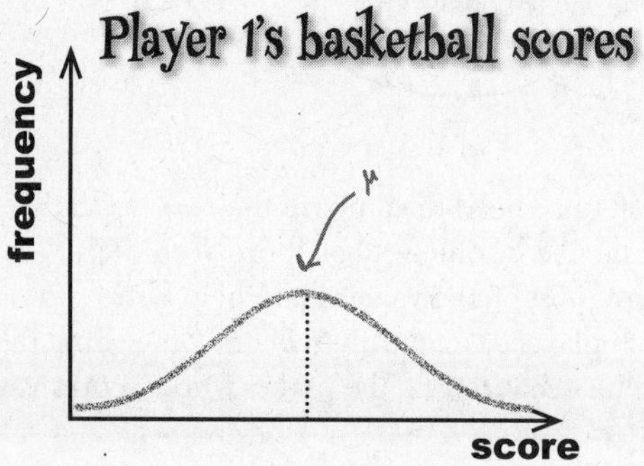

The values here are spread out quite a long way from the mean. If the coach picks this player for the team, he's unlikely to be able to predict how the player will perform on game day. The player may achieve a very high score if he's having a good day. On a bad day, however, he may not score highly at all, and that means he'll potentially lose the game for the team.

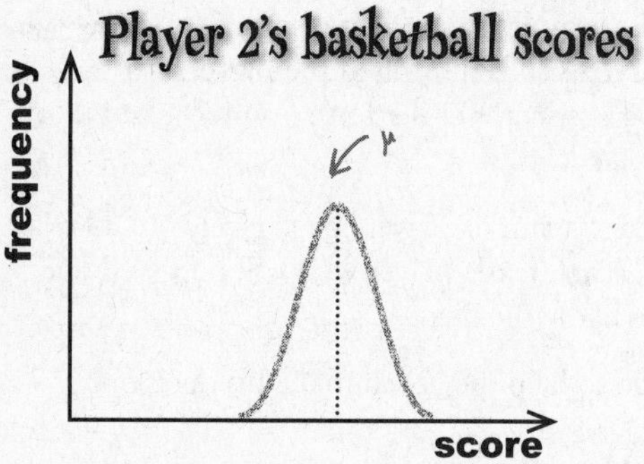

The values for this second set of data are much closer to the mean and vary less. If the coach picks this player, he'll have a good idea of how well the player is likely to perform in each game.

Let's find out.

Calculating average distances

Imagine you have three numbers: 1, 2, and 9. The mean is 4. What happens if we find the average distance of values from the mean?

$$\text{Average distance} = \frac{(1 \text{ to } \mu) + (2 \text{ to } \mu) + (9 \text{ to } \mu)}{3}$$

$$= \frac{3 + 2 + (-5)}{3}$$

$$= 0$$

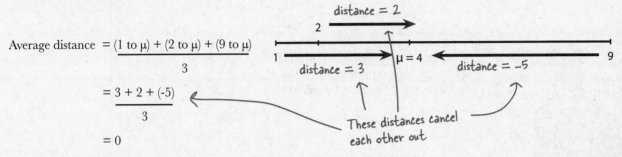

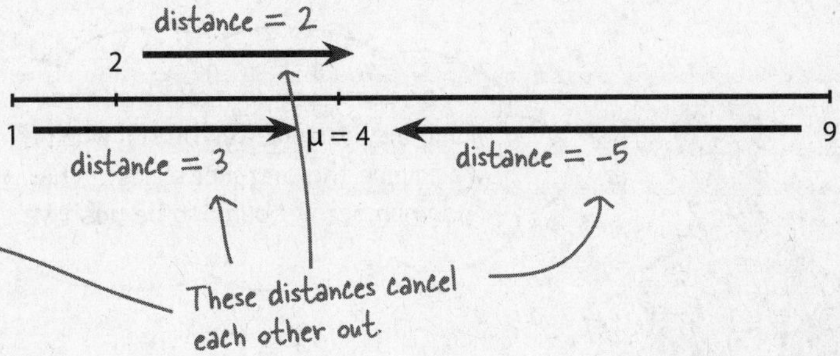

The average distance of values from the mean is always 0. The positive and negative distances cancel each other out. So what can we do now?

there are no Dumb Questions

Q: Why do we have -5 in that equation? I would have thought the distance would be 5. Why is it negative?

A: The distance from 9 to μ is negative because μ is less than 9. 1 and 2 are both less than μ, so the distance is positive for both of them. That's why the distances cancel each other out.

Q: Can't we just take the positive distances and average those?

A: That sounds intuitive, but in practice, statisticians rarely do this. There's another way of making sure that the distances don't cancel out, and you'll see that very soon. This other way of determining how close typical values are to the mean is used a *lot* in statistics, and you'll see it through most of the rest of the book.

Q: Surely the distances don't cancel out for all values. Maybe we were just unlucky.

A: No matter what values you choose, the distances to the mean will always cancel out.

Here's a challenge for you: take a group of numbers, work out the mean, work out the distance of each value from the mean, and then add the distances together. The result will be 0 every time.

Q: Can't I use the interquartile range to see how reliable the scores are?

A: The interquartile range only uses part of the data for measuring spread. If a player has one bad score, this will be excluded by the interquartile range. In order to truly determine reliability and consistency, we need to consider *all* the scores.

Q: The range uses all the scores. Why can't we use that then?

A: The range is only really good for describing the difference between the highest and lowest number. As you saw earlier, this doesn't represent how the values are actually distributed. We need another measure to do this.

The positive and negative distances from the mean cancel each other out.

We can calculate variation with the variance...

We want a way to measure the average distance of values from the mean in a way that stops the distances from cancelling each other out.

We need a way of making all the numbers positive. Maybe it'll work if we square the distances first. Then each number is bound to be positive.

Let's try this with the same three numbers.

Remember that $\mu = 4$.

$$\text{Average (distance)}^2 = \frac{(1 \text{ to } \mu)^2 + (2 \text{ to } \mu)^2 + (9 \text{ to } \mu)^2}{3}$$

$$= \frac{3^2 + 2^2 + (-5)^2}{3}$$

$$= \frac{9 + 4 + 25}{3}$$

$$= 12.67 \text{ (to 2 decimal places)}$$

This time we're adding together three positive numbers.

This time we get a meaningful number, as the distances don't cancel each other out. Every number we add together has to be non-negative because we're squaring the distance from the mean. Adding these numbers together gives us a non-negative result—every time.

This method of measuring spread is called the ***variance***, and it's a very common way of describing the spread of a set of data. Here's a general form of the equation:

The variance is the average of the distances from the mean squared.

$$\textbf{Variance} = \frac{\Sigma(x - \mu)^2}{n}$$

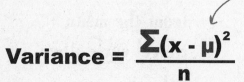

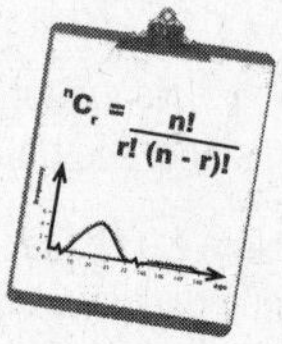

Vital Statistics

Variance

The variance is a way of measuring spread, and it's the average of the distance of values from the mean squared.

$$\frac{\Sigma(x - \mu)^2}{n}$$

...but standard deviation is a more intuitive measure

Statisticians use the variance a lot as a means of measuring the spread of data. It's useful because it uses every value to come up with the result, and it can be thought of as the average of the distances from the mean squared.

But why should I have to think about distances squared? I hardly call that intuitive. Isn't there another way?

What we *really* want is a number that gives the spread in terms of the distance from the mean, not distance squared.

The problem with the variance is that it can be quite difficult to think about spread in terms of distances squared.

There's an easy way to correct this. All we need to do is take the square root of the variance. We call this the ***standard deviation***.

Let's work out the standard deviation for the set of numbers we had before. The variance is 12.67, which means that

$$\text{Standard deviation} = \sqrt{12.67}$$

$$= 3.56 \text{ (to 2 decimal places)}$$

In other words, typical values are a distance of 3.56 away from the mean.

Standard deviation know-how

We've seen that the standard deviation is a way of saying how far typical values are from the mean. The smaller the standard deviation, the closer values are to the mean. The smallest value the standard deviation can take is 0.

Like the mean, the standard deviation has a special symbol, σ. This is the Greek character lowercase Sigma. (We saw uppercase Sigma, Σ, in Chapter 2 to represent summation.)
To find σ, start off by calculating the variance, and then take the square root.

$$\sigma = \sqrt{\text{variance}}$$

$$\updownarrow$$

$$\sigma^2 = \text{variance}$$

I'm the standard deviation. If you need to measure distances from the mean, give me a call.

σ

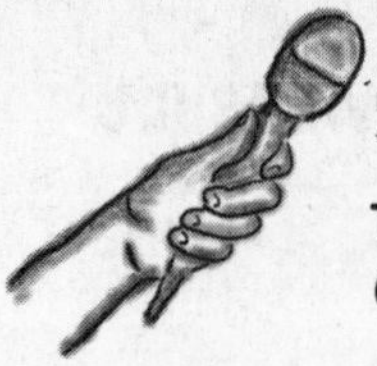

Standard Deviation Exposed

This week's interview:
Getting the measure of Standard Deviation

Head First: Hey, Standard Deviation, great to see you.

Standard Deviation: It's a real pleasure, Head First.

Head First: To start off, I was wondering if you could tell me a bit more about yourself and what you do.

Standard Deviation: I'm really all about measuring the spread of data. Mean does a great job of telling you what's going on at the center, but quite often, that's not enough. Sometimes Mean needs support to give a more complete picture. That's where I come in. Mean gives the average value, and I say how values vary.

Head First: Without meaning to be rude, why should I care about how values vary? Is it really all that important? Surely it's enough to know just the average of a set of values.

Standard Deviation: Let me give you an example. How would you feel if you ordered a meal from the local diner, and when it arrived, you saw that half of it was burnt and the other half raw?

Head First: I'd probably feel unhappy, hungry, and ready to sue the diner. Why?

Standard Deviation: Well, according to Mean, your meal would have been cooked at the perfect temperature. Clearly, that's not the full picture; what you really need to know is the variation. That's where I come in. I look at what Mean thinks is a typical value, and I say how you can expect values to vary from that number.

Head First: I think I get it. Mean gives the average, and you indicate spread. How do you do that, though?

Standard Deviation: That's easy. I just say how far values are from the mean, on average. Suppose the standard deviation of a set of values is 3 cm. You can think of that as saying values are, on average, 3 cm away from the mean. There's a bit more to it than that, but if you think along those lines, you're on the right track.

Head First: Speaking of numbers, Standard Deviation, is it better if you're large or small?

Standard Deviance: Well, that really all depends what you're using me for. If you're manufacturing machine parts, you want me to be small, so you can be sure all the pieces are about the same. If you're looking at wages in a large company, I'll naturally be quite large.

Head First: I see. Tell me, do you have anything to do with Variance?

Standard Deviation: It's funny you should ask that. Variance is just an alter ego of mine. Square me, and I turn into Variance. Take the square root of Variance, and there I am again. We're a bit like Clark Kent and Superman, but without the cape.

Head First: Just one more question. Do you ever feel overshadowed by Mean? After all, he gets a lot more attention than you.

Standard Deviation: Of course not. We're great friends, and we support each other. Besides, that would make me sound negative. I'm never negative.

Head First: Standard Deviation, thank you for your time.

Standard Deviation: It's been a pleasure.

Exercise

It's time for you to flex those standard deviation muscles. Calculate the mean and standard deviation for the following sets of numbers.

1 2 3 4 5 6 7

1 2 3 4 5 6

Exercise Solution

It's time for you to flex those standard deviation muscles. Calculate the mean and standard deviation for the following sets of numbers.

1 2 3 4 5 6 7

Let's start off by calculating the mean.

$$\mu = \frac{1+2+3+4+5+6+7}{7}$$

$$= \frac{28}{7}$$

$$= 4$$

$$\text{Variance} = \frac{(1-4)^2 + (2-4)^2 + (3-4)^2 + (4-4)^2 + (5-4)^2 + (6-4)^2 + (7-4)^2}{7}$$

$$= \frac{3^2 + 2^2 + 1^2 + 0^2 + (-1)^2 + (-2)^2 + (-3)^2}{7}$$

$$= \frac{9+4+1+0+1+4+9}{7}$$

$$= \frac{28}{7}$$

$$= 4 \qquad \sigma = \sqrt{4} = 2$$

1 2 3 4 5 6

$$\mu = \frac{1+2+3+4+5+6}{6}$$

$$= \frac{21}{6}$$

$$= 3.5$$

$$\text{Variance} = \frac{(1-3.5)^2 + (2-3.5)^2 + (3-3.5)^2 + (4-3.5)^2 + (5-3.5)^2 + (6-3.5)^2}{6}$$

$$= \frac{2.5^2 + 1.5^2 + 0.5^2 + (-0.5)^2 + (-1.5)^2 + (-2.5)^2}{6}$$

$$= \frac{6.25 + 2.25 + 0.25 + 0.25 + 2.25 + 6.25}{6}$$

$$= \frac{17.5}{6}$$

$= 2.92$ (to 2 decimal places) $\qquad \sigma = \sqrt{2.92}$

$= 1.71$ (to 2 d.p.)

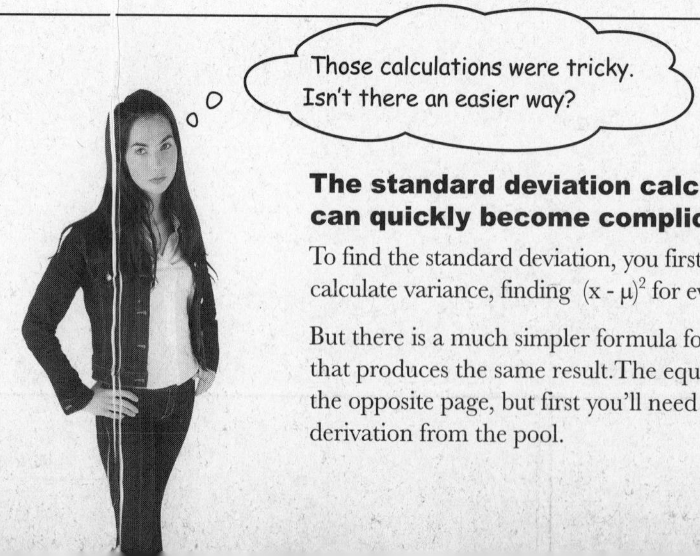
Those calculations were tricky.
Isn't there an easier way?

The standard deviation calculation can quickly become complicated.

To find the standard deviation, you first have to calculate variance, finding $(x - \mu)^2$ for every value of x.

But there is a much simpler formula for variance that produces the same result. The equation's on the opposite page, but first you'll need to rescue the derivation from the pool.

Pool Puzzle

There's an easier calculation for calculating the variance, but what is it? Your **job** is to take equation snippets from the pool, and place them into the blank lines in the derivation. Each snippet will be used **only once**, but you won't need to use every one. Your **goal** is to get to the equation at the end.

Psst – here's a hint. Remember that $\frac{\Sigma x}{n} = \mu$.

$$\frac{\Sigma(x - \mu)^2}{n} = \frac{\Sigma(x - \mu)\,(x - \mu)}{n}$$

$$= \frac{\Sigma(x^2 \ \ldots\ldots\ldots\ldots \ + \mu^2)}{n}$$

See if you can get from here...

$$= \frac{\Sigma x^2}{n} - \frac{2\mu\,\Sigma x}{n} + \frac{\ldots\ldots}{n}$$

to here.

$$= \frac{\ldots\ldots}{\ldots\ldots} - 2\mu \ldots\ldots + \frac{n\mu^2}{n}$$

$$= \frac{\Sigma x^2}{n} - \mu^2$$

Note: each snippet from the pool can only be used once!

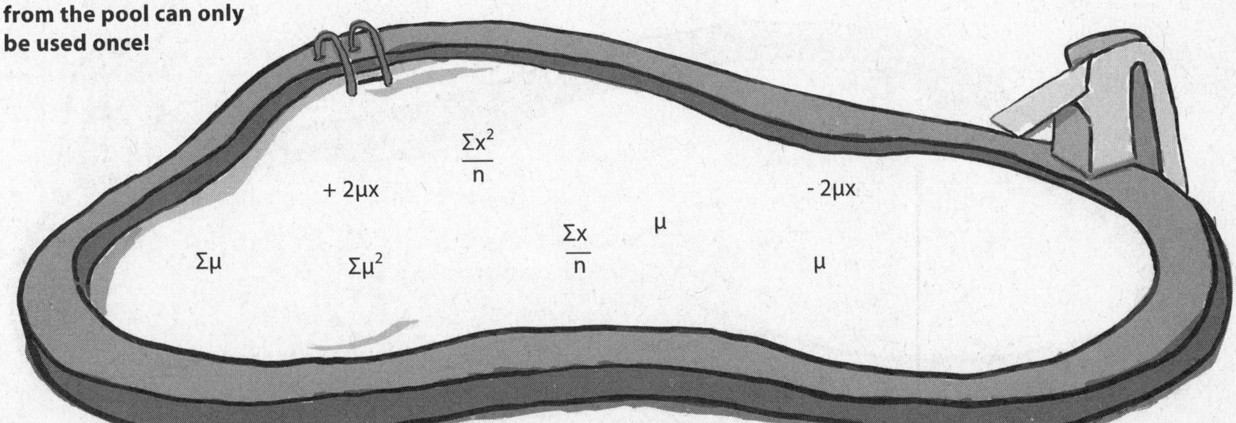

Pool Puzzle Solution

There's an easier calculation for calculating the variance, but what is it? Your **job** is to take equation snippets from the pool, and place them into the blank lines in the derivation. Each snippet will be used **only once**, but you won't need to use every one. Your **goal** is to get to the equation at the end.

$$\frac{\Sigma(x - \mu)^2}{n} = \frac{\Sigma(x - \mu)(x - \mu)}{n}$$

$$= \frac{\Sigma(x^2 - 2\mu x + \mu^2)}{n}$$

$$= \frac{\Sigma x^2}{n} - \frac{2\mu \Sigma x}{n} + \frac{\Sigma x^2}{n}$$

There are n of these.

$$= \frac{\Sigma x^2}{n} - 2\mu\,\mu + \frac{n\mu^2}{n}$$

The n's here cancel each other out.

$$= \frac{\Sigma x^2 - \mu^2}{n}$$

You didn't need these snippets.

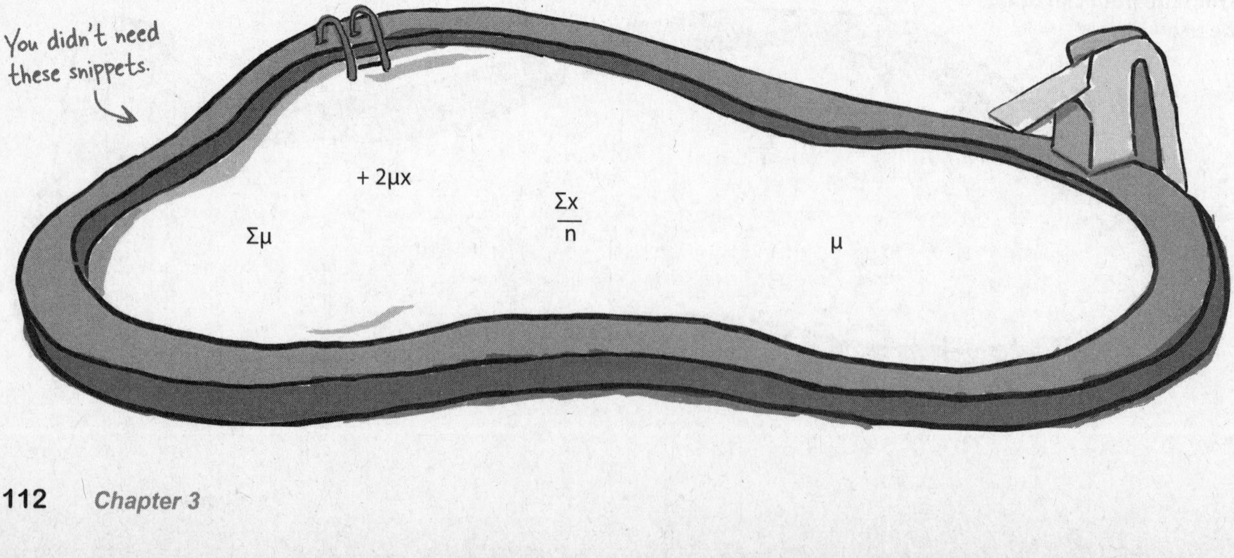

A quicker calculation for variance

As you've seen, the standard deviation is a good way of measuring spread, but the necessary variance calculation quickly becomes complicated. The difficulty lies in having to calculate $(x - \mu)^2$ for every value of x. The more values you're dealing with, the easier it is to make a mistake—particularly if μ is a long decimal number.

Here's a quicker way to calculate the variance:

$$\text{Variance} = \frac{\sum x^2}{n} - \mu^2$$

The advantage of this method is that you don't have to calculate $(x - \mu)^2$. Which means that, in practice, it's less tricky to deal with, and there's less of a chance you'll make mistake.

Vital Statistics

Variance

Here's the quicker way of calculating the variance

$$\frac{\sum x^2}{n} - \mu^2$$

there are no Dumb Questions

Q: So which form of the variance equation should I use?

A: If you're performing calculations, it's generally easier to use the second form, which is:

$$\frac{\sum x^2}{n} - \mu^2$$

This is particularly important if you have a mean with lots of decimals.

Q: How do I work out the standard deviation with this form of the variance equation?

A: Exactly the same way as before. Taking the square root of the variance gives you the standard deviation.

Q: What if I'm told what the standard deviation is, can I find the variance?

A: Yes, you can. The standard deviation is the square root of the variance, which means that the variance is the square of the standard deviation. To find the variance from the standard deviation, square the value of the standard deviation.

Q: I find the standard deviation really confusing. What is it again?

A: The standard deviation is a way of measuring spread. It describes how far typical values are from the mean.

If the standard deviation is high, this means that values are typically a long way from the mean. If the standard deviation is low, values tend to be close to the mean.

Q: Can the standard deviation ever be 0?

A: Yes, it can. The standard deviation is 0 if all of the values are the same. In other words, if each value is a distance of 0 away from the mean, the standard deviation will be 0.

Q: What units is standard deviation measured in?

A: It's measured in the same units as your data. If your measurements are in centimeters, and the standard deviation is 1, this means that values are typically 1 centimeter away from the mean.

Q: I'm sure I've seen formulas for variance where you divide by (n - 1) instead of n. Is that wrong?

A: It's not wrong, but that form of the variance is really used when you're dealing with samples. We'll show you more about this when we talk about sampling later in the book.

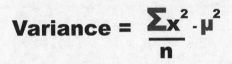

BE the coach

Here are the scores for the three players. The mean for each of them is 10. Your job is to play like you're the coach, and work out the standard deviation for each player. Which player is the most reliable one for your team?

Player 1

Score	7	9	10	11	13
Frequency	1	2	4	2	1

Player 2

Score	7	8	9	10	11	12	13
Frequency	1	1	2	2	2	1	1

Player 3

Score	3	6	7	10	11	13	30
Frequency	2	1	2	3	1	1	1

Exercise

The generous CEO of Starbuzz Coffee wants to give all his employees a pay raise. He's not sure whether to give everyone a straight $2,000 raise or increase salaries by 10%.

a) What happens to the standard deviation if everyone at Starbuzz is given a $2,000 pay raise?

b) What happens to the standard deviation if everyone at Starbuzz is given a 10% pay raise instead?

BE the coach Solution

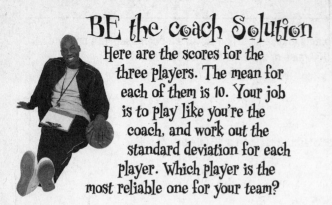

Here are the scores for the three players. The mean for each of them is 10. Your job is to play like you're the coach, and work out the standard deviation for each player. Which player is the most reliable one for your team?

Player 1

Score	7	9	10	11	13
Frequency	1	2	4	2	1

$$\text{Variance} = \frac{7^2 + 2(9^2) + 4(10^2) + 2(11^2) + 13^2}{10} - 100$$

$$= \frac{49 + 162 + 400 + 242 + 169}{10} - 100$$

$$= 2.2$$

$$\text{Standard Deviation} = \sqrt{2.2} = 1.48$$

Player 2

Score	7	8	9	10	11	12	13
Frequency	1	1	2	2	2	1	1

$$\text{Variance} = \frac{7^2 + 8^2 + 2(9^2) + 2(10^2) + 2(11^2) + 12^2 + 13^2}{10} - 100$$

$$= \frac{49 + 64 + 162 + 200 + 242 + 144 + 169}{10} - 100$$

$$= 3$$

$$\text{Standard Deviation} = \sqrt{3} = 1.73$$

Player 3

Score	3	6	7	10	11	13	30
Frequency	2	1	2	3	1	1	1

$$\text{Variance} = \frac{2(3^2) + 6^2 + 2(7^2) + 3(10^2) + 11^2 + 13^2 + 30^2}{11} - 100$$

$$= \frac{18 + 36 + 98 + 300 + 121 + 169 + 900}{11} - 100$$

$$= 49.27$$

$$\text{Standard Deviation} = \sqrt{49.27} = 7.02$$

Player 1 and Player 2 both have small standard deviations, so the values are clustered around the mean. But Player 3 has a standard deviation of 7.02, meaning scores are typically 7.02 points away from the mean. So Player 1 is the most reliable, and Player 3 is the least.

Exercise Solution

The generous CEO of Starbuzz Coffee wants to give all his employees a pay raise. He's not sure whether to give everyone a straight \$2,000 raise or increase salaries by 10%.

a) What happens to the standard deviation if everyone at Starbuzz is given a \$2,000 pay raise?

The standard deviation stays exactly the same. The figures are, in effect, picked up and moved sideways, so the standard deviation doesn't change.

$$\text{standard deviation} = \sqrt{\frac{\sum((x + 2000) - (\mu + 2000))^2}{n}}$$

$$= \sqrt{\frac{\sum(x + 2000 - \mu - 2000)^2}{n}}$$

$$= \sqrt{\frac{\sum(x - \mu)^2}{n}}$$

$$= \text{original standard deviation}$$

b) What happens to the standard deviation if everyone at Starbuzz is given a 10% pay raise instead?

The standard deviation is multiplied by 110%, or 1.1. The figures are stretched, so the standard deviation increases.

$$\text{standard deviation} = \sqrt{\frac{\sum((1.1x) - (1.1\mu))^2}{n}}$$

$$= \sqrt{\frac{\sum 1.1^2 (x - \mu)^2}{n}}$$

$$= 1.1 \sqrt{\frac{\sum(x - \mu)^2}{n}}$$

$$= 1.1 \text{ times original standard deviation}$$

What if we need a baseline for comparison?

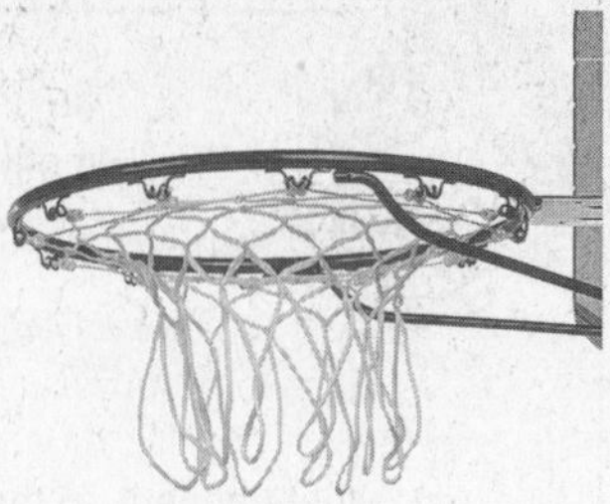

We've seen how the standard deviation can be used to measure how variable a set of values are, and we've used it to pick out the most reliable player for the Statsville All Stars. The standard deviation has other uses, too.

Imagine a situation in which you have two basketball players of different ability. The first player gets the ball into the net an average of 70% of the time, and he has a standard deviation of 20%. The second player has a mean of 40% and a standard deviation of 10%.

In a particular practice session, Player 1 gets the ball into the net 75% of the time, and Player 2 makes a basket 55% of the time. Which player does best against their personal track record?

That's easy—Player 1 does best. Player 1 scores 75% of the time, and Player 2 only scores 55% of the time.

Just looking at the percentages doesn't give the full picture.

75% sounds like a high percentage, but we're not taking into account the mean and standard deviation of each player. Each player has scored more than their personal mean, but which has fared better against their personal track record? How can we compare the two players?

The two players have different means and standard deviations, so how can we compare their personal performance?

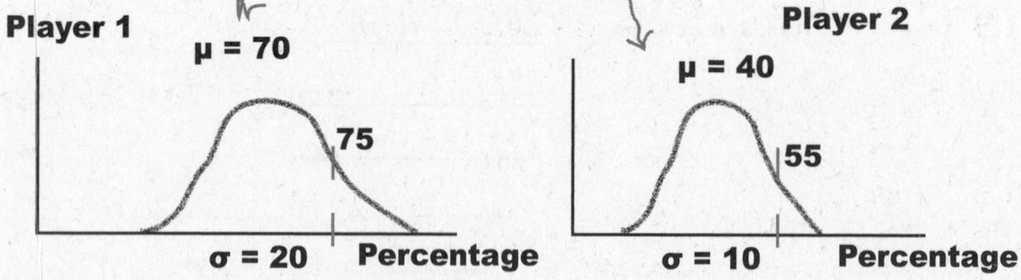

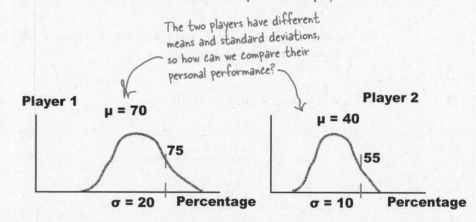

Does this sort of situation sound impossible? Don't worry, we can achieve this with the ***standard score***, or ***z-score***.

Use standard scores to compare values across data sets

Standard scores give you a way of comparing values across different sets of data where the mean and standard deviation differ. They're a way of comparing related data values in different circumstances. As an example, you can use standard scores to compare each player's performance relative to his personal track record—a bit like a personal trainer would.

You find the standard score of a particular value using the mean and standard deviation of the entire data set. The standard score is normally denoted by the letter *z*, and to find the standard score of a particular value *x*, you use the formula:

$$z = \frac{x - \mu}{\sigma}$$

These are the mean and standard deviation of the set of data containing the value x.

Let's calculate the standard scores for each player, and see what those scores tell us.

Calculating standard scores

Let's start by calculating z_1, the standard score of Player 1.

$$z_1 = \frac{75 - 70}{20} = \frac{5}{20} = 0.25$$

So using the mean and standard deviation to standardize the score, Player 1 gets 0.25. What about the score for Player 2?

$$z_2 = \frac{55 - 40}{10} = \frac{15}{10} = 1.5$$

This gives us a standard score of 1.5 for Player 2, compared with a standard score of 0.25 for Player 1. But what does this actually mean?

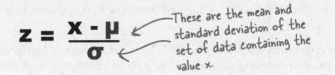

Interpreting standard scores

Standard scores give us a way of comparing values across different data sets even when the sets of data have different means and standard deviations. They're a way of comparing values **as if they came from the same set of data or distribution.**

So what does this mean for our basketball players?

Each player's shooting success rate has a different mean and standard deviation, which makes it difficult to compare how the players are performing relative to their own track record. We can see that in a particular practice, one player got the ball in the net more times than the other. We also notice that both players are scoring at a higher rate than their average. The difficulty lies in comparing performances relative to the personal track record of each player.

The standard score makes such comparisons possible by transforming each set of data into a more generic distribution. We can find the standard score of each player at the practice session, and then transform and compare them.

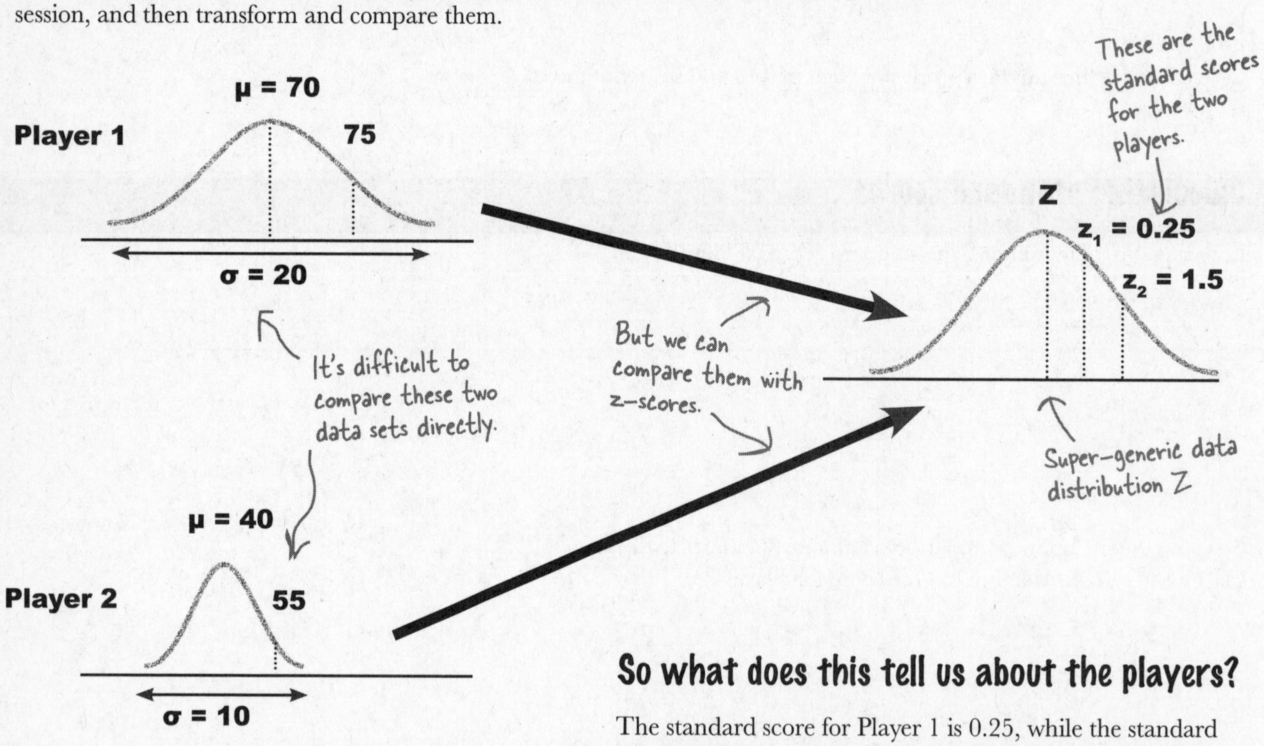

So what does this tell us about the players?

The standard score for Player 1 is 0.25, while the standard score for Player 2 is 1.5. In other words, when we standardize the scores, the score for Player 2 is higher.

This means that even though Player 1 is generally a better shooter and put balls into the net at a higher rate than Player 2, Player 2 performed better relative to his own track record. Player 2 performed better…for him.

Standard Scores Up Close

Standard scores work by transforming sets of data into a new, theoretical distribution with a mean of 0 and a standard deviation of 1. It's a generic distribution that can be used for comparisons. Standard scores effectively transform your data so that it fits this model while making sure it keeps the same basic shape.

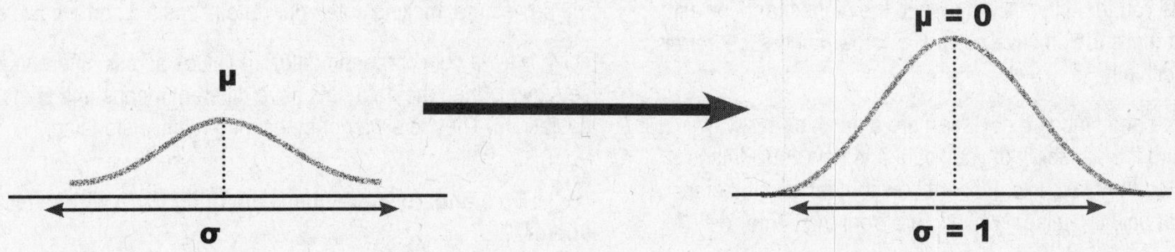
μ
σ
μ = 0
σ = 1

Standard scores can take any value, and they indicate position relative to the mean. Positive z-scores mean that your value is above the mean, and negative z-scores mean that your value is below it. If your z-score is 0, then your value is the mean itself. The size of the number shows how far away the value is from the mean.

Standard deviations from the mean

Sometimes statisticians express the relative position of a particular value in terms of **standard deviations from the mean**. As an example, a statistician may say that a particular value is within 1 standard deviation of the mean. It's really just another way of indicating how close values are to the mean, but what does it mean in practice?

We've seen that using z-scores transforms your data set into a generic distribution with a mean of 0 and a standard deviation of 1. If a value is within 1 standard deviation of the mean, this tells us that the standard score of the value is between -1 and 1. Similarly, if a value is within 2 standard deviations of the mean, the standard score of the value would be somewhere between -2 and 2.

Standard score = number of standard deviations from the mean.

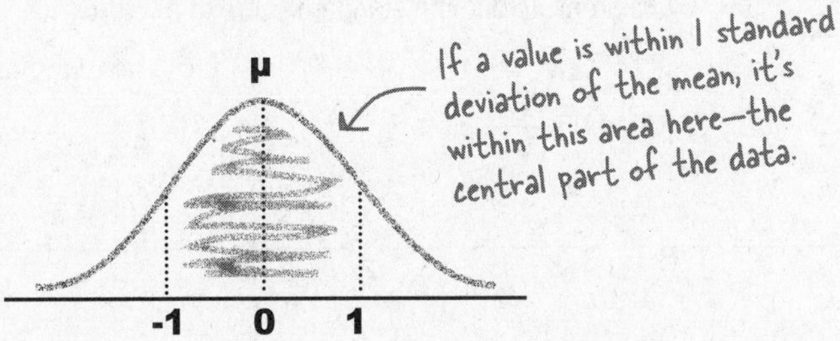
μ
If a value is within 1 standard deviation of the mean, it's within this area here—the central part of the data.
-1
0
1

there are no Dumb Questions

Q: So variance and standard deviation both measure the spread of your data. How are they different from the range?

A: The range is quite a simplistic measure of the spread of your data. It tells you the difference between the highest and lowest values, but that's it. You have no way of knowing how the data is clustered within it.

The variance and standard deviation are a much better way of measuring the variability of your data and how your data is dispersed, as they take into account how the data is clustered. They look at how far values typically are from the center of your data.

Q: And what's the difference between variance and standard deviation? Which one should I use?

A: The standard deviation is the square root of the variance, which means you can find one from the other.

The standard deviation is probably the most intuitive, as it tells you roughly how far your values are, on average, from the mean.

Q: How do standard scores fit into all this?

A: Standard scores use the mean and standard deviation to convert values in a data set to a more generic distribution, while at the same time, making sure your data keeps the same basic shape.

They're a way of comparing different values across different data sets even when the data sets have different means and standard deviations. They're a way of measuring relative standing.

Q: Do standard scores have anything to do with detecting outliers?

A: Good question! Determining outliers can be subjective, but sometimes outliers are defined as being more than 3 standard deviations of the mean. Statisticians have different opinions about this though, so be warned.

BULLET POINTS

- The **variance** and **standard deviation** measure how values are dispersed by looking at how far values are from the mean.
- The variance is calculated using

$$\frac{\sum (x - \mu)^2}{n}$$

- An alternate form is

$$\frac{\sum x^2}{n} - \mu^2$$

- The standard deviation is equal to the square root of the variance, and the variance is the standard deviation squared.
- **Standard scores**, or **z-scores**, are a way of comparing values across different sets of data where the means and standard deviations are different. To find the standard score of a value x, use:

$$z = \frac{x - \mu}{\sigma}$$

Exercise

Complete the table below. Name each type of measure of dispersion we've encountered in the chapter, and show how to calculate it. Try your hardest to fill this out without looking back through the chapter.

Statistic	How to calculate
Range	
	Upper quartile - Lower quartile
Standard Deviation (σ)	
Standard Score	

Complete the table below. Name each type of measure of dispersion we've encountered in the chapter, and show how to calculate it. Try your hardest to fill this out without looking back through the chapter.

Statistic	How to calculate
Range	Upper bound – Lower bound
Interquartile range	Upper quartile - Lower quartile
Standard Deviation (σ)	$\sqrt{\frac{\sum(x-\mu)^2}{n}}$ $\sqrt{\frac{\sum x^2}{n} - \mu^2}$ Both of these give the same result.
Standard Score	$z = \frac{x-\mu}{\sigma}$

Statsville All Stars win the league!

All the basketball matches for the season have now been played, and the Statsville All Stars finished at the top of the league. You clearly helped the coach pick the best player for the team.

Just remember: you owe it all to the friendly neighborhood standard deviation.

Let's hear it for the standard deviation, our new team mascot!

4 calculating probabilities

Taking Chances

What's the probability he's remembered I'm allergic to non-precious metals?

Life is full of uncertainty.

Sometimes it can be impossible to say what will happen from one minute to the next. But certain events are more likely to occur than others, and that's where **probability theory** comes into play. Probability lets you **predict the future** by assessing how likely outcomes are, and knowing what could happen helps you make **informed decisions**. In this chapter, you'll find out more about probability and learn how to take control of the future!

Fat Dan's Grand Slam

FAT DAN'S CASINO

Fat Dan's Casino is the most popular casino in the district. All sorts of games are offered, from roulette to slot machines, poker to blackjack.

It just so happens that today is your lucky day. Head First Labs has given you a whole rack of chips to squander at Fat Dan's, and you get to keep any winnings. Want to give it a try? Go on—you know you want to.

Are you ready to play?
One of Fat Dan's croupiers

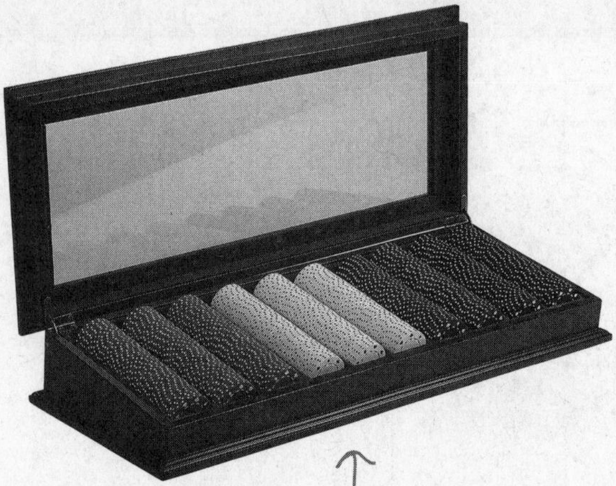

These are all your poker chips; looks like you're in for a fun time.

There's a lot of activity over at the roulette wheel, and another game is just about to start. Let's see how lucky you are.

Roll up for roulette!

You've probably seen people playing roulette in movies even if you've never tried playing yourself. The croupier spins a roulette wheel, then spins a ball in the opposite direction, and you place bets on where you think the ball will land.

Roulette wheel

The roulette wheel used in Fat Dan's Casino has 38 pockets that the ball can fall into. The main pockets are numbered from 1 to 36, and each pocket is colored either red or black. There are two extra pockets numbered 0 and 00. These pockets are both green.

Lightest gray = green,
black = black,
medium gray = red,

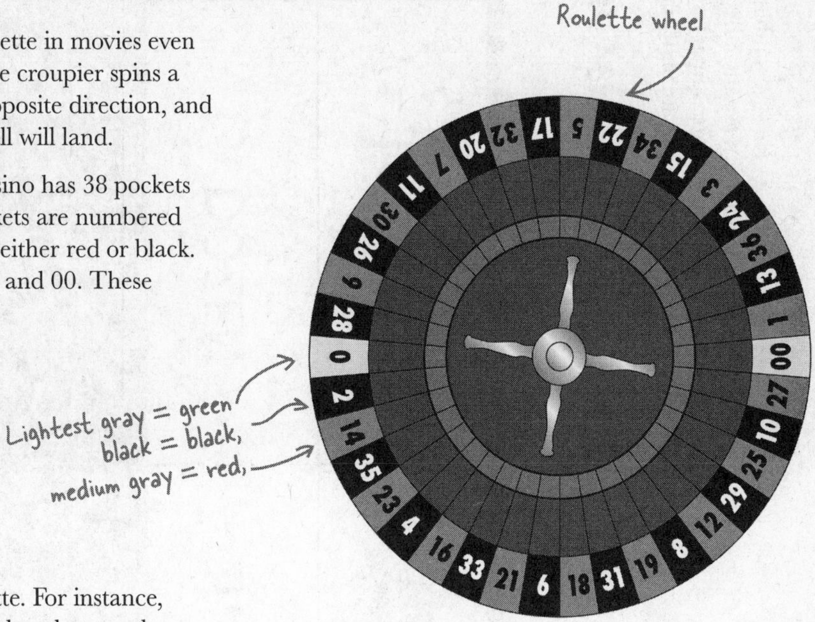

You can place all sorts of bets with roulette. For instance, you can bet on a particular number, whether that number is odd or even, or the color of the pocket. You'll hear more about other bets when you start playing. One other thing to remember: if the ball lands on a green pocket, you lose.

Roulette boards make it easier to keep track of which numbers and colors go together.

Roulette board. (See page 130 for a larger version.)

You place bets on the pocket the ball will fall into on the wheel using the board.

If the ball falls into the 0 or 00 pocket, you lose!

00	3	6	9	12	15	18	21	24	27	30	33	36	2 to 1
	2	5	8	11	14	17	20	23	26	29	32	35	2 to 1
0	1	4	7	10	13	16	19	22	25	28	31	34	2 to 1

1st DOZEN		2nd DOZEN		3rd DOZEN	
1 – 18	EVEN	◆	◆	ODD	19 – 36

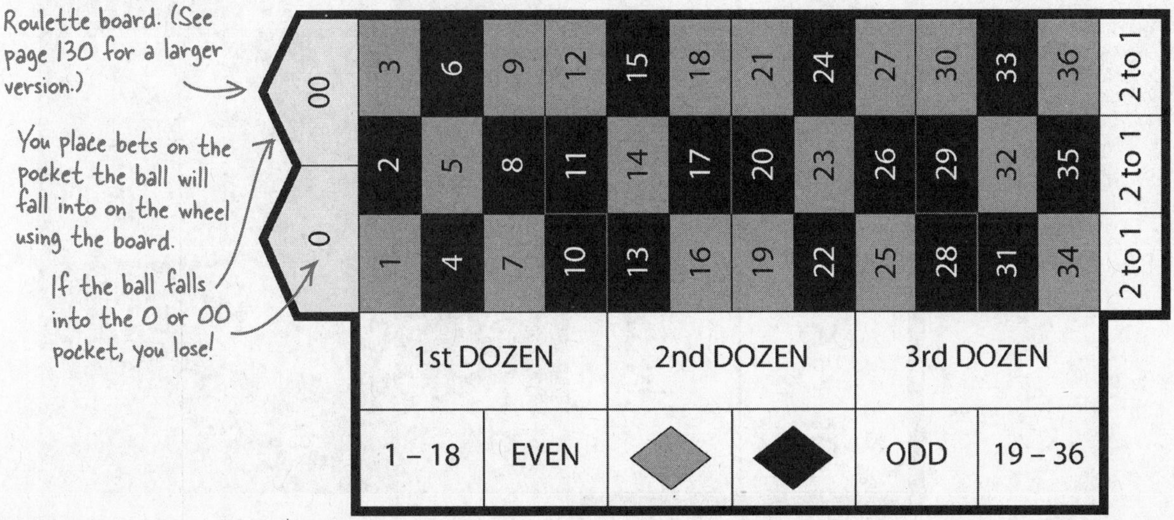

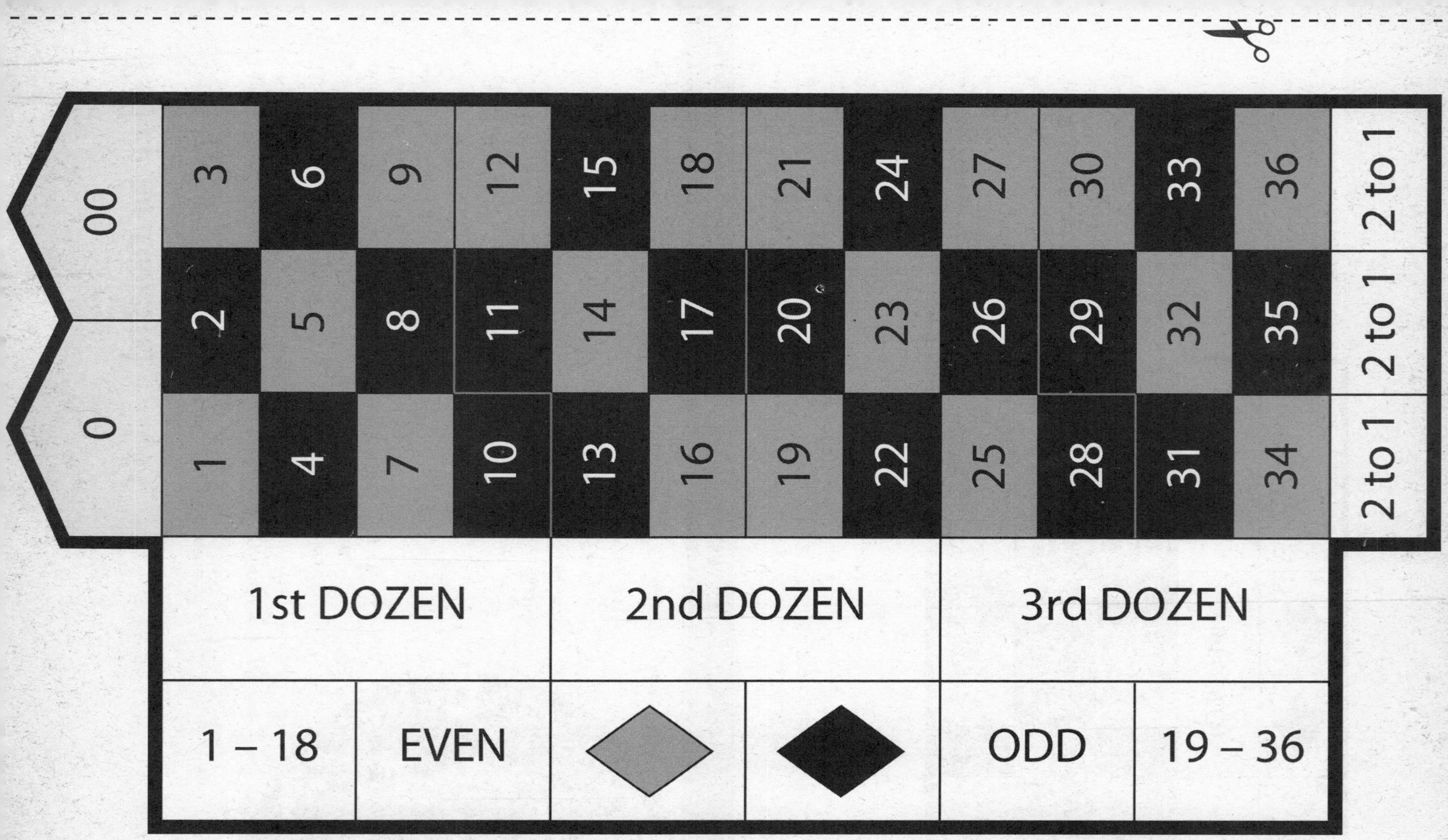

Your very own roulette board

You'll be placing a lot of roulette bets in this chapter. Here's a handy roulette board for you to cut out and keep. You can use it to help work out the probabilities in this chapter.

Just be careful with those scissors.

Place your bets now!

Have you cut out your roulette board? The game is just beginning. Where do you think the ball will land? Choose a number on your roulette board, and then we'll place a bet.

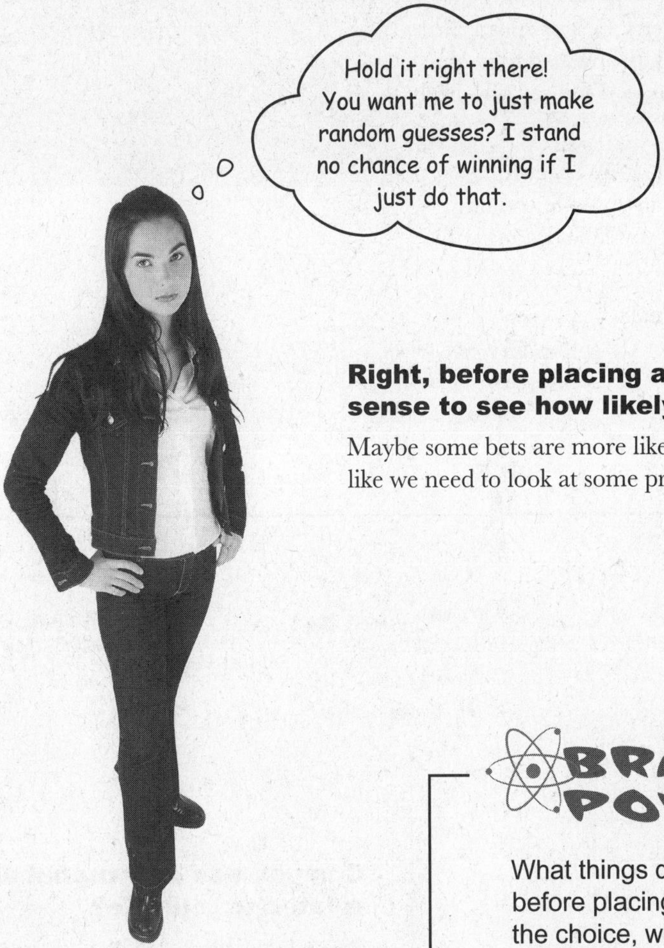

Hold it right there! You want me to just make random guesses? I stand no chance of winning if I just do that.

Right, before placing any bets, it makes sense to see how likely it is that you'll win.

Maybe some bets are more likely than others. It sounds like we need to look at some probabilities...

BRAIN POWER

What things do you need to think about before placing any roulette bets? Given the choice, what sort of bet would you make? Why?

What are the chances?

Have you ever been in a situation where you've wondered "Now, what were the chances of *that* happening?" Perhaps a friend has phoned you at the exact moment you've been thinking about them, or maybe you've won some sort of raffle or lottery.

Probability is a way of measuring the chance of something happening. You can use it to indicate how likely an occurrence is (the probability that you'll go to sleep some time this week), or how unlikely (the probability that a coyote will try to hit you with an anvil while you're walking through the desert). In stats-speak, an ***event*** is any occurrence that has a probability attached to it—in other words, an event is any outcome where you can say how likely it is to occur.

Probability is measured on a scale of 0 to 1. If an event is impossible, it has a probability of 0. If it's an absolute certainty, then the probability is 1. A lot of the time, you'll be dealing with probabilities somewhere in between.

Here are some examples on a probability scale.

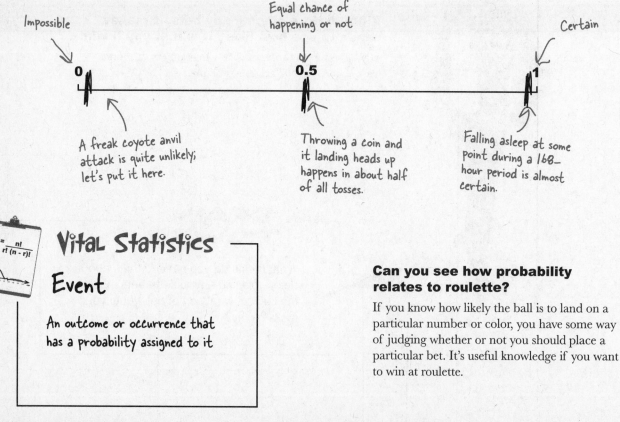

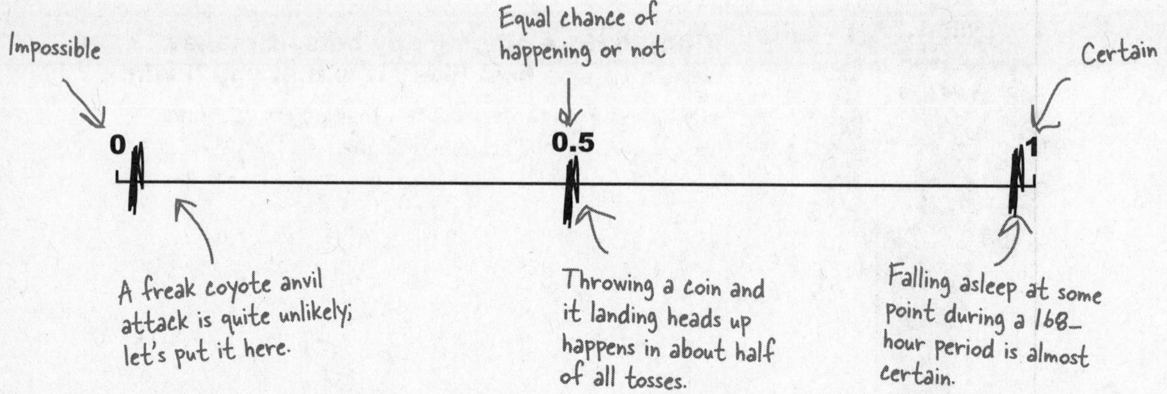

Impossible
Equal chance of happening or not
Certain
0
0.5
1
A freak coyote anvil attack is quite unlikely; let's put it here.
Throwing a coin and it landing heads up happens in about half of all tosses.
Falling asleep at some point during a 168-hour period is almost certain.

Vital Statistics

Event

An outcome or occurrence that has a probability assigned to it

Can you see how probability relates to roulette?

If you know how likely the ball is to land on a particular number or color, you have some way of judging whether or not you should place a particular bet. It's useful knowledge if you want to win at roulette.

Sharpen your pencil

Let's try working out a probability for roulette, the probability of the ball landing on 7. We'll guide you every step of the way.

1. Look at your roulette board. How many pockets are there for the ball to land in?

2. How many pockets are there for the number 7?

3. To work out the probability of getting a 7, take your answer to question 2 and divide it by your answer to question 1. What do you get?

4. Mark the probability on the scale below. How would you describe how likely it is that you'll get a 7?

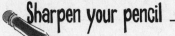

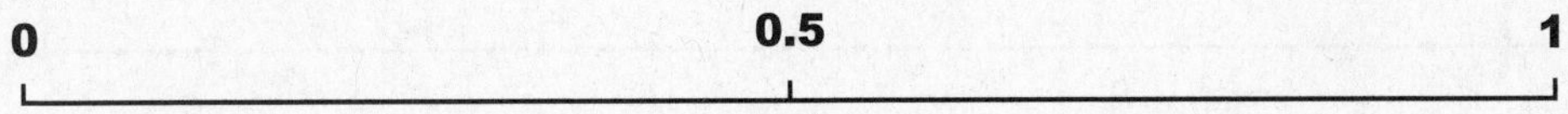

Sharpen your pencil Solution

You had to work out a probability for roulette, the probability of the ball landing on 7. Here's how you calculate the solution, step by step.

1. Look at your roulette board. How many pockets are there for the ball to land in?

There are 38 pockets. ← Don't forget that the ball can land in 0 or 00 as well as the 36 numbers.

2. How many pockets are there for the number 7?

Just 1

3. To work out the probability of getting a 7, take your answer to question 2 and divide it by your answer to question 1. What do you get?

Probability of getting 7 = $\frac{1}{38}$

= 0.026 ← Our answer to 3 decimal places

4. Mark the probability on the scale below? How would you describe how likely it is that you'll get a 7?

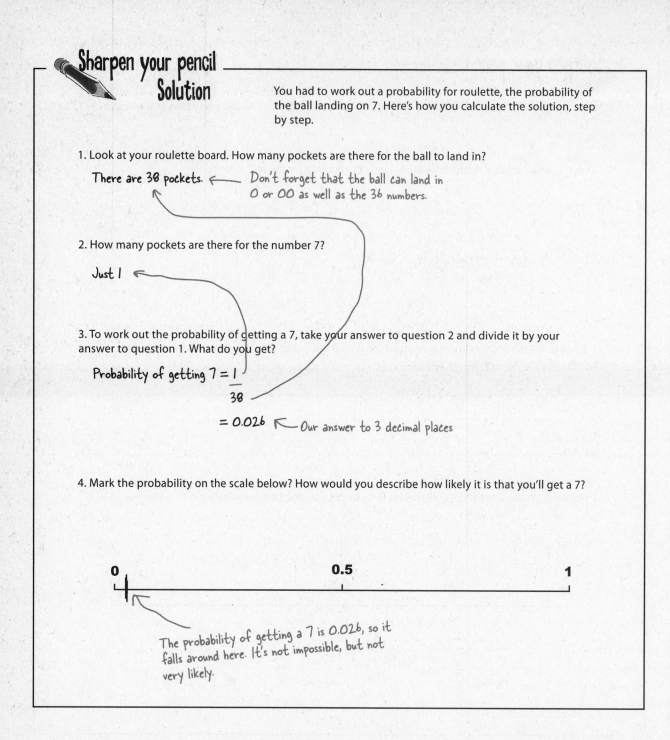

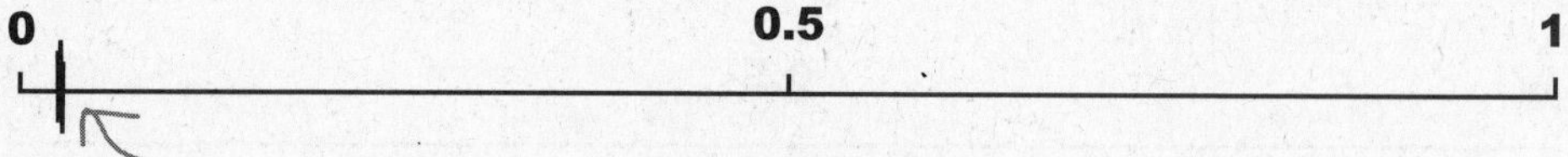

The probability of getting a 7 is 0.026, so it falls around here. It's not impossible, but not very likely.

Find roulette probabilities

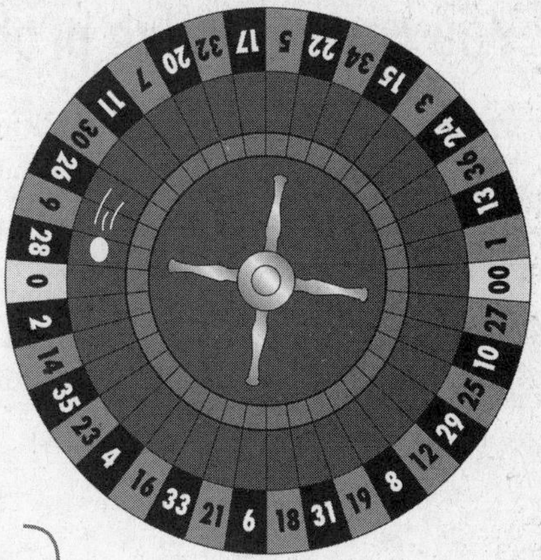

Let's take a closer look at how we calculated that probability.

Here are all the possible outcomes from spinning the roulette wheel. The thing we're really interested in is winning the bet—that is, the ball landing on a 7.

There's just one event we're really interested in: the probability of the ball landing on a 7.

00	3	6	9	12	15	18	21	24	27	30	33	36	2 to 1
0	2	5	8	11	14	17	20	23	26	29	32	35	2 to 1
	1	7	10	13	16	19	22	25	28	31	34		2 to 1

1st DOZEN	2nd DOZEN	3rd DOZEN
1 – 18	EVEN ◆ ◆	ODD	19 – 36

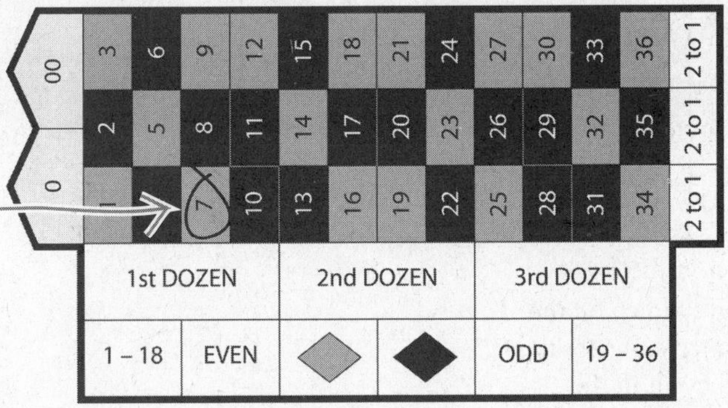

These are all possible outcomes, as the ball could land in any of these pockets.

To find the probability of winning, we take the number of ways of winning the bet and divide by the number of possible outcomes like this:

There's one way of getting a 7, and there are 38 pockets.

$$\text{Probability} = \frac{\text{number of ways of winning}}{\text{number of possible outcomes}}$$

We can write this in a more general way, too. For the probability of any event A:

Probability of event A occurring → $$P(A) = \frac{n(A)}{n(S)}$$

Number of ways of getting an event A

The number of possible outcomes

S is known as the ***possibility space***, or ***sample space***. It's a shorthand way of referring to all of the possible outcomes. Possible events are all subsets of S.

You can visualize probabilities with a Venn diagram

Probabilities can quickly get complicated, so it's often very useful to have some way of visualizing them. One way of doing so is to draw a box representing the possibility space **S**, and then draw circles for each relevant event. This sort of diagram is known as a ***Venn diagram***. Here's a Venn diagram for our roulette problem, where **A** is the event of getting a 7.

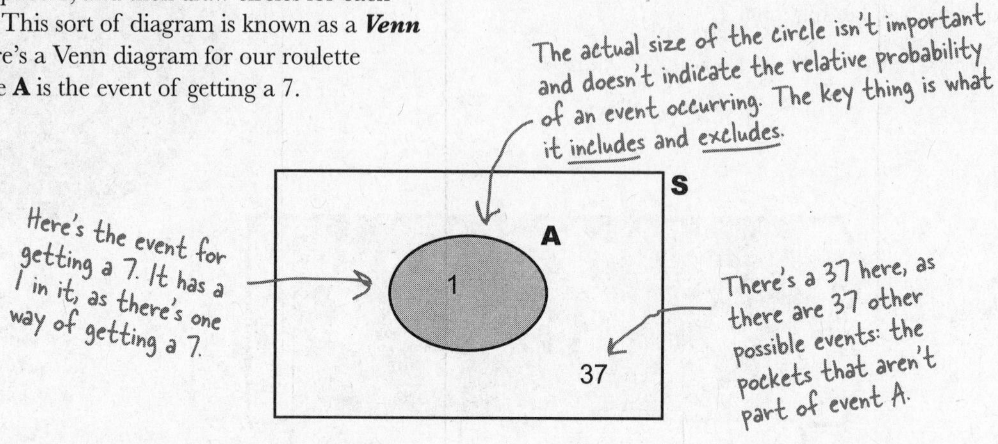
The actual size of the circle isn't important and doesn't indicate the relative probability of an event occurring. The key thing is what it includes and excludes.
S
A
1
37
Here's the event for getting a 7. It has a 1 in it, as there's one way of getting a 7.
There's a 37 here, as there are 37 other possible events: the pockets that aren't part of event A.

Very often, the numbers themselves aren't shown on the Venn diagram. Instead of numbers, you have the option of using the actual probabilities of each event in the diagram. It all depends on what kind of information you need to help you solve the problem.

Complementary events

There's a shorthand way of indicating the event that A does not occur—A'. A' is known as the ***complementary*** event of A.

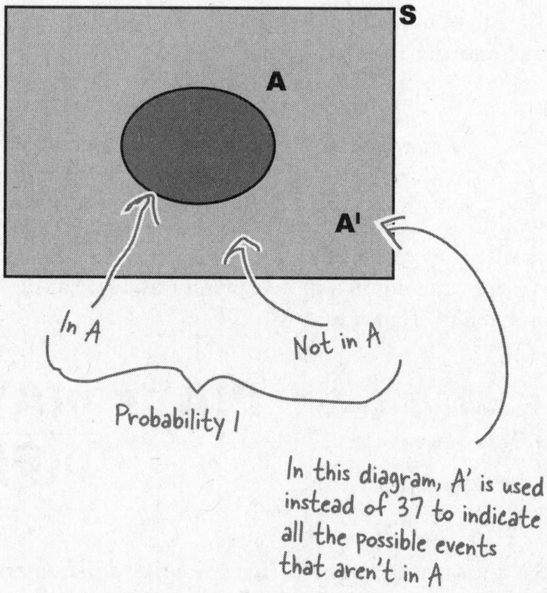
S
A
A'
In A
Not in A
Probability 1
In this diagram, A' is used instead of 37 to indicate all the possible events that aren't in A

There's a clever way of calculating $P(A')$. A' covers every possibility that's not in event A, so between them, A and A' must cover every eventuality. If something's in A, it can't be in A', and if something's not in A, it must be in A'. This means that if you add P(A) and $P(A')$ together, you get 1. In other words, there's a 100% chance that something will be in either A or A'. This gives us

$$P(A) + P(A') = 1$$

or

$$\mathbf{P(A') = 1 - P(A)}$$

BE the croupier

Your job is to imagine you're the croupier and work out the probabilities of various events. For each event below, write down the probability of a successful outcome.

P(9)

P(Green)

P(Black)

P(38)

BE the croupier Solution

Your job was to imagine you're the croupier and work out the probabilities of various events. For each event you should have written down the probability of a successful outcome.

P(9)

The probability of getting a 9 is exactly the same as getting a 7, as there's an equal chance of the ball falling into each pocket.

$$\text{Probability} = \frac{1}{38}$$

$$= 0.026 \text{ (to 3 decimal places)}$$

P(Green)

2 of the pockets are green, and there are 38 pockets total, so:

$$\text{Probability} = \frac{2}{38}$$

$$= 0.053 \text{ (to 3 decimal places)}$$

P(Black)

18 of the pockets are black, and there are 38 pockets, so:

$$\text{Probability} = \frac{18}{38}$$

$$= 0.474 \text{ (to 3 decimal places)}$$

P(38)

This event is actually impossible—there is no pocket labeled 38. Therefore, the probability is 0.

The most likely event out of all these is that the ball will land in a black pocket.

there are no Dumb Questions

Q: Why do I need to know about probability? I thought I was learning about statistics.

A: There's quite a close relationship between probability and statistics. A lot of statistics has its origins in probability theory, so knowing probability will take your statistics skills to the next level. Probability theory can help you make predictions about your data and see patterns. It can help you make sense of apparent randomness. You'll see more about this later.

Q: Are probabilities written as fractions, decimals, or percentages?

A: They can be written as any of these. As long as the probability is expressed in some form as a value between 0 and 1, it doesn't really matter.

Q: I've seen Venn diagrams before in set theory. Is there a connection?

A: There certainly is. In set theory, the possibility space is equivalent to the set of all possible outcomes, and a possible event forms a subset of this. You don't have to already know any set theory to use Venn diagrams to calculate probability, though, as we'll cover everything you need to know in this chapter.

Q: Do I always have to draw a Venn diagram? I noticed you didn't in that last exercise.

A: No, you don't have to. But sometimes they can be a useful tool for visualizing what's going on with probabilities. You'll see more situations where this helps you later on.

Q: Can anything be in both events A and A'?

A: No. A' means everything that isn't in A. If an element is in A, then it can't possibly be in A'. Similarly, if an element is in A', then it can't be in A. The two events are mutually exclusive, so no elements are shared between them.

It's time to play!

A game of roulette is just about to begin.

Look at the events on the previous page. We'll place a bet on the one that's ***most likely*** to occur—that the ball will land in a black pocket.

Let's see what happens.

And the winning number is...

Oh dear! Even though our most likely probability was that the ball would land in a black pocket, it actually landed in the green 0 pocket. You lose some of your chips.

The ball landed in the 0 pocket, so you lost some chips.

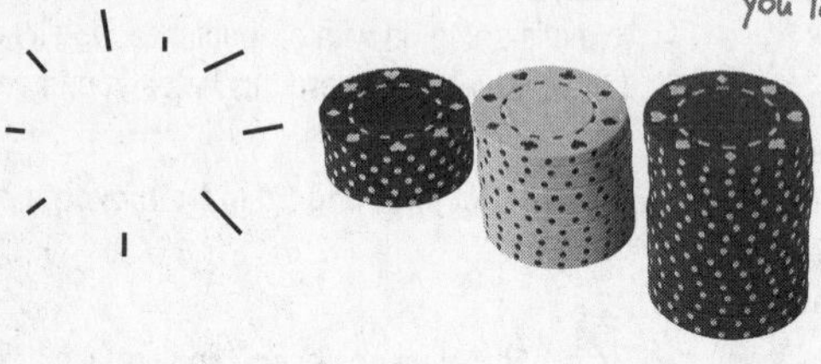

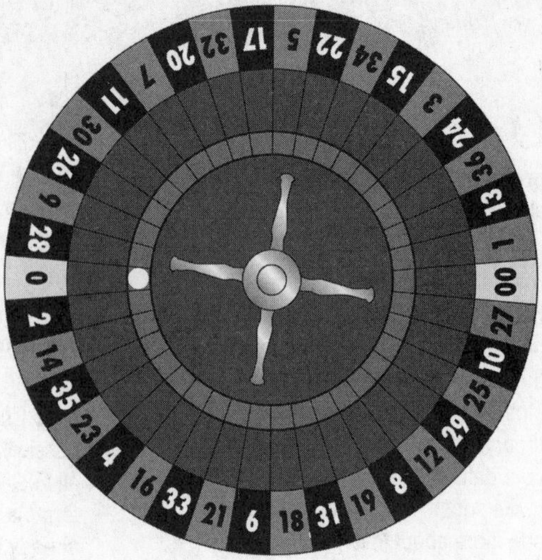

There must be a fix! The probability of getting a black is far higher than getting a green or 0. What went wrong? I want to win!

Probabilities are only indications of how likely events are; they're not guarantees.

The important thing to remember is that a probability indicates a long-term trend only. If you were to play roulette thousands of times, you would expect the ball to land in a black pocket in 18/38 spins, approximately 47% of the time, and a green pocket in 2/38 spins, or 5% of the time. Even though you'd expect the ball to land in a green pocket relatively infrequently, that doesn't mean it can't happen.

No matter how unlikely an event is, if it's not impossible, it can still happen.

Let's bet on an even more likely event

Let's look at the probability of an event that should be more likely to happen. Instead of betting that the ball will land in a black pocket, let's bet that the ball will land in a black *or* a red pocket. To work out the probability, all we have to do is count how many pockets are red or black, then divide by the number of pockets. Sound easy enough?

That's a lot of pockets to count. We've already worked out P(Black) and P(Green). Maybe we can use one of these instead.

We can use the probabilities we already know to work out the one we don't know.

Take a look at your roulette board. There are only three colors for the ball to land on: red, black, or green. As we've already worked out what P(Green) is, we can use this value to find our probability without having to count all those black and red pockets.

$$
\begin{aligned}
P(\text{Black or Red}) &= P(\text{Green}') \\
&= 1 - P(\text{Green}) \\
&= 1 - 0.053 \\
&= 0.947 \text{ (to 3 decimal places)}
\end{aligned}
$$

Sharpen your pencil

Don't just take our word for it. Calculate the probability of getting a black or a red by counting how many pockets are black or red and dividing by the number of pockets.

Sharpen your pencil Solution

Don't just take our word for it. Calculate the probability of getting a black or a red by counting how many pockets are black or red and dividing by the number of pockets.

$$P(\text{Black or Red}) = \frac{36}{38}$$

$$= 0.947 \text{ (to 3 decimal places)}$$

So P(Black or Red) = 1 – P(Green)

You can also add probabilities

There's yet another way of working out this sort of probability. If we know P(Black) and P(Red), we can find the probability of getting a black or red by adding these two probabilities together. Let's see.

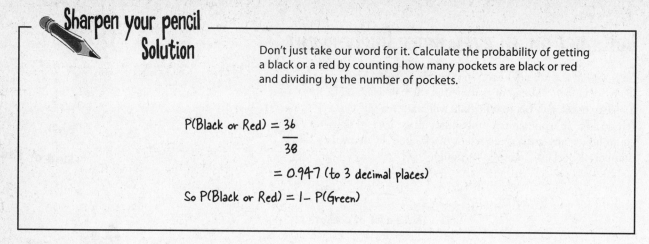

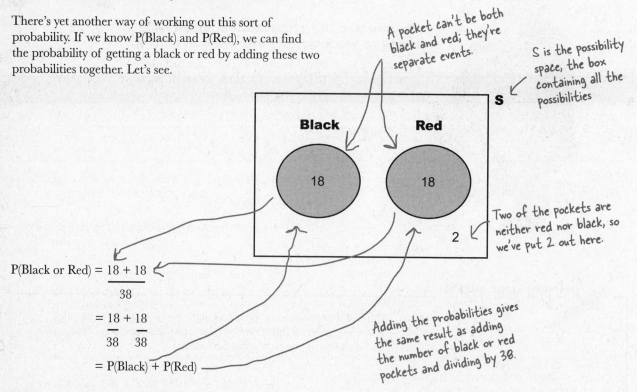

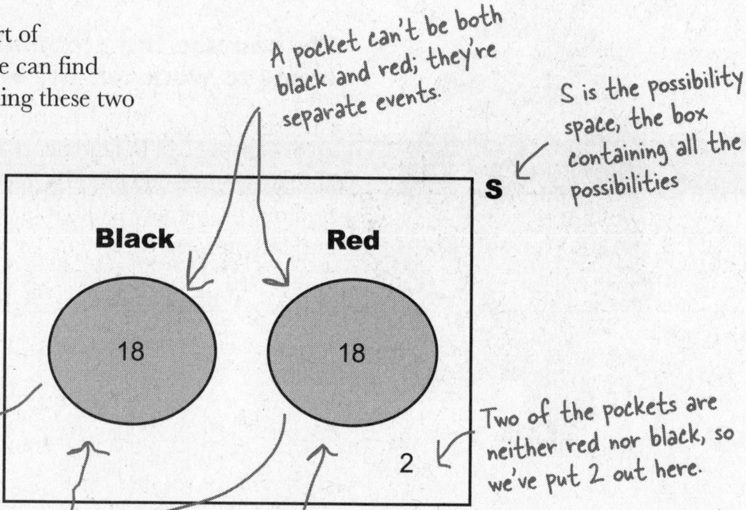
A pocket can't be both black and red; they're separate events.
S is the possibility space, the box containing all the possibilities
S
Black
Red
18
18
2
Two of the pockets are neither red nor black, so we've put 2 out here.

$$P(\text{Black or Red}) = \frac{18 + 18}{38}$$

$$= \frac{18}{38} + \frac{18}{38}$$

$$= P(\text{Black}) + P(\text{Red})$$

Adding the probabilities gives the same result as adding the number of black or red pockets and dividing by 38.

In this case, adding the probabilities gives exactly the same result as counting all the red or black pockets and dividing by 38.

Vital Statistics

Probability

To find the probability of an event A, use

$$P(A) = \frac{n(A)}{n(S)}$$

Vital Statistics

A'

A' is the complementary event of A. It's the probability that event A does not occur.

$$P(A') = 1 - P(A)$$

there are no Dumb Questions

Q: It looks like there are three ways of dealing with this sort of probability. Which way is best?

A: It all depends on your particular situation and what information you are given.

Suppose the only information you had about the roulette wheel was the probability of getting a green. In this situation, you'd have to calculate the probability by working out the probability of not getting a green:

1 - P(Green)

On the other hand, if you knew P(Black) and P(Red) but didn't know how many different colors there were, you'd have to calculate the probability by adding together P(Black) and P(Red).

Q: So I don't have to work out probabilities by counting everything?

A: Often you won't have to, but it all depends on your situation. It can still be useful to double-check your results, though.

Q: If some events are so unlikely to happen, why do people bet on them?

A: A lot depends on the sort of return that is being offered. In general, the less likely the event is to occur, the higher the payoff when it happens. If you win a bet on an event that has a high probability, you're unlikely to win much money. People are tempted to make bets where the return is high, even though the chances of them winning is negligible.

Q: Does adding probabilities together like that always work?

A: Think of this as a special case where it does. Don't worry, we'll go into more detail over the next few pages.

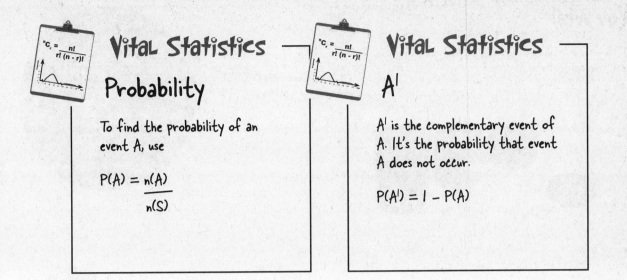

You win!

This time the ball landed in a red pocket, the number 7, so you win some chips.

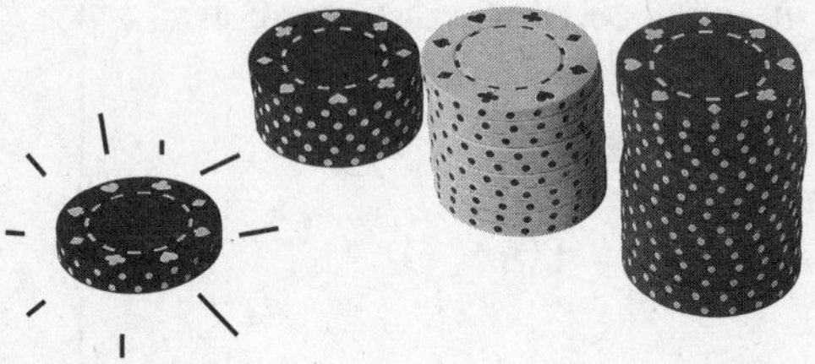
This time, you picked a winning pocket: a red one.

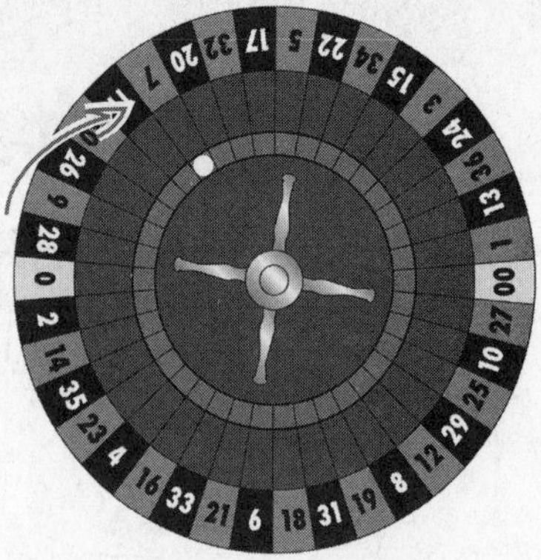

Time for another bet

Now that you're getting the hang of calculating probabilities, let's try something else. What's the probability of the ball landing on a black or even pocket?

Bet:
Black or Even

That's easy. We just add the black and even probabilities together.

Sometimes you can add together probabilities, but it doesn't work in all circumstances.

We might not be able to count on being able to do this probability calculation in quite the same way as the previous one. Try the exercise on the next page, and see what happens.

Sharpen your pencil

Let's find the probability of getting a black or even (assume 0 and 00 are not even).

1. What's the probability of getting a black?

2. What's the probability of getting an even number?

3. What do you get if you add these two probabilities together?

4. Finally, use your roulette board to count all the holes that are either black or even, then divide by the total number of holes. What do you get?

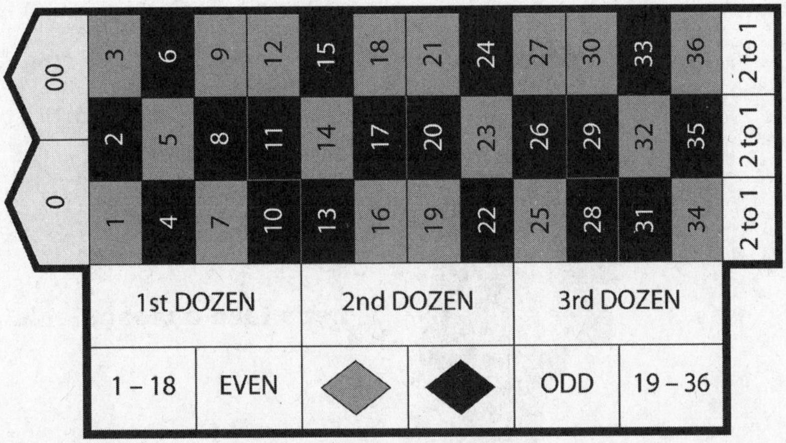

Sharpen your pencil Solution

Let's find the probability of getting a black or even (assume 0 and 00 are not even).

1. What's the probability of getting a black?

18 / 38 = 0.474

2. What's the probability of getting an even number?

18 / 38 = 0.474

3. What do you get if you add these two probabilities together?

0.947

4. Finally, use your roulette board to count all the holes that are either black or even, then divide by the total number of holes. What do you get?

26 / 38 = 0.684

Uh oh! Different answers

I don't get it. Adding probabilities worked OK last time. What went wrong?

Let's take a closer look...

Exclusive events and intersecting events

When we were working out the probability of the ball landing in a black or red pocket, we were dealing with two separate events, the ball landing in a black pocket and the ball landing in a red pocket. These two events are ***mutually exclusive*** because it's impossible for the ball to land in a pocket that's both black and red.

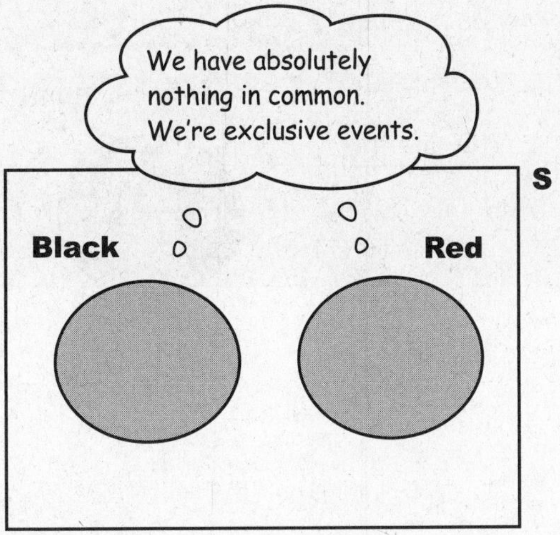
We have absolutely nothing in common. We're exclusive events.
S
Black
Red

If two events are mutually exclusive, only one of the two can occur.

What about the black and even events? This time the events aren't mutually exclusive. It's possible that the ball could land in a pocket that's both black *and* even. The two events ***intersect***.

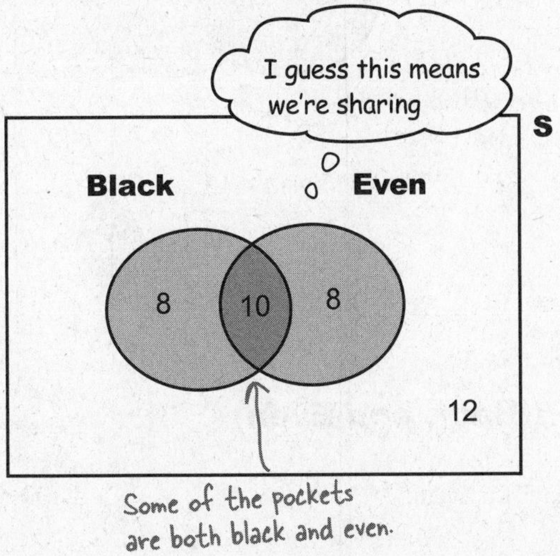
I guess this means we're sharing
S
Black
Even
8
10
8
12
Some of the pockets are both black and even.

If two events intersect, it's possible they can occur simultaneously.

BRAIN POWER

What sort of effect do you think this intersection could have had on the probability?

Problems at the intersection

Calculating the probability of getting a black or even went wrong because we included black and even pockets twice. Here's what happened.

First of all, we found the probability of getting a black pocket and the probability of getting an even number.

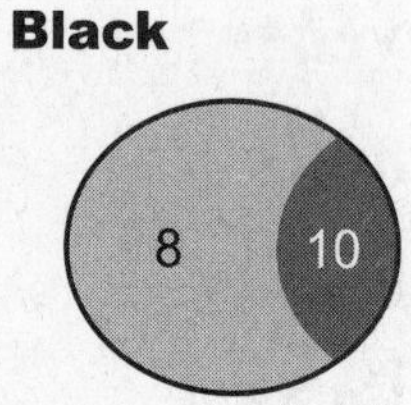

$$P(\text{Black}) = \frac{18}{38} = 0.474$$

$$P(\text{Even}) = \frac{18}{38} = 0.474$$

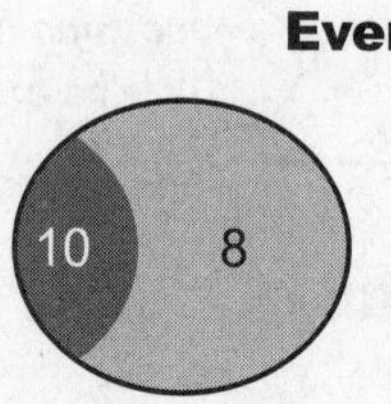

When we added the two probabilities together, we counted the probability of getting a black and even pocket twice.

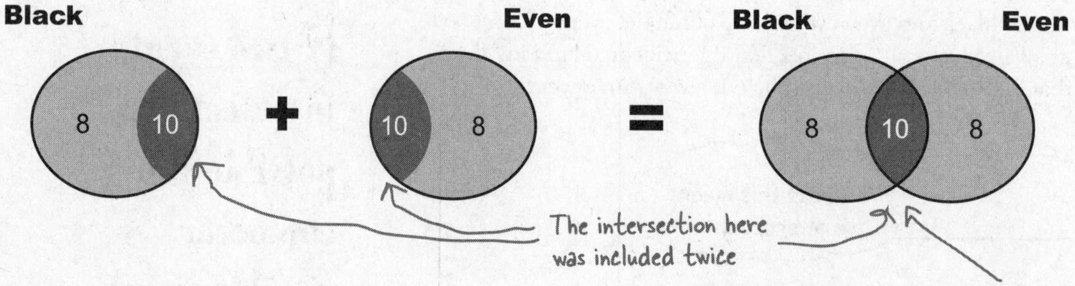

$$P(\text{Black} \cap \text{Even}) = \frac{10}{38} = 0.263$$

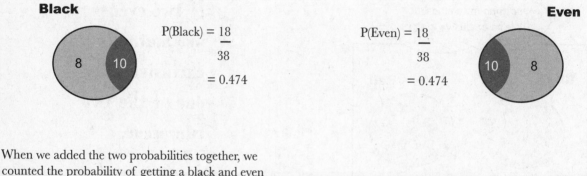

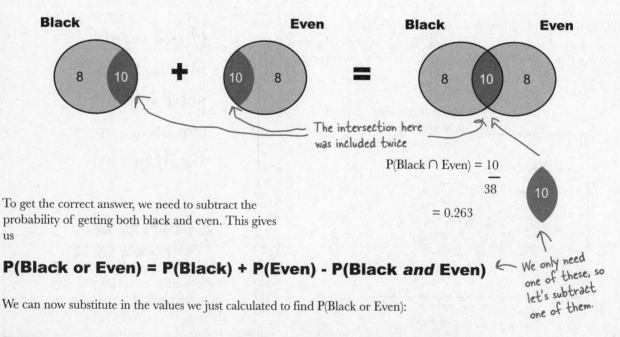

To get the correct answer, we need to subtract the probability of getting both black and even. This gives us

P(Black or Even) = P(Black) + P(Even) - P(Black *and* Even)

We only need one of these, so let's subtract one of them.

We can now substitute in the values we just calculated to find P(Black or Even):

P(Black or Even) = 18/38 + 18/38 - 10/38 = 26/38 = 0.684

Some more notation

There's a more general way of writing this using some more math shorthand.

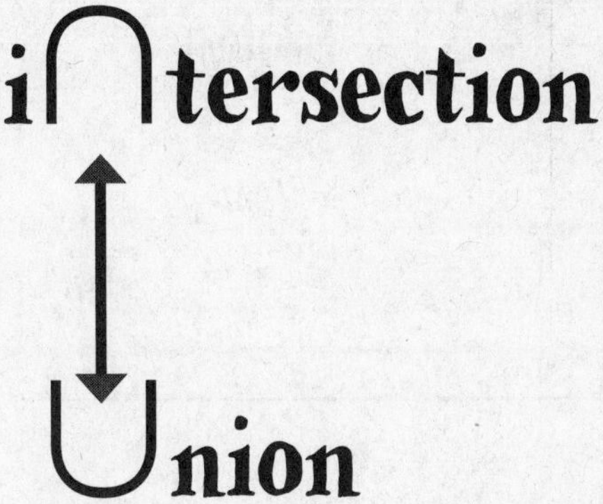

First of all, we can use the notation A ∩ B to refer to the intersection between A and B. You can think of this symbol as meaning "and." It takes the common elements of events.

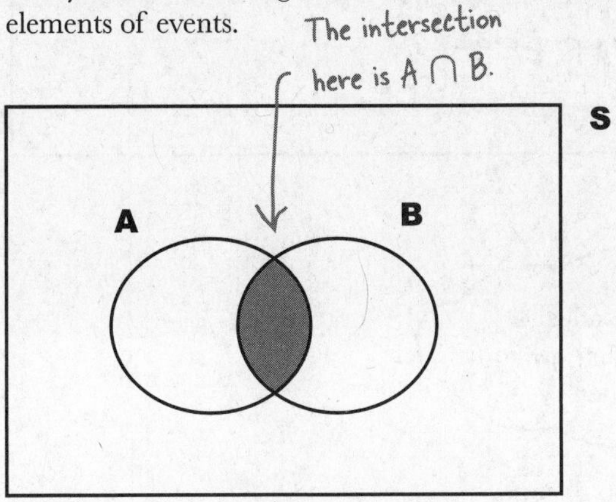

A ∪ B, on the other hand, means the union of A and B. It includes all of the elements in A and also those in B. You can think of it as meaning "or."

If A ∪ B =1, then A and B are said to be ***exhaustive***. Between them, they make up the whole of S. They exhaust all possibilities.

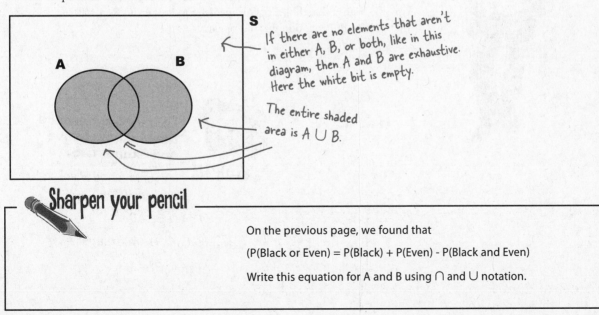

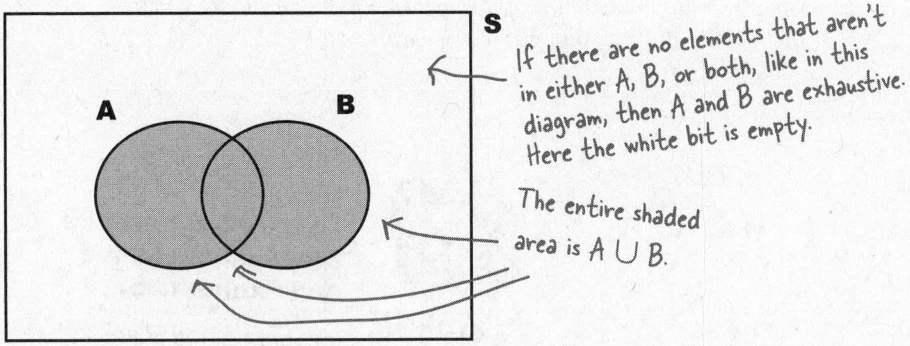

S
A
B
If there are no elements that aren't in either A, B, or both, like in this diagram, then A and B are exhaustive. Here the white bit is empty.
The entire shaded area is A ∪ B.

Sharpen your pencil

On the previous page, we found that

(P(Black or Even) = P(Black) + P(Even) - P(Black and Even)

Write this equation for A and B using ∩ and ∪ notation.

Sharpen your pencil Solution

On the previous page, we found that

(P(Black or Even) = P(Black) + P(Even) - P(Black and Even)

Write this equation for A and B using ∩ and ∪ notation.

P(A or B) →

$$P(A \cup B) = P(A) + P(B) - P(A \cap B)$$

← P(A and B)

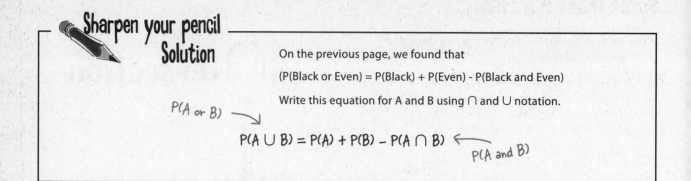

So why is the equation for exclusive events different? Are you just giving me more things to remember?

It's not actually that different.

Mutually exclusive events have no elements in common with each other. If you have two mutually exclusive events, the probability of getting A and B is actually 0—so $P(A \cap B) = 0$. Let's revisit our black-or-red example. In this bet, getting a red pocket on the roulette wheel and getting a black pocket are mutually exclusive events, as a pocket can't be both red and black. This means that $P(\text{Black} \cap \text{Red}) = 0$, so that part of the equation just disappears.

Watch it!

There's a difference between exclusive and exhaustive.

If events A and B are exclusive, then

$$P(A \cap B) = 0$$

If events A and B are exhaustive, then

$$P(A \cup B) = 1$$

BE the probability

Your job is to play like you're the probability and shade in the area that represents each of the following probabilities on the Venn diagrams.

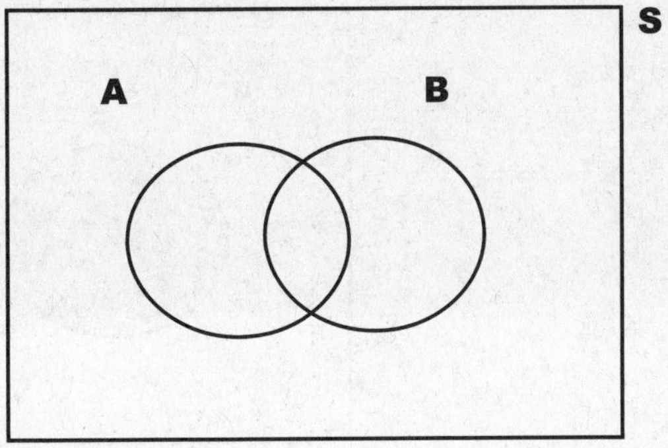

$P(A \cap B) + P(A \cap B')$

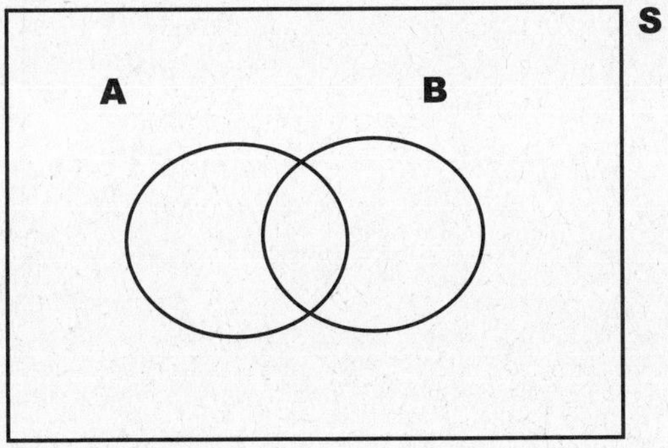

$P(A' \cap B')$

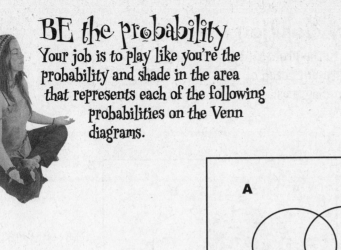

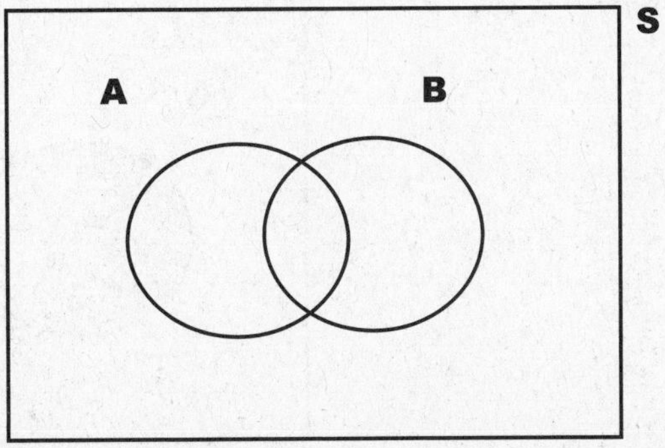

$P(A \cup B) - P(B)$

BE the probability Solution

Your job was to play like you're the probability and shade in the area that represents each of the probabilities on the Venn diagrams.

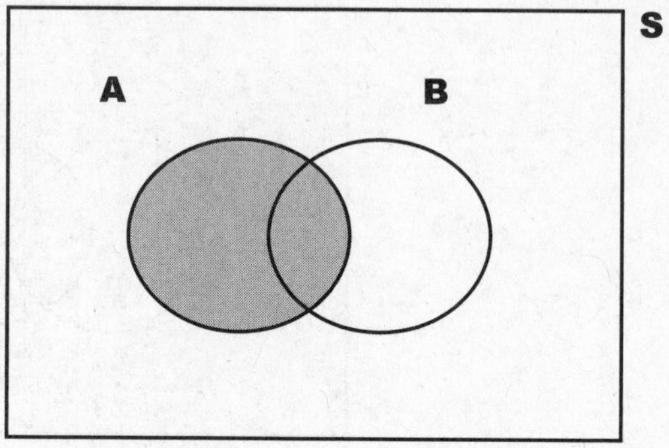

$P(A \cap B) + P(A \cap B')$

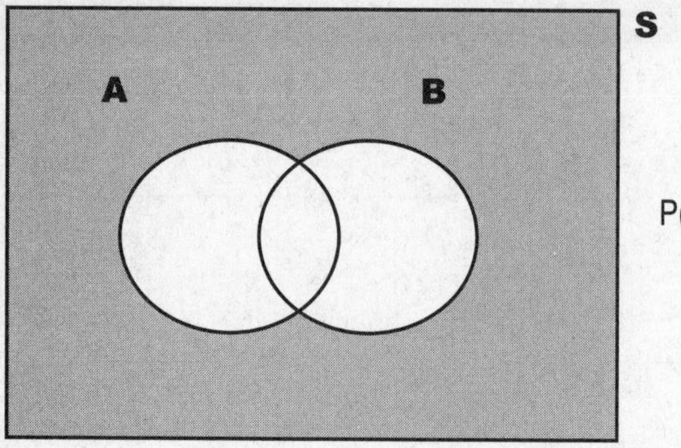

$P(A' \cap B')$

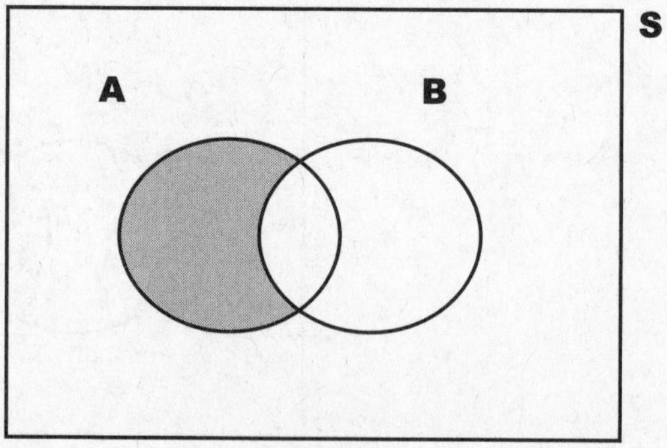

$P(A \cup B) - P(B)$

Exercise

50 sports enthusiasts at the Head First Health Club are asked whether they play baseball, football, or basketball. 10 only play baseball. 12 only play football. 18 only play basketball. 6 play baseball and basketball but not football. 4 play football and basketball but not baseball.

Draw a Venn diagram for this probability space. How many enthusiasts play baseball in total? How many play basketball? How many play football?

Are any sports' rosters mutually exclusive? Which sports are exhaustive (fill up the possibility space)?

Vital Statistics

A or B

To find the probability of getting event A or B, use

$$P(A \cup B) = P(A) + P(B) - P(A \cap B)$$

$\cup$ means OR

$\cap$ means AND

Exercise Solution

50 sports enthusiasts at the Head First Health Club are asked whether they play baseball, football or basketball. 10 only play baseball. 12 only play football. 18 only play basketball. 6 play baseball and basketball but not football. 4 play football and basketball but not baseball.

Draw a Venn diagram for this probability space. How many enthusiasts play baseball in total? How many play basketball? How many play football?

Are any sports' rosters mutually exclusive? Which sports are exhaustive (fill up the possibility space)?

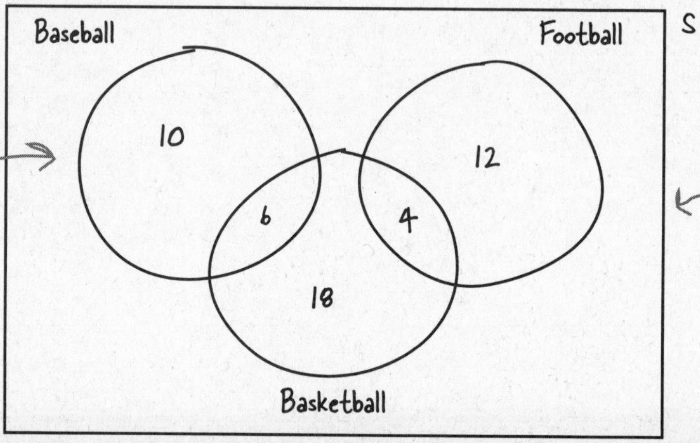

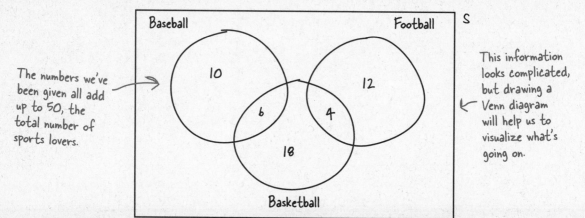

The numbers we've been given all add up to 50, the total number of sports lovers.

This information looks complicated, but drawing a Venn diagram will help us to visualize what's going on.

By adding up the values in each circle in the Venn diagram, we can determine that there are 16 total baseball players, 28 total basketball players, and 16 total football players.

The baseball and football events are mutually exclusive. Nobody plays both baseball and football, so $P(\text{Baseball} \cap \text{Football}) = 0$

The events for baseball, football, and basketball are exhaustive. Together, they fill the entire possibility space, so $P(\text{Baseball} \cup \text{Football} \cup \text{Basketball}) = 1$

there are no Dumb Questions

Q: Are A and A' mutually exclusive or exhaustive?

A: Actually they're both. A and A' can have no common elements, so they are mutually exclusive. Together, they make up the entire possibility space so they're exhaustive too.

Q: Isn't $P(A \cap B) + P(A \cap B')$ just a complicated way of saying P(A)?

A: Yes it is. It can sometimes be useful to think of different ways of forming the same probability, though. You don't always have access to all the information you'd like, so being able to think laterally about probabilities is a definite advantage.

Q: Is there a limit on how many events can intersect?

A: No. When you're referring to the intersection between several events, use more ∩'s. As an example, the intersection of events A, B, and C is $A \cap B \cap C$.

Finding probabilities for multiple intersections can sometimes be tricky. We suggest that if you're in doubt, draw a Venn diagram and take a good, hard look at which probabilities need to be added together and which need to be subtracted.

Another unlucky spin...

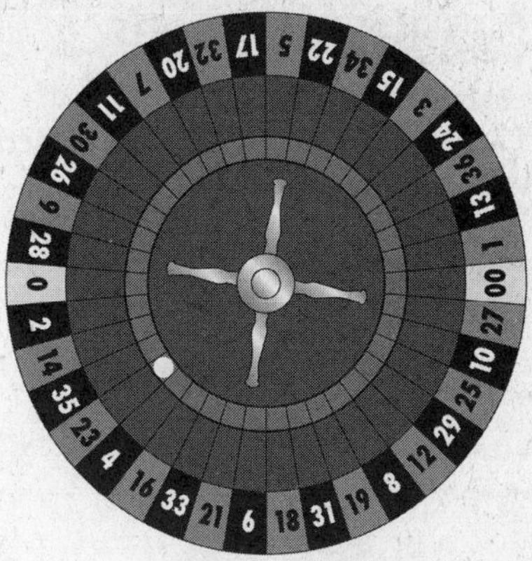

We know that the probability of the ball landing on black or even is 0.684, but, unfortunately, the ball landed on 23, which is red and odd.

...but it's time for another bet

Even with the odds in our favor, we've been unlucky with the outcomes at the roulette table. The croupier decides to take pity on us and offers a little inside information. After she spins the roulette wheel, she'll give us a clue about where the ball landed, and we'll work out the probability based on what she tells us.

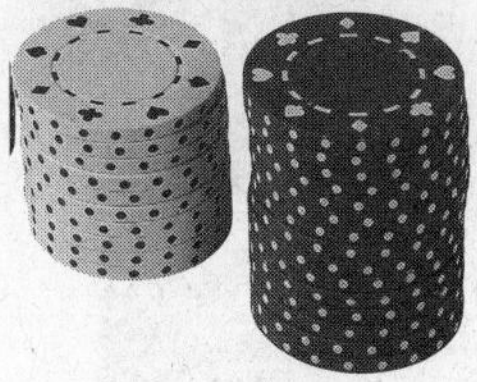

Here's your next bet...and a hint about where the ball landed. Shh, don't tell Fat Dan...
Bet: Even
Clue: The ball landed in a black pocket

Should we take this bet?

How does the probability of getting even given that we know the ball landed in a black pocket compare to our last bet that the ball would land on black or even. Let's figure it out.

Conditions apply

The croupier says the ball has landed in a black pocket. What's the probability that the pocket is also even?

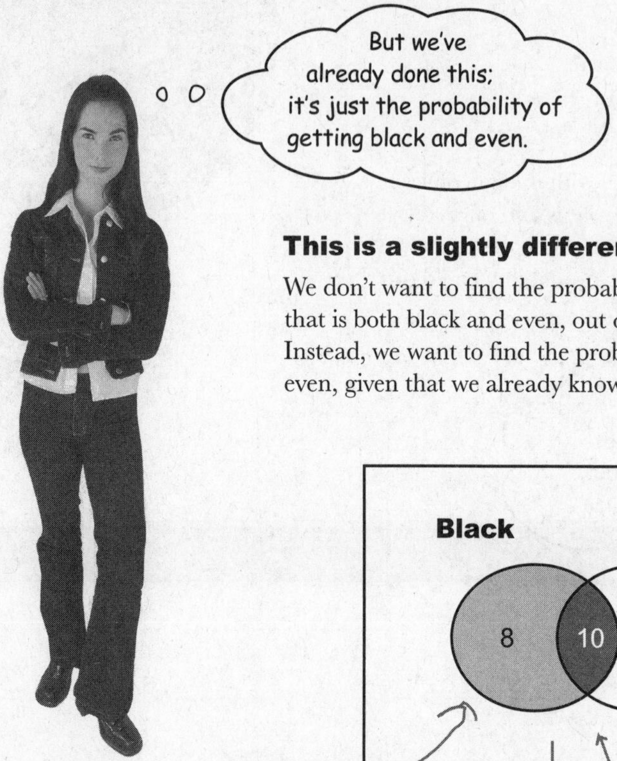
But we've
already done this;
it's just the probability of
getting black and even.

This is a slightly different problem

We don't want to find the probability of getting a pocket that is both black and even, out of all possible pockets. Instead, we want to find the probability that the pocket is even, given that we already know it's black.

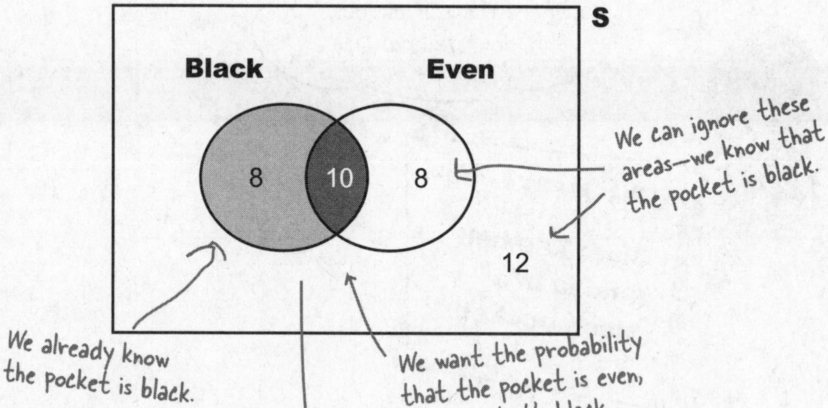
S
Black
Even
8
10
8
12
We can ignore these areas—we know that the pocket is black.
We already know the pocket is black.
We want the probability that the pocket is even, given that it's black.

In other words, we want to find out how many pockets are even out of all the black ones. Out of the 18 black pockets, 10 of them are even, so

$$P(\text{Even given Black}) = \frac{10}{18}$$

$$= 0.556 \text{ (to 3 decimal places)}$$

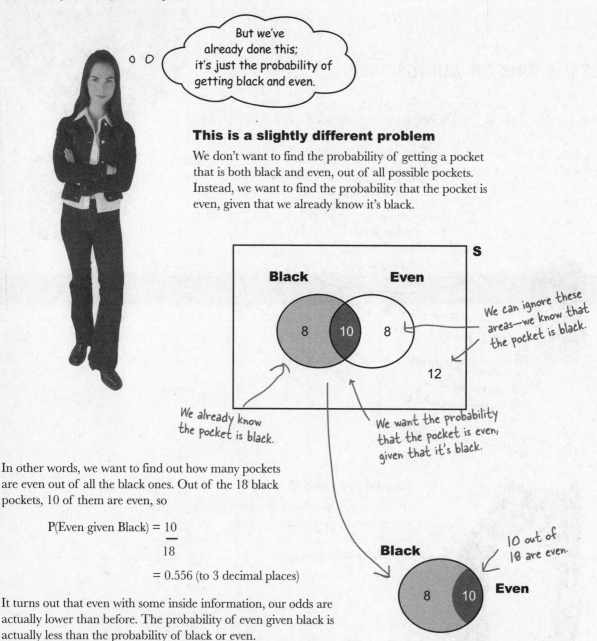

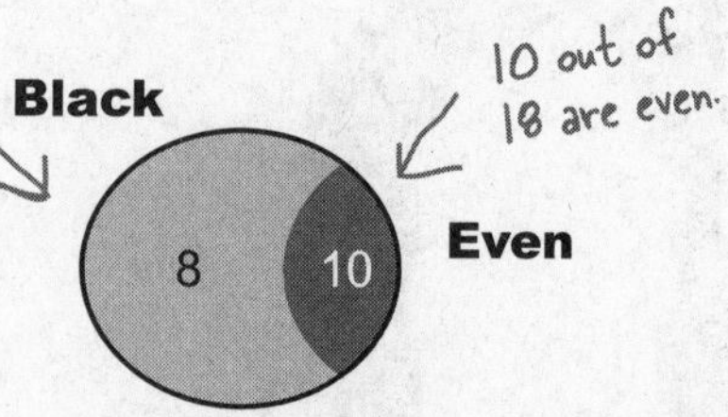
Black
Even
8
10
10 out of 18 are even.

It turns out that even with some inside information, our odds are actually lower than before. The probability of even given black is actually less than the probability of black or even.

However, a probability of 0.556 is still better than 50% odds, so this is still a pretty good bet. Let's go for it.

Find conditional probabilities

I'm a given

So how can we generalize this sort of problem? First of all, we need some more notation to represent ***conditional probabilities***, which measure the probability of one event occurring relative to another occurring.

If we want to express the probability of one event happening given another one has already happened, we use the "|" symbol to mean "given." Instead of saying "the probability of event A occurring given event B," we can shorten it to say

$$P(A \mid B)$$

The probability of A given that we know B has happened.

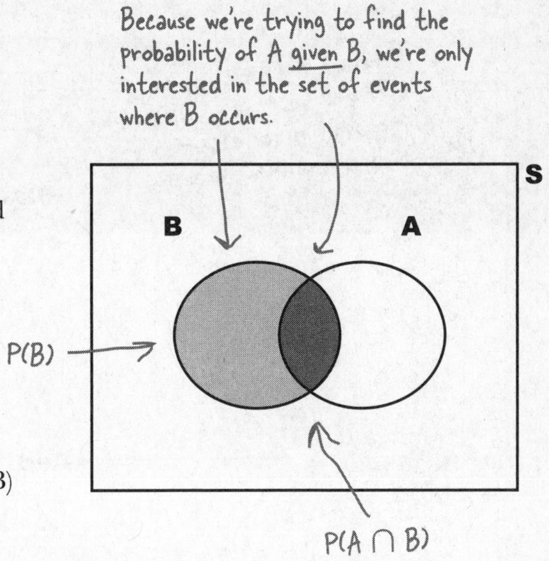
Because we're trying to find the probability of A given B, we're only interested in the set of events where B occurs.
S
B
A
P(B)
P(A ∩ B)

So now we need a general way of calculating P(A | B). What we're interested in is the number of outcomes where both A and B occur, divided by all the B outcomes. Looking at the Venn diagram, we get:

$$P(A \mid B) = \frac{P(A \cap B)}{P(B)}$$

We can rewrite this equation to give us a way of finding $P(A \cap B)$

$$P(A \cap B) = P(A \mid B) \times P(B)$$

It doesn't end there. Another way of writing $P(A \cap B)$ is $P(B \cap A)$. This means that we can rewrite the formula as

$$P(B \cap A) = P(B \mid A) \times P(A)$$

In other words, just flip around the A and the B.

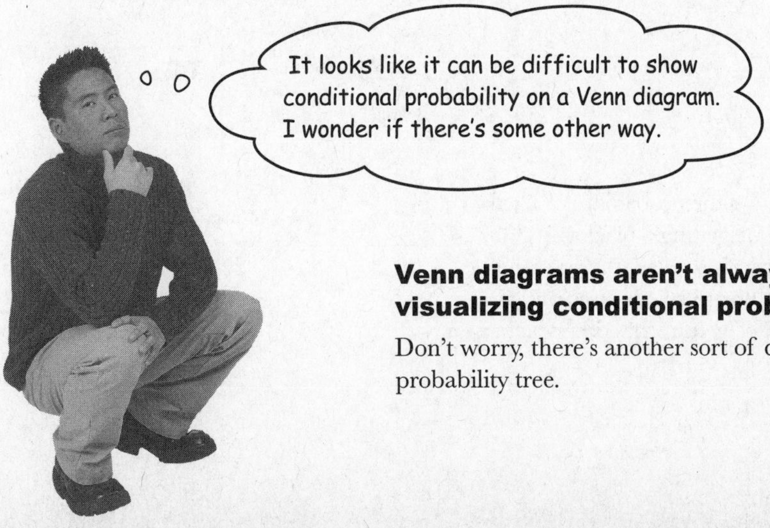
It looks like it can be difficult to show conditional probability on a Venn diagram. I wonder if there's some other way.

Venn diagrams aren't always the best way of visualizing conditional probability.

Don't worry, there's another sort of diagram you can use—a probability tree.

You can visualize conditional probabilities with a probability tree

It's not always easy to visualize conditional probabilities with Venn diagrams, but there's another sort of diagram that really comes in handy in this situation—the ***probability tree***. Here's a probability tree for our problem with the roulette wheel, showing the probabilities for getting different colored and odd or even pockets.

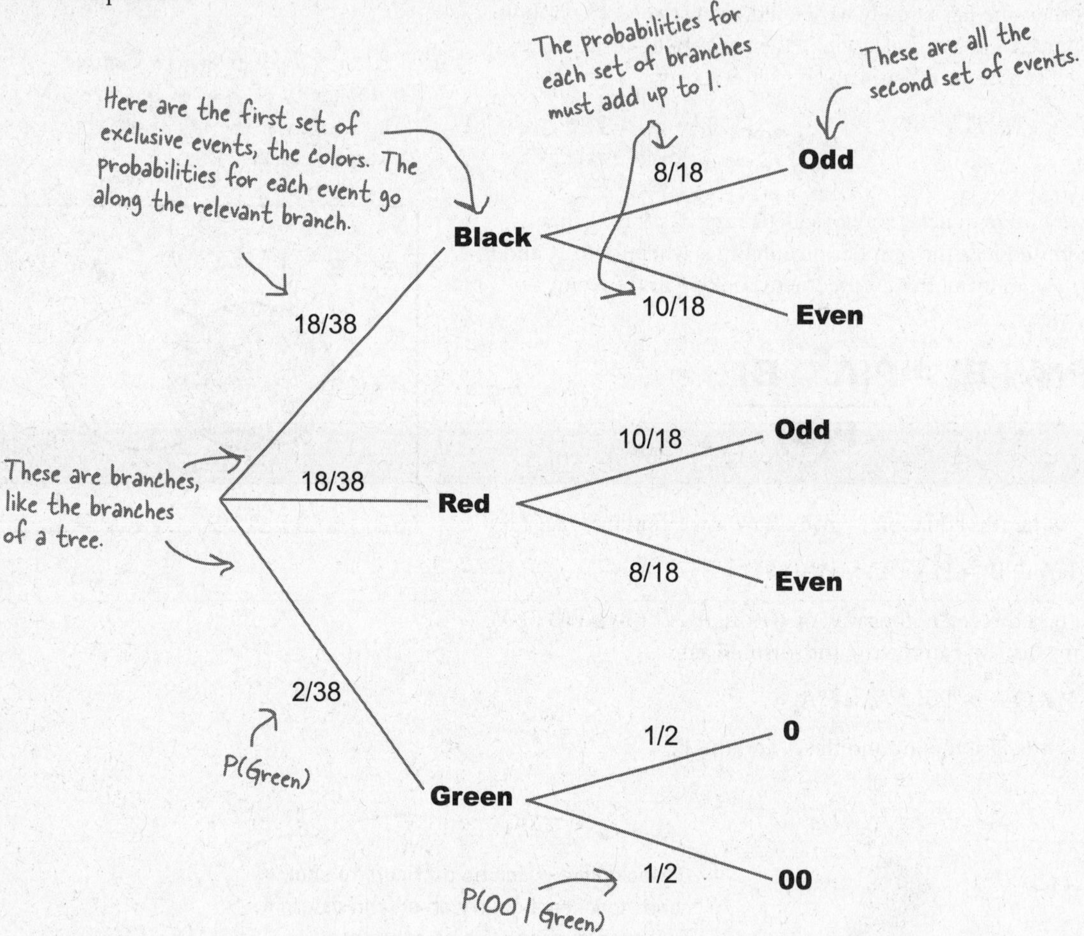

The first set of branches shows the probability of each outcome, so the probability of getting a black is 18/38, or 0.474. The second set of branches shows the probability of outcomes **given the outcome of the branch it is linked to**. The probability of getting an odd pocket given we know it's black is 8/18, or 0.444.

Trees also help you calculate conditional probabilities

Probability trees don't just help you visualize probabilities; they can help you to calculate them, too.

Let's take a general look at how you can do this. Here's another probability tree, this time with a different number of branches. It shows two levels of events: A and A¹ and B and B¹. A¹ refers to every possibility not covered by A, and B¹ refers to every possibility not covered by B.

You can find probabilities involving intersections by multiplying the probabilities of linked branches together. As an example, suppose you want to find $P(A \cap B)$. You can find this by multiplying $P(B)$ and $P(A \mid B)$ together. In other words, you multiply the probability on the first level B branch with the probability on the second level A branch.

To find $P(A \cap B)$, just multiply the probabilities for these two branches together.

This is the same equation you saw earlier—just multiply the adjoining branches together.

$P(A \cap B) = P(A \mid B) \times P(B)$

$P(A^{I} \cap B) = P(A^{I} \mid B) \times P(B)$

$P(A \cap B^{I}) = P(A \mid B^{I}) \times P(B^{I})$

$P(A^{I} \cap B^{I}) = P(A^{I} \mid B^{I}) \times P(B^{I})$

The probability of not getting event B

The probability of not getting A given that B hasn't happened

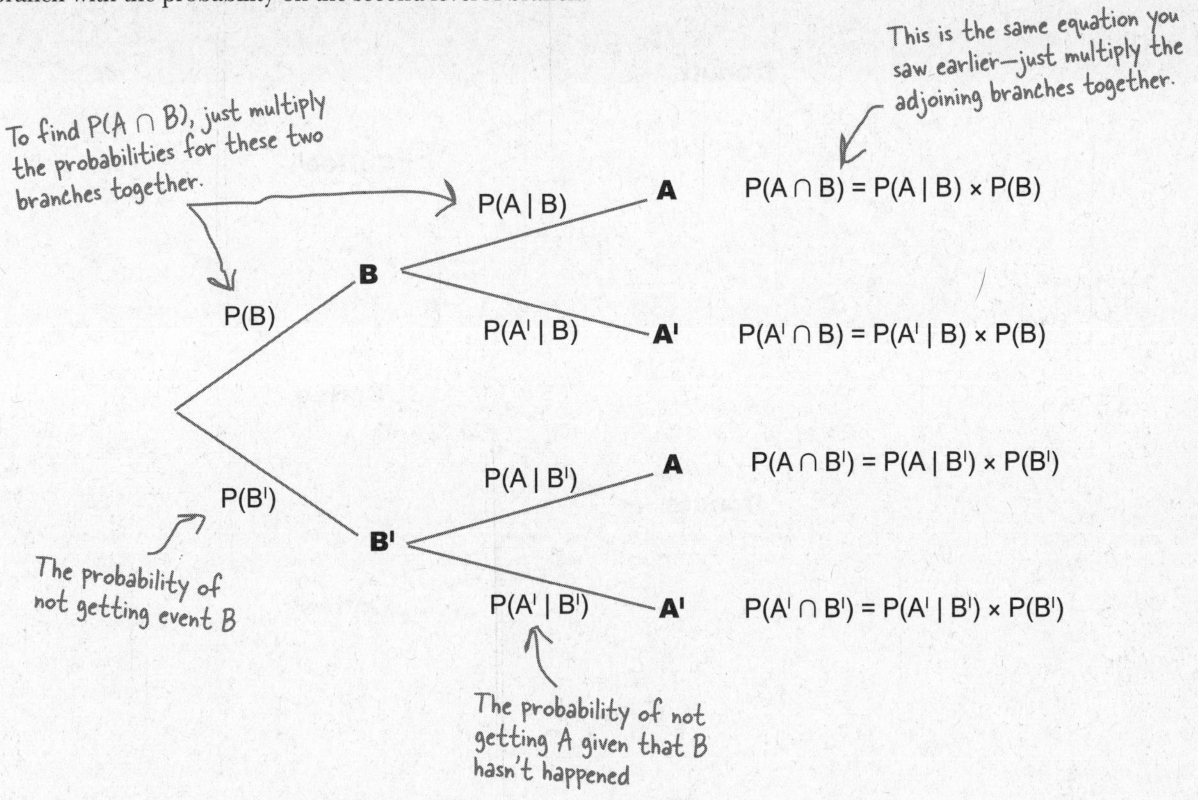

Using probability trees gives you the same results you saw earlier, and it's up to you whether you use them or not. Probability trees can be time-consuming to draw, but they offer you a way of visualizing conditional probabilities.

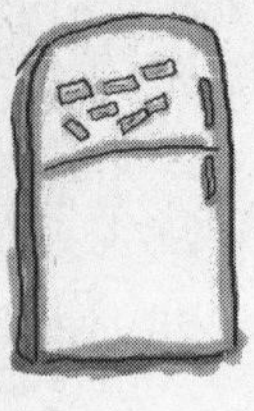

Probability Magnets

Duncan's Donuts are looking into the probabilities of their customers buying donuts *and* coffee. They drew up a probability tree to show the probabilities, but in a sudden gust of wind, they all fell off. Your task is to pin the probabilities back on the tree. Here are some clues to help you.

$P(\text{Donuts}) = 3/4$ $P(\text{Coffee} \mid \text{Donuts}') = 1/3$ $P(\text{Donuts} \cap \text{Coffee}) = 9/20$

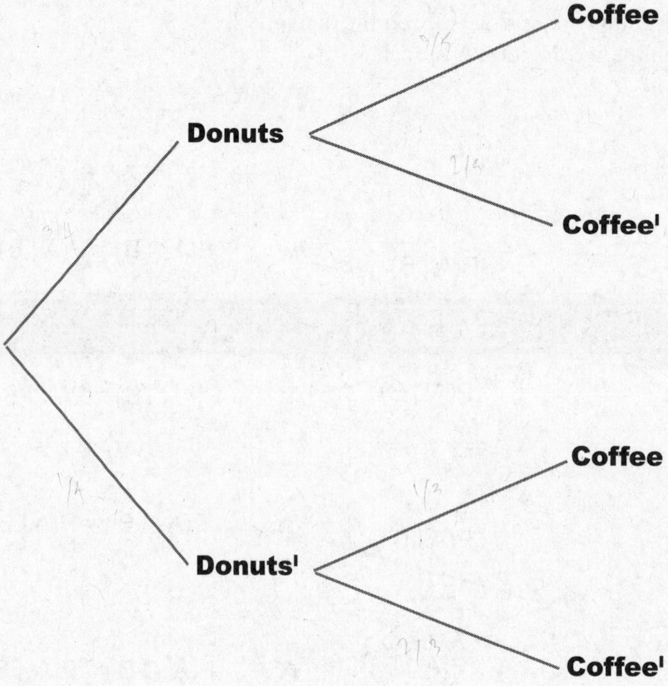

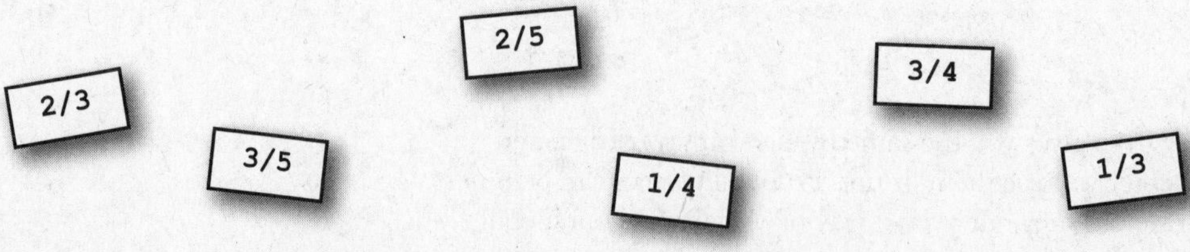

Handy hints for working with trees

1. Work out the levels

Try and work out the different levels of probability that you need. As an example, if you're given a probability for P(A | B), you'll probably need the first level to cover B, and the second level A.

2. Fill in what you know

If you're given a series of probabilities, put them onto the tree in the relevant position.

3. Remember that each set of branches sums to 1

If you add together the probabilities for all of the branches that fork off from a common point, the sum should equal 1. Remember that $P(A) = 1 - P(A')$.

4. Remember your formula

You should be able to find most other probabilities by using

$$P(A \mid B) = \frac{P(A \cap B)}{P(B)}$$

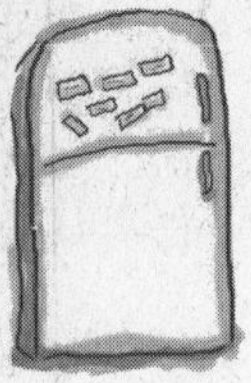

Probability Magnets Solution

Duncan's Donuts are looking into the probabilities of their customers buying Donuts *and* Coffee. They drew up a probability tree to show the probabilities, but in a sudden gust of wind they all fell off. Your task is to pin the probabilities back on the tree. Here are some clues to help you.

P(Donuts) = 3/4 P(Coffee | Donuts') = 1/3 P(Donuts ∩ Coffee) = 9/20

$$P(\text{Coffee} \mid \text{Donuts}) = \frac{P(\text{Coffee} \cap \text{Donuts})}{P(\text{Donuts})} = \frac{9/20}{3/4} = 3/5$$

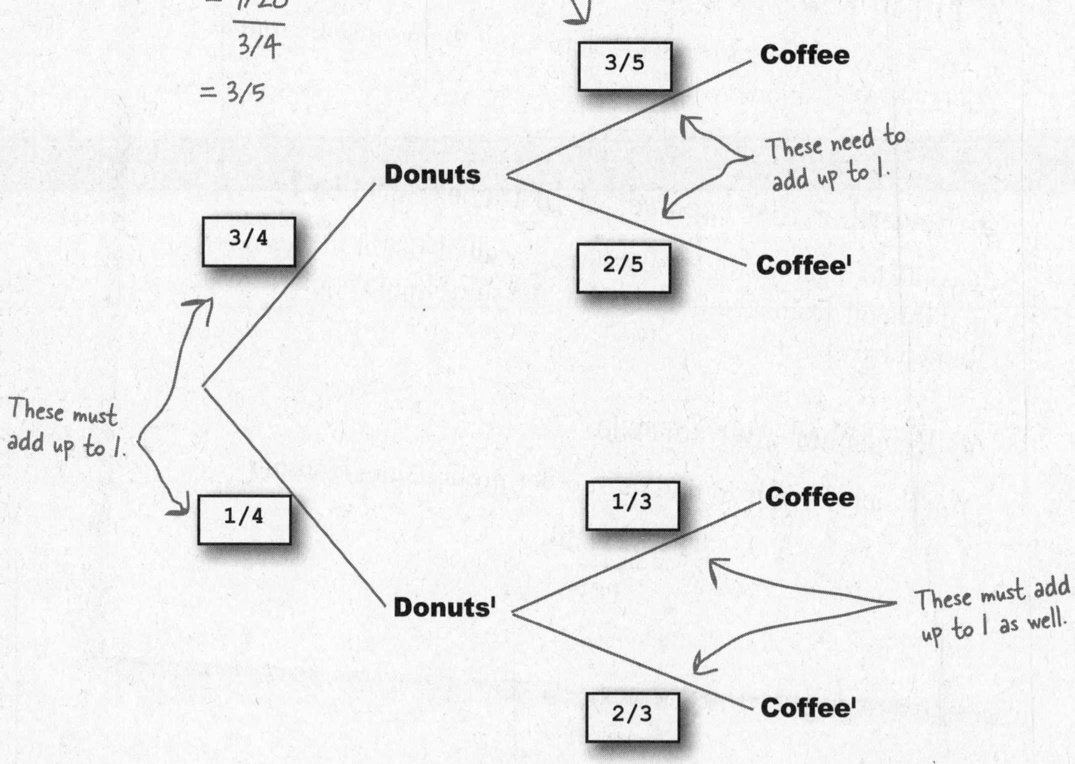

Exercise

We haven't quite finished with Duncan's Donuts! Now that you've completed the probability tree, you need to use it to work out some probabilities.

1. $P(\text{Donuts}^I)$

2. $P(\text{Donuts}^I \cap \text{Coffee})$

3. $P(\text{Coffee}^I \mid \text{Donuts})$

4. $P(\text{Coffee})$ ← Hint: How many ways are there of getting coffee? (You can get coffee with or without donuts.)

Hint: maybe some of your other answers can help you.

5. $P(\text{Donuts} \mid \text{Coffee})$

Exercise Solution

Your job was to use the completed probability tree to work out some probabilities.

1. P(Donuts')

1/4

We can read this one off the tree. We were given P(Donuts) = 3/4, so P(Donuts') must be 1/4.

2. P(Donuts' ∩ Coffee)

1/12

We can find this by multiplying together P(Donuts') and P(Coffee | Donuts'). We've just found P(Donuts') = 1/4, and looking at the tree, P(Coffee | Donuts') = 1/3. Multiplying these together gives 1/12.

3. P(Coffee' | Donuts)

2/5

We can read this off the tree.

4. P(Coffee)

8/15

This probability is tricky, so don't worry if you didn't get it.

To get P(Coffee), we need to add together P(Coffee ∩ Donuts) and P(Coffee ∩ Donuts'). This gives us 1/12 + 9/20 = 8/15.

5. P(Donuts | Coffee)

27/32

You'll only be able to do this if you found P(Coffee).

P(Donuts | Coffee) = P(Donuts ∩ Coffee) / P(Coffee). This gives us (9/20) / (8 / 15) = 27/32.

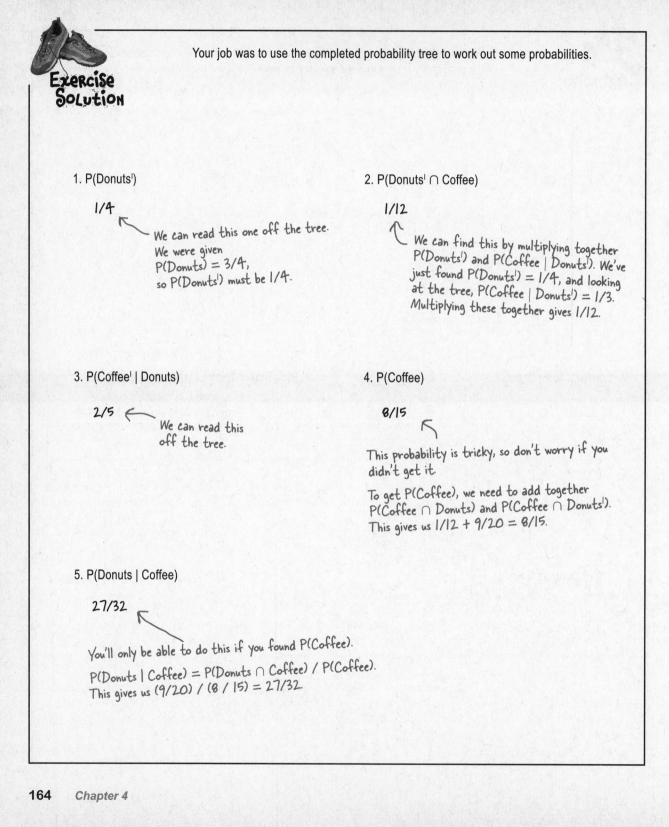

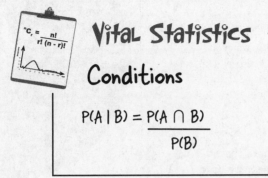

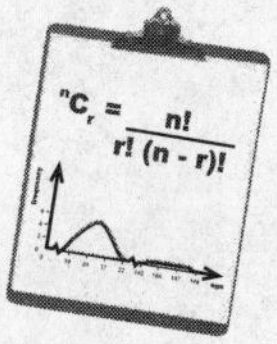

Vital Statistics

Conditions

$$P(A \mid B) = \frac{P(A \cap B)}{P(B)}$$

there are no Dumb Questions

Q: I still don't get the difference between P(A ∩ B) and P(A | B).

A: P(A ∩ B) is the probability of getting both A and B. With this probability, you can make no assumptions about whether one of the events has already occurred. You have to find the probability of both events happening without making any assumptions.

P(A | B) is the probability of event A given event B. In other words, you make the assumption that event B has occurred, and you work out the probability of getting A under this assumption.

Q: So does that mean that P(A | B) is just the same as P(A)?

A: No, they refer to different probabilities. When you calculate P(A | B), you have to assume that event B has already happened. When you work out P(A), you can make no such assumption.

Q: Is P(A | B) the same as P(B | A)? They look similar.

A: It's quite a common mistake, but they are very different probabilities. P(A | B) is the probability of getting event A given event B has already happened. P(B | A) is the probability of getting event B given event A occurred. You're actually finding the probability of a different event under a different set of assumptions.

Q: Are probability trees better than Venn diagrams?

A: Both diagrams give you a way of visualizing probabilities, and both have their uses. Venn diagrams are useful for showing basic probabilities and relationships, while probability trees are useful if you're working with conditional probabilities. It all depends what type of problem you need to solve.

Q: Is there a limit to how many sets of branches you can have on a probability tree?

A: In theory there's no limit. In practice you may find that a very large probability tree can become unwieldy, but you may still find it easier to draw a large probability tree than work through complex probabilities without it.

Q: If A and B are mutually exclusive, what is P(A | B)?

A: If A and B are mutually exclusive, then P(A ∩ B) = 0 and P(A | B) = 0. This makes sense because if A and B are mutually exclusive, it's impossible for both events to occur. If we assume that event B has occurred, then it's impossible for event A to happen, so P(A | B) = 0.

Bad luck!

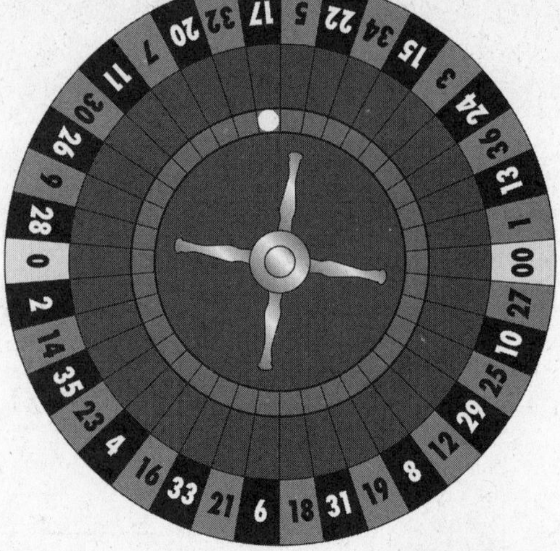

You placed a bet that the ball would land in an even pocket given we've been told it's black. Unfortunately, the ball landed in pocket 17, so you lose a few more chips.

Maybe we can win some chips back with another bet. This time, the croupier says that the ball has landed in an even pocket. What's the probability that the pocket is also black?

This is the opposite of the previous bet.

But that's a similar problem to the one we had before. Do you mean we have to draw another probability tree and work out a whole new set of probabilities? Can't we use the one we had before?

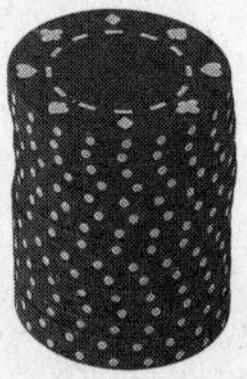

We can reuse the probability calculations we already did.

Our previous task was to figure out P(Even | Black), and we can use the probabilities we found solving that problem to calculate P(Black | Even). Here's the probability tree we used before:

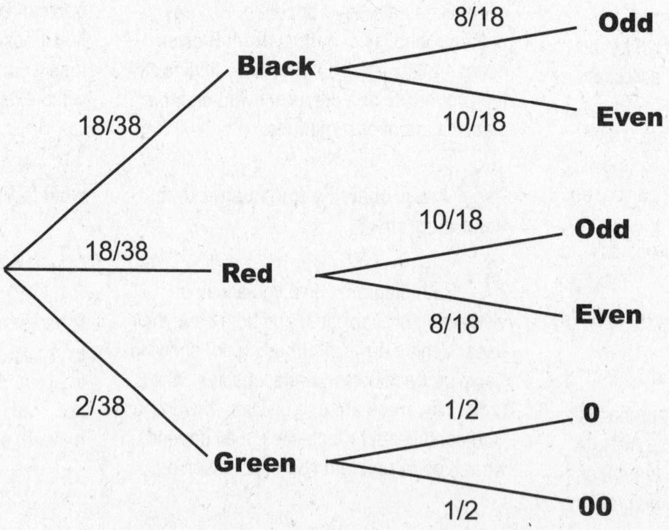

We can find P(Black | Even) using the probabilities we already have

So how *do* we find P(Black | Even)? There's still a way of calculating this using the probabilities we already have even if it's not immediately obvious from the probability tree. All we have to do is look at the probabilities we already have, and use these to somehow calculate the probabilities we don't yet know.

Let's start off by looking at the overall probability we need to find, P(Black | Even).

Using the formula for finding conditional probabilities, we have

$$P(\text{Black} \mid \text{Even}) = \frac{P(\text{Black} \cap \text{Even})}{P(\text{Even})}$$

If we can find what the probabilities of P(Black ∩ Even) and P(Even) are, we'll be able to use these in the formula to calculate P(Black | Even). All we need is some mechanism for finding these probabilities.

Sound difficult? Don't worry, we'll guide you through how to do it.

Use the probabilities you have to calculate the probabilities you need

Step 1: Finding P(Black ∩ Even)

Let's start off with the first part of the formula, P(Black ∩ Even).

Sharpen your pencil

Take a look at the probability tree on the previous page. How can you use it to find P(Black ∩ Even)?

Hint: P(Black ∩ Even) = P(Even ∩ Black)

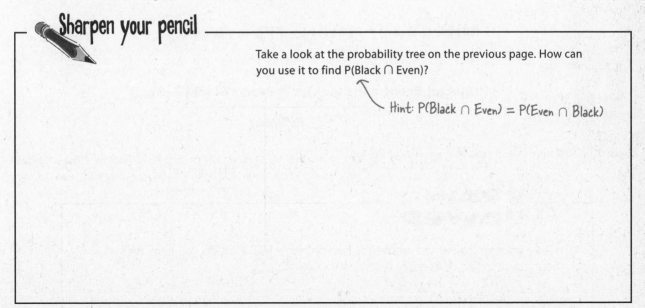

Sharpen your pencil Solution

Take a look at the probability tree opposite. How can you use it to find P(Black ∩ Even)?

You can find P(Black ∩ Even) by multiplying together P(Black) and P(Even | Black). This gives us

$$P(Black \cap Even) = P(Black) \times P(Even \mid Black)$$
$$= \frac{18}{38} \times \frac{10}{18}$$
$$= \frac{10}{38}$$
$$= \frac{5}{19}$$

So where does this get us?

We want to find the probability P(Black | Even). We can do this by evaluating

$$P(Black \mid Even) = \frac{P(Black \cap Even)}{P(Even)}$$

These two quantities are equivalent...

So far we've only looked at the first part of the formula, P(Black ∩ Even), and you've seen that you can calculate this using

$$P(Black \cap Even) = P(Black) \times P(Even \mid Black)$$

This gives us

$$P(Black \mid Even) = \frac{P(Black) \times P(Even \mid Black)}{P(Even)}$$

...so we can substitute P(Black) × P(Even | Black) for P(Black ∩ Even) in our original formula.

So how do we find the next part of the formula, P(Even)?

BRAIN POWER

Take another look at the probability tree on page 168. How do you think we can use it to find P(Even)?

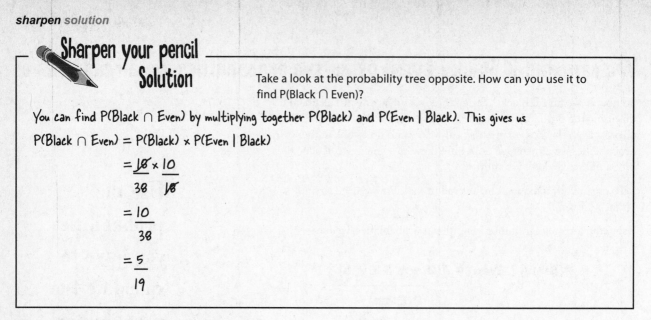

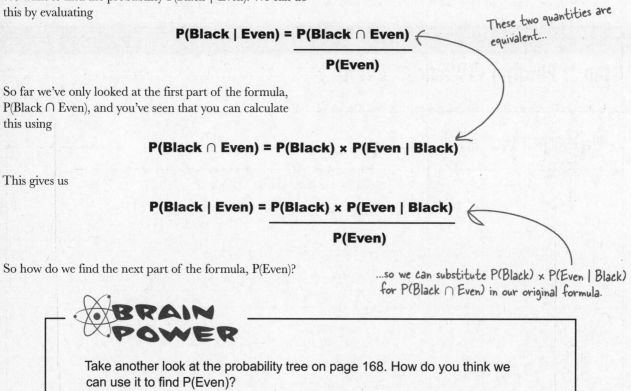

Step 2: Finding P(Even)

The next step is to find the probability of the ball landing in an even pocket, P(Even). We can find this by considering all the ways in which this could happen.

A ball can land in an even pocket by landing in either a pocket that's both black and even, or in a pocket that's both red and even. These are all the possible ways in which a ball can land in an even pocket.

This means that we find P(Even) by adding together P(Black ∩ Even) and P(Red ∩ Even). In other words, we add the probability of the pocket being both black and even to the probability of it being both red and even. The relevant branches are highlighted on the probability tree.

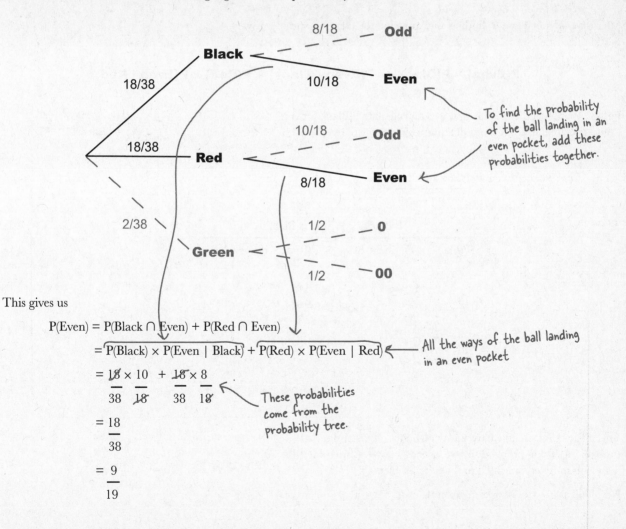

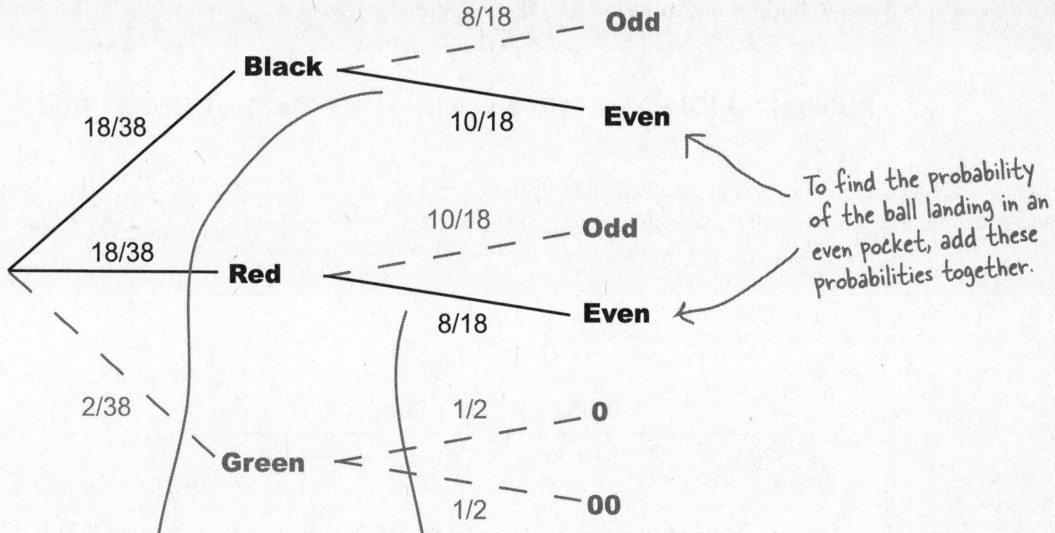

This gives us

$$P(Even) = P(Black \cap Even) + P(Red \cap Even)$$

$$= P(Black) \times P(Even \mid Black) + P(Red) \times P(Even \mid Red)$$

All the ways of the ball landing in an even pocket

$$= \frac{18}{38} \times \frac{10}{18} + \frac{18}{38} \times \frac{8}{18}$$

These probabilities come from the probability tree.

$$= \frac{18}{38}$$

$$= \frac{9}{19}$$

Step 3: Finding P(Black | Even)

Can you remember our original problem? We wanted to find P(Black | Even) where

$$\textbf{P(Black | Even)} = \frac{\textbf{P(Black} \cap \textbf{Even)}}{\textbf{P(Even)}}$$

We started off by finding an expression for P(Black ∩ Even)

$$\textbf{P(Black} \cap \textbf{Even) = P(Black)} \times \textbf{P(Even | Black)}$$

After that we moved on to finding an expression for P(Even), and found that

$$\textbf{P(Even) = P(Black)} \times \textbf{P(Even | Black) + P(Red)} \times \textbf{P(Even | Red)}$$

This is what we just calculated using the probability tree.

Putting these together means that we can calculate P(Black | Even) using probabilities from the probability tree

$$\begin{aligned} P(\text{Black} \mid \text{Even}) &= \frac{P(\text{Black} \cap \text{Even})}{P(\text{Even})} \\ &= \frac{P(\text{Black}) \times P(\text{Even} \mid \text{Black})}{P(\text{Black}) \times P(\text{Even} \mid \text{Black}) + P(\text{Red}) \times P(\text{Even} \mid \text{Red})} \\ &= \frac{5}{19} \div \frac{9}{19} \\ &= \frac{5}{19} \times \frac{19}{9} \\ &= \frac{5}{9} \end{aligned}$$

We calculated these earlier, so we can substitute in our results.

This means that we now have a way of finding *new* conditional probabilities using probabilities we already know—something that can help with more complicated probability problems.

Let's look at how this works in general.

These results can be generalized to other problems

Imagine you have a probability tree showing events A and B like this, and assume you know the probability on each of the branches.

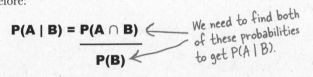

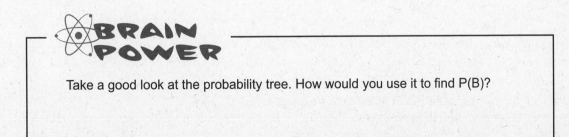

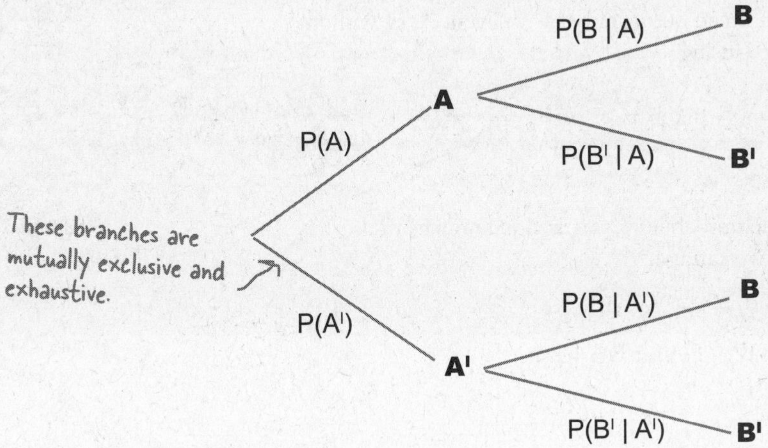

Now imagine you want to find P(A | B), and the information shown on the branches above is all the information that you have. How can you use the probabilities you have to work out P(A | B)?

We can start with the formula we had before:

$$P(A \mid B) = \frac{P(A \cap B)}{P(B)}$$

We need to find both of these probabilities to get P(A | B).

Now we can find P(A ∩ B) using the probabilities we have on the probability tree. In other words, we can calculate P(A ∩ B) using

$$P(A \cap B) = P(A) \times P(B \mid A)$$

But how do we find P(B)?

BRAIN POWER

Take a good look at the probability tree. How would you use it to find P(B)?

Use the Law of Total Probability to find P(B)

To find P(B), we use the same process that we used to find P(Even) earlier; we need to add together the probabilities of all the different ways in which the event we want can possibly happen.

There are two ways in which even B can occur: either with event A, or without it. This means that we can find P(B) using:

$$P(B) = P(A \cap B) + P(A' \cap B)$$

Add together both of the intersections to get P(B).

We can rewrite this in terms of the probabilities we already know from the probability tree. This means that we can use:

$$P(A \cap B) = P(A) \times P(B \mid A)$$

$$P(A' \cap B) = P(A') \times P(B \mid A')$$

This gives us:

$$\mathbf{P(B) = P(A) \times P(B \mid A) + P(A') \times P(B \mid A')}$$

This is sometimes known as the ***Law of Total Probability***, as it gives a way of finding the total probability of a particular event based on conditional probabilities.

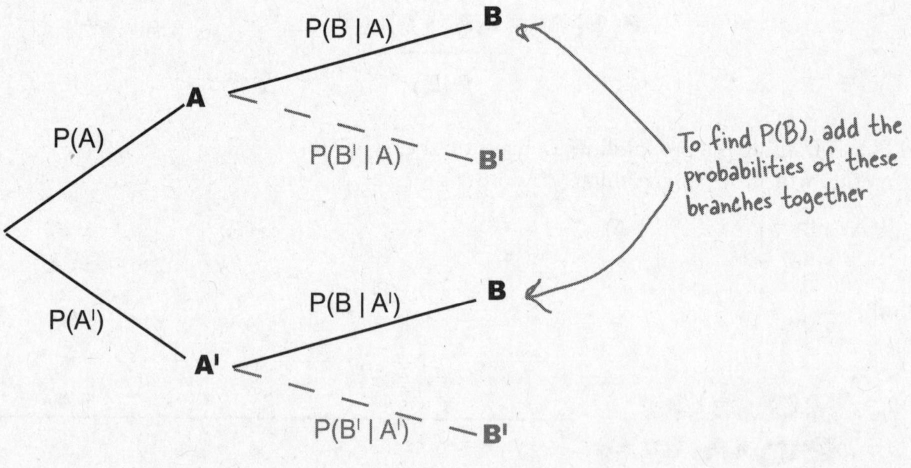

Now that we have expressions for P(A ∩ B) and P(B), we can put them together to come up with an expression for P(A | B).

Introducing Bayes' Theorem

We started off by wanting to find P(A | B) based on probabilities
we already know from a probability tree. We already know P(A),
and we also know P(B | A) and P(B | A'). What we need is a
general expression for finding conditional probabilities that are
the *reverse* of what we already know, in other words P(A | B).

We started off with:

$$P(A \mid B) = \frac{P(A \cap B)}{P(B)}$$

With substitution, this formula...

On page 127, we found P(A ∩ B) = P(A) × P(B | A). And on the
previous page, we discovered P(B) = P(A) × P(B | A) + P(A') × P(B |
A').

If we substitute these into the formula, we get:

$$P(A \mid B) = \frac{P(A) \times P(B \mid A)}{P(A) \times P(B \mid A) + P(A') \times P(B \mid A')}$$

...becomes this formula.

This is called **Bayes' Theorem**. It gives you a means of finding reverse
conditional probabilities, which is really useful if you don't know every
probability up front.

> **Relax**
>
> **Bayes' Theorem is
> one of the most
> difficult aspects of
> probability.**
>
> Don't worry if it looks complicated—this
> is as tough as it's going to get. And even
> though the formula is tricky, visualizing the
> problem can help.

...divide the probability
of this branch...

...by the probability
of these two
branches added
together.

Here's A. To find
P(A | B)...

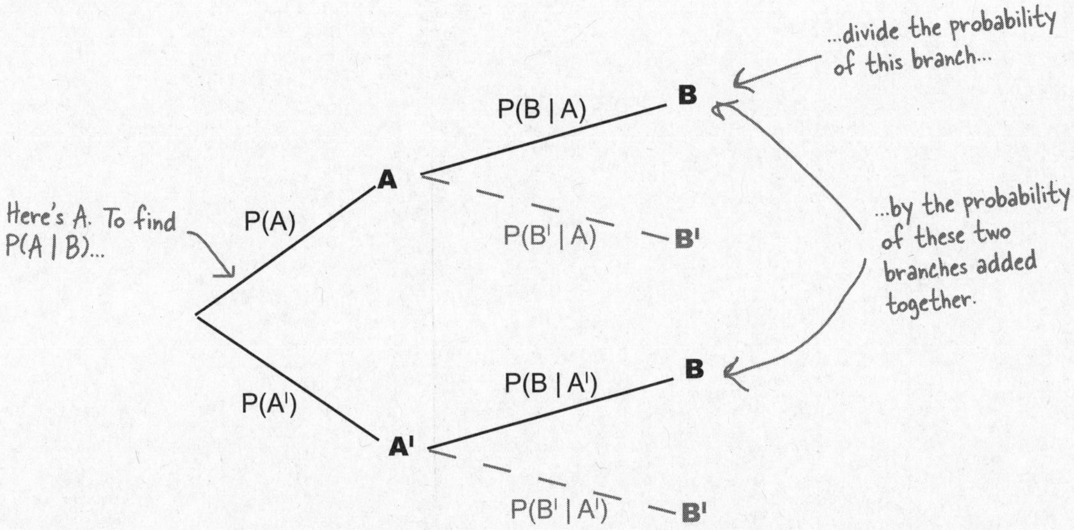

Long Exercise

The Manic Mango games company is testing two brand-new games. They've asked a group of volunteers to choose the game they most want to play, and then tell them how satisfied they were with game play afterwards.

80 percent of the volunteers chose Game 1, and 20 percent chose Game 2. Out of the Game 1 players, 60 percent enjoyed the game and 40 percent didn't. For Game 2, 70 percent of the players enjoyed the game and 30 percent didn't.

Your first task is to fill in the probability tree for this scenario.

Manic Mango selects one of the volunteers at random to ask if she enjoyed playing the game, and she says she did. Given that the volunteer enjoyed playing the game, what's the probability that she played game 2? Use Bayes' Theorem.

Hint: What's the probability of someone choosing game 2 and being satisfied? What's the probability of someone being satisfied overall? Once you've found these, you can use Bayes Theorem to obtain the right answer.

Long Exercise Solution

The Manic Mango games company is testing two brand-new games. They've asked a group of volunteers to choose the game they most want to play, and then tell them how satisfied they were with game play afterwards.

80 percent of the volunteers chose Game 1, and 20 percent chose Game 2. Out of the Game 1 players, 60 percent enjoyed the game and 40 percent didn't. For Game 2, 70 percent of the players enjoyed the game and 30 percent didn't.

Your first task is to fill in the probability tree for this scenario.

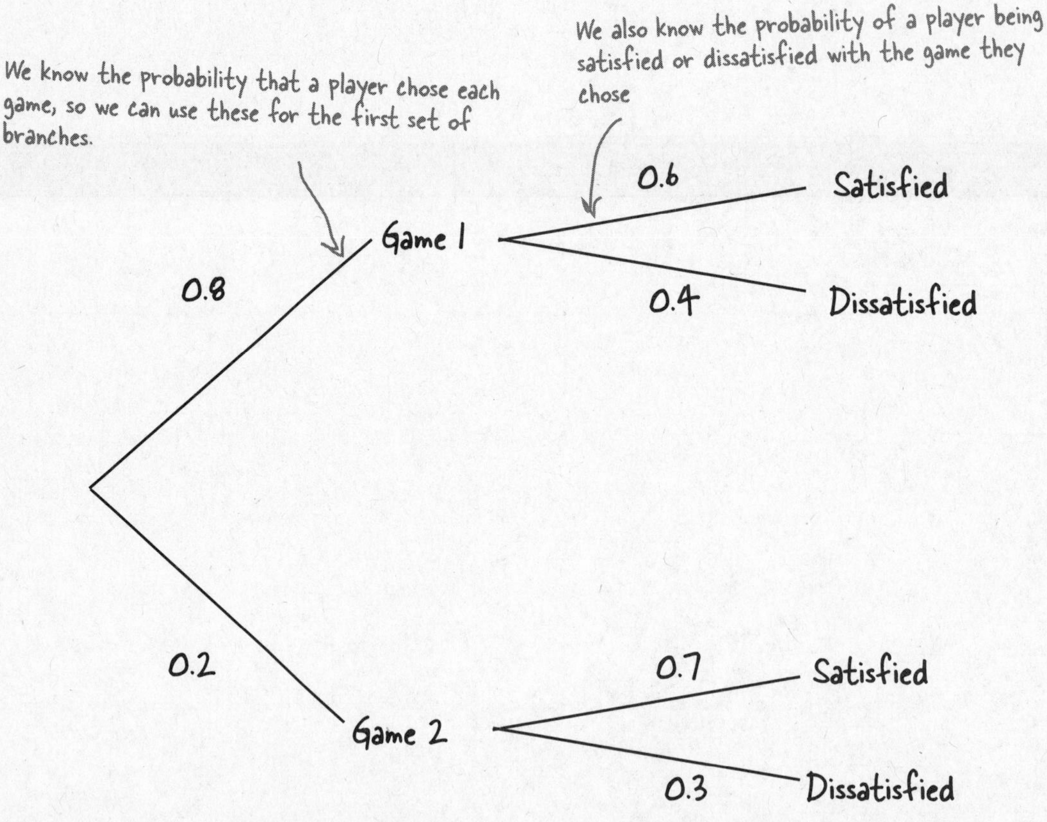

We know the probability that a player chose each game, so we can use these for the first set of branches.
We also know the probability of a player being satisfied or dissatisfied with the game they chose
0.8
Game 1
0.6
Satisfied
0.4
Dissatisfied
0.2
Game 2
0.7
Satisfied
0.3
Dissatisfied

Manic Mango selects one of the volunteers at random to ask if she enjoyed playing the game, and she says she did. Given that the volunteer enjoyed playing the game, what's the probability that she played game 2? Use Bayes' Theorem.

We need to use Bayes' Theorem to find P(Game 2 | Satisfied). This means we need to use

$$P(\text{Game 2} \mid \text{Satisfied}) = \frac{P(\text{Game 2})\, P(\text{Satisfied} \mid \text{Game 2})}{P(\text{Game 2})\, P(\text{Satisfied} \mid \text{Game 2}) + P(\text{Game 2})\, P(\text{Dissatisfied} \mid \text{Game 2})}$$

Let's start with P(Game 2) P(Satisfied | Game 2)

We've been told that P(Game 2) = 0.2 and P(Satisfied | Game 2) = 0.7. This means that

$$P(\text{Game 2})\, P(\text{Satisfied} \mid \text{Game 2}) = 0.2 \times 0.7$$
$$= 0.14$$

The next thing we need to find is P(Game 2) P(Dissatisfied | Game 2). We've been told that P(Dissatisfied | Game 2) = 0.3, and we've already seen that P(Game 2) = 0.2. This gives us

$$P(\text{Game 2})\, P(\text{Dissatisfied} \mid \text{Game 2}) = 0.2 \times 0.3$$
$$= 0.06$$

Substituting this into the formula for Bayes' Theorem gives us

$$P(\text{Game 2} \mid \text{Satisfied}) = \frac{P(\text{Game 2})\, P(\text{Satisfied} \mid \text{Game 2})}{P(\text{Game 2})\, P(\text{Satisfied} \mid \text{Game 2}) + P(\text{Game 2})\, P(\text{Dissatisfied} \mid \text{Game 2})}$$

$$= \frac{0.14}{0.14 + 0.06}$$

$$= \frac{0.14}{0.2}$$

$$= 0.7$$

Vital Statistics

Law of Total Probability

If you have two events A and B, then

$$P(B) = P(B \cap A) + P(B \cap A')$$
$$= P(A)\,P(B \mid A) + P(A')\,P(B \mid A')$$

The Law of Total Probability is the denominator of Bayes' Theorem.

Vital Statistics

Bayes' Theorem

If you have n mutually exclusive and exhaustive events, A_1 through to A_n, and B is another event, then

$$P(A \mid B) = \frac{P(A)\,P(B \mid A)}{P(A)\,P(B \mid A) + P(A')\,P(B \mid A')}$$

there are no Dumb Questions

Q: So when would I use Bayes' Theorem?

A: Use it when you want to find conditional probabilities that are in the opposite order of what you've been given.

Q: Do I have to draw a probability tree?

A: You can either use Bayes' Theorem right away, or you can use a probability tree to help you. Using Bayes' Theorem is quicker, but you need to make sure you keep track of your probabilities. Using a tree is useful if you can't remember Bayes' Theorem. It will give you the same result, and it can keep you from losing track of which probability belongs to which event.

Q: When we calculated P(Black | Even) in the roulette wheel problem, we didn't include any probabilities for the ball landing in a green pocket. Did we make a mistake?

A: No, we didn't. The only green pockets on the roulette board are 0 and 00, and we don't classify these as even. This means that P(Even | Green) is 0; therefore, it has no effect on the calculation.

Q: The probability P(Black|Even) turns out to be the same as P(Even|Black): they're both 5/9. Is that always the case?

A: True, it happens here that P(Black | Even) and P(Even | Black) have the same value, but that's not necessarily true for other scenarios.

If you have two events, A and B, you can't assume that P(A | B) and P(B | A) will give you the same results. They are two separate probabilities, and making this sort of assumption could actually cost you valuable points in a statistics exam. You need to use Bayes' Theorem to make sure you end up with the right result.

Q: How useful is Bayes' Theorem in real life?

A: It's actually pretty useful. For example, it can be used in computing as a way of filtering emails and detecting which ones are likely to be junk. It's sometimes used in medical trials too.

We have a winner!

Congratulations, this time the ball landed on 10, a pocket that's both black and even. You've won back some chips.

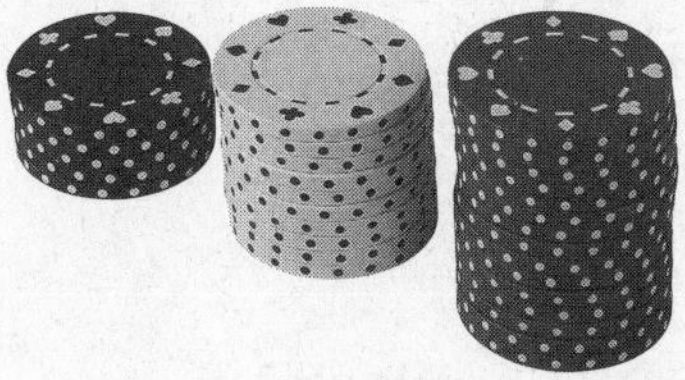

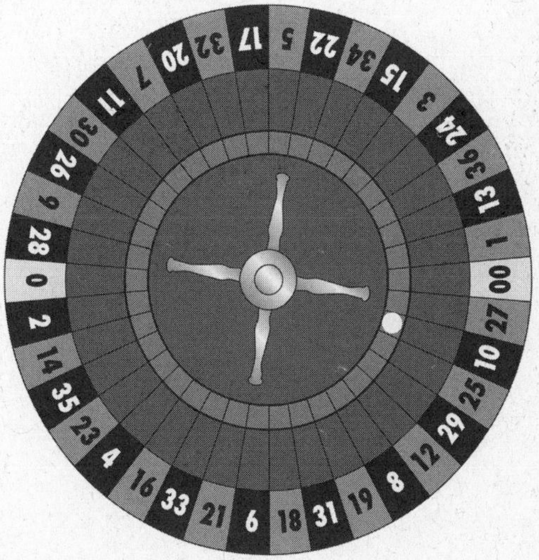

It's time for one last bet

Before you leave the roulette table, the croupier has offered you a great deal for your final bet, triple or nothing. If you bet that the ball lands in a black pocket twice in a row, you could win back all of your chips.

Here's the probability tree. Notice that the probabilities for landing on two black pockets in a row are a bit different than they were in our probability tree on page 166, where we were trying to calculate the likelihood of getting an even pocket given that we knew the pocket was black.

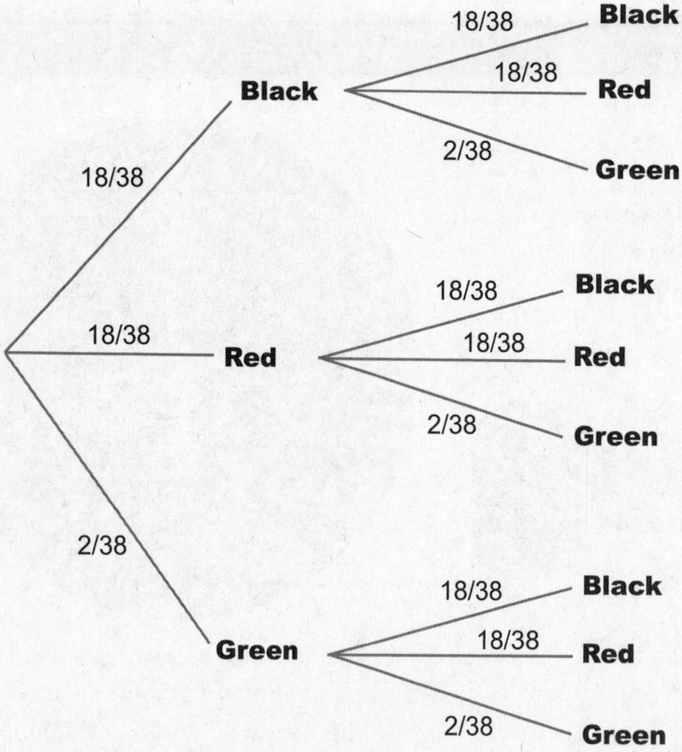

If events affect each other, they are dependent

The probability of getting black followed by black is a slightly different problem from the probability of getting an even pocket given we already know it's black. Take a look at the equation for this probability:

P(Even | Black) = 10/18 = 0.556

For P(Even | Black), the probability of getting an even pocket is affected by the event of getting a black. We already know that the ball has landed in a black pocket, so we use this knowledge to work out the probability. We look at how many of the pockets are even out of all the black pockets.

If we didn't know that the ball had landed on a black pocket, the probability would be different. To work out P(Even), we look at how many pockets are even out of all the pockets

P(Even) = 18/38 = 0.474

These two probabilities are different

P(Even | Black) gives a different result from P(Even). In other words, the knowledge we have that the pocket is black changes the probability. These two events are said to be ***dependent***.

In general terms, events A and B are said to be dependent if P(A | B) is different from P(A). It's a way of saying that the probabilities of A and B are affected by each other.

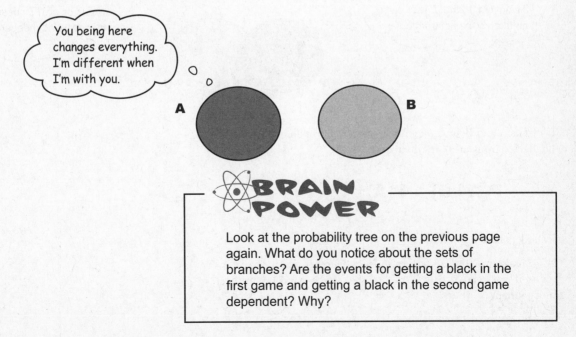

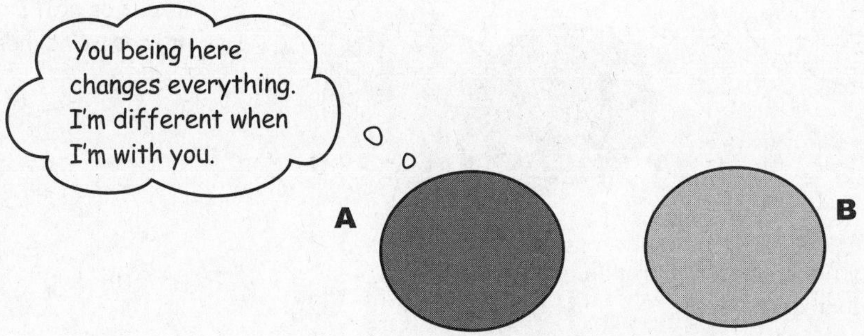

You being here changes everything. I'm different when I'm with you.
A
B

BRAIN POWER

Look at the probability tree on the previous page again. What do you notice about the sets of branches? Are the events for getting a black in the first game and getting a black in the second game dependent? Why?

If events do not affect each other, they are independent

Not all events are dependent. Sometimes events remain completely unaffected by each other, and the probability of an event occurring remains the same irrespective of whether the other event happens or not. As an example, take a look at the probabilities of P(Black) and P(Black | Black). What do you notice?

P(Black) = 18/38 = 0.474

P(Black | Black) = 18/38 = 0.474

These probabilities are the same. The events are independent.

These two probabilities have the same value. In other words, the event of getting a black pocket in this game has no bearing on the probability of getting a black pocket in the next game. These events are ***independent***.

Independent events aren't affected by each other. They don't influence each other's probabilities in any way at all. If one event occurs, the probability of the other occurring remains exactly the same.

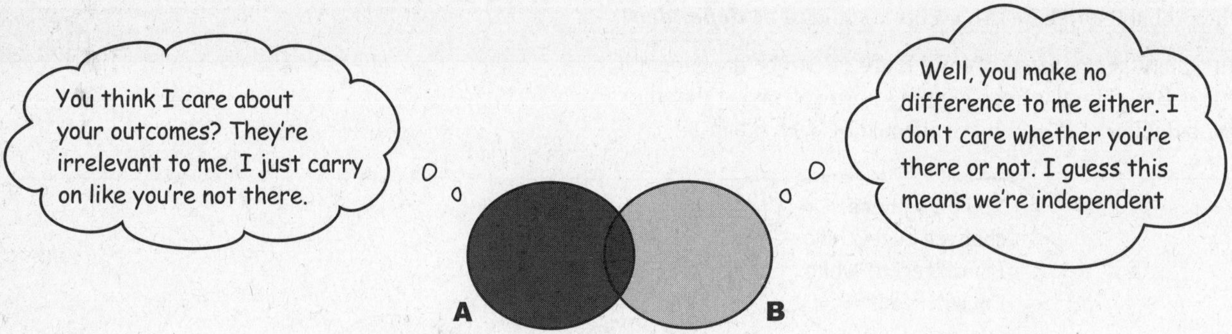

If events A and B are independent, then the probability of event A is unaffected by event B. In other words

$$P(A \mid B) = P(A)$$

for independent events.

We can also use this as a test for independence. If you have two events A and B where P(A | B) = P(A), then the events A and B must be independent.

More on calculating probability for independent events

It's easier to work out other probabilities for independent events too, for example $P(A \cap B)$.

We already know that

$$P(A \mid B) = \frac{P(A \cap B)}{P(B)}$$

If A and B are independent, $P(A \mid B)$ is the same as $P(A)$. This means that

$$P(A) = \frac{P(A \cap B)}{P(B)}$$

or

$$\mathbf{P(A \cap B) = P(A) \times P(B)}$$

for independent events. In other words, if two events are independent, then you can work out the probability of getting both events A and B by multiplying their individual probabilities together.

Watch it!

If A and B are mutually exclusive, they can't be independent, and if A and B are independent, they can't be mutually exclusive.

If A and B are mutually exclusive, then if event A occurs, event B cannot. This means that the outcome of A affects the outcome of B, and so they're dependent.

Similarly if A and B are independent, they can't be mutually exclusive.

Sharpen your pencil

It's time to calculate another probability. What's the probability of the ball landing in a black pocket twice in a row?

Sharpen your pencil Solution

It's time to calculate another probability. What's the probability of the ball landing in a black pocket twice in a row?

We need to find P(Black in game 1 ∩ Black in game 2). As the events are independent, the result is

$$18/38 \times 18/38 = 324/1444$$
$$= 0.224 \text{ (to 3 decimal places)}$$

there are no Dumb Questions

Q: What's the difference between being independent and being mutually exclusive?

A: Imagine you have two events, A and B.

If A and B are mutually exclusive, then if event A happens, B cannot. Also, if event B happens, then A cannot. In other words, it's impossible for both events to occur.

If A and B are independent, then the outcome of A has no effect on the outcome of B, and the outcome of B has no effect on the outcome of A. Their respective outcomes have no effect on each other.

Q: Do both events have to be independent? Can one event be independent and the other dependent?

A: No. The two events are independent *of each other*, so you can't have two events where one is dependent and the other one is independent.

Q: Are all games on a roulette wheel independent? Why?

A: Yes, they are. Separate spins of the roulette wheel do not influence each other. In each game, the probabilities of the ball landing on a red, black, or green remain the same.

Q: You've shown how a probability tree can demonstrate independent events. How do I use a Venn diagram to tell if events are independent?

A: A Venn diagram really isn't the best way of showing dependence. Venn diagrams are great if you need to examine intersections and show mutually exclusive events. They're not great for showing independence though.

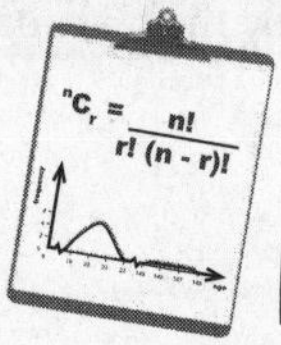

Vital Statistics

Independence

If two events A and B are independent, then

$$P(A \mid B) = P(A)$$

If this holds for any two events, then the events must be independent. Also

$$P(A \cap B) = P(A) \times P(B)$$

Five Minute Mystery

The Case of the Two Classes

The Head First Health Club prides itself on its ability to find a class for everyone. As a result, it is extremely popular with both young and old.

The Health Club is wondering how best to market its new yoga class, and the Head of Marketing wonders if someone who goes swimming is more likely to go to a yoga class. "Maybe we could offer some sort of discount to the swimmers to get them to try out yoga."

The CEO disagrees. "I think you're wrong," he says. "I think that people who go swimming and people who go to yoga are independent. I don't think people who go swimming are any more likely to do yoga than anyone else."

They ask a group of 96 people whether they go to the swimming or yoga classes. Out of these 96 people, 32 go to yoga and 72 go swimming. 24 people are exceptionally eager and go to both.

So who's right? Are the yoga and swimming classes dependent or independent?

Fireside Chats

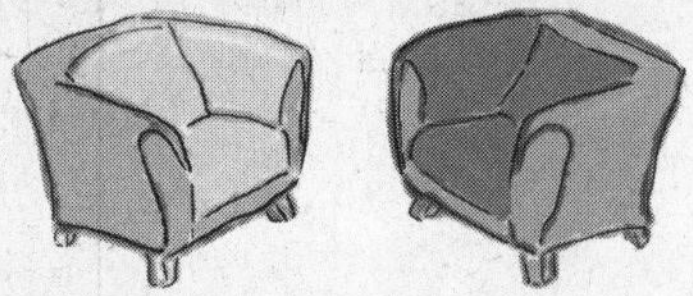

Tonight's talk: **Dependent and Independent discuss their differences**

Dependent: Independent, glad you could show up. I've been wanting to catch up with you for some time.

Independent: Really, Dependent? How come?

Dependent: Well, I hear you keep getting fledgling statisticians into trouble. They're doing fine until you show up, and then, whoa, wrong probabilities all over the place! That ∩ guy has a particularly poor opinion of you.

Independent: I'm a little hurt that ∩'s been saying bad things about me; I thought I made life easy for him. You want to work out the probability of getting two independent events? Easy! Just multiply the probabilities for the two events together and job done.

Dependent: It's that simplistic attitude of yours that gets people into trouble. They think, "Hey, that Independent guy looks easy. I'll just use him for this probability." The next thing you know, ∩ has his probabilities all in a twist. That's just not the right way of dealing with dependent events.

Independent: You're blowing this all out of proportion. Even if people do decide to use me instead of you, I don't see that it can make all that much difference.

Dependent: You don't understand the seriousness of the situation. If people use your way of calculating ∩'s probability, and the events are dependent, they're *guaranteed* to get the wrong answer. That's just not good enough. For dependent events, you only get the right answer if you take that | guy into account—he's a given.

Independent: I can't say I pay all that much attention to him. With independent events, probabilities just turn out the same.

Dependent: You're doing it again; you're oversimplifying things. Well, I've had enough. I think that people need to think of me first instead of you; that would sort out all of these problems.

Independent: Yeah? Like how?

Dependent: By really thinking through whether events are dependent or not. Let me give you an example. Suppose you have a deck of 52 cards, and thirteen of them are diamonds. Imagine you choose a card at random and it's a diamond. What would be the probability of that happening?

Independent: That's easy. It's 13/52, or 1/4.

Dependent: What if you pick out a second card? What's the probability of pulling out a second diamond?

Independent: It's the same isn't it? 1/4.

Dependent: No! The events are dependent. You can no longer say there are 13 diamonds in a pack of 52 cards. You've just removed one diamond, so there are 12 diamonds left out of 51 cards. The probability drops to 12/51, or 4/17.

Independent: Not fair, I assumed you put the first card back! That would have meant the probability of getting a diamond would have been the same as before, and I would have been right. The events would have been independent.

Dependent: But they weren't. When people think about you first, it leads them towards making all sorts of inappropriate assumptions. No wonder ∩ gets so messed up.

Independent: Well, thanks for the chat, Dependent, I'm glad we had a chance to sort things out.

Dependent: Think nothing of it. Just make sure you think things through a bit more carefully next time.

Solved: The Case of the Two Classes

Are the yoga and swimming classes dependent or independent?

The CEO's right—the classes are independent. Here's how he knows.

32 people out of 96 go to yoga classes, so

$$P(\text{Yoga}) = 1/3$$

72 people go swimming, so

$$P(\text{Swimming}) = 3/4$$

24 people go to both classes, so

$$P(\text{Yoga} \cap \text{Swimming}) = 1/4$$

So how do we know the classes are independent? Let's multiply together P(Yoga) and P(Swimming) and see what we get.

$$P(\text{Yoga}) \times P(\text{Swimming}) = 1/3 \times 3/4$$

$$= 1/4$$

As this is the same as $P(\text{Yoga} \cap \text{Swimming})$, we know that the classes are independent.

Dependent or Independent?

Here are a bunch of situations and events. Your task is to say which of these are dependent, and which are independent.

	Dependent	Independent
Throwing a coin and getting heads twice in a row.	☐	☐
Removing socks from a drawer until you find a matching pair.	☐	☐
Choosing chocolates at random from a box and picking dark chocolates twice in a row.	☐	☐
Choosing a card from a deck of cards, and then choosing another one.	☐	☐
Choosing a card from a deck of cards, putting the card back in the deck, and then choosing another one.	☐	☐
The event of getting rain given it's a Thursday.	☐	☐

Dependent or Independent? Solution

Here are a bunch of situations and events. Your task was to say which of these are dependent, and which are independent.

	Dependent	Independent
Throwing a coin and getting heads twice in a row. *(The second coin throw isn't affected by the first.)*	☐	☑
Removing socks from a drawer until you find a matching pair. *(When you remove one sock, there are fewer socks to choose from the next time, and this affects the probability.)*	☑	☐
Choosing chocolates at random from a box and picking dark chocolates twice in a row.	☑	☐
Choosing a card from a deck of cards, and then choosing another one.	☑	☐
Choosing a card from a deck of cards, putting the card back in the deck, and then choosing another one.	☐	☑
The event of getting rain given it's a Thursday. *(It's no more or less likely to rain just because it's Thursday, so these two events are independent.)*	☐	☑

Winner! Winner!

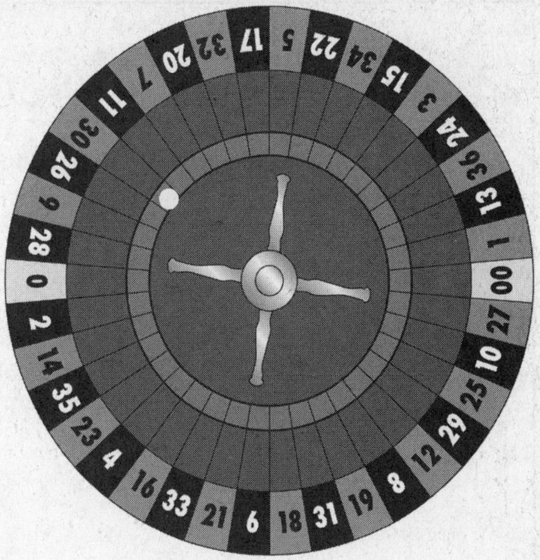

On both spins of the wheel, the ball landed on 30, a red square, and you doubled your winnings.

You've learned a lot about probability over at Fat Dan's roulette table, and you'll find this knowledge will come in handy for what's ahead at the casino. It's a pity you didn't win enough chips to take any home with you, though.

[Note from Fat Dan: That's a relief.]

It's great that we know our chances of winning all these different bets, but don't we need to know more than just probability to make smart bets?

Besides the chances of winning, you also need to know how much you stand to win in order to decide if the bet is worth the risk.

Betting on an event that has a very low probability may be worth it if the payoff is high enough to compensate you for the risk. In the next chapter, we'll look at how to factor these payoffs into our probability calculations to help us make more informed betting decisions.

The Absent-Minded Diners

Three absent-minded friends decide to go out for a meal, but they forget where they're going to meet. Fred decides to throw a coin. If it lands heads, he'll go to the diner; tails, and he'll go to the Italian restaurant. George throws a coin, too; heads, it's the Italian restaurant; tails, it's the diner. Ron decides he'll just go to the Italian restaurant because he likes the food.

What's the probability all three friends meet? What's the probability one of them eats alone?

Exercise

Here are some more roulette probabilities for you to work out.

1. The probability of the ball having landed on the number 17 given the pocket is black.

2. The probability of the ball landing on pocket number 22 twice in a row.

3. The probability of the ball having landed in a pocket with a number greater than 4 given that it's red.

4. The probability of the ball landing in pockets 1, 2, 3, or 4.

The Absent-Minded Diners solution

Three absent-minded friends decide to go out for a meal, but they forget where they're going to meet. Fred decides to throw a coin. If it lands heads, he'll go to the diner; tails, and he'll go to the Italian restaurant. George throws a coin, too; heads, it's the Italian restaurant; tails, it's the diner. Ron decides he'll just go to the Italian restaurant because he likes the food.

What's the probability all three friends meet? What's the probability one of them eats alone?

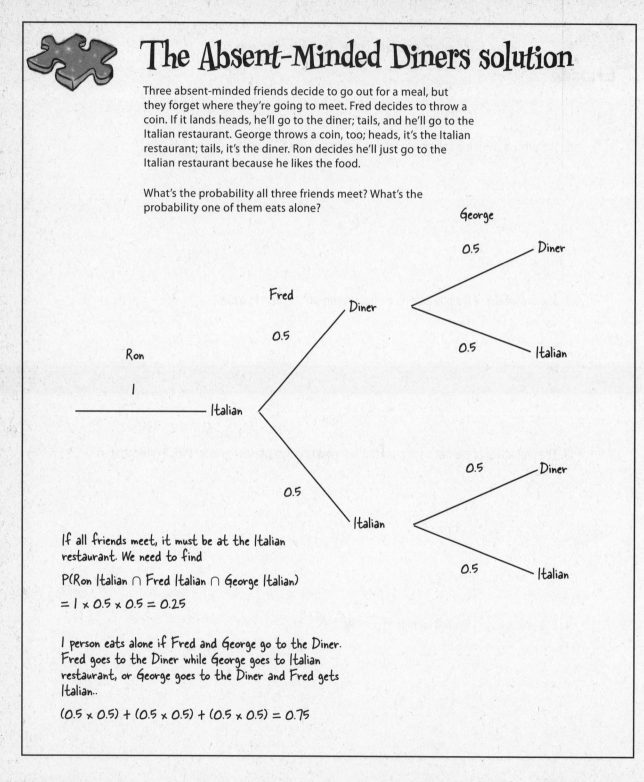

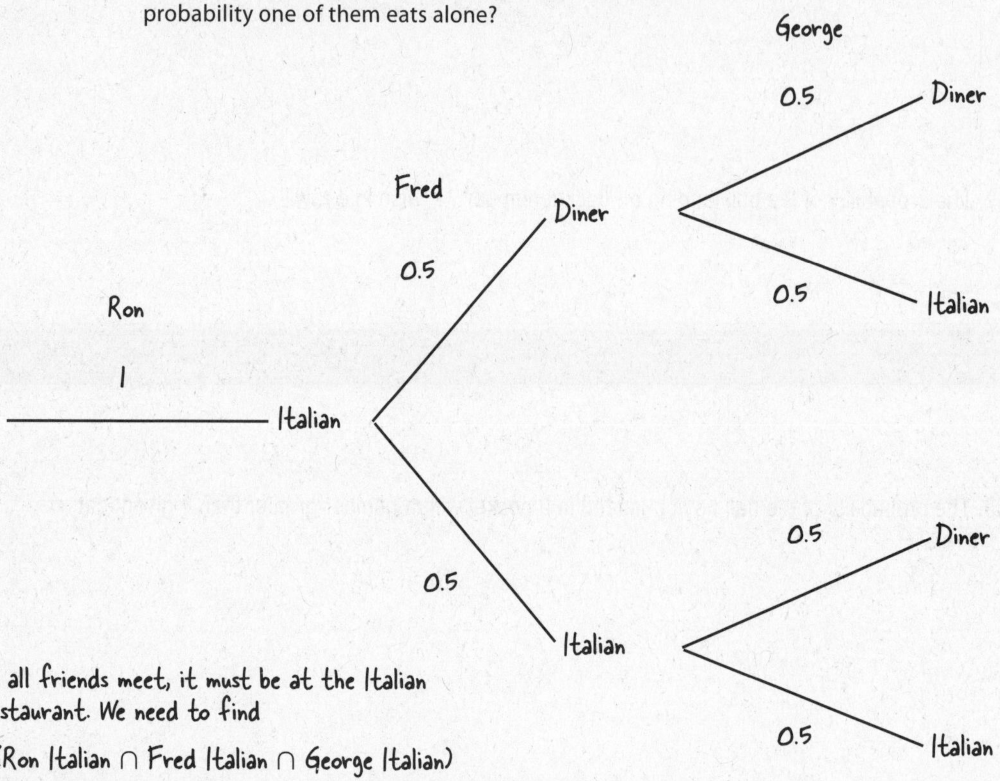

If all friends meet, it must be at the Italian restaurant. We need to find

$P(\text{Ron Italian} \cap \text{Fred Italian} \cap \text{George Italian})$

$= 1 \times 0.5 \times 0.5 = 0.25$

1 person eats alone if Fred and George go to the Diner. Fred goes to the Diner while George goes to Italian restaurant, or George goes to the Diner and Fred gets Italian..

$(0.5 \times 0.5) + (0.5 \times 0.5) + (0.5 \times 0.5) = 0.75$

Exercise Solution

Here are some more roulette probabilities for you to work out.

1. The probability of the ball having landed on the number 17 given the pocket is black.

There are 18 black pockets, and one of them is numbered 17.
P(17 | Black) = 1/18 = 0.0556 (to 3 decimal places)

2. The probability of the ball landing on pocket number 22 twice in a row.

We need to find $P(22 \cap 22)$. As these events are independent, this is equal to $P(22) \times P(22)$. The probability of getting a 22 is 1/38, so
$P(22 \cap 22) = 1/38 \times 1/38 = 1/1444 = 0.00069$ (to 5 decimal places)

3. The probability of the ball having landed in a pocket with a number greater than 4 given that it's red.

P(Above 4 | Red) = 1 - P(4 or below | Red)

There are 2 red numbers below 4, so this gives us

1 - (1/18 + 1/18) = 8/9 = 0.889 (to 3 decimal places)

4. The probability of the ball landing in pockets 1, 2, 3, or 4.

The probability of each pocket is 1/38, so the probability of this event is $4 \times 1/38 = 4/38 = 0.105$ (to 3 decimal places)

5 using discrete probability distributions

Manage Your Expectations

OK, so falling out the tree was unexpected, but you have to take a long-term view of these things.
Taking Chances... Safely

Unlikely events happen, but what are the consequences?

So far we've looked at how probabilities tell you how likely certain events are. What probability *doesn't* tell you is the **overall impact** of these events, and what it means to you. Sure, you'll sometimes make it big on the roulette table, but is it really worth it with all the money you lose in the meantime? In this chapter, we'll show you how you can use probability to **predict long-term outcomes**, and also **measure the certainty** of these predictions.

Back at Fat Dan's Casino

Have you ever felt mesmerized by the flashing lights of a slot machine? Well, you're in luck. At Fat Dan's Casino, there's a full row of shiny slot machines just waiting to be played. Let's play one of them, which costs $1 per ***game*** (pull of the lever). Who knows, maybe you'll hit jackpot!

The slot machine has three windows, and if all three windows line up in the right way, the cash will come cascading out.

The amount of money you can win looks tempting, but I'd like to know the probability of getting any of these combinations before playing.

This sounds like something we can calculate. Here are the probabilities of a particular image appearing in a particular window:

$	Cherry	Lemon	Other
0.1	0.2	0.2	0.5

The three windows are independent of each other, which means that the image that appears in one of the windows has no effect on the images that appear in any of the others.

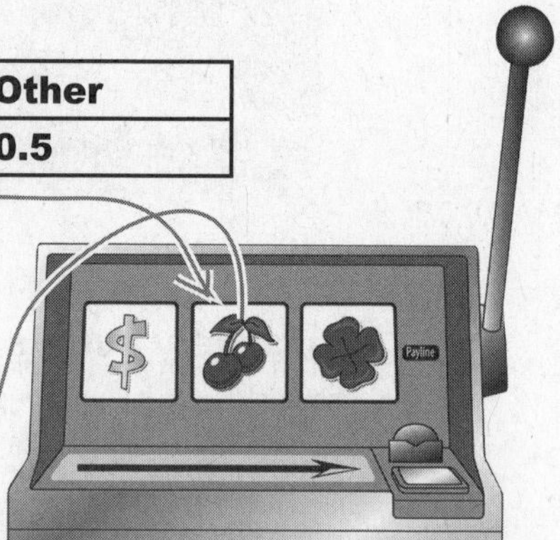

The probability of a cherry appearing in this window is 0.2.

BE the gambler

Take a look at the poster for the slot machine on the facing page. Your job is to play like you're the gambler and work out the probability of getting each combination on the poster. What's the probability of not winning anything?

probability of 

probability of  **(any order)**

probability of 

probability of 

probability of winning nothing

BE the gambler solution

Take a look at the poster for the slot machine on the facing page. Your job is to play like you're the gambler and work out the probability of getting each combination on the poster. What's the probability of not winning anything?

probability of \$ \$ \$

$$P(\$, \$, \$) = P(\$) \times P(\$) \times P(\$)$$

$$= 0.1 \times 0.1 \times 0.1$$

$$= 0.001$$

The probability of a dollar sign appearing in a window is 0.1

probability of \$ \$  (any order)

There are three ways of getting this:

$$P(\$, \$, \text{cherry}) + P(\$, \text{cherry}, \$) + P(\text{cherry}, \$, \$)$$

$$= (0.1^2 \times 0.2) + (0.1^2 \times 0.2) + (0.1^2 \times 0.2)$$

$$= 0.006$$

probability of 

$$P(\text{lemon, lemon, lemon}) = P(\text{lemon}) \times P(\text{lemon}) \times P(\text{lemon})$$

$$= 0.2 \times 0.2 \times 0.2$$

$$= 0.008$$

A lemon appearing in a window is independent of ones appearing in the other two windows, so you multiply the three probabilities together.

probability of 

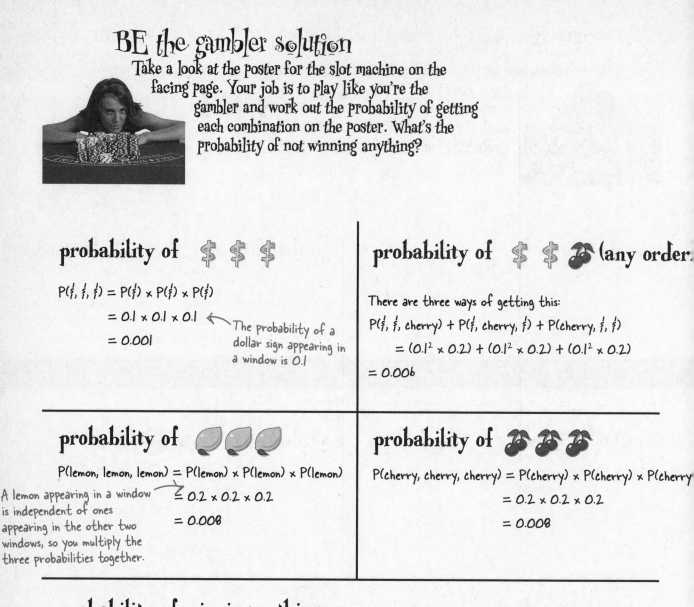

$$P(\text{cherry, cherry, cherry}) = P(\text{cherry}) \times P(\text{cherry}) \times P(\text{cherry})$$

$$= 0.2 \times 0.2 \times 0.2$$

$$= 0.008$$

probability of winning nothing

This means we get none of the winning combinations.

Rather than work out all the possible ways in which you could lose, you can say P(losing) = 1 – P(winning).

$$P(\text{losing}) = 1 - P(\$, \$, \$) - P(\$, \$, \text{cherry (any order)}) - P(\text{cherry, cherry, cherry}) - P(\text{lemon, lemon, lemon})$$

$$= 1 - 0.001 - 0.006 - 0.008 - 0.008$$

$$= 0.977$$

These are the four probability values we calculated above.

We can compose a probability distribution for the slot machine

Here are the probabilities of the different winning combinations on the slot machine.

This is just a summary of the probabilities we just worked out.

Combination	None	Lemons	Cherries	Dollars/cherry	Dollars
Probability	0.977	0.008	0.008	0.006	0.001

This looks useful, but I wonder if we can take it one step further. We've found the probabilities of getting each of the winning combinations, but what we're **really** interested in is how much we'll win or lose.

We don't just want to know the probability of winning, we want to know how *much* we stand to win.

The probabilities are currently written in terms of combinations of symbols, which makes it hard to see at a glance what out gain will be.

We don't have to write them like this though. Instead of writing the probabilities in terms of slot machine images, we can write them in terms of how much we win or lose on each game. All we need to do is take the amount we'll win for each combination, and subtract the amount we've paid for the game.

Combination	None	Lemons	Cherries	$s/cherry	Dollars
Gain	**-$1**	**$4**	**$9**	**$14**	**$19**
Probability	0.977	0.008	0.008	0.006	0.001

We lose $1 if we don't hit a winning combination.

These are the same probabilities, just written in terms of how much we'll gain.

Our gain for hitting each winning combination: the payoff minus the $1 we paid to play.

The table gives us the ***probability distribution*** of the winnings, a set of the probabilities for every possible gain or loss for our slot machine.

Probability Distributions Up Close

When you derived the probabilities of the slot machine, you calculated the probability of making each gain or loss. In other words, you calculated the probability distribution of a ***random variable***, which is a variable that can takes on a set of values, where each value is associated with a specific probability. In the case of Fat Dan's slot machine, the random variable represents the amount we'll gain in each game.

When we want to refer to a random variable, it's usual to represent it by a capital letter, like X or Y. The particular *values* that the variable can take are represented by a lowercase letter—for example, x or y. Using this notation, P(X = x) is a way of saying "the probability that the variable X takes a particular value x."

Here's our slot machine probability distribution written using this notation:

The value of each combination's winnings is represented by x.

Here x is 19.

Combination	None	Lemons	Cherries	$s/cherry	Dollars
x	**-1**	**4**	**9**	**14**	**19**
P(X = x)	0.977	0.008	0.008	0.006	0.001

X is the variable.

The probability that the variable X is 9—in other words, that the value of the winnings is $9.

The variable is ***discrete***. This means that it can only take exact values.

As well as giving a table of the probability distribution, we can also show the distribution on a chart to help us visualize it. Here is a bar chart showing the slot machine probabilities.

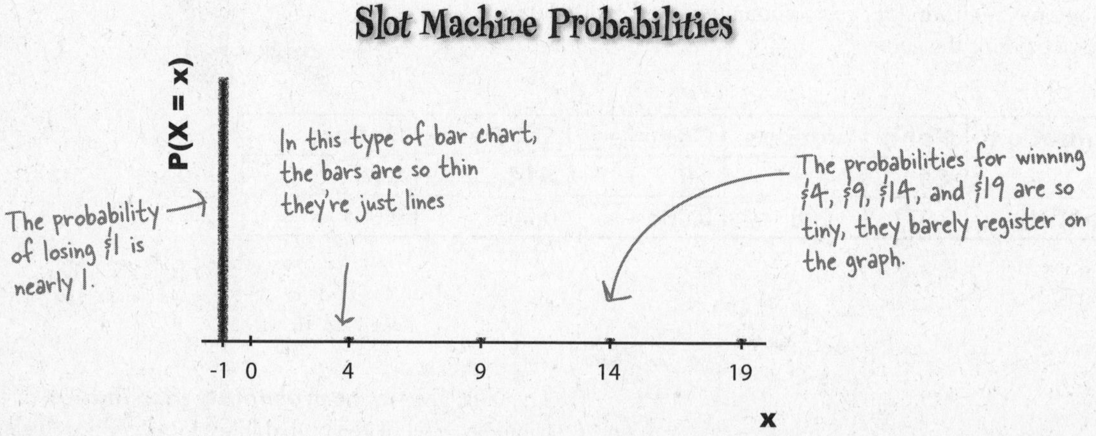
Slot Machine Probabilities
P(X = x)
In this type of bar chart, the bars are so thin they're just lines
The probability of losing $1 is nearly 1.
The probabilities for winning $4, $9, $14, and $19 are so tiny, they barely register on the graph.
-1
0
4
9
14
19
x

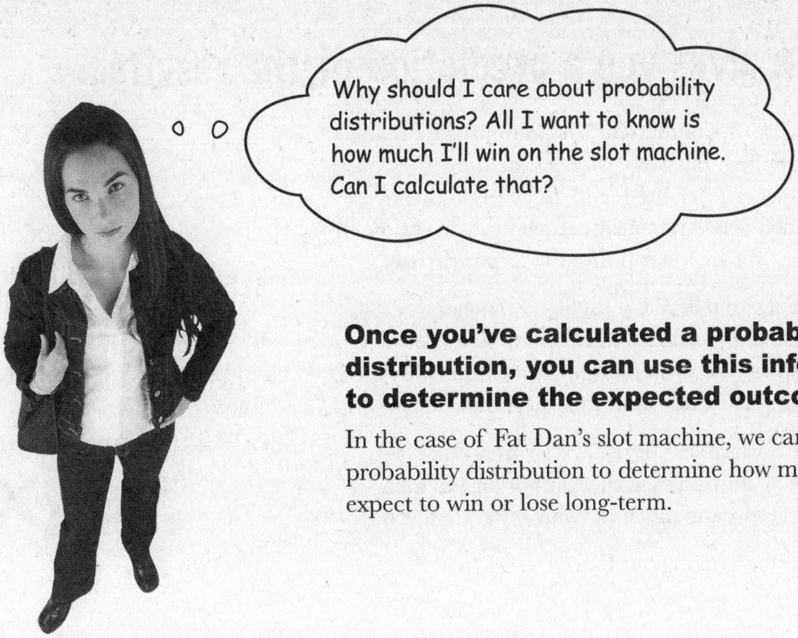

Why should I care about probability distributions? All I want to know is how much I'll win on the slot machine. Can I calculate that?

Once you've calculated a probability distribution, you can use this information to determine the expected outcome.

In the case of Fat Dan's slot machine, we can use our probability distribution to determine how much you can expect to win or lose long-term.

there are no Dumb Questions

Q: Why couldn't we have just used the symbols instead of winnings? I'm not sure we've really gained that much.

A: We could have, but we can do more things if we have numeric data because we can use it in calculations. You'll see shortly how we can use numeric data to work out how much we can expect to win on each game, for instance. We couldn't have done that if we had just used symbols.

Q: What if I want to show probability distributions on a Venn diagram?

A: It's not that appropriate to show probability distributions like that. Venn diagrams and probability trees are useful if you want to calculate probabilities. With a probability distribution, the probabilities have already been calculated.

Q: Can you use any letter to represent a variable?

A: Yes, you can, as long as you don't confuse it with anything else. It's most common to use letters towards the end of the alphabet, though, such as X and Y.

Q: Should I use the same letter for the variable and the values? Would I ever use X for the variable and y for the values?

A: Theoretically, there's nothing to stop you, but in practice you'll find it more confusing if you use different letters. It's best to stick to using the same letter for each.

Q: You said that a discrete random variable is one where you can say precisely what the values are. Isn't that true of every variable?

A: No, it's not. With the slot machine winnings, you know precisely what the winnings are going to be for each symbol combination. You can't get any more precise, and it wouldn't matter how many times you played. For each game the possible values remain the same.

Sometimes you're given a range of values where any value within the range is possible. As an example, suppose you were asked to measure pieces of string that are between 10 inches and 11 inches long. The length could be literally any value within that range.

Don't worry about the distinction too much for now; we'll look at this in more detail later on in the book. For now, every random variable we look at will be discrete.

Expectation gives you a prediction of the results...

You have a probability distribution for the amount you could gain on the slot machines, but now you need to know how much you can expect to win or lose long-term. You can do this by calculating how much you can typically expect to win or lose in each game. In other words, you can find the ***expectation***.

The expectation of a variable X is a bit like the mean, but for probability distributions. You even calculate it in a similar way. To find the expectation, you multiply each value x by the probability of getting that value, and then sum the results.

The expectation of a variable X is usually written E(X), but you'll sometimes see it written as μ, the symbol for the mean. Think of the expectation and mean as twins separated at birth.

I'm the expectation. Treat me like I'm mean.

$$E(X) = \mu$$

Here's the equation for working out E(X):

Multiply each value by its probability.

$$E(X) = \sum xP(X = x)$$

E(X) is the expectation of X

Once you've done multiplying, add the whole lot up together.

Let's use this to calculate the expectation of the slot machine gain. Here's a reminder of our probability distribution:

x	-1	4	9	14	19
P(X = x)	0.977	0.008	0.008	0.006	0.001

$$E(X) = (-1 \times 0.977) + (4 \times 0.008) + (9 \times 0.008) + (14 \times 0.006) + (19 \times 0.001)$$

$$= -0.977 + 0.032 + 0.072 + 0.084 + 0.019$$

$$= -0.77$$

This is the amount in $'s you can expect to gain on each pull of the lever—and it's negative!

In other words, over a large number of games, you can expect to lose $0.77 for each game. This means that if you played the slot machine 100 times, you could expect to lose $77.

...and variance tells you about the spread of the results

The expectation tells you how much on average you can expect to win or lose with each game. If you lost this amount every single time, where would the fun be, and who would play?

Just because you can expect to lose each time you play doesn't mean there isn't a small chance you'll win big. Just like the mean, the expectation doesn't give the full story as the amount you stand to gain on each game could vary a lot. How do you think we can measure this?

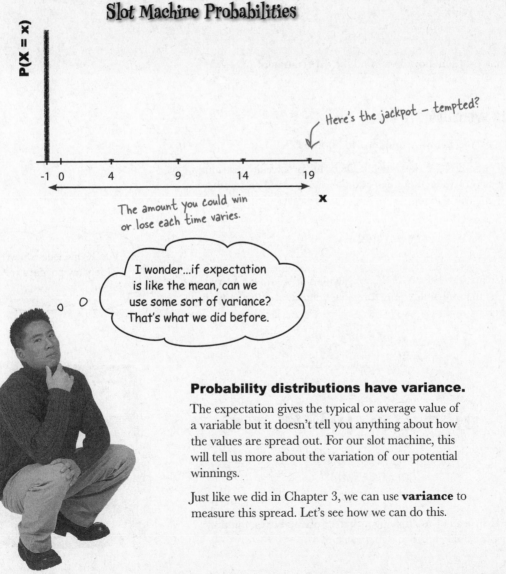

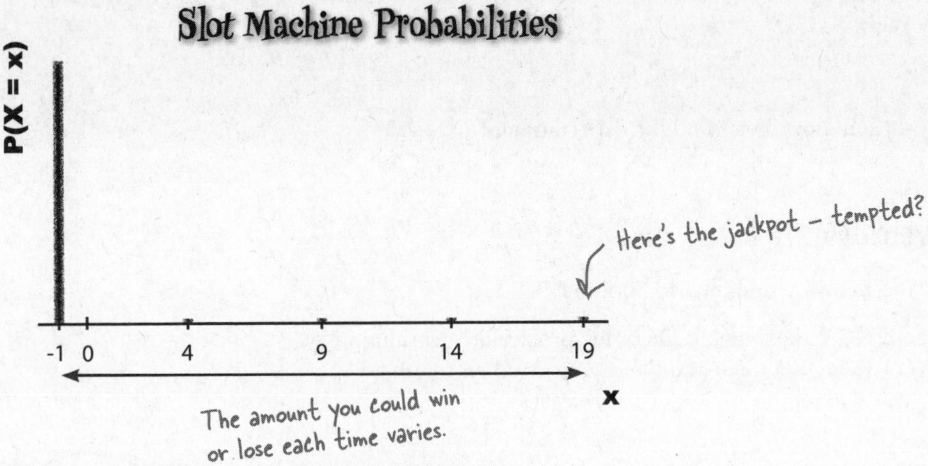
Slot Machine Probabilities
P(X = x)
Here's the jackpot - tempted?
-1
0
4
9
14
19
x
The amount you could win or lose each time varies.

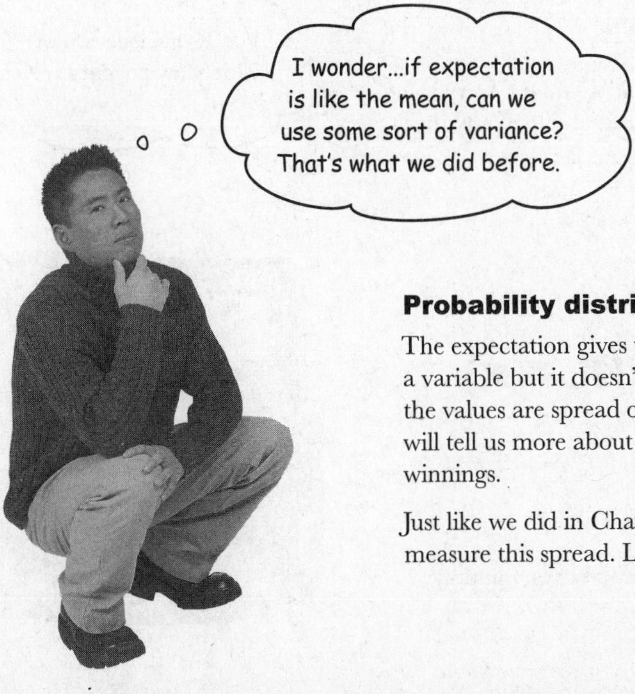
I wonder...if expectation is like the mean, can we use some sort of variance? That's what we did before.

Probability distributions have variance.

The expectation gives the typical or average value of a variable but it doesn't tell you anything about how the values are spread out. For our slot machine, this will tell us more about the variation of our potential winnings.

Just like we did in Chapter 3, we can use **variance** to measure this spread. Let's see how we can do this.

Variances and probability distributions

Back in Chapter 3, we calculated the variance of a set of numbers. We worked out $(x - \mu)^2$ for each number, and then we took the average of these results.

We can do something similar to work out the variance of a variable X. Instead of finding the **average** of $(X - \mu)^2$, we find its **expectation**. We use this formula:

$$\text{Var}(X) = E(X - \mu)^2$$

This is the variance. A shorthand way of referring to the variance of X is Var(X).

μ is the alternative way of writing E(X).

We need to find the expectation of $(X - \mu)^2$—but how?

There's just one problem: how do we find the expectation of $(X - \mu)^2$?

So how do we calculate $E(X - \mu)^2$?

Finding $E(X - \mu)^2$ is actually quite similar to finding E(X).

When we calculate E(X), we take each value in the probability distribution, multiply it by its probability, and then add the results together. In other words, we use the calculation

$$E(X) = \Sigma xP(X = x)$$

When we calculate the variance of X, we calculating $(x - \mu)^2$ for every value x, multiply it by the probability of getting that value x, and then add the results together.

Go through each value x and work out what $(x - \mu)^2$ is. Then multiply it by the probability of getting x...

$$E(X - \mu)^2 = \sum(x - \mu)^2P(X = x)$$

...and then add these results together.

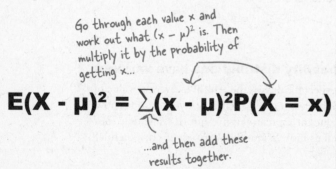

Var(X) measures how widely my payouts vary.
Payline

In other words, instead of multiplying x by its probability, you multiply $(x - \mu)^2$ by the probability of getting that value of x.

Let's calculate the slot machine's variance

Let's see if we can use this to calculate the variance of the slot machine. To do this, we subtract μ from each value, square the result, and then multiply each one by the probability. As a reminder, E(X) or μ is -0.77.

Here's a reminder of the slot machine probabilities.

x	-1	4	9	14	19
P(X = x)	0.977	0.008	0.008	0.006	0.001

We found E(X) = -0.77 back on page 204

$$\text{Var}(X) = E(X - \mu)^2$$

$$= (-1+0.77)^2 \times 0.977 + (4+0.77)^2 \times 0.008 + (9+0.77)^2 \times 0.008 + (14+0.77)^2 \times 0.006 + (19+0.77)^2 \times 0.001$$

$$= (-0.23)^2 \times 0.977 + 4.77^2 \times 0.008 + 9.77^2 \times 0.008 + 14.77^2 \times 0.006 + 19.77^2 \times 0.001$$

$(X-\mu)^2 \times P(X=x)$

$$= 0.0516833 + 0.1820232 + 0.7636232 + 1.3089174 + 0.3908529$$

This means that while the expectation of our winnings is -0.77, the variance is 2.6971.

What about the standard deviation? Can we calculate that too?

As well as having a variance, probability distributions have a standard deviation.

It serves a similar function to the standard deviation of a set of values. It's a way of measuring how far away from the center you can expect your values to be.

As before, the standard deviation is calculated by taking the square root of the variance like this:

$$\sigma = \sqrt{\text{Var}(X)}$$

We can use the same symbol for standard deviation as before.

This means that the standard deviation of the slot machine winnings is √2.6971, or 1.642. This means that on average, our winnings per game will be 1.642 away from the expectation of -0.77.

BRAIN POWER

Would you prefer to play on a slot machine with a high or low variance? Why?

there are no Dumb Questions

Q: So expectation is a lot like the mean. Is there anything for probability distributions that's like the median or mode?

A: You can work out the most likely probability, which would be a bit like the mode, but you won't normally have to do this. When it comes to probability distributions, the measure that statisticians are most interested in is the expectation.

Q: Shouldn't the expectation be one of the values that X can take?

A: It doesn't have to be. Just as the mean of a set of values isn't necessarily the same as one of the values, the expectation of a probability distribution isn't necessarily one of the values X can take.

Q: Are the variance and standard deviation the same as we had before when we were dealing with values?

A: They're the same, except that this time we're dealing with probability distributions. The variance and standard deviation of a set of values are ways of measuring how far values are spread out from the mean. The variance and standard deviation of a probability distribution measure how the probabilities of particular values are dispersed.

Q: I find the concept of $E(X - \mu)^2$ confusing. Is it the same as finding $E(X - \mu)$ and then squaring the end result?

A: No, these are two different calculations. $E(X - \mu)^2$ means that you find the square of $X - \mu$ for each value of X, and then find the expectation of all the results. If you calculate $E(X - \mu)$ and then square the result, you'll get a completely different answer.

Technically speaking, you're working out $E((X - \mu)^2)$, but it's not often written that way.

Q: So what's the difference between a slot machine with a low variance and one with a high variance?

A: A slot machine with a high variance means that there's a lot more variability in your overall winnings. The amount you could win overall is less predictable.

In general, the smaller the variance is, the closer your average winnings per game are likely to be to the expectation. If you play on a slot machine with a larger variance, your overall winnings will be less reliable.

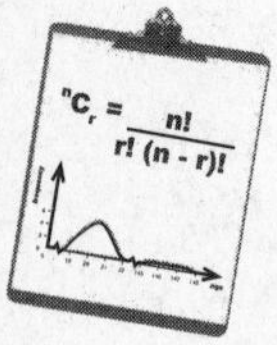

Vital Statistics

Expectation

Use the following formula to find the expectation of a variable X:

$$E(X) = \sum xP(X = x)$$

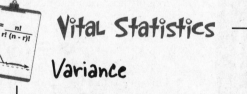

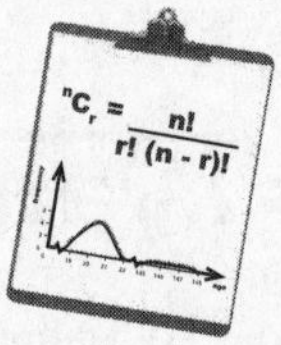

Vital Statistics

Variance

Use the following formula to calculate the variance

$$Var(X) = E(X - \mu)^2$$

Exercise

Here's the probability distribution of a random variable X:

x	**1**	**2**	**3**	**4**	**5**
P(X = x)	0.1	0.25	0.35	0.2	0.1

1. What's the value of E(X)?

2. What's the value of Var(X)?

Exercise Solution

Here's the probability distribution of a random variable X:

x	1	2	3	4	5
P(X = x)	0.1	0.25	0.35	0.2	0.1

1. What's the value of E(X)?

$$E(X) = \sum xP(X=x)$$

← Multiply each value by the probability of it occurring, and take the sum of all the results.

$$= 1\times0.1 + 2\times0.25 + 3\times0.35 + 4\times0.2 + 5\times0.1$$

$$= 0.1 + 0.5 + 1.05 + 0.8 + 0.5$$

$$= 2.95$$

2. What's the value of Var(X)?

Go through each value x and work out what $(x - \mu)^2$ is. Then multiply it by the probability of getting x. Once you've done that, add the whole lot up together.

$$Var(X)^2 = E(X - \mu)^2$$

$$= \sum(x - \mu)^2P(X=x)$$

$$= (1-2.95)^2\times0.1 + (2-2.95)^2\times0.25 + (3-2.95)^2\times0.35 + (4-2.95)^2\times0.2 + (5-2.95)^2\times0.1$$

$$= (-1.95)^2\times0.1 + (-0.95)^2\times0.25 + (0.05)^2\times0.35 + (1.05)^2\times0.2 + (2.05)^2\times0.1$$

$$= 3.8025\times0.1 + 0.9025\times0.25 + 0.0025\times0.35 + 1.1025\times0.2 + 4.2025\times0.1$$

$$= 0.38025 + 0.225625 + 0.000875 + 0.2205 + 0.42025$$

$$= 1.2475$$

The Case of the Moving Expectation

Statsville broadcasts a number of popular quiz shows, and among these is Seal or No Seal. In this show, the contestant is shown a number of boxes containing different amounts of money, and they have to choose one of them, without looking inside. The remaining boxes are opened one by one, and with each one that's opened, the contestant is offered the chance to keep the money in the box they've chosen, sight unseen, or accept another offer based on the amount of money contained in the rest of the unopened boxes. The Statsville Seal Sanctuary get a donation based on any winnings the contestant gets.

The latest contestant is an amateur statistician, and he figures he'll be in a better position to win if he knows what the expectation is of all the boxes. He's just finished calculating the expectation when the producer comes over to him.

"You're on in three minutes," says the producer, "and we've changed all the values in the boxes. They're now worth twice as much, minus $10."

The contestant stares at the producer in horror. Are all his calculations for nothing? He can't possibly work out the expectation from scratch in three minutes. What should he do?

How can the contestant figure out the new expectation in record time?

Fat Dan changed his prices

In the past few minutes, Fat Dan has changed the cost and prizes of the slot machine. Here's the new lineup.

Instead of paying $1 for each game, the price has now gone up to $2.

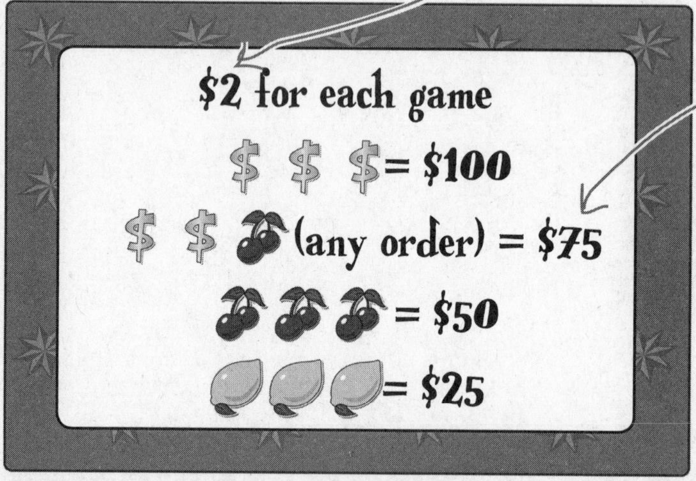

The prizes are 5 times the original.

The cost of one game (pull of the lever) on the slot machine is now $2 instead of $1, but the prizes are now five times greater. If we win, we'll be able to make a lot more money than before.

Here's the new probability distribution.

y	-2	23	48	73	98
P(Y = y)	0.977	0.008	0.008	0.006	0.001

This time we're using Y, not X.

If we knew what the expectation and variance were, we'd have an idea of how much we could win long-term.

Sharpen your pencil

What's the expectation and variance of the new probability distribution? How do these values compare to the previous payout distribution's expectation of -0.77 and variance of 2.6971?

y	**-2**	**23**	**48**	**73**	**98**
P(Y = y)	0.977	0.008	0.008	0.006	0.001

Sharpen your pencil Solution

What's the expectation and variance of the new probability distribution? How do these values compare to the previous payout distribution's expectation of -0.77 and variance of 2.6971?

y	-2	23	48	73	98
P(Y = y)	0.977	0.008	0.008	0.006	0.001

$$E(Y) = (-2) \times 0.977 + 23 \times 0.008 + 48 \times 0.008 + 73 \times 0.006 + 98 \times 0.001$$

$$= -1.954 + 0.184 + 0.384 + 0.438 + 0.098$$

$$= -0.85$$

$$Var(Y) = E(Y - \mu)^2$$

$$= \sum(y - \mu)^2 P(Y=y)$$

$$= (-2+0.85)^2 \times 0.977 + (23+0.85)^2 \times 0.008 + (48+0.85)^2 \times 0.008 + (73+0.85)^2 \times 0.006 + (98+0.85)^2 \times 0.001$$

$$= (-1.15)^2 \times 0.977 + (23.85)^2 \times 0.008 + (48.85)^2 \times 0.008 + (73.85)^2 \times 0.006 + (98.85)^2 \times 0.001$$

$$= 1.3225 \times 0.977 + 568.8225 \times 0.008 + 2386.3225 \times 0.008 + 5453.8225 \times 0.006 + 9771.3225 \times 0.001$$

$$= 1.2920825 + 4.55058 + 19.09058 + 32.722935 + 9.7713225$$

$$= 67.4275$$

The expectation is slightly lower, so in the long term, we can expect to lose \$0.85 each game. The variance is much larger. This means that we stand to lose more money in the long term on this machine, but there's less certainty.

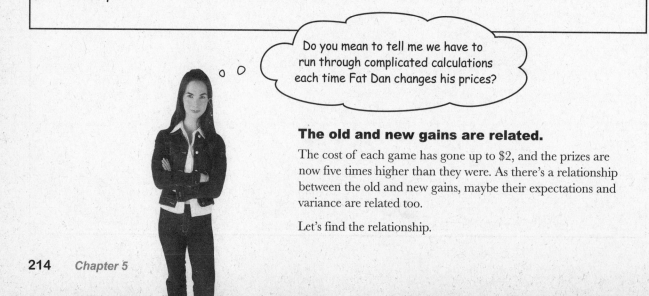

Do you mean to tell me we have to run through complicated calculations each time Fat Dan changes his prices?

The old and new gains are related.

The cost of each game has gone up to \$2, and the prizes are now five times higher than they were. As there's a relationship between the old and new gains, maybe their expectations and variance are related too.

Let's find the relationship.

Pool Puzzle

It's time for a bit of algebra. Your **job** is to take numbers from the pool and place them into the blank lines in the calculations. You may **not** use the same number more than once, and you won't need to use all the numbers. Your **goal** is to come up with an expression for the new gains on the slot machine in terms of the old. X represents the old gains, Y the new.

X = (original win) - (original cost)

= (original win) - ....

(original win) = ... + ...

Y = 5 (original win) - (new cost)

= 5(............... + ...) - ....

= 5 .... + ... - ....

= .... + ...

Note: each thing from the pool can only be used once!

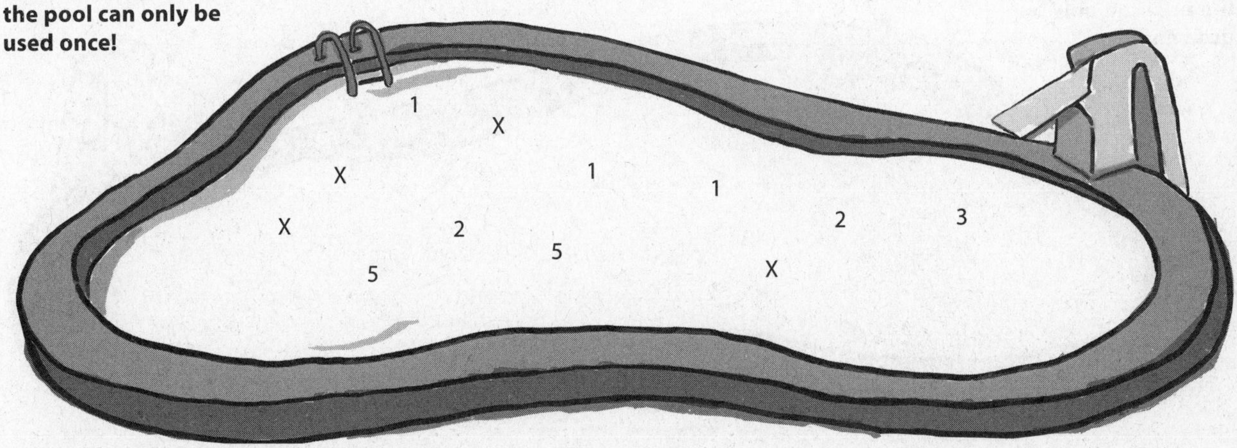

Pool Puzzle Solution

It's time for a bit of algebra. Your **job** is to take numbers from the pool and place them into the blank lines in the calculations. You may **not** use the same number more than once, and you won't need to use all the numbers. Your **goal** is to come up with an expression for the new gains on the slot machine in terms of the old. X represents the old gains, Y the new.

X = (original win) - (original cost)

= (original win) - 1 ← The original game cost $1

(original win) = X + 1 ← This gives us the winnings of the original game in terms of X.

We can substitute in our expression for the original winnings.

Y = 5 (original win) - (new cost)

= 5(X + 1) - 2

= 5 X + 5 - 2

= 5 X + 3

So Y = 5X + 3. There's a definite relationship between X and Y.

Note: each thing from the pool can only be used once!

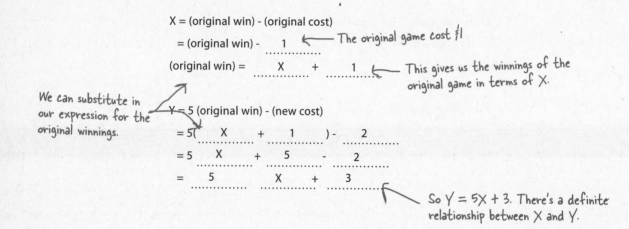

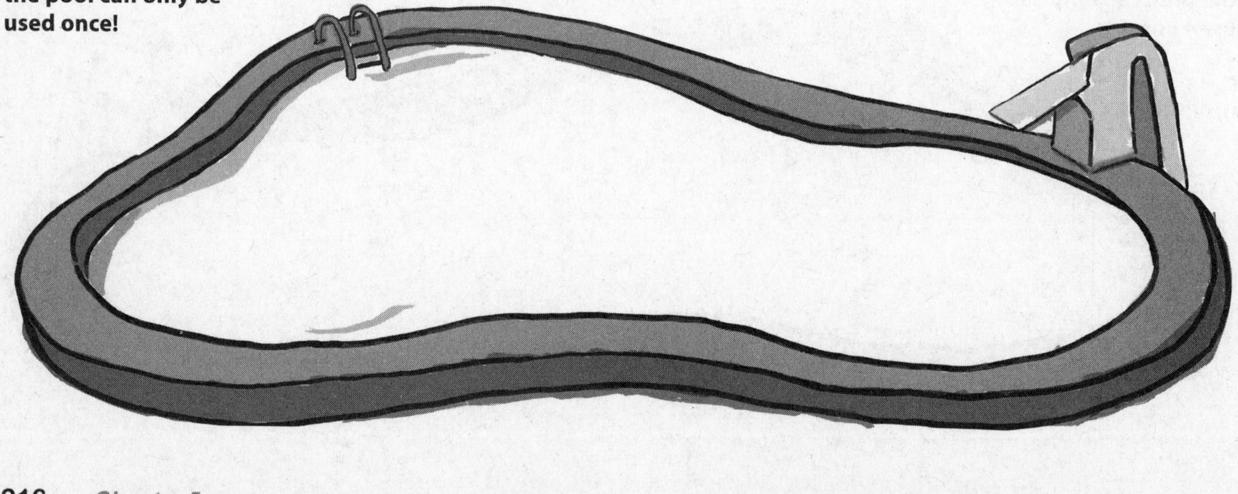

There's a linear relationship between E(X) and E(Y)

We've found that we can relate the new gains to the old using $Y = 5X + 3$, where Y refers to the new gains, and X refers to the old. What we want to do now is see if there's a relationship between E(X) and E(Y), and Var(X) and Var(Y).

If there is a relationship, this will save us lots of time if Fat Dan changes his prices again. As long as we know what the relationship is between the old and the new, we'll be able to quickly calculate the new expectation and variance.

Sharpen your pencil

Let's see whether there's a pattern in the relationship between E(X) and E(Y), and Var(X) and Var(Y).

1. E(X) is -0.77 and E(Y) = -0.85. What is 5 x E(X)? What is 5 x E(X) + 3? How does this relate to E(Y)?

2. Var(X) = 2.6971 and Var(Y) = 67.4275. What is 5 x Var(X)? What is 5^2 x Var(X)? How does this relate to Var(Y)?

3. How could you generalize this for any probability distribution where Y = aX + b?

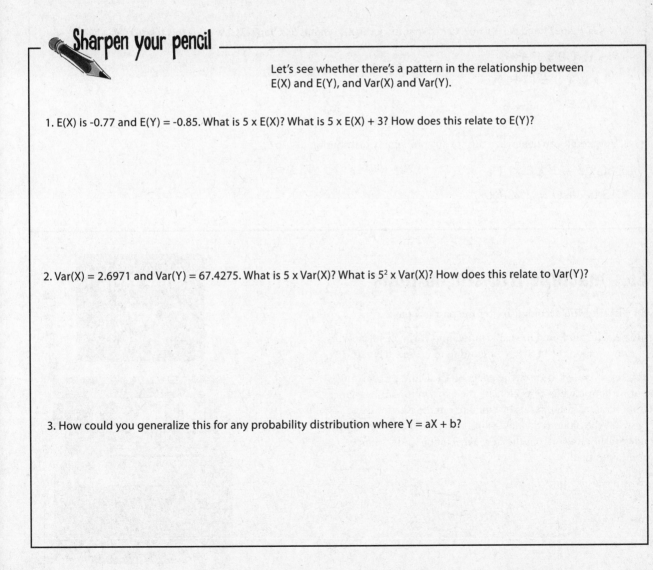

Sharpen your pencil Solution

Let's see whether there's a pattern in the relationship between E(X) and E(Y), and Var(X) and Var(Y).

1. E(X) is -0.77 and E(Y) = -0.85. What is 5 x E(X)? What is 5 x E(X) + 3? How does this relate to E(Y)?

$5 \times E(X) = -3.85$

$5 \times E(X) + 3 = -0.85$

$E(Y) = 5 \times E(X) + 3.$

2. Var(X) = 2.6971 and Var(Y) = 67.4275. What is 5 x Var(X)? What is 5^2 x Var(X)? How does this relate to Var(Y)?

$5 \times Var(X) = 13.4855$

$5^2 \times Var(X) = 67.4275$

$Var(Y) = 5^2 \times Var(X)$

3. How could you generalize this for any probability distribution aX + b?

$E(aX + b) = a\, E(X) + b$

$Var(aX + b) = a^2\, Var(X)$

Slot machine transformations

So what did you accomplish over the past few pages?

First of all, you found the expectation and variance of X, where X is the amount of money you stand to make in each game.

You then wanted to know the effect of Fat Dan's price changes but without having to recalculate the expectation and variance from scratch. You did this by working out the relationship between the old and the new gains, and then using the relationship to work out the new expectation and variance. You found that:

$$E(5X + 3) = 5E(X) + 3$$

$$Var(5X + 3) = 5^2Var(X)$$

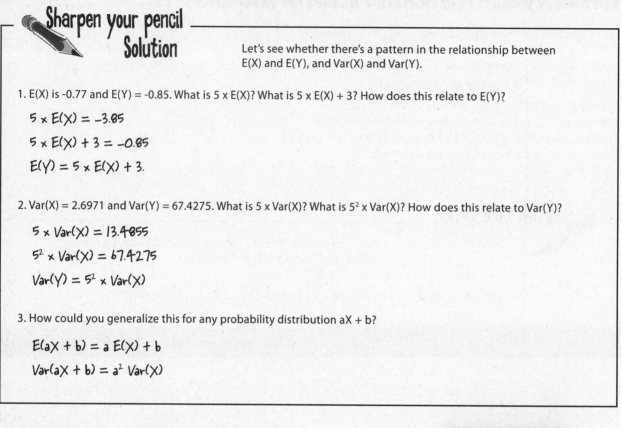

Now pays 5 times more!
Payline

General formulas for linear transforms

We can generalize this for any random variable. For any random variable X

$$E(aX + b) = aE(X) + b$$

Multiply the expectation by a, and then add b.

$$Var(aX + b) = a^2Var(X)$$

Square the a and multiply it by the variance of X (drop the b).

This is called a linear transform, as we are dealing with a linear change to X. In other words, the underlying probabilities stay the same but the values are changed into new values of the form aX + b.

there are no Dumb Questions

Q: Do a and b have to be constants?

A: Yes they do. If a and b are variables, then this result won't hold true.

Q: Where did the b go in the variance?

A: Adding a constant value to the distribution makes no difference to the overall variance, only to the expectation.

When you add a constant to a variable, it in effect moves the distribution along while keeping the same basic shape. This means that the expectation shifts along by b, but as the shape remains unchanged, the variance says the same.

Q: I'm surprised I have to multiply the variance by a^2. Why's that?

A: When you multiply a variable by a constant, you multiply all its underlying values by that constant.

When you calculate the variance, you perform calculations based on the square of the underlying values. And as these have been multiplied by a, the end result is that you multiply the variance by a^2.

Q: Do I really have to remember how to do linear transforms? Are they important?

A: Yes, they are. They can save you a lot of time in the long run, as they eliminate the need for you to have to calculate the expectation and variance of a probability distribution every time the values change. Rather than calculating a new probability distribution, then calculating the expectation and variance from scratch, you can just plug the expectation and variance you already calculated into the equations above.

Knowing linear transforms can also help you out in exams. First of all, you can save valuable time if you know what shortcuts you can take. Furthermore, exam papers don't always give you the underlying probability distribution. You might be told the expectation of variable, and you may have to transform it based on very basic information.

Q: I tried calculating the expectation and variance the long way round and came up with a different answer. Why?

A: You've seen by now that it's easy to make mistakes when you calculate expectations and variances. If you calculate these longhand, there's a good chance you made a mistake somewhere along the line. You're always better off using statistical shortcuts where possible.

Solved: The Case of the Moving Expectation.

How can the contestant figure out the new expectation in record time?

The contestant looks around in panic for a brief moment and then relaxes. The change in values isn't such a big problem after all.

The contestant has already spent time calculating the expectation of the original values of all the boxes, and this has given him an idea of how much money is available for him to win.

The producer has told him that the new prizes are ten dollars less than twice the original prizes. In other words, this is a linear transform. If X represents the original prize money and Y the new, the values are transformed using $Y = 2X - 10$.

The contestant finds $E(Y)$ using $E(2X - 10) = 2E(X) - 10$. This means that all he has to do to find the new expectation is double his original expectation and subtract 10.

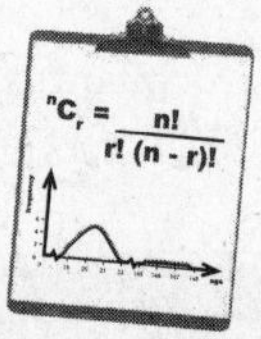

Vital Statistics

Linear Transforms

If you have a variable X and numbers a and b, then:

$$E(aX + b) = aE(X) + b$$

$$Var(aX + b) = a^2Var(X)$$

BULLET POINTS

- **Probability distributions** describe the probability of all possible outcomes of a given variable.
- The **expectation** is the expected average long-term outcome. It's represented as either $E(X)$ or μ, and is calculated using $E(X) = \Sigma xP(X=x)$.
- The **expectation of a function of X** is given by $E(f(X)) = \Sigma f(x)P(X=x)$
- The **variance of a probability distribution** is given by $Var(X) = E(X - \mu)^2$
- The **standard deviation of a probability distribution** is given by $\sigma = \sqrt{Var(X)}$
- **Linear transforms** are when a variable X is transformed into $aX + b$, where a and b are constants. The expectation and variance are given by:
  $E(aX + b) = aE(X) + b$
  $Var(aX + b) = a^2Var(X)$

So do linear transforms give me a quick way of calculating the expectation and variance when I want to play multiple games?

There's a difference between using linear transforms and playing multiple games.

With linear transforms, all of the probabilities stay the same, but the possible values change. The values are transformed, but not the probabilities. There are still the same number of possible values.

When you play multiple games, both the values and the probabilities are different, and even the number of possible values can change. It's not possible to just transform the values, and working out the probabilities can quickly become complicated.

Let's look at a simple example. Imagine you were playing on a very simple slot machine with probability distribution X.

x	-1	5
P(X = x)	0.9	0.1

To find the probability distribution of 2X, you just need to multiply the x values by 2. The underlying values change because the potential gains have doubled.

The amounts here are multiplied by 2. The probabilities stay the same as before.

2x	-2	10
P(2X = 2x)	0.9	0.1

What if you were going to play two games on the slot machine? You'd need to work out the probability distribution from scratch by considering all the possible outcomes from both games.

This is like pulling the handle twice. The possible gains and probabilities are different.

y=-2 if you lose both games.

W represents the outcome of two games

w	-2	4	10
P(W = w)	0.81	0.18	0.01

y=10 if you win both games.

y=4 when you by get -1 in one game and 5 in the other.

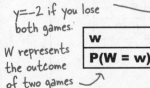

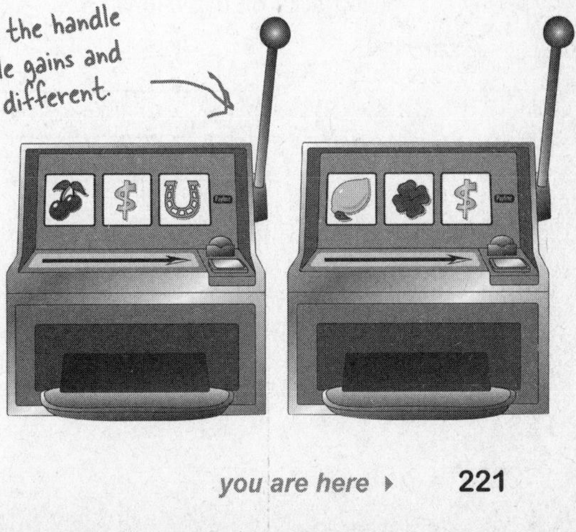

This time, both the probabilities and values have changed. So how can we find the expectation and variance for this situation?

Every pull of the lever is an independent observation

When we play multiple games on the slot machine, each game is called an event, and the outcome of each game is called an ***observation***. Each observation has the same expectation and variance, but their outcomes can be different. You may not gain the same amount in each game.

We need some way of differentiating between the different games or observations. If the probability distribution of the slot machine gains is represented by X, we call the first observation X_1 and the second observation X_2.

Each game is called an event. The outcome of each game is called an observation.

We have the same expectation and variance, but we're separate events.
Observation for game 1
Observation for game 2
X_1
X_2

X_1 and X_2 have the same probabilities, possible values, expectation and variance as X. In other words, they have the same probability distribution, even though they are separate observations and their outcomes can be different.

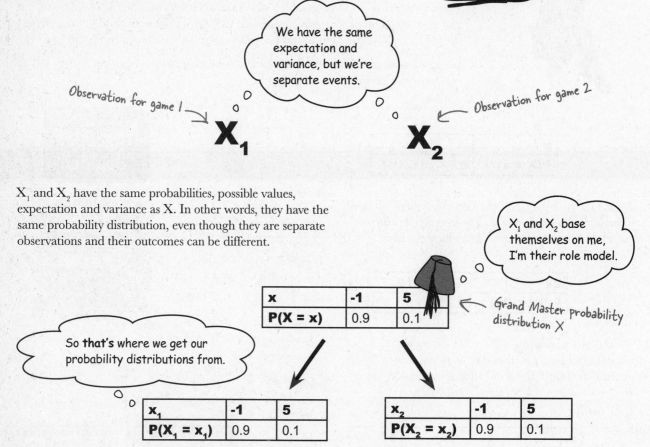

X_1 and X_2 base themselves on me, I'm their role model.
Grand Master probability distribution X

x	-1	5
P(X = x)	0.9	0.1

So that's where we get our probability distributions from.

x_1	-1	5
$P(X_1 = x_1)$	0.9	0.1

x_2	-1	5
$P(X_2 = x_2)$	0.9	0.1

When we want to find the expectation and variance of two games on the slot machine, what we really want to find is the expectation and variance of $X_1 + X_2$. Let's take a look at some shortcuts.

Observation shortcuts

Let's find the expectation and variance of $X_1 + X_2$.

$X_1 + X_2$ is not the same as 2X.

$X_1 + X_2$ means you are considering two observations of X. 2X means you have one observation, but the possible values have doubled.

Expectation

First of all, let's deal with $E(X_1 + X_2)$.

$$E(X_1 + X_2) = E(X_1) + E(X_2)$$
$$= E(X) + E(X)$$
$$= 2E(X)$$

$E(X_1)$ and $E(X_2)$ are both equal to E(X) as X_1 and X_2 follow the same probability distribution as X

In other words, if we have the expectation of two observations, we multiply E(X) by 2. This means that if we were to play two games on a slot machine where E(X) = -0.77, the expectation would be -0.77×2, or -1.54.

We can extend this to deal with multiple observations. If we want to find the expectation of *n* observations, we can use

If there are n observations, we just multiply E(X) by n.

$$\mathbf{E(X_1 + X_2 + \ldots X_n) = nE(X)}$$

Variance

So what about $Var(X_1 + X_2)$? Here's the calculation.

$$Var(X_1 + X_2) = Var(X_1) + Var(X_2)$$
$$= Var(X) + Var(X)$$
$$= 2Var(X)$$

$Var(X_1)$ and $Var(X_2)$ are the same as Var(X) as X_1 and X_2 follow the same probability distribution as X.

This means that if we were to play two games on a slot machine where Var(X) = 2.6971, the variance would be 2.6971×2, or 5.3942.

We can extend this for any number of independent observations. If we have *n* independent observations of X

Multiply Var(X) by n, the number of observations.

$$\mathbf{Var(X_1 + X_2 + \ldots X_n) = nVar(X)}$$

In other words, to find the expectation and variance of multiple observations, just multiply E(X) and Var(X) by the number of observations.

there are no Dumb Questions

Q: Isn't $E(X_1 + X_2)$ the same as $E(2X)$?

A: They look similar but they're actually two different concepts.

With E(2X), you want to find the expectation of a variable where the underlying values have been doubled. In other words, there's only one variable, but the values are twice the size.

With $E(X_1 + X_2)$, you're looking at two separate instances of X, and you're looking at the joint expectation. As an example, if X represents the distribution of a game, then $X_1 + X_2$ represents the distribution of two games.

Q: So are X_1 and X_2 the same?

A: They follow the same distribution, but they're different instances or observations. As an example, X_1 could refer to game 1, and X_2 to game 2. They both have the same probability distribution, but the actual outcome of each might be different.

Q: I see that the new variance is nVar(X) and not n^2Var(X) like we had for linear transforms. Why's that?

A: This time we have a series of independent observations, all distributed the same way. This means that we can find the overall variance by adding the variance of each one together. If we have *n* independent observations, then this gives us nVar(X).

When we calculate the variance of Var(nX), we multiply the underlying values by *n*. As the variance is formed by squaring the underlying values, this means that the resulting variance is n^2Var(X).

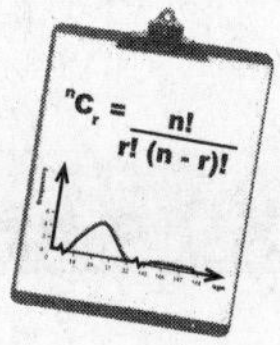

Vital Statistics

Independent Observations

Use the following formula to calculate the variance

$$E(X_1 + X_2 + \ldots + X_n) = nE(X)$$

$$Var(X_1 + X_2 + \ldots + X_n) = nVar(X)$$

BULLET POINTS

- Probability distributions describe the probability of all possible outcomes of a given random variable.
- The expectation of a random variable X is the expected long-term average. It's represented as either E(X) or μ. It's calculated using

$$E(X) = \Sigma xP(X=x)$$

- The variance of a random variable X is given by

$$Var(X) = E(X - \mu)^2$$

- The standard deviation σ is the square root of the variance.
- Linear transforms are when a random variable X is transformed into aX + b, where a and b are numbers. The expectation and variance are given by

$$E(aX + b) = aE(X) + b$$
$$Var(aX + b) = a^2Var(X)$$

LINEAR TRANSFORM OR INDEPENDENT OBSERVATION?

Below are a series of scenarios. Assuming you know the distribution of each X, and your task is to say whether you can solve each problem using linear transforms or independent observations.

	Linear transform	Independent observation
The amount of coffee in an extra large cup of coffee; X is the amount of coffee in a normal-sized cup.	☐	☐
Drinking an extra cup of coffee per day; X is the amount of coffee in a cup.	☐	☐
Finding the net gain from buying 10 lottery tickets; X is the net gain of buying 1 lottery ticket.	☐	☐
Finding the net gain from a lottery ticket after the price of tickets goes up; X is the net gain of buying 1 lottery ticket.	☐	☐
Buying an extra hen to lay eggs for breakfast; X is the number of eggs laid per week by a certain breed of hen.	☐	☐

Linear Transform or Independent Observation? Solution

Below are a series of scenarios. Assuming you know the distribution of each X, and your task is to say whether you can solve each problem using linear transforms or independent observations.

	Linear transform	Independent observation
The amount of coffee in an extra large cup of coffee; X is the amount of coffee in a normal-sized cup.	☑	☐
Drinking an extra cup of coffee per day; X is the amount of coffee in a cup.	☐	☑
Finding the net gain from buying 10 lottery tickets; X is the net gain of buying 1 lottery ticket.	☐	☑
Finding the net gain from a lottery ticket after the price of tickets goes up; X is the net gain of buying 1 lottery ticket.	☑	☐
Buying an extra hen to lay eggs for breakfast; X is the number of eggs laid per week by a certain breed of hen.	☐	☑

The winnings from each lottery ticket are independent of the others.

Changing the price of a ticket changes the expected winnings, but not the probability of winning, so this can be solved with linear transforms.

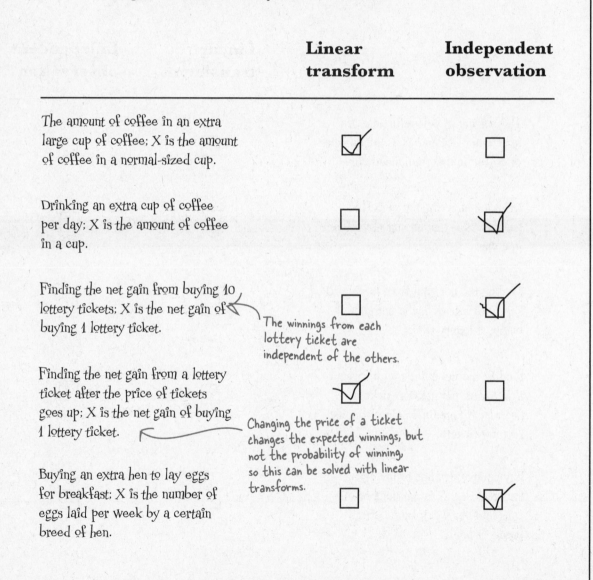

Exercise

The local diner has started selling fortune cookies at $0.50 per cookie. Hidden within each cookie is a secret message. Most messages predict a good future for the buyer, but others offer money off at the diner. The probability of getting $2 off is 0.1, the probability of getting $5 off is 0.07, and the probability of getting $10 off is 0.03.

If X is the net gain, what's the probability distribution of X? What are the values of E(X) and Var(X)?

The diner decides to put the price of the cookies up to $1. What are the new expectation and variance?

Exercise Solution

The local diner has started selling fortune cookies at \$0.50 per cookie. Hidden within each cookie is a secret message. Most messages predict a good future for the buyer, but others offer money off at the diner. The probability of getting \$2 off is 0.1, the probability of getting \$5 off is 0.07, and the probability of getting \$10 off is 0.03.

If X is the net gain, what's the probability distribution of X? What are the values of E(X) and Var(X)?

Here's the probability distribution of X:

x	-0.5	1.5	4.5	9.5
P(X = x)	0.8	0.1	0.07	0.03

$$E(X) = (-0.5)\times 0.8 + 1.5\times 0.1 + 4.5\times 0.07 + 9.5\times 0.03$$
$$= -0.4 + 0.15 + 0.315 + 0.285$$
$$= 0.35$$

$$Var(X) = E(X - \mu)^2$$
$$= \sum(x - \mu)^2 P(X=x)$$
$$= (-0.5-0.35)^2\times 0.8 + (1.5-0.35)^2\times 0.1 + (4.5-0.35)^2\times 0.07 + (9.5-0.35)^2\times 0.03$$
$$= (-0.85)^2\times 0.8 + (1.15)^2\times 0.1 + (4.15)^2\times 0.07 + (9.15)^2\times 0.03$$
$$= 0.7225\times 0.8 + 1.3225\times 0.1 + 17.2225\times 0.07 + 83.7225\times 0.03$$
$$= 0.578 + 0.13225 + 1.205575 + 2.511675$$
$$= 4.4275$$

The diner decides to put the price of the cookies up to \$1. What are the new expectation and variance?

The diner puts the price of the cookies up by \$0.50, which means that the new net gains are modelled by X - 0.5

$$E(X - 0.5) = E(X) - 0.5$$
$$= 0.35 - 0.5$$
$$= -0.15$$

$$Var(X - 0.5) = Var(X)$$
$$= 4.4275$$

New slot machine on the block

Fat Dan has brought in a new model slot machine. Each game costs more, but if you win you'll win big. Here's the probability distribution:

Each game costs more than the other slot machine, but just look at the jackpot!

x	-5	395
P(X = x)	0.99	0.01

We've looked at the expectation and variance of playing a single machine, and also for playing several independent games on the same machine. What happens if we play two different machines at once?

In this situation, we have two different, independent probability distributions for our machines:

x	-5	395
P(X = x)	0.99	0.01

These are the current gains of Fat Dan's new slot machine.

y	-2	23	48	73	98
P(Y = y)	0.977	0.008	0.008	0.006	0.001

These are the current gains of our original slot machine.

So how can we find the expectation and variance of playing one game each on both machines?

We could work out the probability distribution of X + Y, but that would be time-consuming, and we might make a mistake. I wonder if we can take another shortcut?

Add E(X) and E(Y) to get E(X + Y)...

We want to find the expectation and variance of playing one game each on both of the slot machines. In other words, we want to find E(X + Y) and Var(X + Y) where X and Y are random variables representing the two machines. X and Y are independent.

One way of doing this would be to calculate the probability distribution of X + Y, and then calculate the expectation and variance.

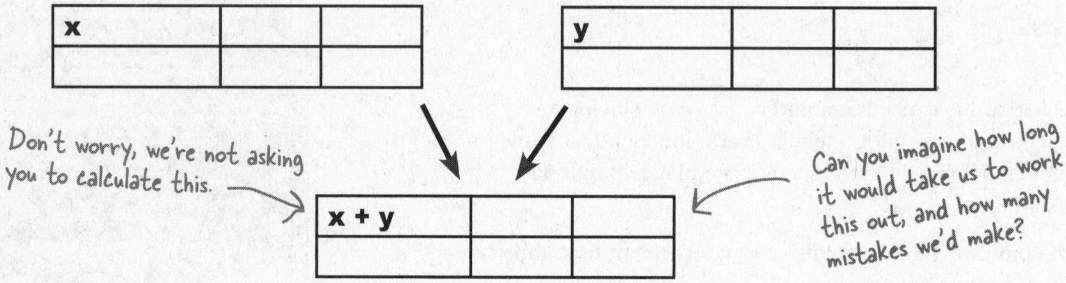

x
y
x + y
Don't worry, we're not asking you to calculate this.
Can you imagine how long it would take us to work this out, and how many mistakes we'd make?

Fortunately we don't have to do this. To find E(X + Y), all we need to do is add together E(X) and E(Y).

$$E(X + Y) = E(X) + E(Y)$$

Intuitively this makes sense. If, for example, you were playing two games where you would expect to win \$5 in one game and \$10 in the other, you would expect to win \$15 overall—\$5 + \$10.

We can do something similar with the variance. To find Var(X +Y), we add the two variances together. This works for all independent random variables.

$$Var(X + Y) = Var(X) + Var(Y)$$

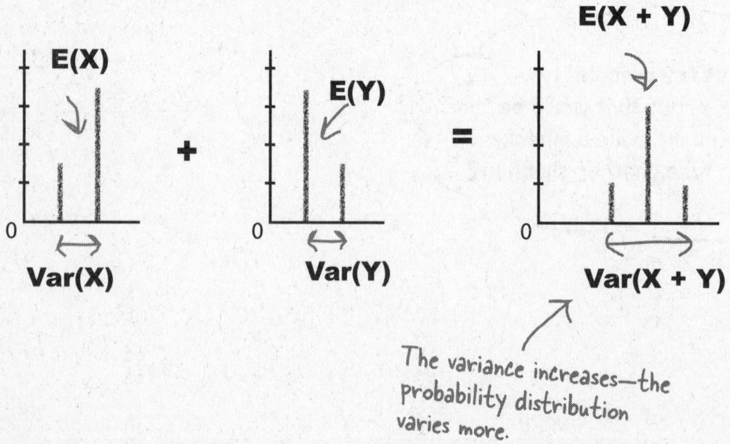

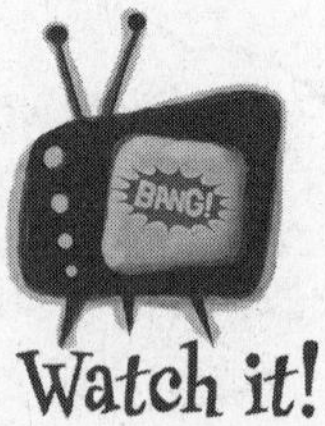

Adding the variances together only works for independent random variables

If X and Y are not independent, then Var(X + Y) is no longer equal to Var(X) + Var(Y).

...and subtract E(X) and E(Y) to get E(X - Y)

You're not just limited to adding random variables; you can also subtract one from the other. Instead of using the probability distribution of X + Y, we can use X – Y.

If you're dealing with the difference between two random variables, it's easy to find the expectation. To find E(X – Y), we subtract E(Y) from E(X).

Finding the variance of X – Y is less intuitive. To find Var(X – Y), we add the two variances together.

E(X - Y) = E(X) – E(Y)

Var(X - Y) = Var(X) + Var(Y)

We add the variances, so be careful!

But that doesn't make sense. Why should we add the variances?

Because the variability increases.

When we subtract one random variable from another, the variance of the probability distribution still increases.

Watch it!

If you're subtracting two random variables, add the variances.

It's easy to make this mistake as at first glance it seems counterintuitive. Just remember that if the two variables are independent, Var(X - Y) = Var(X) + Var(Y)

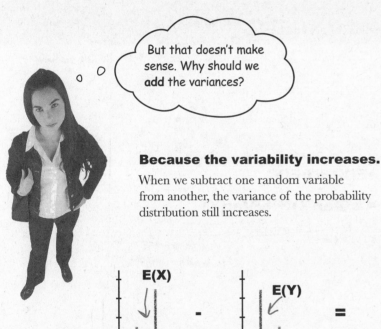

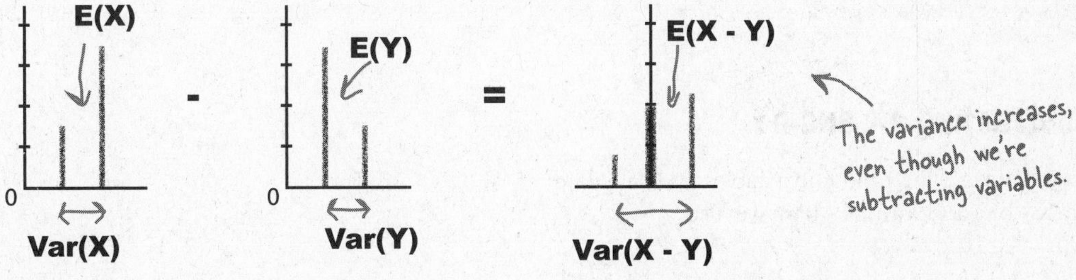

When we subtract independent random variables, the variance is exactly the same as if we'd added them together. The amount of variability can only increase.

Subtracting independent random variables still increases the variance.

You can also add and subtract linear transformations

It doesn't stop there. As well as adding and subtracting random variables, we can also add and subtract their linear transforms.

Imagine what would happen if Fat Dan changed the cost and prizes on both machines, or even just one of them. The last thing we'd want to do is work out the entire probability distribution in order to find the new expectations and variances.

Fortunately, we can take another shortcut.

Suppose the gains on the X and Y slot machines are changed so that the gains for X become aX, and the gains for Y become bY. a and b can be any number.

$$X \rightarrow aX$$

$$Y \rightarrow bY$$

a and b can be any number.

To find the expectation and variance for combinations of aX and bY, we can use the following shortcuts.

Adding aX and bY

If we want to find the expectation and variance of aX + bY, we use

$$E(aX + bY) = aE(X) + bE(Y)$$

$$Var(aX + bY) = a^2Var(X) + b^2Var(Y)$$

It's a linear transform, so we square the numbers here.

We square the numbers because it's a linear transform, just like before.

Subtracting aX and bY

If we subtract the random variables and calculate E(aX - bY) and Var(aX - bY), we use

$$E(aX - bY) = aE(X) - bE(Y)$$

$$Var(aX - bY) = a^2Var(X) + b^2Var(Y)$$

Remember to add the variances.

Just as before, we add the variances, even though we're subtracting the random variables.

there are no Dumb Questions

Q: So if X and Y are games, does aX + bY mean *a* games of X and *b* games of Y?

A: aX + bY actually refers to two linear transforms added together. In other words, the underlying values of X and Y are changed. This is different from independent observations, where each game would be an independent observation.

Q: I can't see when I'd ever want to use X – Y. Does it have a purpose?

A: X – Y is really useful if you want to find the difference between two variables. E(X – Y) is a bit like saying "What do you expect the difference between X and Y to be", and Var(X – Y) tells you the variance.

Q: Why do you add the variances for X – Y? Surely you'd subtract them?

A: At first it sounds counterintuitive, but when you subtract one variable from another, you actually increase the amount of variability, and so the variance increases. The variability of subtracting a variable is actually the same as adding it.

Another way of thinking of it is that calculating the variance squares the underlying values. Var(X + bY) is equal to Var(X) + b^2Var(Y), and if b is -1, this gives us Var(X - Y). As $(-1)^2 = 1$, this means that Var(X - Y) = Var(X) + Var(Y).

Q: Can we do this if X and Y aren't independent?

A: No, these rules only apply if X and Y are independent. If you need to find the variance of X + Y where there's dependence, you'll have to calculate the probability distribution from scratch.

Q: It looks like the same rules apply for X + Y as $X_1 + X_2$. Is this correct?

A: Yes, that's right, as long as X, Y, X_1 and X_2 are all independent.

BULLET POINTS

- **Independent observations of X** are different instances of X. Each observation has the same probability distribution, but the outcomes can be different.
- If $X_1, X_2, ..., X_n$ are independent observations of X then:

  $E(X_1 + X_2 + ... + X_n) = nE(X)$
  $Var(X_1 + X_2 + ... X_n) = nVar(X)$

- If X and Y are independent random variables, then:

  E(X + Y) = E(X) + E(Y)
  E(X - Y) = E(X) - E(Y)
  Var(X + Y) = Var(X) + Var(Y)
  Var(X - Y) = Var(X) + Var(Y)

- The expectation and variance of linear transforms of X and Y are given by

  E(aX + bY) = aE(X) + bE(Y)
  E(aX - bY) = aE(X) - bE(Y)
  Var(aX + bY) = a2Var(X) + b2Var(Y)
  Var(aX - bY) = a2Var(X) + b2Var(Y)

Exercise

Below you'll see a table containing expectations and variances. Write the formula or shortcut for each one in the table. Where applicable, assume variables are independent.

Statistic	Shortcut or formula
$E(aX + b)$	
$Var(aX + b)$	
$E(X)$	
$E(f(X))$	
$Var(aX - bY)$	
$Var(X)$	
$E(aX - bY)$	
$E(X_1 + X_2 + X_3)$	
$Var(X_1 + X_2 + X_3)$	
$E(X^2)$	
$Var(aX - b)$	

Exercise

A restaurant offers two menus, one for weekdays and the other for weekends. Each menu offers four set prices, and the probability distributions for the amount someone pays is as follows:

Weekday:

x	**10**	**15**	**20**	**25**
P(X = x)	0.2	0.5	0.2	0.1

Weekend:

y	**15**	**20**	**25**	**30**
P(Y = y)	0.15	0.6	0.2	0.05

Who would you expect to pay the restaurant most: a group of 20 eating at the weekend, or a group of 25 eating on a weekday?

Exercise Solution

Below you'll see a table containing expectations and variances. Write the formula or shortcut for each one in the table. Where applicable, assume variables are independent.

Statistic	Shortcut or formula
E(aX + b)	$aE(X) + b$
Var(aX + b)	$a^2Var(X)$
E(X)	$\sum xP(X = x)$
E(f(X))	$\sum f(x)P(X = x)$
Var(aX - bY)	$a^2Var(X) + b^2Var(Y)$
Var(X)	$E(X - \mu)^2 = E(X^2) - \mu^2$
E(aX - bY)	$aE(X) - bE(Y)$
$E(X_1 + X_2 + X_3)$	$3E(X)$
$Var(X_1 + X_2 + X_3)$	$3Var(X)$
$E(X^2)$	$\sum x^2P(X = x)$
Var(aX - b)	$a^2Var(X)$

Exercise Solution

A restaurant offers two menus, one for weekdays and the other for weekends. Each menu offers four set prices, and the probability distributions for the amount someone pays is as follows:

Weekday:

x	**10**	**15**	**20**	**25**
P(X = x)	0.2	0.5	0.2	0.1

Weekend:

y	**15**	**20**	**25**	**30**
P(Y = y)	0.15	0.6	0.2	0.05

Who would you expect to pay the restaurant most: a group of 20 eating at the weekend, or a group of 25 eating on a weekday?

Let's start by finding the expectation of a weekday and a weekend. X represents someone paying on a weekday, and Y represents someone paying at the weekend.

$$E(X) = 10 \times 0.2 + 15 \times 0.5 + 20 \times 0.2 + 25 \times 0.1$$
$$= 2 + 7.5 + 4 + 2.5$$
$$= 16$$

$$E(Y) = 15 \times 0.15 + 20 \times 0.6 + 25 \times 0.2 + 30 \times 0.05$$
$$= 2.25 + 12 + 5 + 1.5$$
$$= 20.75$$

Each person eating at the restaurant is an independent observation, and to find the amount spent by each group, we multiply the expectation by the number in each group.

25 people eating on a weekday gives us $25 \times E(X) = 25 \times 16 = 400$

20 people eating at the weekend gives us $20 \times E(Y) = 20 \times 20.75 = 415$

This means we can expect 20 people eating at the weekend to pay more than 25 people eating on a weekday.

Jackpot!

You've covered a lot of ground in this chapter. You learned how to use probability distributions, expectation, and variance to predict how much you stand to win by playing a specific slot machine.

And you discovered how to use linear transforms and independent observations to anticipate how much you'll win when the payout structure changes or when you play multiple games on the same machine.

Exercise

Sam likes to eat out at two restaurants. Restaurant A is generally more expensive than restaurant B, but the food quality is generally much better.

Below you'll find two probability distributions detailing how much Sam tends to spend at each restaurant. As a general rule, what would you say is the difference in price between the two restaurants? What's the variance of this?

Restaurant A:

x	**20**	**30**	**40**	**45**
P(X = x)	0.3	0.4	0.2	0.1

Restaurant B:

y	**10**	**15**	**18**
P(Y = y)	0.2	0.6	0.2

Exercise Solution

Sam likes to eat out at two restaurants. Restaurant A is generally more expensive than restaurant B, but the food quality is generally much better.

Below you'll find two probability distributions detailing how much Sam tends to spend at each restaurant. As a general rule, what would you say is the difference in price between the two restaurants? What's the variance of this?

Restaurant A:

x	**20**	**30**	**40**	**45**
P(X = x)	0.3	0.4	0.2	0.1

Restaurant B:

y	**10**	**15**	**18**
P(Y = y)	0.2	0.6	0.2

Let's start by finding the expectation and variance of X and Y.

$$E(X) = 20\times0.3 + 30\times0.4 + 40\times0.2 + 45\times0.1$$
$$= 6 + 12 + 8 + 4.5$$
$$= 30.5$$

$$Var(X) = (20-30.5)^2\times0.3 + (30-30.5)^2\times0.4 + (40-30.5)^2\times0.2 + (45-30.5)^2\times0.1$$
$$= (-10.5)^2\times0.3 + (-0.5)^2\times0.4 + 9.5^2\times0.2 + 14.5^2\times0.1$$
$$= 110.25\times0.3 + 0.25\times0.4 + 90.25\times0.2 + 210.25\times0.1$$
$$= 33.075 + 0.1 + 18.05 + 21.025$$
$$= 72.25$$

$$E(Y) = 10\times0.2 + 15\times0.6 + 18\times0.2$$
$$= 2 + 9 + 3.6$$
$$= 14.6$$

$$Var(Y) = (10-14.6)^2\times0.2 + (15-14.6)^2\times0.6 + (18-14.6)^2\times0.2$$
$$= (-4.6)^2\times0.2 + 0.4^2\times0.6 + 3.4^2\times0.2$$
$$= 21.16\times0.2 + 0.16\times0.6 + 11.56\times0.2$$
$$= 4.232 + 0.096 + 2.312$$
$$= 6.64$$

The difference between X and Y is modeled by X – Y.

$$E(X - Y) = E(X) - E(Y)$$
$$= 30.5 - 14.6$$
$$= 15.9$$

$$Var(X - Y) = Var(X) + Var(Y)$$
$$= 72.25 + 6.64$$
$$= 78.89$$

6 permutations and combinations

Making Arrangements

If I try every permutation, sooner or later I'll get through to Tom's Tattoo Parlor.

Sometimes, order is important.

Counting **all the possible ways** in which you can order things is time consuming, but the trouble is, this sort of information is **crucial** for calculating some probabilities. In this chapter, we'll show you a **quick way** of deriving this sort of information without you having to figure out what all of the possible outcomes are. Come with us and we'll show you how to **count the possibilities**.

The Statsville Derby

One of the biggest sporting events in Statsville is the Statsville Derby. Horses and jockeys travel from far and wide to see which horse can complete the track in the shortest time, and you can place bets on the outcome of each race. There's a lot of money to be made if you can predict the top three finishers in each race.

The opening set of races is for rookies, horses that have never competed in a race before. This time, no statistics are available for previous races to help you anticipate how well each horse will do. This means you have to assume that each horse has an equal chance of winning, and it all comes down to simple probability.

The first race of the day, the three-horse race, is just about to begin, and the Derby is taking bets. You have $500 of winnings from Fat Dan's Casino to spend at the Derby. If you can correctly predict the order in which the three horses finish, the payout is 7:1, which means you'll win 7 times your bet, or $3,500.

Should we take this bet? Let's work out some probabilities and find out.

Want to join in with the fun? If you know a thing or two about probability, you could do very well indeed.

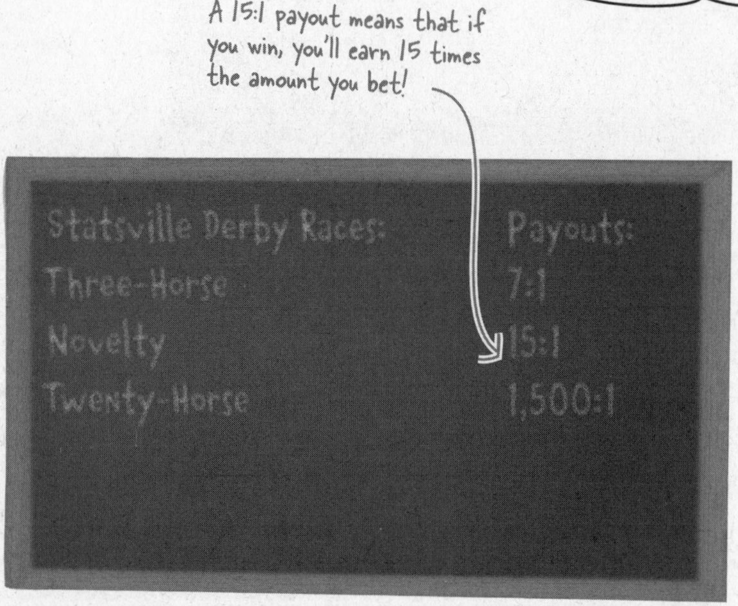

A 15:1 payout means that if you win, you'll earn 15 times the amount you bet!
Statsville Derby Races:
Payouts:
Three-Horse
7:1
Novelty
15:1
Twenty-Horse
1,500:1

It's a three-horse race

The first race is a very simple one between three horses, and in order to make the most amount of money, you need to predict the exact order in which horses finish the race. Here are the contenders.

Cheeky Sherbet

Ruby Toupee

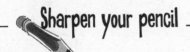

Frisky Funboy

Sharpen your pencil

How many different ways are there in which the horses can finish the race? (Assume there are no ties and that every horse finishes.) What's the probability of winning a bet on the correct finishing order?

Calculate your expected winnings for this bet.

Hint: Find the probability distribution for this event. Then use this to calculate the expectation.

Sharpen your pencil Solution

How many different ways are there in which the horses can finish the race? (Assume there are no ties and that every horse finishes.) What's the probability of winning a bet on the correct finishing order?

Calculate your expected winnings for this bet.

There are 6 different ways in which the race can be finished:

Cheeky Sherbet, Ruby Toupee, Frisky Funboy

Cheeky Sherbet, Frisky Funboy, Ruby Toupee

Ruby Toupee, Cheeky Sherbet, Frisky Funboy

Ruby Toupee, Frisky Funboy, Cheeky Sherbet

Frisky Funboy, Cheeky Sherbet, Ruby Toupee

Frisky Funboy, Ruby Toupee, Cheeky Sherbet

The probability of getting the order right is therefore 1/6.

Here's the probability distribution for amount of money you can expect to win if you bet \$500 with odds of 7:1

Three-horse race:

x	-500	3,500
P(X = x)	0.833	0.167

$E(X) = -500 \times 0.833 + 3{,}500 \times 0.167$

$= 168$

We can expect to win \$168 each time this race is won.

Yes, you can expect to win $168 on this bet, but the house is still going to win 5/6 times you play. Do you feel lucky?

A three-horse race? How likely is that? Most races will have far more horses taking part.

Exactly, most races will have more than three horses.

So what we need is some quick way of figuring out how many finishing orders there are for each race, one that works irrespective of how many horses are racing.

Working out the number of ways in which three horses can finish a race is straightforward; there are only 6 possibilities. The trouble is, the more horses there are taking part in the race, the harder and more time consuming it is to work out every possible finishing order.

Let's take a closer look at the different ways of ordering the three horses we have for the race and see if we can spot a pattern. We can do this by looking at each position, one by one.

How many ways can they cross the finish line?

Let's start by looking at the first position of the race.

One of the horses has to win the race, and this can be any one of the three horses taking part. This means that there are three ways of filling the number one position.

So what about the second position in the race?

If one of the horses has finished the race, this means there are two horses left. Either of these can come second in the race. This means that there are two ways of filling the number two position, no matter which horse came first.

2 ways
One horse has already finished the race, so there are only two horses that can finish second.

Once two horses have finished the race, there's only one position left for the final horse—third place.

1 way
Only one horse hasn't finished the race, so there's only one position left for him: last.

So how does this help us calculate all the possible finishing orders?

Calculate the number of arrangements

We just saw that there were 3 ways of filling the first position, and for each of these, there are 2 ways of filling the second position. And no matter how those first two slots are filled, there's only one way of filling the last position. In other words, the number of ways in which we can fill all three positions is:

$$3 \times 2 \times 1 = 6$$

2 ways of filling the 2nd position
3 ways of filling the 1st position
6 ways of filling all 3 positions
1 way of filling the 3rd position

This means that we can tell there are 6 different ways of ordering the three horses, without us having to figure out each of the arrangements.

So what if there are n horses?

You've seen that there are $3 \times 2 \times 1$ ways of ordering 3 horses. You can generalize this for any number n. If you want to work out the number of ways there are of ordering n separate objects, you can get the right result by calculating:

$$n \times (n - 1) \times (n - 2) \times \ldots \times 3 \times 2 \times 1$$

This means that if you have to work out the number of ways in which you can order n separate objects, you can come up with a precise figure without having to figure out every possible arrangement.

This type of calculation is called the ***factorial*** of a number. In math notation, factorials are represented as an exclamation point. For example, the factorial of 3 is written as 3!, and the factorial of n is n!. You pronounce it "n factorial."

So when we write n!, this is just a shorthand way of saying "take all the numbers from n down to 1, and multiply them together." In other words, perform the following calculation:

$$n! = n \times (n - 1) \times (n - 2) \times \ldots \times 3 \times 2 \times 1$$

The advantage of n! is that a lot of calculators have this as an available function. If, for example, you want to find the number of arrangements of 4 separate objects, all you have to do is calculate 4!, giving you $4 \times 3 \times 2 \times 1 = 24$ separate arrangements.

Going round in circles

There's one exception to this rule, and that's if you're arranging objects in a circle.

Here's an example. Imagine you want to stand four horses in a circle, and you want to find the number of possible ways in which you can order them. Now, let's focus on arrangements where Frisky Funboy has Ruby Toupee on his immediate right, and Cheeky Sherbet on his immediate left. Here are two of the four possible arrangements of this.

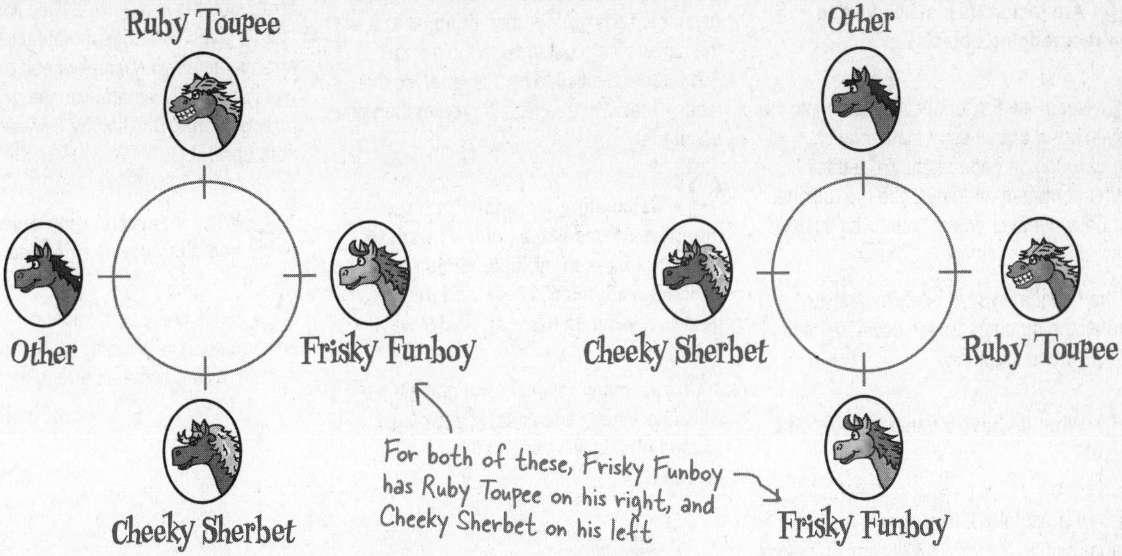
Ruby Toupee
Other
Frisky Funboy
Cheeky Sherbet
Other
Cheeky Sherbet
Ruby Toupee
Frisky Funboy
For both of these, Frisky Funboy has Ruby Toupee on his right, and Cheeky Sherbet on his left

At first glance, these two arrangements look different, but they're actually the same. The horses are in exactly the same positions relative to each other, the only difference is that in the second arrangement, the horses have walked a short distance round the circle. This means that some of the ways in which you can order the horses are actually the same.

So how do we solve this sort of problem?

The key here is to fix the position of one of the horses, say Frisky Funboy. With Frisky Funboy standing in a fixed position, you can count the number of ways in which the remaining 3 horses can be ordered, and this will give you the right result without any duplicates.

In general, if you have n objects you need to arrange in a circle, the number of possible arrangements is given by

$$(n - 1)!$$

← The number of ways of arranging n objects in a circle

there are no Dumb Questions

Q: How do I pronounce n!?

A: You pronounce it as "n factorial." The ! symbol is used to indicate a mathematical operation, and not to indicate any sort of exclamation.

Q: Are factorials just used when you're arranging objects?

A: Not at all. Factorials also come into play in other branches of mathematics, like calculus. In general, they're a useful math shorthand, and you'll see the factorial symbol whenever you're faced with this sort of multiplication task.

All the factorial symbol really means is "take all the numbers from n down to 1 and multiply them together."

Q: What if I have a value 0? How do I find 0!?

A: 0! is actually 1. This may seem like a strange result, but it's a bit like saying there's only one way to arrange 0 objects.

Q: What about if you want to find the factorial of a negative number? Or one that's not an integer?

A: Factorials only work with positive integers, so you can't find the factorial of a negative number, or one that's not an integer.

One way of looking at this is that it doesn't make sense to arrange bits of objects. Each thing you're arranging is classed as a whole object. Equally, you can't have a negative number of objects.

Q: Can the result of a factorial ever be an odd number?

A: There are only two occasions where this can be true, when n is 0 or when n is 1. In both these cases, n! = 1.

For all other values of n, n! is even. This is because if n is greater than or equal to 2, the calculation must include the number 2. Any integer multiplied by 2 is even, so this means that n! is even if n is greater than or equal to 2.

Q: Calculating factorials for large numbers seems like a pain. If I want to find 10!, I have to multiply 10 numbers (10×9×8×7×6×5×4×3×2×1), and the result gets really big. Is there an easier way.

A: Yes, many scientific and graphing calculators have a factorial key (typically labeled n!) that will perform this calculation for you.

Q: If I'm arranging n objects in a circle, there are (n - 1)! arrangements. What if clockwise and counterclockwise arrangements are considered to be the same?

A: In this case, the number of arrangements is (n - 1)!/2. Calculating (n - 1)! gives you twice the number of arrangements you actually need as it gives you both clockwise and counterclockwise arrangements. Dividing by 2 gives you the right answer.

Q: What if I'm arranging objects in a circle and absolute position matters?

A: In this case the number of arrangements is given by n!. In that situation, it's exactly the same as arranging n objects.

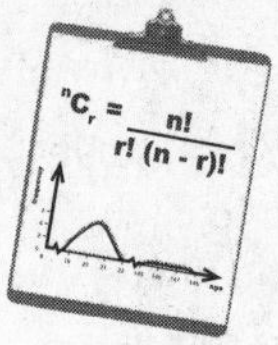

Vital Statistics

Formulas for arrangements

If you want to find the number of possible arrangements of n objects, use n! where

$n! = n \times (n-1) \times \ldots \times 3 \times 2 \times 1$

In other words, multiply together all the numbers from n down to 1.

If you are arranging n objects in a circle, then there are (n – 1)! possible arrangements.

Exercise

Paula wants to telephone the Statsville Health Club, but she has a very poor memory. She knows that the telephone number contains the numbers 1,2,3,4,5,6 and 7, but she can't remember the order. What's the probability of getting the right number at random?

Paula has just been reminded that the first three numbers is some arrangement of the numbers 1, 2 and 3, and the last four numbers is some arrangement of the numbers 4, 5, 6, and 7. She can't remember the order of each set of numbers though. What's the probability of getting the right telephone number now?

Hint: This time you need to arrange two groups of numbers.

Sharpen your pencil

The Statsville Derby is organizing a parade for the end of the season. 10 horses are taking part, and they will parade round the race track in a circle. The exact horse order will be chosen at random, and if you guess the horse order correctly, you win a prize.

What's the probability that if you make a guess on the exact horse order, you'll win the prize?

Exercise Solution

Paula wants to telephone the Statsville Health Club, but she has a very poor memory. She knows that the telephone number contains the numbers 1,2,3,4,5,6 and 7, but she can't remember the order. What's the probability of getting the right number at random?

There are 7 numbers so there are 7! possible arrangements. $7! = 7\times6\times5\times4\times3\times2\times1 = 5040$.
The probability of getting the right number is therefore $1/5040 = 0.0002$

Paula has just been reminded that the first three numbers is some arrangement of the numbers 1, 2 and 3, and the last four numbers is some arrangement of the numbers 4, 5, 6, and 7. She can't remember the order of each set of numbers though. What's the probability of getting the right telephone number now? Hint: This time you need to arrange two groups of numbers.

We start by splitting the numbers into two groups, one for the first three numbers (1, 2, 3), and another for the last four (4, 5, 6, 7). This gives us

Number of ways of arranging 1, 2, 3 is $3! = 3\times2\times1 = 6$
Number of ways of arranging 4, 5, 6, 7 is $4! = 4\times3\times2\times1 = 24$

To find the total number of possible arrangements, we multiply together the number of ways of arranging each group. This gives

Total number of possible arrangements is $3!\times4! = 6\times24 = 144$

The probability of getting the right number is therefore $1/144 = 0.0069$

Sharpen your pencil Solution

The Statsville Derby are organizing a parade for the end of the season. 10 horses are taking part, and they will parade round the race track in a circle. The exact horse order will be chosen at random, and if you guess the horse order correctly, you win a prize.

What's the probability that if you make a guess on the exact horse order, you'll win the prize?

10 horses will be parading in a circle, which means there are 9! possible orders for the horses.
$9! = 362880$, which means there are 362880 possible orders for the parade.

The probability of guessing correctly is $1/9!$ – which is a number very close to 0.

It's time for the novelty race

The Statsville Derby is unusual in that not all of the animals taking part in the races have to be horses. In the next race, three of the contenders are zebras, and they're racing against three horses.

In this race, it's the *type* of animal that matters rather than the particular animal itself. In other words, all we're interested in is which sort of animal finishes the race in which position. The question is, how many ways are there of ordering all the animals by species?

The Derby's offering a special bet: if you can predict whether a horse or zebra will finish in each place, the payout is 15:1. The question is, should you make this bet?

In the last race, you had a 1/6 probability of predicting the top finishers correctly. But let's see how you fare in the novelty race; it's a Statsville tradition.

BRAIN POWER

How would you go about solving this sort of problem? Write down your ideas in the space below.

Arranging by individuals is different than arranging by type

So if there are three horses and three zebras in today's novelty race, how can we calculate how many different orderings there are of horses and zebra.

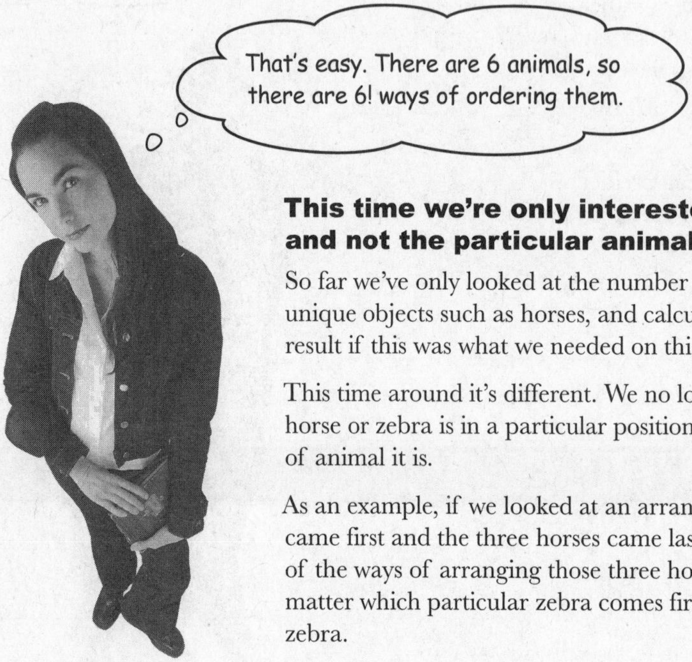

This time we're only interested in the *type* of animal, and not the particular animal itself.

So far we've only looked at the number of ways in which we can order unique objects such as horses, and calculating 6! would be the correct result if this was what we needed on this occasion.

This time around it's different. We no longer care about which particular horse or zebra is in a particular position; we only care about what type of animal it is.

As an example, if we looked at an arrangement where the three zebras came first and the three horses came last, we wouldn't want to count all of the ways of arranging those three horses and three zebras. It doesn't matter which particular zebra comes first; it's enough to know it's a zebra.

I'll knock the stripes off those zebras.
For this sort of problem, we care about which type of animal is in which position, but we don't care about the name of the animal itself.

We need to arrange animals by type

There are 6! ways of ordering the 6 animals, but the problem with this result is that it assumes we want to know all possible arrangements of individual horses and zebras.

Let's start by looking at the zebras. There are 3! ways of arranging the three zebras, and the result 6! includes each of these 3! arrangements. But since we're not concerned about which individual zebra goes where, these arrangements are all the same. So, to eliminate these repetitions, we can just divide the total number of arrangements by 3!

We're classing the 3 zebras as being alike, so we divide the number of arrangements by 3!

Next, let's take the horses. There are 3! ways of arranging the three horses, and the number of arrangements we have so far includes each of these 3! arrangements. As with the zebras, we divide the end result by 3! to eliminate duplicate orderings.

This time we're classing the 3 horses as being alike. There are 3! ways of arranging the horses, so we divide the total number of arrangements by 3!

This means that the number of ways of arranging the 6 animals according to species is

$$\frac{6!}{3!3!} = \frac{720}{6 \times 6} = \frac{720}{36} = 20$$

There are 6! animals altogether...

...but the 3 zebras and 3 horses are alike, so we divide by the number of ways we can arrange these like animals

In other words, the probability of betting correctly on the right order in which the different species finish the race is 1/20.

Turn the page and we'll look at this in more detail.

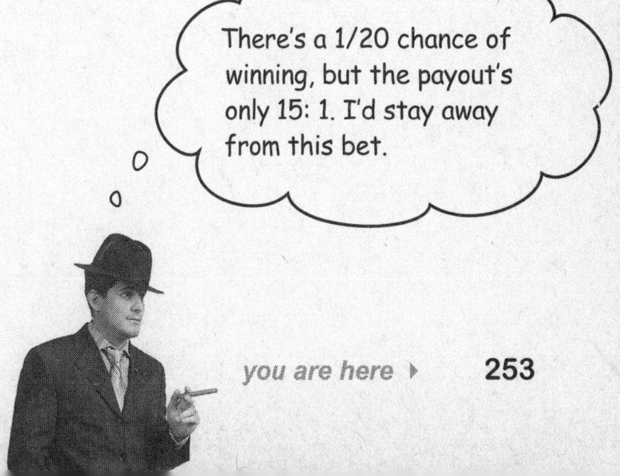
There's a 1/20 chance of winning, but the payout's only 15: 1. I'd stay away from this bet.

Generalize a formula for arranging duplicates

Imagine you need to count the number of ways in which n objects can be arranged. Then imagine that k of the objects are alike.

To find the number of arrangements, start off by calculating the number of arrangements for the n objects as if they were all unique. Then divide by the number of ways in which the k objects (the ones that are alike) can be arranged. This gives you:

$$\frac{n!}{k!}$$

There are n objects in total. (→ n!)

k of the objects are alike. (→ k!)

If you have n objects where k are alike the number of arrangements is given by n!/k!

We can take this further.

Imagine you want to arrange *n* objects, where *k* of one type are alike, and *j* of another type are alike, too. You can find the number of possible arrangements by calculating:

$$\frac{n!}{j!k!}$$

There are n objects in total. (→ n!)

j of one type of object are alike, and so are k of another type. (→ j!k!)

The number of ways of arranging n objects where j of one type are alike, and so are k of another type.

In general, when calculating arrangements that include duplicate objects, divide the total number of arrangements (n!) by the number of arrangements of each set of alike objects (j!, k!, and so on).

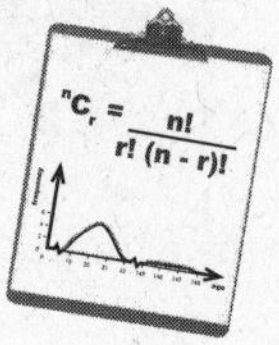

Vital Statistics

Arranging by type

If you want to arrange n objects where j of one type are alike, k of another type are alike, so are m of another type and so on, the number of arrangements is given by

$$\frac{n!}{j!k!m!\ldots}$$

Exercise

The Statsville Derby have decided to experiment with their races. They've decided to hold a race between 3 horses, 2 zebras and 5 camels, where all the animals are equally likely to finish the race first.

1. How many ways are there of finishing the race if we're interested in individual animals?

2. How many ways are there of finishing the race if we're just interested in the species of animal in each position?

3. What's the probability that all 5 camels finish the race consecutively if each animal has an equal chance of winning? (Assume we're interested in the species in each position, not the individual animals themselves.)

Exercise Solution

The Statsville Derby have decided to experiment with their races. They've decided to hold a race between 3 horses, 2 zebras and 5 camels, where all the animals are equally likely to finish the race first.

1. How many ways are there of finishing the race if we're interested in individual animals?

There are 10 animals so there are $10! = 3,628,800$ different arrangements.

2. How many ways are there of finishing the race if we're just interested in the species of animal in each position?

There are 3 horses, 2 zebras and 5 camels.

$$\text{Number of arrangements} = \frac{10!}{3!2!5!}$$

← There are 10 animals.

← We treat the 3 horses as being alike, and the 2 zebras, and also the 5 camels.

$$= \frac{3,628,800}{6 \times 2 \times 120}$$

$$= \frac{3,628,800}{1,440}$$

$$= 252$$

3. What's the probability that all 5 camels finish the race consecutively if each animal has an equal chance of winning? (Assume we're interested in the species in each position, not the individual animals themselves.)

First of all, let's find the number of ways in which the 5 camels can finish the race together. To do this, we class the 5 camels as one single object. That way, we're guaranteed to keep them together. This means that if we add our 1 group of camels to the 3 horses and 2 zebras, we actually need to arrange 6 objects

$$\text{Number of arrangements} = \frac{6!}{3!2!}$$

← 1 group of camels + 3 horses + 2 zebras

← We treat the 3 horses as being alike, and the 2 zebras. We don't need to divide by 5! for the 5 camels, as we're counting them as 1 object.

$$= \frac{720}{6 \times 2}$$

$$= \frac{720}{12}$$

$$= 60$$

Then, to find the probability of this occurring, we just need to divide the number of ways the camels finish together by all the possible ways the animal types can finish the race, which we calculated above.

The probability of all 5 camels finishing together is therefore $60/252 = 5/21$

there are no Dumb Questions

Q: Why did you treat the 5 camels as one object in the last part of the exercise? Surely they're individual camels.

A: They're individual camels, but in the last part of that problem we need to make sure we keep the camels together. To do this, we bundle all the camels together and treat them as one object.

Q: It seems like the number of arrangements for the different objects has a lot to do with how you group them into like groups.

A: That's right. Mastering arrangements is a skill, but a lot depends on how you think things through.

The key thing is to think really carefully about what sort of problem you're actually trying to solve and to get lots of practice.

Q: Are there many races where horses, zebras and camels all race together?

A: It's unlikely. But hey, this is Statsville, and the Statsville Derby runs its own events.

It's time for the twenty-horse race

The novelty race is over, with the zebras taking the lead.

The next race is between 20 horses.

Think you can predict
the top three horses?
If you can, the payout's
a whopping 1,500 to 1.

BRAIN POWER

How would you go about finding the number of ways in which you can pick three horses out of twenty?

How many ways can we fill the top three positions?

The main race is about to begin. There are twenty horses racing, and we need to find the number of possible arrangements of the top three horses. This way, we can work out the probability of guessing the exact order correctly.

We can work out the solution the same way we did earlier, by looking at how many ways there are of filling the first three positions.

Let's start with the first position. There are 20 horses in total, so this means there are 20 different ways of filling the first position. Once this position has been filled, that leaves 19 ways of filling the second position and 18 ways of filling the third.

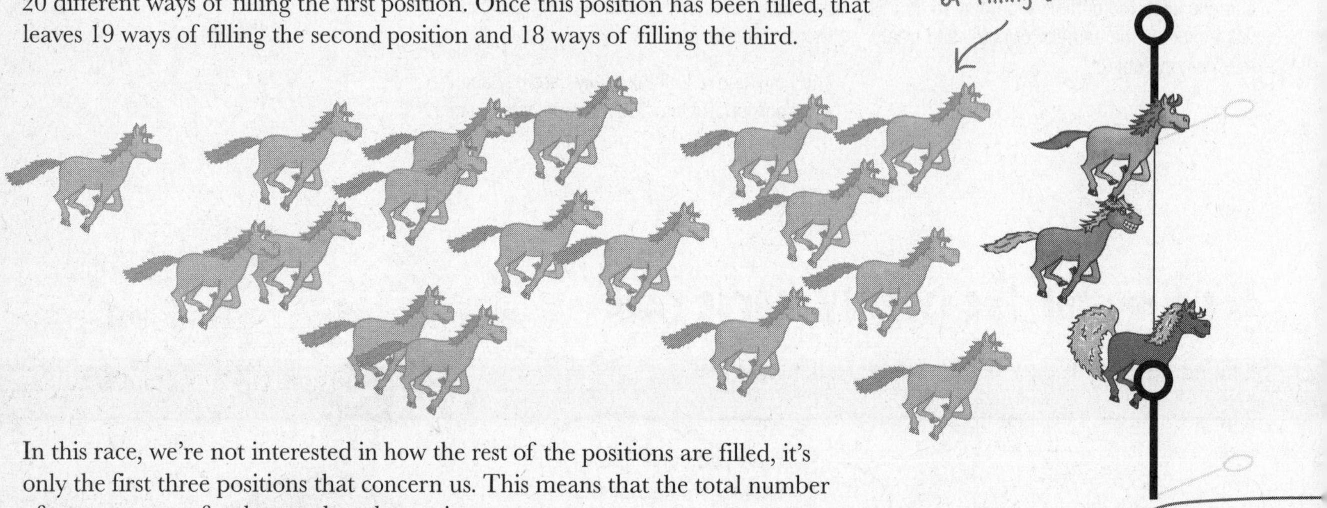

In this race, we're not interested in how the rest of the positions are filled, it's only the first three positions that concern us. This means that the total number of arrangements for the top three horses is

$$20 \times 19 \times 18 = 6,840$$

So the probability of guessing the precise order in which the top three horses finish the race is 1/6,840.

That gives us the right answer, but it could get complicated if there were more horses, or if we wanted to fill more positions.

We need a more concise way of solving this sort of problem.

At the moment we only have three numbers to multiply together, but what if there were more?

We need to generalize a formula that will allow us to find the total number of arrangements of a certain number of horses, drawn from a larger pool of horses.

Examining permutations

So how can we rewrite the calculation in terms of factorials?

The number of arrangements is 20 × 19 × 18. Let's rewrite it and see where it gets us.

$$20 \times 19 \times 18 = \frac{20 \times 19 \times 18 \times (17 \times 16 \times ... \times 3 \times 2 \times 1)}{(17 \times 16 \times ... \times 3 \times 2 \times 1)}$$

If we multiply it by 17!/17!, this will still give us the same answer.

$$= \frac{20!}{17!}$$

This is the same expression written in terms of factorials.

This is the same expression that we had before, but this time written in terms of factorials.

The number of arrangements of 3 objects taken from 20 is called the number of ***permutations***. As you've seen, this is calculated using

$$\frac{20!}{(20 - 3)!}$$

$$= \frac{2,432,902,008,176,640,000}{355,687,428,096,000}$$

$$= 6,840$$

This is the same answer we got earlier

Permutations give the total number of ways you can order a certain number of objects (r), drawn from a larger pool of objects (n).

In general, the number of permutations of r objects taken from n is the number of possible way in which each set of r objects can be ordered. It's generally written ^nP_r, where

$$^nP_r = \frac{n!}{(n - r)!}$$

This is the total number of objects (n)

This is the number of positions we want to fill (r)

So if you want to know how many ways there are of ordering r objects taken from a pool of n, permutations are the key.

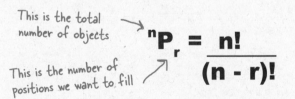

I never said anything about the horse order. Just guess which horses are in the top three and I'll make it worth your while...

What if horse order doesn't matter

So far we've found the number of permutations of ordering three horses taken from a group of twenty. This means that we know how many exact arrangements we can make.

This time around, we don't want to know how many different permutations there are. We want to know the number of ***combinations*** of the top three horses instead. We still want to know how many ways there are of filling the top three positions, but this time the exact arrangement doesn't matter.

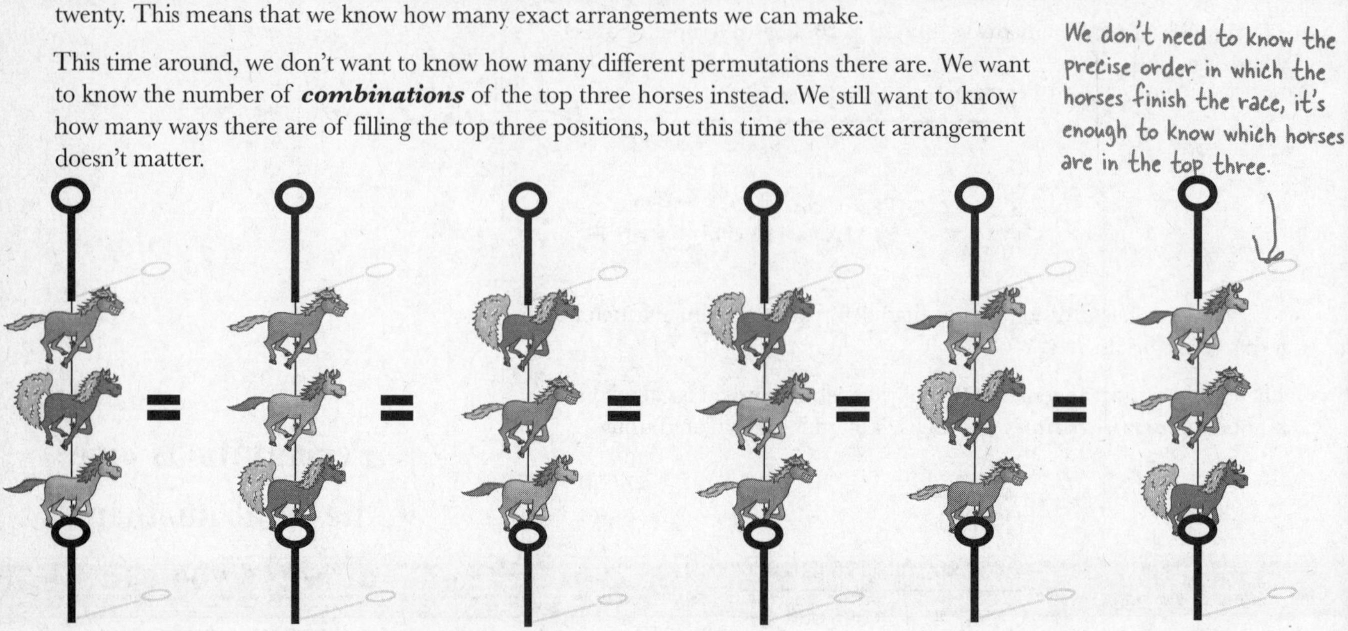

So how can we solve this sort of problem?

At the moment, the number of permutations includes the number of ways of arranging the 3 horses that are in the top three. There are 3! ways of arranging each set of 3 horses, so let's divide the number of permutations by 3!. This will give us the number of ways in which the top three positions can be filled but *without* the exact order mattering.

The result is

$$\frac{20!}{3!17!} = \frac{6,840}{3!}$$

$$= 1,140$$

This means that there are 6,840 permutations for filling the first three places in the race, but if you're not concerned about the order, there are 1,140 combinations.

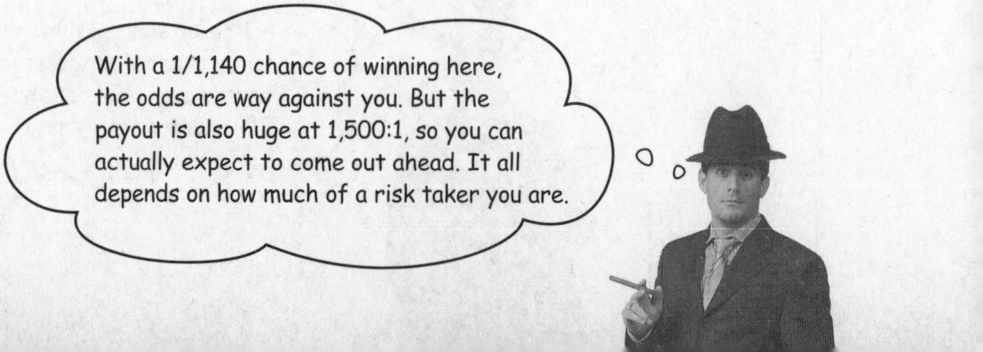

With a 1/1,140 chance of winning here, the odds are way against you. But the payout is also huge at 1,500:1, so you can actually expect to come out ahead. It all depends on how much of a risk taker you are.

Examining combinations

Earlier on we found a general way of calculating permutations. Well, there's a way of doing this for combinations too.

In general, the number of combinations is the number of ways of choosing r objects from n, without needing to know the exact order of the objects. The number of combinations is written ^nC_r, where

$$^nC_r = \frac{n!}{r!\,(n-r)!}$$

This is the total number of objects. (n)

This is the number of positions we want to fill. (r)

This bit is calculated in the same way as a permutation. (n!)

You divide by an extra r! if it's a combination.

So what's the difference between a combination and a permutation?

Permutations	Combinations
A ***permutation*** is the number of ways in which you can choose objects from a pool, and where the order in which you choose them counts. It's a lot more specific than a combination as you want to count the number of ways in which you fill each position.	A ***combination*** is the number of ways in which you can choose objects from a pool, without caring about the exact order in which you choose them. It's a lot more general than a permutation as you don't need to know how each position has been filled. It's enough to know which objects have been chosen.

Permutation: order matters.

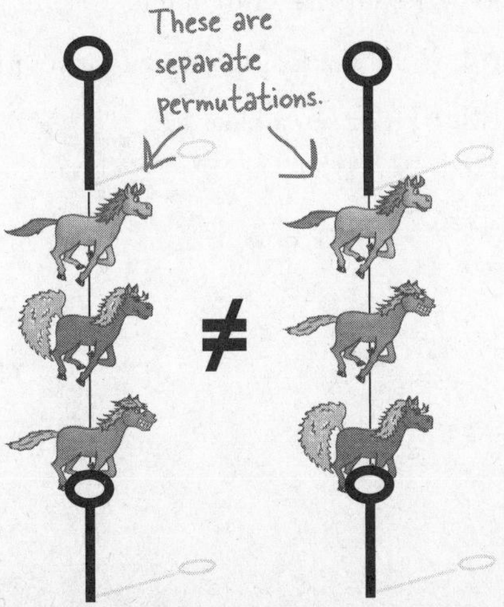

These are separate permutations.
≠

Combination: order doesn't matter.

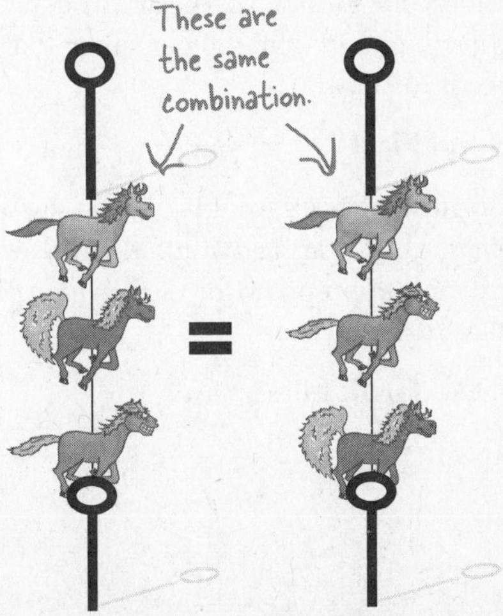

These are the same combination.
=

Combination Exposed

This week's interview:
Does order really matter?

Head First: Combination, great to have you on the show.

Combination: Thanks for inviting me, Head First.

Head First: Now, let's get straight onto business. A lot of people have noticed that you and Permutation are very similar to each other. Is that something you'd agree with?

Combination: I can see why people might think that because we deal with very similar situations. We're both very much concerned with choosing a certain number of objects from a pool. Having said that, I'd say that's where the similarity ends.

Head First: So what makes you different?

Combination: Well, for starters we both have very different attitudes. Permutation is very concerned about order, and really cares about the exact order in which objects are picked. Not only does he want to select objects, he wants to arrange them too. I mean, come on!

Head First: I take it you don't?

Combination: No way! I'm sure permutation shows a lot of dedication and all that, but quite frankly, life's too short. As far as I'm concerned, if an object's picked from the pool, then that's all anyone needs to know.

Head First: So are you better than permutation?

Combination: I wouldn't like to say that either one of us is better as such; it just depends which of us is the most appropriate for the situation. Take music players, for instance.

Head First: Music players?

Combination: Yes. Lots of music players have playlists where you can choose which songs you want to play.

Head First: I think I see where you're headed...

Combination: Now, both Permutation and I are both interested in what's on the playlist, but in different ways. I'm happy just knowing what songs are on it, but Permutation takes it way further. He doesn't just want to know what songs are on the playlist, he wants to know the exact order too. Change the order of the songs, and it's the same Combination, but a different Permutation.

Head First: Tell me a bit about your calculation. Is calculating a Combination similar to how you'd calculate a Permutation?

Combination: It's similar, but there's a slight difference. With Permutation, you find n!, and then divide it by (n-r)!. My calculation is similar, except that you divide by an extra r!. This makes me generally smaller—which makes sense because I'm not as fussy as Permutation.

Head First: Generally smaller?

Combination: I'll phrase that differently. Permutation is never smaller than me.

Head First: Combination, thank you for your time.

Combination: It's been a pleasure.

there are no Dumb Questions

Q: I've heard of something called "choose." What's that?

A: It's another term for the combination. ^nC_r basically means "you have n objects, choose r," so it's sometimes called the choose function.

Q: Can a permutation ever be smaller than a combination?

A: Never. To calculate a combination, you divide by an extra number, so the end result is smaller.

The closest you get to this is when a permutation and combination are identical. This is only ever the case when you're choosing 0 objects or 1.

Q: Which is a permutation and which is a combination? I get confused.

A: A **permutation** is when you care about the number of possible arrangements of the objects you've chosen. A **combination** is when you don't mind about their precise order; it's enough that you've chosen them.

Q: I get confused. If I want to find the number of combinations of choosing r objects from n, do I write that ^nC_r or ^rC_n?

A: It's ^nC_r. One way of remembering this is that the higher of the two numbers is higher up in the shorthand.

Q: Are there other ways of writing this? I think I've seen combinations somewhere else, but they didn't look like that.

A: There are different ways of writing combinations. We've used the shorthand ^nC_r, but an alternative is

$$\binom{n}{r}$$

Q: Are permutations and combinations really important?

A: They are, particularly combinations. You'll see more of these a bit later on in the book, so look out for when you might need them.

Q: Dealing with permutations and combinations looks similar to when you're dealing with like objects. Is that right?

A: It's a similar process. When you're dealing with like objects, you divide the total number of arrangements by the number of ways in which you can divide the like objects.

For permutations, it's as though you're treating all the objects you don't choose as being alike, so you divide n! by (n-r)!. For combinations, it's as though the objects you pick are alike, too. This means you divide the number of permutations by r!.

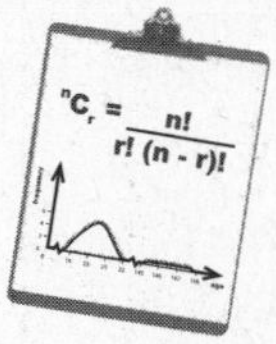

Vital Statistics

Permutations

If you choose r objects from a pool of n, the number of permutations is given by

$$^nP_r = \frac{n!}{(n-r)!}$$

Combinations

If you choose r objects from a pool of n, the number of combinations is given by

$$^nC_r = \frac{n!}{r!(n-r)!}$$

Exercise

The Statsville All Stars are due to play a basketball match. There are 12 players in the roster, and 5 are allowed on the court at any one time.

1. How many different arrangements are there for choosing who's on the court at the same time?

2. The coach classes 3 of the players as expert shooters. What's the probability that all 3 of these players will be on the court at the same time, if they're chosen at random?

Exercise

It's time for you to work out some poker probabilities. See how you get on.

A poker hand consists of 5 cards and there are 52 cards in a pack. How many different arrangements are there?

A royal flush is a hand that consists of a 10, Jack, Queen, King and Ace, all of the same suit. What's the probability of getting this combination of cards? Use your answer above to help you.

Four of a kind is when you have four cards of the same denomination. Any extra card makes up the hand. What's the probability of getting this combination?

A flush is where all 5 cards belong to the same suit. What's the probability of getting this?

Exercise Solution

The Statsville All Stars are due to play a basketball match. There are 12 players in the roster, and 5 are allowed on the court at any one time.

1. How many different arrangements are there for choosing who's on the court at the same time?

There are 12 players in the roster, and we need to count the number of ways of choosing 5 of them. We don't need to consider the order in which we pick the players, so we can work this out using combinations.

$$^{12}C_5 = \frac{12!}{5!(12-5)!}$$

$$= \frac{12!}{5!7!}$$

$$= 792$$

2. The coach classes 3 of the players as expert shooters. What's the probability that all 3 of these players will be on the court at the same time, if they're chosen at random?

Let's start by finding the number of ways in which the three shooters can be on the court at the same time.

If the three expert shooters are on the court at the same time, this means that there are 2 more places left for the other players. We need to find the number of combinations of filling these 2 places from the remaining 9 players.

$$^{9}C_2 = \frac{9!}{2!(9-2)!}$$

$$= \frac{9!}{2!7!}$$

$$= 36$$

This means that the probability of all 3 shooters being on the court at the same time is 36/792 = 1/22.

Exercise Solution

It's time for you to work out some poker probabilities. See how you get on.

A poker hand consists of 5 cards and there are 52 cards in a pack. How many different arrangements are there?

There are 52 cards in a pack, and we want to choose 5.

$$^{52}C_5 = \frac{52!}{47!5!} = 2,598,960$$

A royal flush is a hand that consists of a 10, Jack, Queen, King and Ace, all of the same suit. What's the probability of getting this combination of cards? Use your answer above to help you.

There's one way of getting this combination for each suit, and there are 4 suits. This means that the number of ways of getting a royal flush is 4.

$$P(\text{Royal Flush}) = \frac{4}{2,598,960}$$

$$= 1/649,740$$

$$= 0.0000015$$

Four of a kind is when you have four cards of the same denomination. Any extra card makes up the hand. What's the probability of getting this combination?

Let's start with the 4 cards of the same denomination. There are 13 denominations in total, which means there are 13 ways of combining these 4 cards. Once these 4 cards have been chosen, there are 48 cards left. This means the number of ways of getting this hand is $13 \times 48 = 624$.

$$P(\text{Four of a Kind}) = \frac{624}{2,598,960}$$

$$= 1/4165$$

$$= 0.00024$$

A flush is where all 5 cards belong to the same suit. What's the probability of getting this?.

To find the number of possible combinations, find the number of ways of choosing a suit, and then choose 5 cards from the suit. There are 13 cards in each suit. This means the number of combinations is

$$4 \times {}^{13}C_5 = 4 \times \frac{13!}{8!5!}$$

$$= 4 \times 1287 = 5148$$

$$P(\text{Flush}) = \frac{5148}{2,598,960}$$

$$= 33/16660$$

$$= 0.00198$$

It's the end of the race

The race between the twenty horses is over, and the overall winner is Ruby Toupee, followed by Cheeky Sherbet and Frisky Funboy. If you decided to bet on these three horses, you just won big!

Winner of this year's Statsville Derby: Ruby Toupee

2nd place: Cheeky Sherbet

3rd place: Frisky Funboy

In this chapter, you've learned how to cope with different arrangements, and how to quickly count the number of possible combinations and permutations *without* having to work out each and every possibility.

The sort of knowledge you've gained gives you enormous probability and statistical power. Keep reading, and we'll show you how to gain even greater mastery.

7 geometric, binomial, and poisson distributions

Keeping Things Discrete

Calculating probability distributions takes time.

So far we've looked at how to calculate and use probability distributions, but wouldn't it be nice to have something **easier to work with**, or just **quicker to calculate**? In this chapter, we'll show you some **special probability distributions** that follow very definite patterns. Once you know these patterns, you'll be able to use them to **calculate probabilities, expectations, and variances in record time**. Read on, and we'll introduce you to the geometric, binomial and Poisson distributions.

Meet Chad, the hapless snowboarder

Chad likes to snowboard, but he's accident-prone. If there's a lone tree on the slopes, you can guarantee it will be right in his path. Chad wishes he didn't keep hitting trees and falling over; his insurance is costing him a fortune.

Chad

Chad's about here—just follow the tree damage to see how well his first run went.
Ouch! Rock! Ouch! Flag! Ouch! Tree!

There's a lot riding on Chad's performance on the slopes: his ego, his success with the ski bunnies on the trail, his insurance premiums. If it's likely he'll make it down the slopes in less than 10 tries, he's willing to risk embarrassment, broken bones, and a high insurance deductible to try out some new snowboarding tricks.

The probability of Chad making a clear run down the slope is 0.2, and he's going to keep on trying until he succeeds. After he's made his first successful run down the slopes, he's going to stop snowboarding, and head back to the lodge triumphantly.

Sharpen your pencil

It's time to exercise your probability skills. The probability of Chad making a successful run down the slopes is 0.2 for any given trial (assume trials are independent). What's the probability he'll need two trials? What's the probability he'll make a successful run down the slope in one **or** two trials? Remember, when he's had his first successful run, he's going to stop.

Chad is remarkably resilient, and any collisions in a given run don't affect his performance in future trials.

Hint: You may want to draw a probability tree to help visualize the problem.

Sharpen your pencil Solution

It's time to exercise your probability skills. The probability of Chad making a successful run down the slopes is 0.2 for any given trial (assume trials are independent). What's the probability he'll need two trials? What's the probability he'll make a successful run down the slope in one **or** two trials? Remember, when he's had his first successful run, he's going to stop..

Here's a probability tree for the first two trials, as these are all that's needed to work out the probabilities.

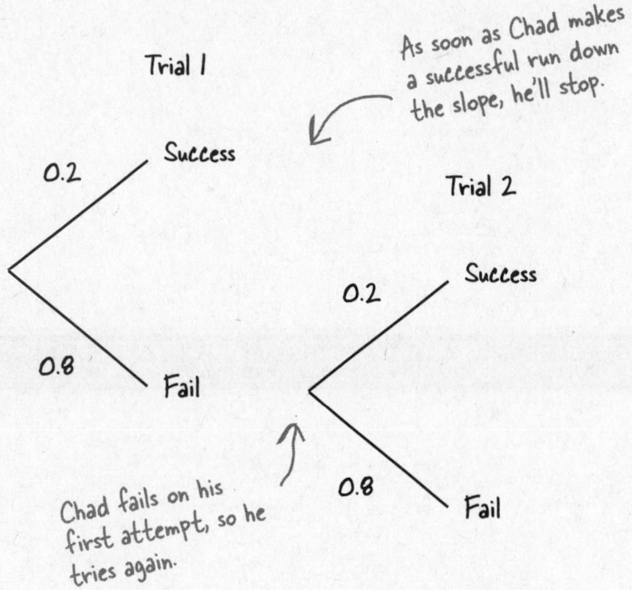

Trial 1
Trial 2
0.2
Success
0.8
Fail
0.2
Success
0.8
Fail
As soon as Chad makes a successful run down the slope, he'll stop.
Chad fails on his first attempt, so he tries again.

If we say X is the number of trials needed to get down the slopes, then

$P(X = 1) = P(\text{Success in trial 1})$

$= 0.2$

$P(X = 2) = P(\text{Success in trial 2} \cap \text{Failure in trial 1})$

$= 0.2 \times 0.8$

$= 0.16$

$P(X \leq 2) = P(X = 1) + P(X = 2)$ ← We can add these probabilities because they're independent.

$= 0.2 + 0.16$

$= 0.36$

We need to find Chad's probability distribution

So far you've found the probability that Chad will need fewer than three attempts to make it down the slope. But what if you needed to look at the probability of him needing fewer than 10 attempts (for insurance reasons), or even 20 or 100?

Rather than work out the probabilities from scratch every time, it would be useful if we could use a probability distribution. To do this, we need to work out the probability for every single possible number of attempts Chad needs to get down the slope.

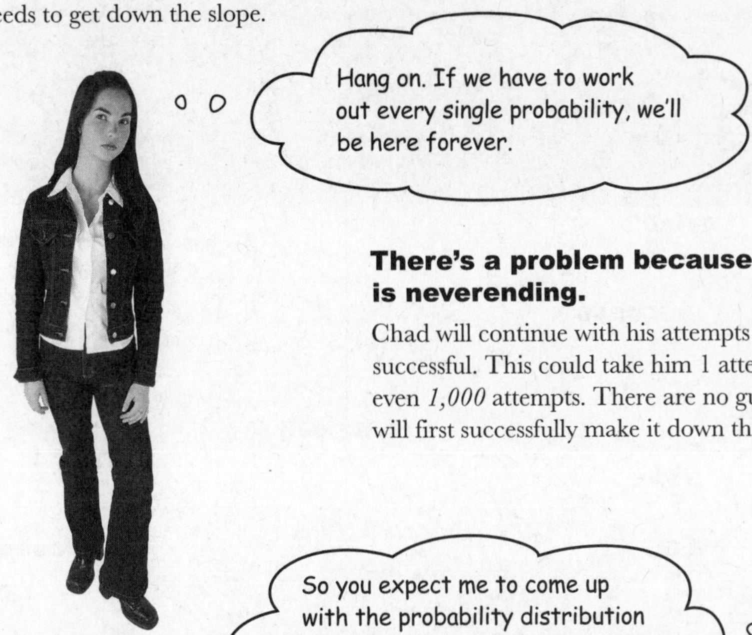

Hang on. If we have to work out every single probability, we'll be here forever.

There's a problem because the number of possibilities is neverending.

Chad will continue with his attempts to make it down the slope until he is successful. This could take him 1 attempt, 10 attempts, 100 attempts, or even *1,000* attempts. There are no guarantees about exactly when Chad will first successfully make it down the slopes.

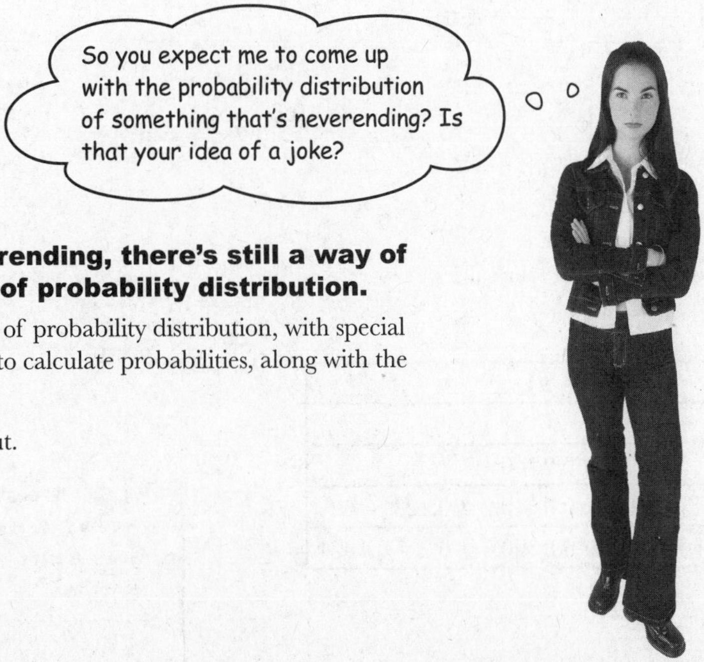

So you expect me to come up with the probability distribution of something that's neverending? Is that your idea of a joke?

Even though it's neverending, there's still a way of figuring out this type of probability distribution.

This is actually a special kind of probability distribution, with special properties that makes it easy to calculate probabilities, along with the expectation and variance.

Let's see if we can figure it out.

There's a pattern to this probability distribution

Let's define the variable X to be the number of trials needed for Chad to make a successful run down the slope. Chad only needs to make one successful run, and then he'll stop.

Let's start off by examining the first four trials so that we can calculate probabilities for the first four values of X. By doing this, we can see if there's some sort of pattern that will help us to easily work out the probabilities of other values.

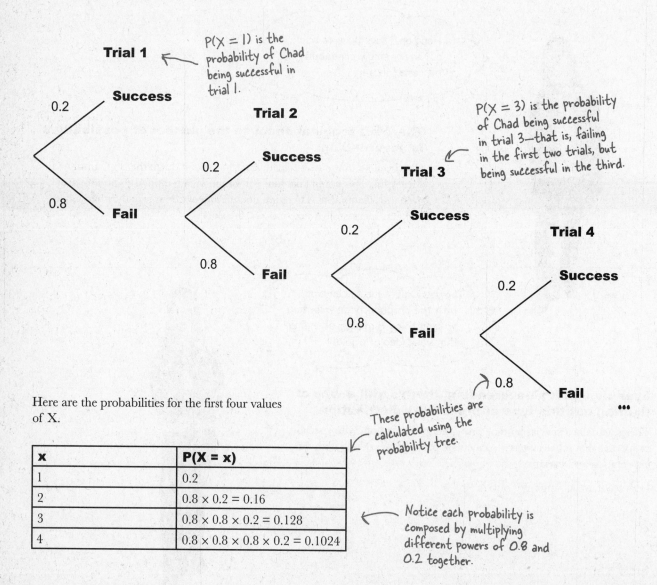

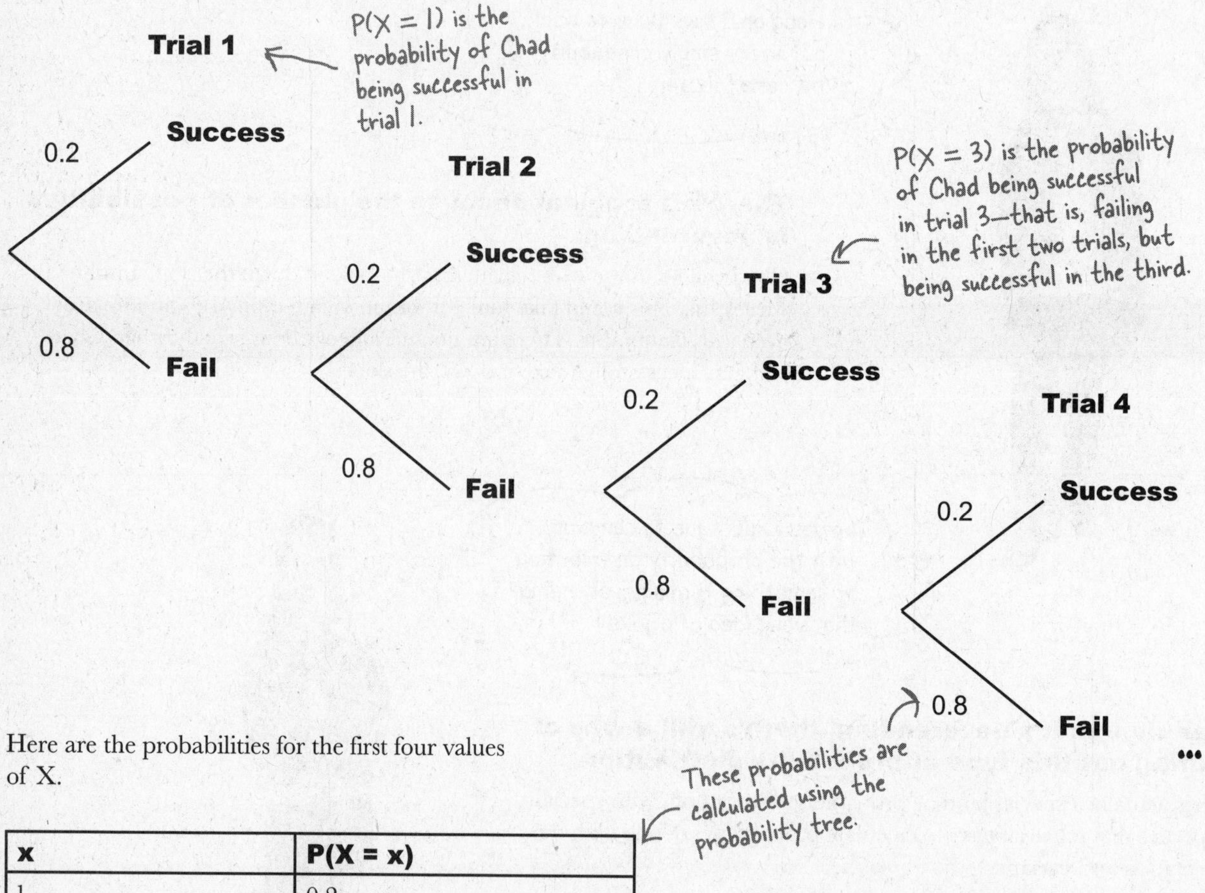

Trial 1
P(X = 1) is the probability of Chad being successful in trial 1.
Success
0.2
0.8
Fail
Trial 2
Success
0.2
0.8
Fail
Trial 3
P(X = 3) is the probability of Chad being successful in trial 3—that is, failing in the first two trials, but being successful in the third.
Success
0.2
0.8
Fail
Trial 4
Success
0.2
0.8
Fail
...
These probabilities are calculated using the probability tree.

Here are the probabilities for the first four values of X.

x	P(X = x)
1	0.2
2	0.8 × 0.2 = 0.16
3	0.8 × 0.8 × 0.2 = 0.128
4	0.8 × 0.8 × 0.8 × 0.2 = 0.1024

Notice each probability is composed by multiplying different powers of 0.8 and 0.2 together.

Here's a table containing the the probabilities of X for different values. Complete the table, filling out the probability that there will be *x* number of trials, and indicating what the power of 0.8 and 0.2 are in each case (the number of times.0.8 and 0.2 appear in P(X = x)).

x	P(X = x)	Power of 0.8	Power of 0.2
1	0.2	0	1
2	0.8 × 0.2	1	1
3	$0.8^2 \times 0.2$	2	
4			
5			
r			

r is a particular value of x but we're not saying which one. Can you guess what the probability will be in terms of r?

We've left extra space here for your calculations.

Exercise Solution

Here's a table containing the the probabilities of X for different values. Complete the table, filling out the probability that there will be *x* number of trials, and indicating what the power of 0.8 and 0.2 are in each case (the number of times.0.8 and 0.2 appear in P(X = x)).

x	P(X = x)	Power of 0.8	Power of 0.2
1	0.2	0	1
2	0.8×0.2	1	1
3	$0.8^2 \times 0.2$	2	1
4	$0.8^3 \times 0.2$	3	1
5	$0.8^4 \times 0.2$	4	1
r	$0.8^{r-1} \times 0.2$	r – 1	1

For X = 4, Chad fails three times and succeeds on his fourth attempt.
P(X = 4) is therefore $0.8 \times 0.8 \times 0.8 \times 0.2$, as the probability of failing on a particular run is 0.8 and the probability of success is 0.2.

For X = 5, Chad fails on his first four attempts. and succeeds on his fifth. This means
$P(X = 5) = 0.8 \times 0.8 \times 0.8 \times 0.8 \times 0.2$.

So what if P(X = r)? For Chad to be successful on his r'th attempt, he must have failed in his first (r–1) attempts, before succeeding in his r'th. Therefore
$P(X = r) = 0.8 \times 0.8 \times \ldots \times 0.8 \times 0.2$, which means that in our expression, 0.8 is taken to the (r–1)th power.

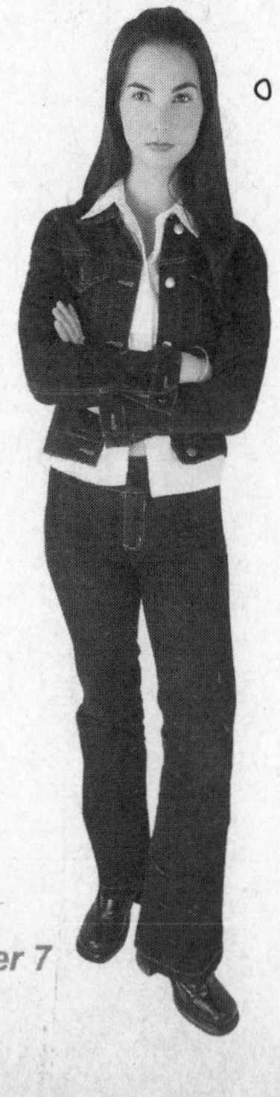

First you say P(X = x), then you say P(X = r). I wish you'd make your mind up.

They refer to two different things.

When we use P(X = x), we're using it to demonstrate *x* taking on any value in the probability distribution. In the table above, we show various values of *x*, and we calculate the probability of getting each of these values.

When we use P(X = r), *x* takes on the particular value *r*. We're looking for the probability of getting this specific value. It's just that we haven't specified what the value of *r* is so that we can come up with a generalized calculation for the probability.

It's a bit like saying that *x* can take on any value, including the fixed value *r*.

The probability distribution can be represented algebraically

As you can see, the probabilities of Chad's snowboarding trials follow a particular pattern. Each probability consists of multiples of 0.8 and 0.2. You can quickly work out the probabilities for any value *r* by using:

$$P(X = r) = 0.8^{r-1} \times 0.2$$

In other words, if you want to find P(X = 100), you don't have to draw an enormous probability tree to work out the probability, or think your way through exactly what happens in every trial. Instead, you can use:

$$P(X = 100) = 0.8^{99} \times 0.2$$

We can generalize this even further. If the probability of success in a trial is represented by ***p*** and the probability of failure is ***1 - p***, which we'll call ***q***, we can work out any probability of this nature by using:

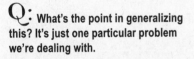

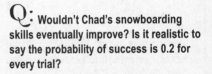

$$P(X = r) = q^{r-1}p$$

(r – 1) failures and 1 success. In our case, p = 0.2 and q = 0.8.

q is equal to 1 – p. If p represents the probability of success, then q represents the probability of failure.

This formula is called the ***geometric distribution***.

there are no Dumb Questions

Q: What's the point in generalizing this? It's just one particular problem we're dealing with.

A: We're generalizing it so that we can apply the results to other similar problems. If we can generalize the results for this kind of problem, it will be quicker to use it for other similar situations in the future.

Q: You said we needed to find an expression for P(X = r). What's r?

A: P(X = r) means "the probability that X is equal to value r," where r is the number of trials we need to get the first success.

If you wanted to find, say, P(X = 20), you could substitute r for 20. This would give you a quick way of finding the probability.

Q: Why is it the letter r? Why not some other letter?

A: We used the letter r so that we could generalize the result for any particular number. We could have used practically any other letter, but using r is common.

Q: How can we have a probability distribution if the number of possibilities is endless?

A: We don't *have* to specify a probability distribution by physically listing the probability of every possible outcome. The key thing is that we need a way of *describing* every possibility, which we can do with a formula for computing the probability.

Q: Wouldn't Chad's snowboarding skills eventually improve? Is it realistic to say the probability of success is 0.2 for every trial?

A: That may be a fair assumption. But in this problem, Chad is truly hapless when it comes to snowboarding, and we have to assume that his skills *won't* improve—which means his probability of success on the slopes will follow the geometric distribution.

Geometric Distribution Up Close

We said that Chad's snowboarding exploits are an example of the ***geometric distribution***. The geometric distribution covers situations where:

1. You run a series of independent trials.

2. There can be either a success or failure for each trial, and the probability of success is the same for each trial.

3. The main thing you're interested in is how many trials are needed in order to get the first successful outcome.

So if you have a situation that matches this set of criteria, you can use the geometric distribution to help you take a few shortcuts. The important thing to be aware of is that we use the word "success" to mean that the event we're interested in happens. If we're looking for an event that has negative connotations, in statistical terms it's still counted as a success.

Let's use the variable X to represent **the number of trials needed to get the first successful outcome**—in other words, the number of trials needed for the event we're interested in to happen.

To find the probability of X taking a particular value *r*, you can get a quick result by using:

$$P(X = r) = p\,q^{r-1}$$

where *p* is the probability of success, and $q = 1 - p$, the probability of failure. In other words, to get a success on the *r*th attempt, there must first have been $(r - 1)$ failures.

The geometric distribution has a distinctive shape.

$P(X = r)$ is at its highest when $r = 1$, and it gets lower and lower as r increases. Notice that the probability of getting a success is highest for the first trial. This means that **the mode of any geometric distribution is always 1**, as this is the value with the highest probability.

This may sound counterintuitive, but it's most likely that only one attempt will be needed for a successful outcome.

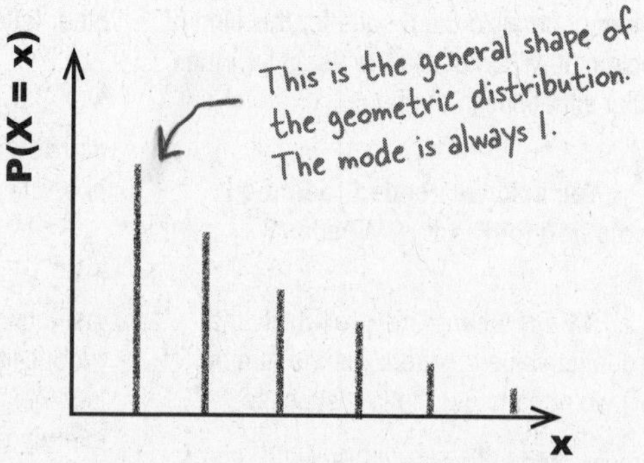
P(X = x)
x
This is the general shape of the geometric distribution. The mode is always 1.

The geometric distribution also works with inequalities

As well as finding exact probabilities for the geometric distribution, there's also a quick way of finding probabilities that deal with inequalities.

Let's start with $P(X > r)$.

$P(X > r)$ is the probability that more than r trials will be needed in order to get the first successful outcome. In order for more than *r* trials to be needed, this means that the first *r* trials must have ended in failure. This means that you find the probability by multiplying the probability of failure together *r* times.

$$\mathbf{P(X > r) = q^r}$$

For the number of trials needed for a success to be greater than r, there must have been r failures.

We don't need p in this formula because we don't need to know exactly which trial was successful, just that there must be more than r trials.

We can use this to find $P(X \leq r)$, the probability that r or fewer trials are needed in order for there to be a successful outcome.

If we add together $P(X \leq r)$ and $P(X > r)$, the total must be 1. This means that

$$P(X \leq r) + P(X > r) = 1$$

or

$$P(X \leq r) = 1 - P(X > r)$$

This is because $P(X \leq r)$ is the opposite of $P(X > r)$. $P(X \leq r) = 1 - P(X > r)$.

This gives us

$$\mathbf{P(X \leq r) = 1 - q^r}$$

From above, we know that $P(X > r) = q^r$ so we substitute in q^r for $P(X > r)$ to get this formula.

If a variable X follows a geometric distribution where the probability of success in a trial is p, this can be written as

$$\mathbf{X \sim Geo(p)}$$

This is a quick way of saying "X follows a geometric distribution where the probability of success is p."

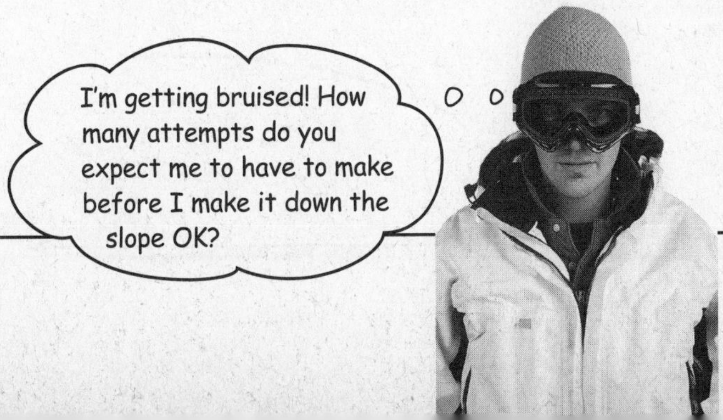

I'm getting bruised! How many attempts do you expect me to have to make before I make it down the slope OK?

The pattern of expectations for the geometric distribution

So far we've found probabilities for the number of attempts Chad needs to make before successfully makes it down the slope, but what if we want to find the expectation and variance? If we know the expectation, for instance, we'll be able to say how many attempts we expect Chad to make before he's successful.

As a reminder, expectation is the average value that you expect to get, a bit like the mean but for probability distributions.

Variance is a measure of how much you can expect this to varies by.

Can you remember how we found expectations earlier in the book? We find E(X) by calculating $\sum xP(X = x)$. The probabilities in this case go on forever, but let's start by working out the first few values to see if there's some sort of pattern.

Here are the first few values of x, where X ~ Geo(0.2)

This is the running total of xP(X = x)

x	P(X = x)	xP(X = x)	xP(X ≤ x)
1	0.2	0.2	0.2
2	$0.8 \times 0.2 = 0.16$	0.32	0.52
3	$0.8^2 \times 0.2 = 0.128$	0.384	0.904
4	$0.8^3 \times 0.2 = 0.1024$	0.4096	1.3136
5	$0.8^4 \times 0.2 = 0.08192$	0.4096	1.7232
6	$0.8^5 \times 0.2 = 0.065536$	0.393216	2.116416
7	$0.8^6 \times 0.2 = 0.0524288$	0.3670016	2.4834176
8	$0.8^7 \times 0.2 = 0.04194304$	0.33554432	2.81894608

Can you see what happens to the values of xP(X = x)?

The values of xP(X = x) start off small, and then they get larger until x = 5. When x is larger than 5, the values start decreasing again, and keep on decreasing as x gets larger. As x gets larger, xP(X = x) becomes smaller and smaller until it makes virtually no difference to the running total.

We can see this more clearly if we chart the cumulative total of xP(X = x):

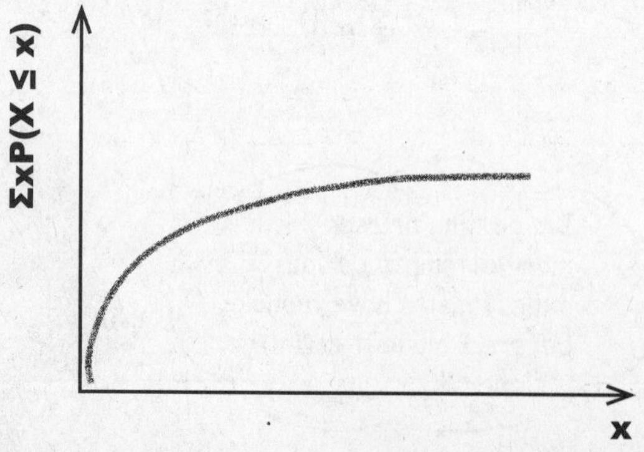

Expectation is 1/p

Drawing the chart for the running total of $xP(X = x)$ shows you that as x gets larger, the running total gets closer and closer to a particular value, 5. In fact, the running total of $xP(X = x)$ for an infinite number of trials is 5 itself. This means that

$$E(X) = 5$$

This makes intuitive sense. The probability of a successful outcome is 0.2. This is a bit like saying that 1 in 5 attempts tend to be successful, so we can expect Chad to make 5 attempts before he is successful.

I can expect to make it down in 5 tries? Not bad!

We can generalize this for any value p. If $X \sim Geo(p)$ then

$$\mathbf{E(X) = \frac{1}{p}}$$

The expectation is 1 divided by the probability of success.

We're not just limited to finding the expectation of the geometric distribution, we can find the variance too.

Sharpen your pencil

Let's see if we can find an expression for the variance of the geometric distribution in the same way that we did for the expectation. Complete the table below. What do you notice?

x	P(X = x)	$x^2P(X = x)$	$x^2P(X \le x)$
1	0.2		
2	$0.8 \times 0.2 = 0.16$		
3	$0.8^2 \times 0.2 = 0.128$		
4	$0.8^3 \times 0.2 = 0.1024$		
5	$0.8^4 \times 0.2 = 0.08192$		
6	$0.8^5 \times 0.2 = 0.065536$		
7	$0.8^6 \times 0.2 = 0.0524288$		
8	$0.8^7 \times 0.2 = 0.04194304$		
9	$0.8^8 \times 0.2 = 0.033554432$		
10	$0.8^9 \times 0.2 = 0.0268435456$		

Remember, the variance is given by $E(X^2) - E^2(X)$.

Sharpen your pencil Solution

Let's see if we can find an expression for the variance of the geometric distribution in the same way that we did for the expectation. Complete the table below. What do you notice?

x	P(X = x)	$x^2P(X = x)$	$x^2P(X \leq x)$
1	0.2	0.2	0.2
2	0.8 × 0.2 = 0.16	0.64	0.84
3	0.8^2 × 0.2 = 0.128	1.152	1.992
4	0.8^3 × 0.2 = 0.1024	1.6384	3.6304
5	0.8^4 × 0.2 = 0.08192	2.048	5.6784
6	0.8^5 × 0.2 = 0.065536	2.359296	8.037696
7	0.8^6 × 0.2 = 0.0524288	2.5690112	10.6067072
8	0.8^7 × 0.2 = 0.04194304	2.68435456	13.29106176
9	0.8^8 × 0.2 = 0.033554432	2.717908992	16.00897075
10	0.8^9 × 0.2 = 0.0268435456	2.68435456	18.69332531

This time $x^2P(X = x)$ increases until x reaches 10. When x reaches 10 it starts to go down again.

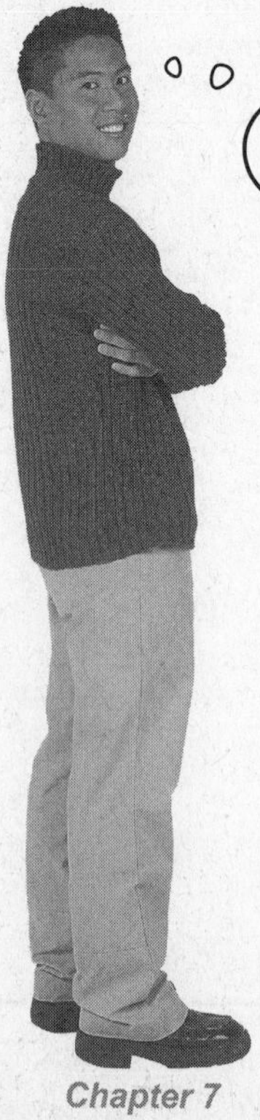

I get it, so $x^2P(X = x)$ gets larger for a while, but after that, it gets smaller and smaller as x gets larger and larger.

That's right.

$x^2P(X = x)$ gets larger and larger up until a certain point, and then it starts decreasing again. Eventually it becomes very close to 0.

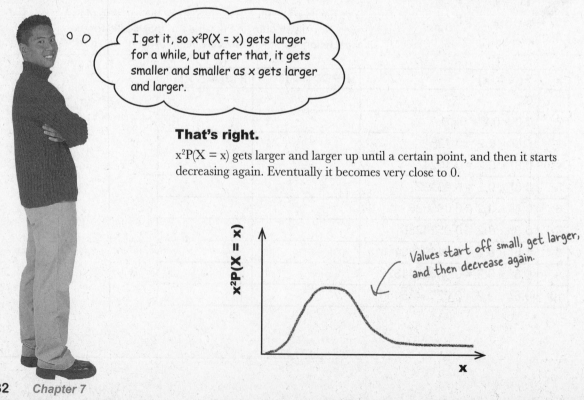

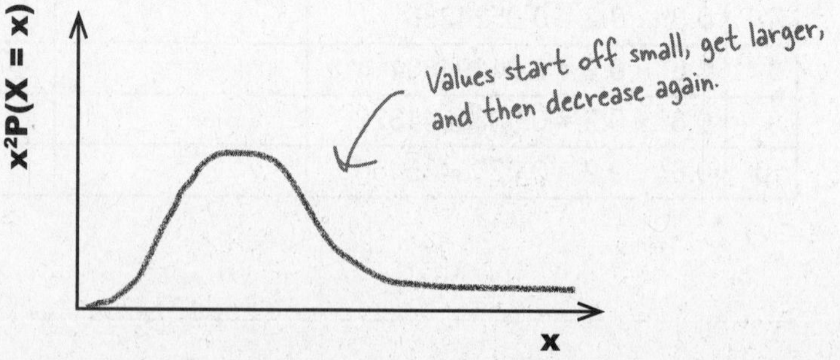

x²P(X = x)
Values start off small, get larger, and then decrease again.
x

Finding the variance for our distribution

So how does this help us find the variance of the number of trials it takes Chad to make a successful run down the slopes?

We find the variance of a probability distribution by calculating

$$Var(X) = E(X^2) - E^2(X)$$

This means that we calculate $\sum x^2 P(X = x)$, and then subtract E(X) squared. By graphing the resulting values against the values of x, you can see the pattern of Var(X) as x increases. Here's the graph of $x^2P(X \leq x) - E^2(X)$

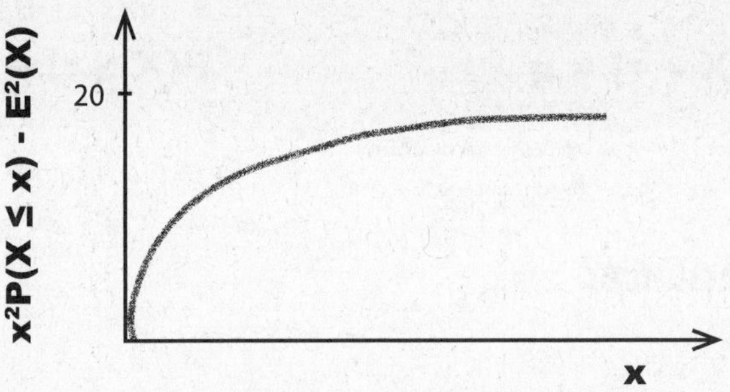

x²P(X ≤ x) - E²(X)
20
x

As x gets larger, the value of $x^2P(X \leq x) - E^2(X)$ gets closer and closer to a particular value, this time 20.

As with the expectation, we can generalize this. If $X \sim Geo(p)$ then

$$\mathbf{Var(X) = \frac{q}{p^2}}$$

Even though there's no fixed number of trials, you can still work out what the expectation and variance are.

A quick guide to the geometric distribution

Here's a quick summary of everything you could possibly need to know about the Geometric distribution

When do I use it?

Use the Geometric distribution if you're running independent trials, each one can have a success or failure, and you're interested in how many trials are needed to get the first successful outcome

How do I calculate probabilities?

Use the following handy formulae. p is the probability of success in a trial, q = 1 - p, and X is the number of trials needed in order to get the first successful outcome. We say X ~ Geo(p).

$$P(X = r) = p\,q^{r-1}$$

The probability of the first success being in the r'th trial

$$P(X > r) = q^r$$

The probability you'll need more than r trials to get your first success

$$P(X \leq r) = 1 - q^r$$

The probability you'll need r trials or less to get your first success

What about the expectation and variance?

Just use the following

$$E(X) = 1/p \qquad Var(X) = q/p^2$$

there are no Dumb Questions

Q: Can I trust these formulae? Can I use them any time I need to find probabilities and expectations?

A: You can use these shortcuts whenever you're dealing with the geometric distribution, as they're shortcuts for that probability distribution. If you're dealing with a situation that can't be modelled by the geometric distribution, don't use these shortcuts.

Remember, the geometric distribution is used for situations where you're running independent trials (so the probability stays the same for each one), each trial ends in either success or failure, and the thing you're interested in is how many trials are needed to get the first successful outcome.

Q: What about if my circumstances are different? What if I have a fixed number of trials and I want to find the number of successful outcomes?

A: You can't use the geometric distribution to model this sort of situation, but don't worry, there are other methods.

Q: Do I have to learn all of these shortcuts?

A: If you have to deal with the geometric distribution, knowing the formulae will save you a lot of time. If you're sitting for a statistics exam, check whether your exam syllabus covers it.

Q: Why does the distribution use the letters p and q?

A: The letter p stands for probability. In this case, it's the probability of getting a successful outcome in one trial.

The letter q is often used in statistics to represent 1 - p, or p'. You'll see quite a lot of it through the rest of this chapter and the rest of the book.

BE the snowboarder

The probability that another snowboarder will make it down the slope without falling over is 0.4. Your job is to play like you're the snowboarder and work out the following probabilities for your slope success.

1. The probability that you will be successful on your second attempt, while failing on your first.

2. The probability that you will be successful in 4 attempts or fewer.

3. The probability that you will need more than 4 attempts to be successful.

4. The number of attempts you expect you'll need to make before being successful.

5. The variance of the number of attempts.

BE the snowboarder solution

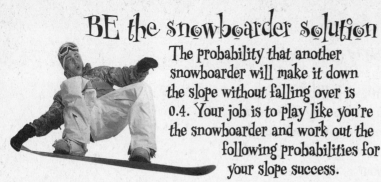

The probability that another snowboarder will make it down the slope without falling over is 0.4. Your job is to play like you're the snowboarder and work out the following probabilities for your slope success.

Let's use $X \sim Geo(0.4)$, where X is the number of trials needed by this second snowboarder to make a clean run down the slope.

1. The probability that you will be successful on your second attempt, while failing on your first.

$$P(X = 2) = p \times q$$
$$= 0.4 \times 0.6$$
$$= 0.24$$

2. The probability that you will be successful in 4 attempts or fewer.

$$P(X \leq 4) = 1 - q^4$$
$$= 1 - 0.6^4$$
$$= 1 - 0.1296$$
$$= 0.8704$$

3. The probability that you will need more than 4 attempts to be successful.

$$P(X > 4) = q^4$$
$$= 0.6^4$$
$$= 0.1296$$

Or you could have found this by using $P(X > 4) = 1 - P(X \leq 4) = 1 - 0.8704 = 0.1296$

4. The number of attempts you expect you'll need to make before being successful.

$$E(X) = 1/p$$
$$= 1/0.4$$
$$= 2.5$$

4. The variance of the number of attempts.

$$Var(X) = q/p^2$$
$$= 0.6/0.4^2$$
$$= 0.6/0.16$$
$$= 3.75$$

Let's hit the slopes!

You've mastered the geometric distribution

Thanks to your skills with the geometric distribution, Chad not only knows the probability of him making a clear run down the slopes after any number of tries, but also how many times he can expect it to take to get down the hill successfully, and how much variability there is.

With an expectation of 5 tries to make it down the slopes, and a variance of 20, he feels much more confident he can impress the ladies without serious bodily harm.

Now let's move on to…

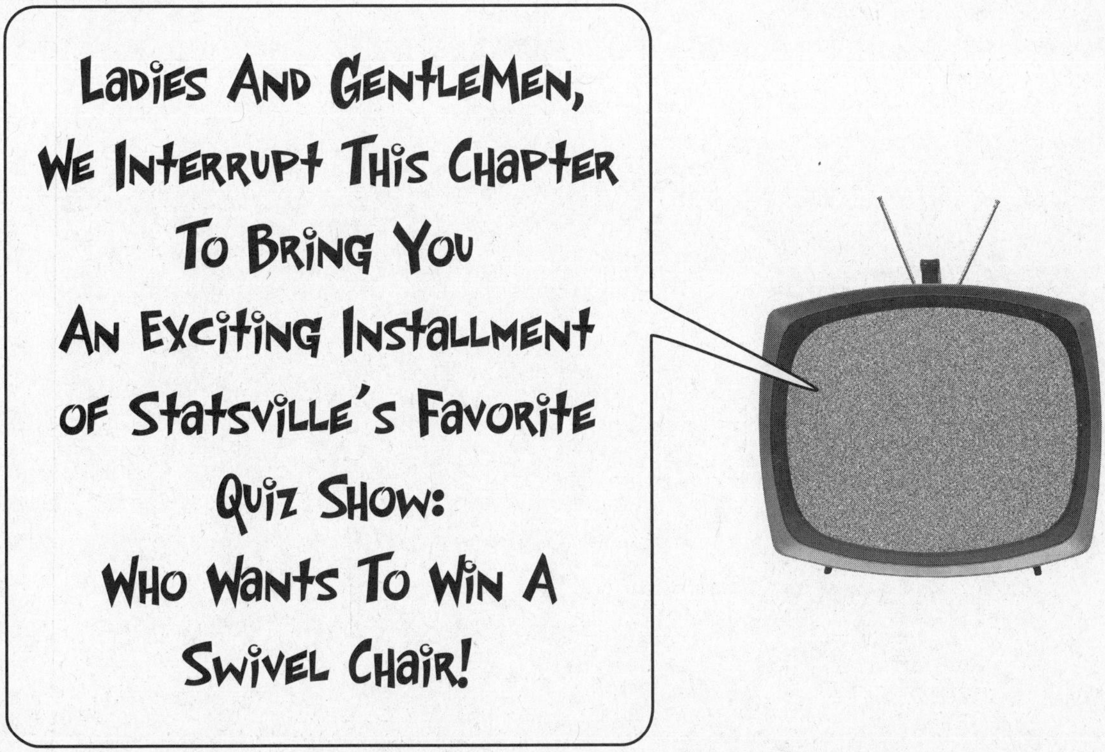
Ladies And Gentlemen,
We Interrupt This Chapter
To Bring You
An Exciting Installment
of Statsville's Favorite
Quiz Show:
Who Wants To Win A
Swivel Chair!

WHO WANTS TO WIN A
SWIVEL CHAIR
Hello, and welcome to Who Wants To Win A Swivel Chair, Statville's favorite quiz show. We've got some fiendishly difficult questions on the show tonight. I hope you're feeling lucky.

We've got some great questions for you today, so let's get started. In Round One I'm going to ask you three questions, and for each question there are four possible answers. You can quit now and walk away with the consolation prize, but if you play on and beat your competitors, you'll move on to the next round and be one step closer to winning a swivel chair. The title of Round One is "All About Me." Good luck!

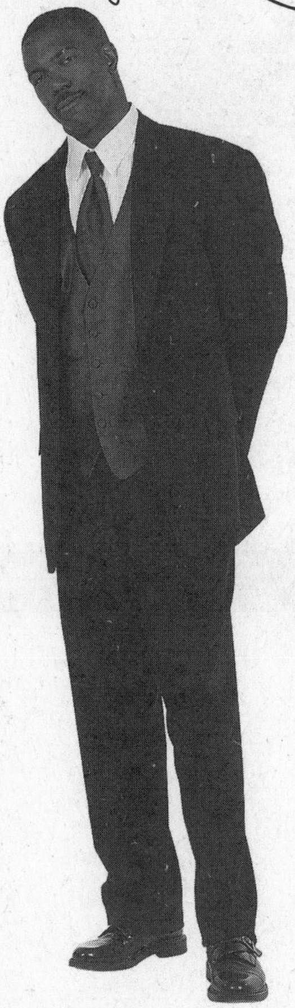

Sharpen your pencil

Here are the questions for Round One. The questions are all about the game show host. Put a check mark next to the correct answer.

1. What's his favorite color?

- **A:** Red
- **B:** Blue
- **C:** Green
- **D:** Yellow

2. In what month is his birthday?

- **A:** January
- **B:** February
- **C:** March
- **D:** April

3. What do people like most about him?

- **A:** Good looks
- **B:** Charm
- **C:** Sense of Humor
- **D:** Intelligence

there are no Dumb Questions

Q: What's a quiz show doing in the middle of my chapter? I thought we were talking about probability distributions.

A: We still are. This situation is ideal for another sort of probability distribution. Keep reading and everything will become clear.

Q: I don't know the answers to these questions. What should I do?

A: If you don't know the answers you'll have to answer them at random. Give it your best shot - you might win a swivel chair.

Should you play, or walk away?

It's unlikely you'll know the game show host well enough to answer these questions, so let's see if we can find the probability distribution for the number of questions you'll get correct if you choose answers at random. That should help you decide whether or not to play on.

Here's a probability tree for the three questions:

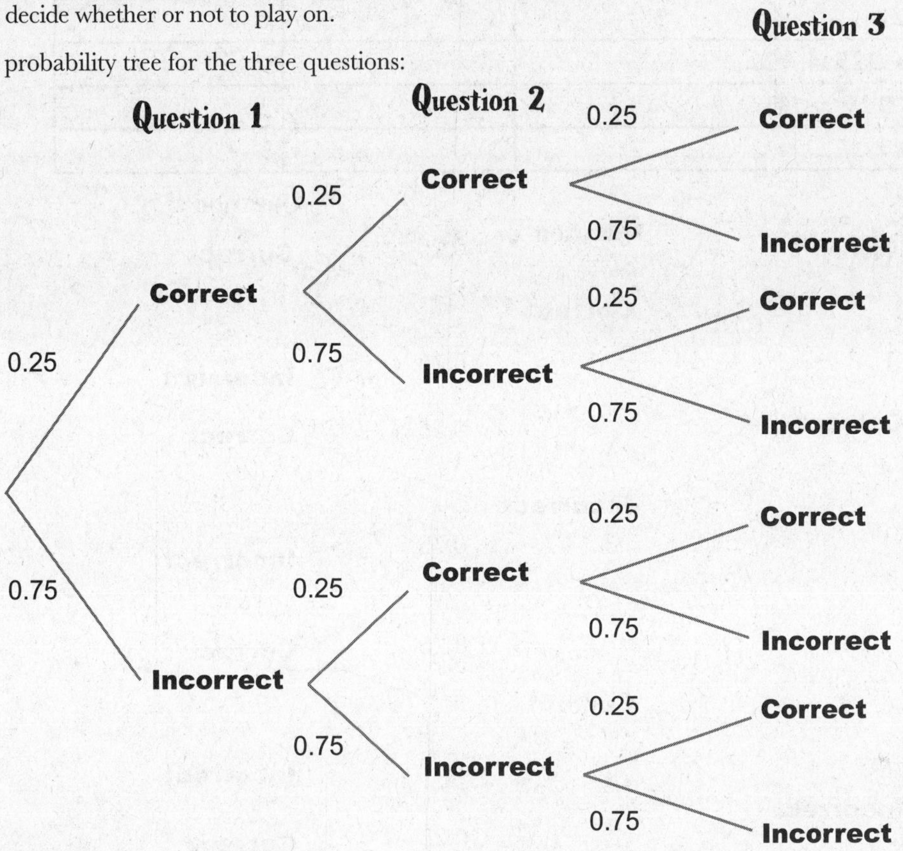

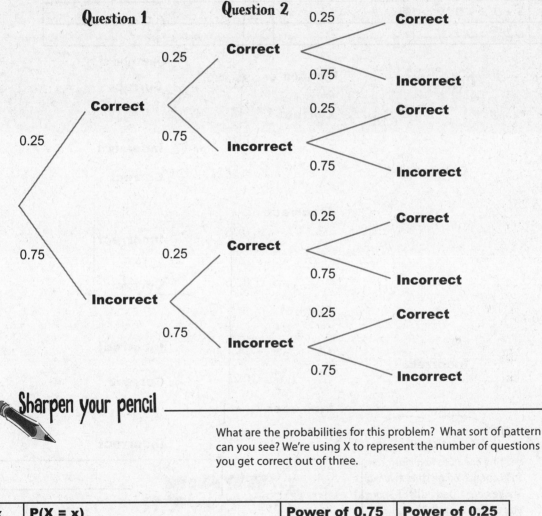

Sharpen your pencil

What are the probabilities for this problem? What sort of pattern can you see? We're using X to represent the number of questions you get correct out of three.

x	P(X = x)	Power of 0.75	Power of 0.25
0	0.75^3	3	0
1			
2			
3			

Sharpen your pencil Solution

What are the probabilities for this problem? What sort of pattern can you see? We're using X to represent the number of questions you get correct out of three.

x	P(X = x)	Power of 0.75	Power of 0.25
0	$0.75^3 = .422$	3	0
1	$3 \times 0.75^2 \times 0.25 = .422$	2	1
2	$3 \times 0.75 \times 0.25^2 = .141$	1	2
3	$0.25^3 = .015$	0	3

There are 3 different ways you can get one question right, and all of them have a probability of $0.75^2 \times 0.25$.

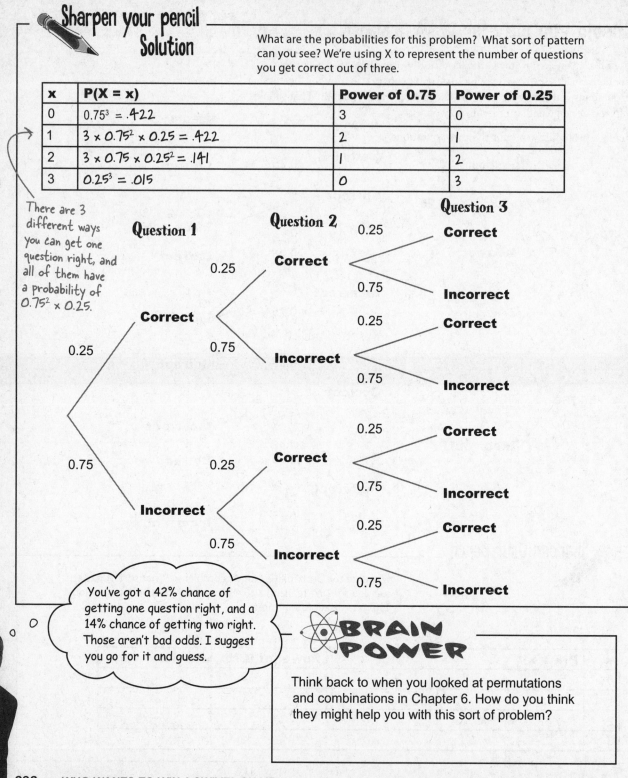

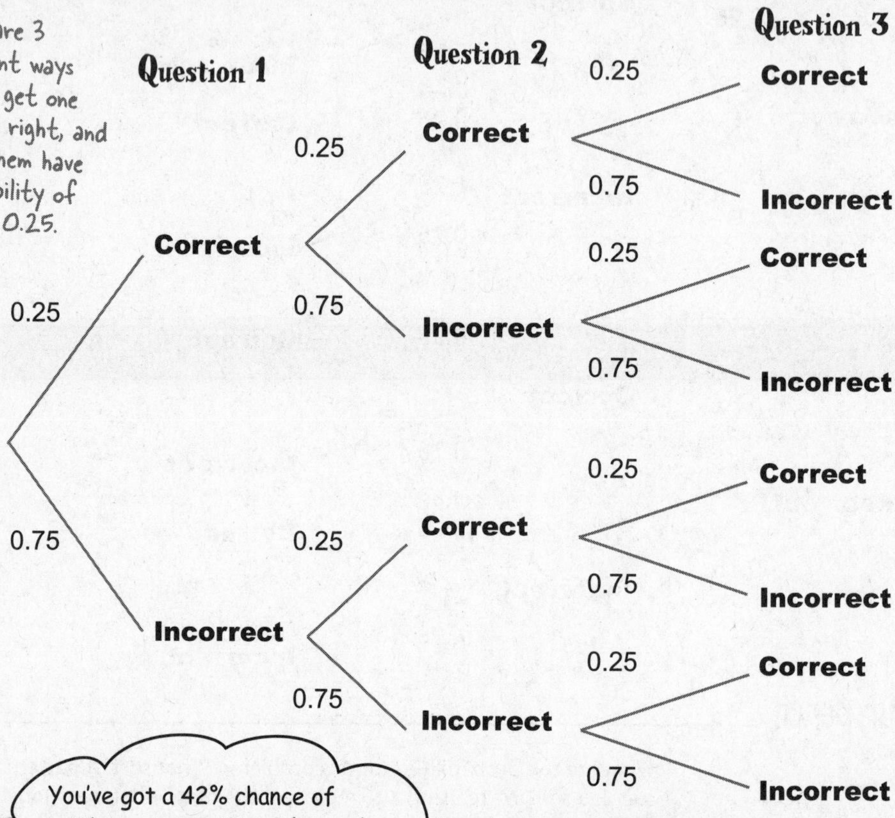

You've got a 42% chance of getting one question right, and a 14% chance of getting two right. Those aren't bad odds. I suggest you go for it and guess.

BRAIN POWER

Think back to when you looked at permutations and combinations in Chapter 6. How do you think they might help you with this sort of problem?

Generalizing the probability for three questions

So far we've looked at the probability distribution of X, the number of questions we answer correctly out of three.

Just as with the geometric distribution, there seems to be a pattern in the way the probabilities are formed. Each probability contains different powers of 0.75 and 0.25. As x increases, the power of 0.75 decreases while the power of 0.25 increases.

In general, P(X = r) is given by:

$$P(X = r) = \,? \times 0.25^r \times 0.75^{3-r}$$

The probability of getting a question right

There are 3 questions

r is the number of questions we get right

What's this?

The probability of getting a question wrong

In other words, to find the probability of getting exactly r questions right, we calculate 0.25^r, multiply it by 0.75^{3-r}, and then multiply the whole lot by some number. But what?

What's the missing number?

For each probability, we need to answer a certain number of questions correctly, and there are different ways of achieving this. As an example, there are three different ways of answering exactly one question correctly out of three questions. Another way of looking at this is that there are 3 different **combinations**.

Just to remind you, a combination ${}^nC_r$ is the number of ways of choosing r objects from n, without needing to know the exact order. This is exactly the situation we have here. We need to choose r correct questions from 3.

We covered this back in Chapter 6; look back if you need a reminder.

This means that the probability of getting r questions correct out of 3 is given by

$$P(X = r) = {}^3C_r \times 0.25^r \times 0.75^{3-r}$$

So, by this formula, the probability of getting 1 question correct is:

$$P(X = r) = {}^3C_1 \times 0.25 \times 0.75^{3-1}$$
$$= 3!/(3-1)! \times 0.25 \times 0.5625$$
$$= 6/2 \times 0.0625 \times 0.75$$
$$=0.422$$

This is the same result we got using our chart on the previous page.

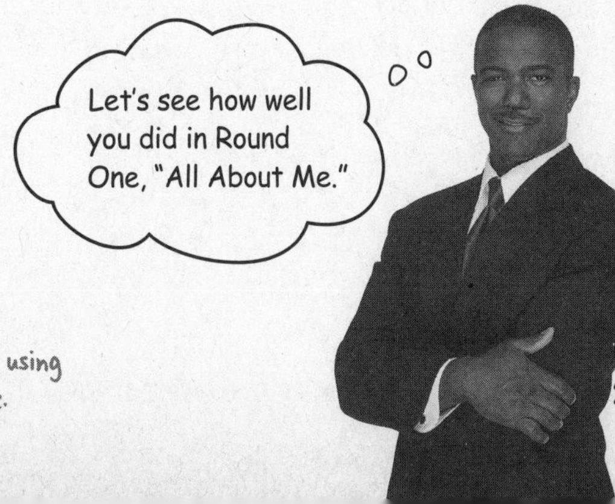

Let's see how well you did in Round One, "All About Me."

Sharpen your pencil Solution

Here are the questions for Round One. The questions are all about the game show host.

1. What's his favorite color?

- **A:** Red
- **B:** Blue
- **C:** Green
- ✓ **D:** Yellow

2. In what month is his birthday?

- **A:** January
- **B:** February
- ✓ **C:** March
- **D:** April

3. What do people like most about him?

- **A:** Good looks
- **B:** Charm
- **C:** Sense of Humor
- ✓ **D:** Intelligence

Looks like you tied with another contender. Congratulations, you're through to the next round.

Round Two of Who Wants To Win A Swivel Chair is called "More About Me." This time I'll ask you five questions. As before, there are four possible answers to each question. Do you want to play on?

Sharpen your pencil

Here are the questions for Round Two. The questions are all about the game show host.

1. What was the name of his first girlfriend?

- **A:** Mary
- **B:** Marie
- **C:** Maggie
- **D:** May

2. What would be an ideal gift for him?

- **A:** A statue
- **B:** A tin dog
- **C:** A horse
- **D:** A hovercraft

3. What is his greatest achievement?

- **A:** Hosting a quiz show
- **B:** Winning Mr Statsville 2008
- **C:** Raising $1000 for the seal sanctuary
- **D:** Releasing an album

4. What is his secret ambition?

- **A:** To launch a range of sports equipment
- **B:** To release an exercise DVD
- **C:** To launch his own range of menswear
- **D:** To have his own hair care range

5. In what year was he abducted by aliens?

- **A:** 2005
- **B:** 2006
- **C:** 2007
- **D:** 2008

It looks like these questions are just as obscure as the ones in the previous round, so you'll have to answer questions at random again.

Let's see if we can work out the probability distribution for this new set of questions.

Let's generalize the probability further

So far you've seen that the probability of getting r questions correct out of 3 is given by

$$P(X = r) = {}^{3}C_{r} \times 0.25^{r} \times 0.75^{3-r}$$

where the probability of answering a question correctly is 0.25, and the probability of answering incorrectly is 0.75.

The next round of Who Wants To Win A Swivel Chair has 5 questions instead of 3. Rather than rework this probability for 5 questions, let's rework it for n questions instead. That way we'll be able to use the same formula for every round of Who Wants To Win A Swivel Chair.

Just replace the 3 with n.

So what's the formula for the probability of getting r questions right out of n? It's actually

$$P(X = r) = {}^{n}C_{r} \times 0.25^{r} \times 0.75^{n-r}$$

What if the probability of getting a question right changes? I wonder if we can generalize this further.

Yes, we can generalize this further.

Imagine the probability of getting a question right is given by p, and the probability of getting a question wrong is given by 1 – p, or q. The probability of getting r questions right out of n is given by

$$P(X = r) = {}^{n}C_{r} \times p^{r} \times q^{n-r}$$

This sort of problem is called the ***binomial distribution***. Let's take a closer look.

Binomial Distribution Up Close

Guessing the answers to the questions on Who Wants To Win A Swivel Chair is an example of the ***binomial distribution***. The binomial distribution covers situations where

These two are like the Geometric distribution.

1. You're running a series of independent trials.

2. There can be either a success or failure for each trial, and the probability of success is the same for each trial.

3. There are a finite number of trials.

This is different.

Just like the geometric distribution, you're running a series of independent trials, and each one can result in success or failure. The difference is that this time you're interested in the number of successes.

Let's use the variable X to represent **the number of successful outcomes out of n trials**. To find the probability there are r successes, use:

$$P(X = r) = {}^nC_r\, p^r q^{n-r} \quad \text{where} \quad {}^nC_r = \frac{n!}{r!\,(n-r)!}$$

p is the probability of a successful outcome in each trial, and n is the number of trials. We can write this as

$$X \sim B(n, p)$$

The exact shape of the binomial distribution varies according to the values of n and p. The closer to 0.5 p is, the more symmetrical the shape becomes. In general it is skewed to the right when p is below 0.5, and skewed to the left when p is greater than 0.5.

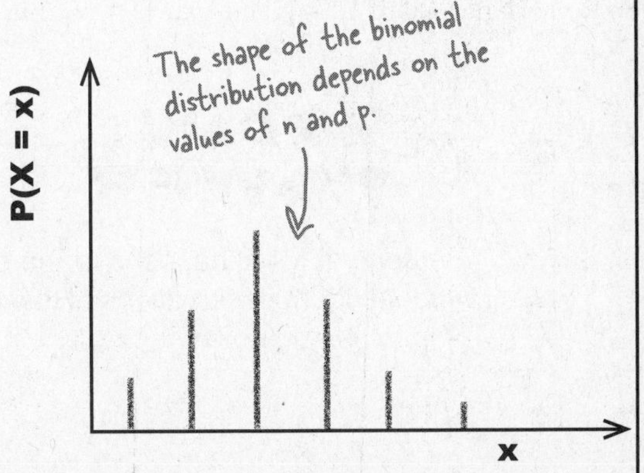

The shape of the binomial distribution depends on the values of n and p.
P(X = x)
x

What's the expectation and variance?

So far we've looked at how to use the binomial distribution to find basic probabilities, which allows us to calculate the probability of getting a certain number of questions correct. But how many questions can we actually expect to get right if we choose the answers at random? That will help you better decide whether we should answer the next round of questions.

Let's see if we can find a general expression for the expectation and variance. We'll start by working out the expectation and variance for a single trial, and then see if we can extend it to n independent trials.

Let's look at one trial

Suppose we conduct just one trial. Each trial can only result in success or failure, so in one trial, it's possible to have 0 or 1 successes. If $X \sim B(1, p)$, the probability of 1 success is p, and the probability of 0 successes is q.

This is the probability distribution of X where $X \sim B(1, p)$.

x	0	1
P(X = x)	q	p

We can use this to find the expectation and variance of X. Let's start with the expectation.

$$E(X) = 0q + 1p$$
$$= p$$

$$Var(X) = E(X^2) - E(X)^2$$
$$= (0q + 1p) - p^2$$
$$= p - p^2$$
$$= p(1 - p)$$
$$= pq$$

$E(X) = p$, so $E(X)^2 = p^2$

$(0q + 1p)$ is $E(X^2)$

So for a single trial, $E(X) = p$ and $Var(X) = pq$. But what if there are n trials?

BRAIN POWER

In general, what happens to the expectation and variance when there are n independent observations? How can this help us now?

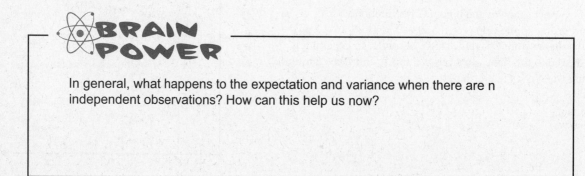

Pool Puzzle

Let's see if you can derive the expectation and variance for Y ~ B(n, p). Your **job** is to take elements from the pool and place them into the blank lines of the calculations. You may **not** use the same element more than once, and you won't need to use all the elements.

Hint: Each X_i is a separate trial. $E(X_i) = p$, and $Var(X_i) = pq$

You need to find the expectation and variance of n independent trials.

$E(X) = E(X_1) + E(X_2) + ... + E(X_n)$

= .... $E(X_i)$

= ....

$Var(X) = Var(X_1) + Var(X_2) + ... + Var(X_n)$

= .... $Var(X_i)$

= ....

Note: each element in the pool can only be used once!

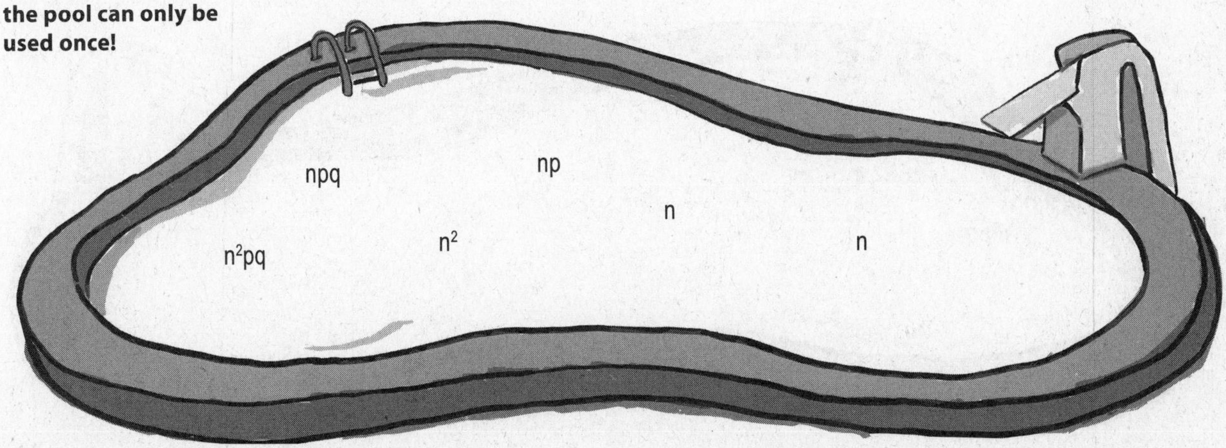

Pool Puzzle Solution

Let's see if you can derive the expectation and variance for $Y \sim B(n, p)$. Your **job** is to take elements from the pool and place them into the blank lines of the calculations. You may **not** use the same element more than once, and you won't need to use all the elements.

Hint: Each X_i is a separate trial. $E(X_i) = p$, and $Var(X_i) = pq$

You need to find the expectation and variance of n independent trials.

$$E(X) = E(X_1) + E(X_2) + \dots + E(X_n)$$
$$= \underline{n}\; E(X_i)$$
$$= \underline{np}$$

Since the trials are independent, $E(X_1) = E(X_2) = E(X_3)$, and so on.

If $X \sim B(n, p)$, then

$E(X) = np$

$Var(X) = npq$

$$Var(X) = Var(X_1) + Var(X_2) + \dots + Var(X_n)$$
$$= \underline{n}\; Var(X_i)$$
$$= \underline{npq}$$

Since the trials are independent, $Var(X_1) = Var(X_2) = Var(X_3)$, and so on.

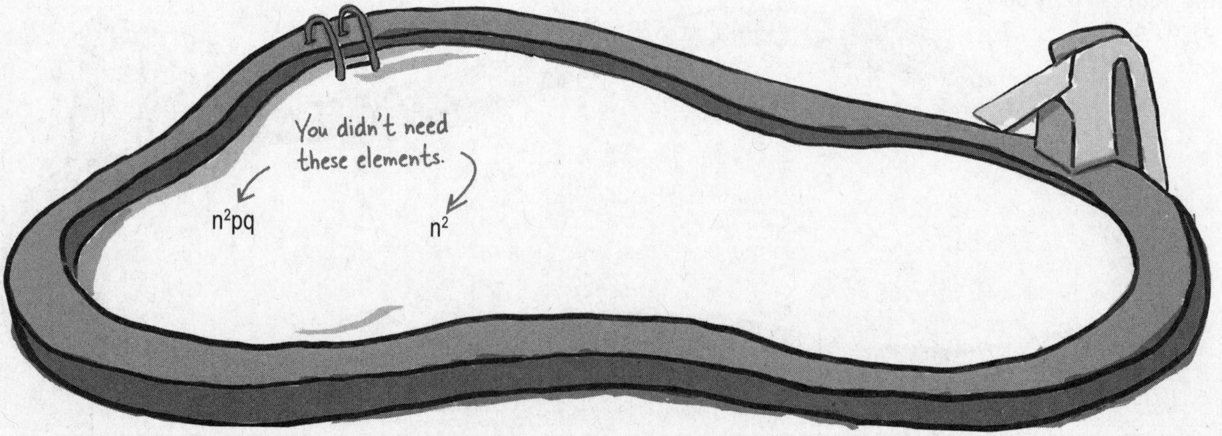
You didn't need these elements.
n²pq
n²

Binomial expectation and variance

Let's summarize what we just did. First of all, we took at one trial, where the probability of success is *p*, and where the distribution is binomial. Using this, we found the expectation and variance of a single trial.

We then considered n independent trials, and used shortcuts to find the expectation and variance of *n* trials. We found that if X ~ B(n, p),

$$E(X) = np$$

$$Var(X) = npq$$

These formulae work for any binomial distribution.

This is useful to know as it gives us a quick way of finding the expectation and variance of any probability distribution, without us having to work out lots of individual probabilities.

there are no Dumb Questions

Q: The geometric distribution and the binomial distribution seem similar. What's the difference between them? Which one should I use when?

A: The geometric and binomial distributions do have some things in common. Both of them deal with independent trials, and each trial can result in success or failure. The difference between them lies in what you actually need to find out, and this dictates which probability distribution you need to use.

If you have a fixed number of trials and you want to know the probability of getting a certain number of successes, you need to use the binomial distribution. You can also use this to find out how many successes you can expect to have in your n trials.

If you're interested in how many trials you'll need before you have your first success, then you need to use the geometric distribution instead.

Q: The geometric distribution has a mode. Does the binomial distribution?

A: Yes, it does. The mode of a probability distribution is the value with the highest probability. If p is 0.5 and n is even, the mode is np. If p is 0.5 and n is odd it has two modes, the two values either side of np. For other values of n and p, finding the mode is a matter of trial and error, but it's generally fairly close to np.

Q: So for both the geometric and the binomial distributions you run a series of trials. Does the probability of success have to be the same for each trial?

A: In order for the geometric or binomial distribution to be applicable, the probability of success in each trial must be the same. If it's not, then neither the geometric nor binomial distribution is appropriate.

Q: I've tried calculating E(X) and it's not a value that's in the probability distribution. Did I do something wrong?

A: When you calculate E(X), the result may not be a possible value in your probability distribution. It may not be a value that can actually occur. If you get a result like this, it doesn't mean that you've made a mistake, so don't worry.

Q: Are there any other sorts of probability distribution?

A: Yes, there are. Keep reading and you'll find out more.

Your quick guide to the binomial distribution

Here's a quick summary of everything you could possibly need to know about the binomial distribution

When do I use it?

Use the binomial distribution if you're running a fixed number of independent trials, each one can have a success or failure, and you're interested in the number of successes or failures

How do I calculate probabilities?

Use

$$P(X = r) = {}^{n}C_{r}\, p^{r}\, q^{n-r}$$

$${}^{n}C_{r} = \frac{n!}{r!\,(n-r)!}$$

where p is the probability of success in a trial, q = 1 - p, n is the number of trials, and X is the number of successes in the n trials.

What about the expectation and variance?

$$E(X) = np \qquad Var(X) = npq$$

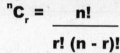

Exercise Solution

In the latest round of Who Wants To Win A Swivel Chair, there are 5 questions. The probability of getting a successful outcome in a single trial is 0.25

1. What's the probability of getting exactly two questions right?

2. What's the probability of getting exactly three questions right?

3. What's the probability of getting two or three questions right?

4. What's the probability of getting no questions right?

5. What are the expectation and variance?

Exercise Solution

In the latest round of Who Wants To Win A Swivel Chair, there are 5 questions. The probability of getting a successful outcome in a single trial is 0.25

1. What's the probability of getting exactly two questions right?

If X represents the number of questions answered correctly, then $X \sim B(n, p)$

$$
\begin{aligned}
P(X = 2) &= {}^5C_2 \times 0.25^2 \times 0.75^3 \\
&= \frac{5!}{3!2!} \times 0.0625 \times 0.421875 \\
&= 10 \times 0.0264 \\
&= 0.264
\end{aligned}
$$

2. What's the probability of getting exactly three questions right?

$$
\begin{aligned}
P(X = 3) &= {}^5C_3 \times 0.25^3 \times 0.75^2 \\
&= \frac{5!}{2!3!} \times .0.015625 \times 0.5625 \\
&= 10 \times 0.00879 \\
&= 0.0879
\end{aligned}
$$

3. What's the probability of getting two or three questions right?

$$
\begin{aligned}
P(X = 2 \text{ or } X = 3) &= P(X = 2) + P(X = 3) \\
&= 0.264 + 0.0879 \\
&= 0.3519
\end{aligned}
$$

4. What's the probability of getting no questions right?

$$
\begin{aligned}
P(X = 0) &= 0.75^5 \\
&= 0.237
\end{aligned}
$$

5. What are the expectation and variance?

$$
\begin{aligned}
E(X) &= np \\
&= 5 \times 0.25 \\
&= 1.25
\end{aligned}
$$

$$
\begin{aligned}
Var(X) &= npq \\
&= 5 \times 0.25 \times 0.75 \\
&= 0.9375
\end{aligned}
$$

So, you can expect to get less than 2 questions correct? I think now's about time to quit. Sorry you won't win the swivel chair, though.

Sharpen your pencil Solution

Here are the questions for Round Two. The questions are all about the game show host.

1. What was the name of his first girlfriend?

- A: Mary
- B: Marie
- ✓ C: Maggie
- D: May

2. What would be an ideal gift for him?

- A: A statue
- ✓ B: A tin dog
- C: A horse
- D: A hovercraft

3. What is his greatest achievement?

- A: Hosting a quiz show
- B: Winning Mr Statsville 2008
- ✓ C: Raising $1000 for the seal sanctuary
- D: Releasing an album

4. What is his secret ambition?

- A: To launch a range of sports equipment
- B: To release an exercise DVD
- C: To launch his own range of menswear
- ✓ D: To have his own hair care range

5. In what year was he abducted by aliens?

- ✓ A: 2005
- B: 2006
- C: 2007
- D: 2008

It's been great having you as a contestant on the show, and we'd love to have you back later on. But we've just had a phone call from the Statsville cinema. Some problem about popcorn...?

The Statsville Cinema has a problem

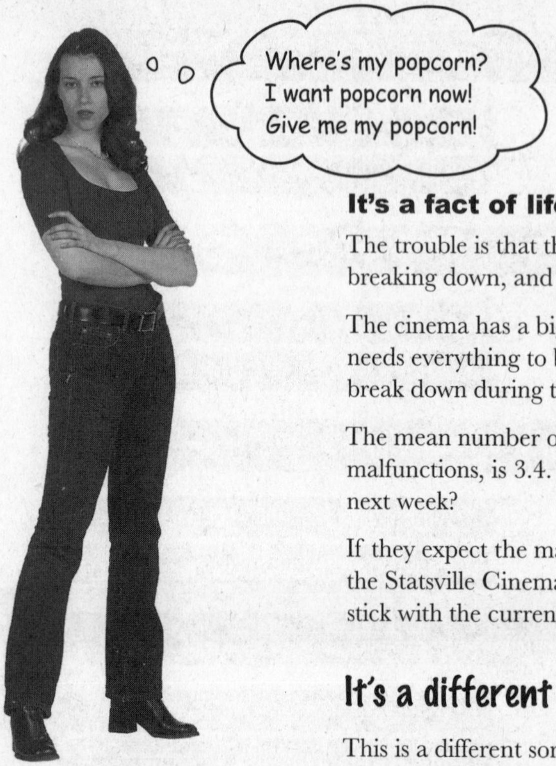

Where's my popcorn?
I want popcorn now!
Give me my popcorn!

It's a fact of life that cinemagoers like popcorn.

The trouble is that the popcorn machine at the Statsville Cinema keeps breaking down, and the customers aren't happy.

The cinema has a big promotion on next week, and the cinema manager needs everything to be perfect. He doesn't want the popcorn machine to break down during the week, or people won't come back.

The mean number of popcorn machine malfunctions per week, or rate of malfunctions, is 3.4. What's the probability that it won't break down at all next week?

If they expect the machine to break down more than a few times next week, the Statsville Cinema will buy a new popcorn machine, but if not, they'll stick with the current one and run the risk of a breakdown.

It's a different sort of distribution

This is a different sort of problem from the ones we've encountered so far.

This time there's no series of attempts or trials. Instead, we have a situation where we know the rate at which malfunctions happen, and where malfunctions occur at random.

So how do we find probabilities?

The trouble with this sort of problem is that while we know the mean number of popcorn machine malfunctions per week, the actual number of breakdowns varies each week. On the whole we can expect 3 or 4 malfunctions per week, but in a bad week there'll be far more, and in a good week there might be none at all.

We need to find the probability that the popcorn machine won't break down next week.

Sound difficult? Don't worry, there's a probability distribution that's designed for just this sort of situation. It's called the ***Poisson distribution***.

Poisson Distribution Up Close

The Poisson distribution covers situations where:

1. Individual events occur at random and independently in a given interval. This can be an interval of time or space—for example, during a week, or per mile.

2. You know the mean number of occurrences in the interval or the rate of occurrences, and it's finite. The mean number of occurrences is normally represented by the Greek letter λ (lambda).

Let's use the variable X to represent **the number of occurrences in the given interval**, for instance the number of breakdowns in a week. If X follows a Poisson distribution with a mean of λ occurrences per interval or rate, we write this as:

$$X \sim Po(\lambda)$$

We're not going to derive it here, but to find the probability that there are r occurrences in a specific interval, use the formula:

$$P(X = r) = \frac{e^{-\lambda} \lambda^r}{r!}$$

Don't let appearances put you off. It's pretty straightforward to calculate in practice.

e is a mathematical constant. It always stands for 2.718, so you can just substitute in this number for e in the Poisson formula. Many scientific calculators have an e^x key that will calculate powers of e for you.

The formula for the probability uses the exponential function e^x, where x is some number. It's a standard function available on most calculators, so even though the formula might look daunting at first, it's actually quite straightforward to use in practice.

As an example, if X ~ Po(2)

$$P(X = 3) = \frac{e^{-2} \times 2^3}{3!}$$

Use the formula and substitute in r = 3 and λ = 2.

$$= \frac{e^{-2} \times 8}{6}$$

$$= e^{-2} \times 1.333$$

$$= 0.180$$

So if X follows a Poisson distribution, what's its expectation and variance? It's easier than you might think...

Expectation and variance for the Poisson distribution

Finding the expectation and variance for the Poisson distribution is a lot easier than finding it for other distributions.

If $X \sim Po(\lambda)$, $E(X)$ is the number of occurrences we can expect to have in a given intervals, so for the popcorn machine, it's the number of breakdowns we can expect to have in a typical week. In other words, $E(X)$ is the mean number of occurrences in the given interval.

Now, if $X \sim Po(\lambda)$, then the mean number of occurrences is given by λ. In other words, $E(X)$ is equal to λ, the parameter that defines our Poisson distribution.

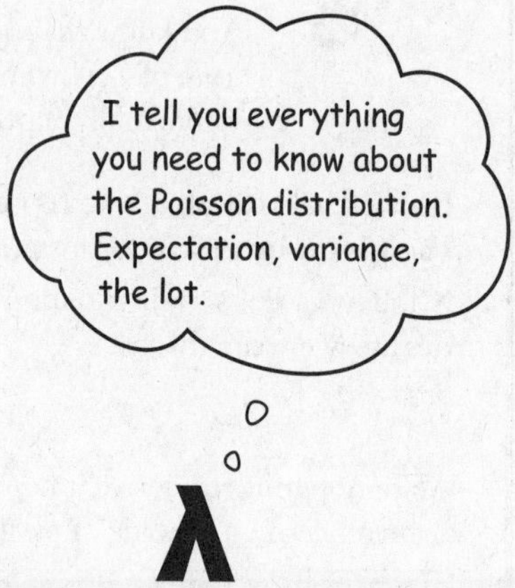

To make things even simpler, the variance of the Poisson distribution is also given by λ, so if $X \sim Po(\lambda)$,

$$E(X) = \lambda \qquad Var(X) = \lambda$$

In other words, if you're given a Poisson distribution $Po(\lambda)$, you don't have to calculate anything at all to find the expectation and variance. It's the parameter of the Poisson distribution itself.

I tell you everything you need to know about the Poisson distribution. Expectation, variance, the lot.

What does the Poisson distribution look like?

The shape of the Poisson distribution varies depending on the value of λ. If λ is small, then the distribution is skewed to the right, but it becomes more symmetrical as λ gets larger.

If λ is an integer, then there are two modes, λ and $\lambda - 1$. If λ is not an integer, then the mode is λ.

The shape of the Poisson distribution depends on the value of λ.

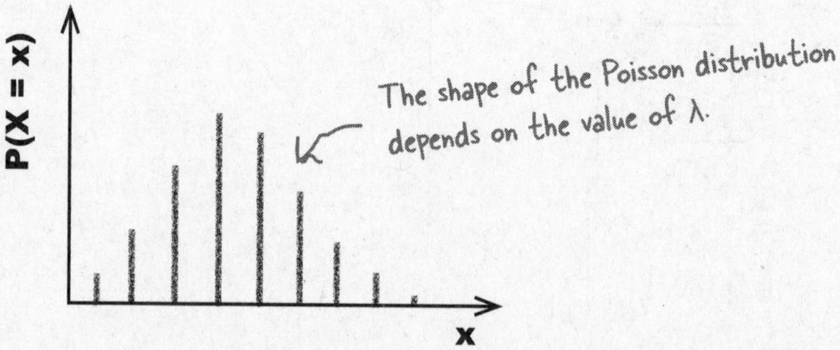

BE the popcorn machine

Your job is to play like you're the popcorn machine and say what the probability is of you malfunctioning a particular number of times next week. Remember, the mean number of times you break down in a week is 3.4.

1. What's the probability of the machine not malfunctioning next week?

2. What's the probability of the machine malfunctioning three times next week?

3. What's the expectation and variance of the machine malfunctions?

BE the popcorn machine solution

Your job is to play like you're the popcorn machine and say what the probability is of you malfunctioning a particular number of times next week. Remember, the mean number of times you break down in a week is 3.4.

Let's use X to represent the number of times the popcorn machine malfunctions in a week. We have

$X \sim Po(3.4)$

1. What's the probability of the machine not malfunctioning next week?

If there are no malfunctions, then X must be 0.

$$P(X = 0) = \frac{e^{-\lambda}\lambda^r}{r!}$$

$$= \frac{e^{-3.4} \times 3.4^0}{0!}$$

$$= \frac{e^{-3.4} \times 1}{1}$$

$$= 0.033$$

Looks like we can expect the machine to break down only 3.4 times next week, so we'll risk it and skip that new machine. Don't tell the moviegoers.

2. What's the probability of the machine malfunctioning three times next week?

$$P(X = 3) = \frac{e^{-3.4} \times 3.4^3}{3!}$$

$$= \frac{e^{-3.4} \times 39.304}{6}$$

$$= 0.033 \times 6.55$$

$$= 0.216$$

3. What's the expectation and variance of the machine malfunctions?

$$E(X) = \lambda$$
$$= 3.4$$

$$Var(X) = \lambda$$
$$= 3.4$$

there are no Dumb Questions

Q: How come we use λ to represent the mean for the Poisson distribution? Why not use μ like we do elsewhere?

A: We use λ because for the Poisson distribution, the parameter of the distribution, expectation and variance are all the same. It's a way of making sure we keep everything neutral.

Q: Where does the formula for the Poisson distribution come from?

A: It can actually be derived from the other distributions, but the mathematics are quite involved. In practice it's best to just accept the formula, and remember the situations in which it's useful.

Q: What's the difference between the Poisson distribution and the other probability distributions?

A: The key difference is that the Poisson distribution doesn't involve a series of trials. Instead, it models the number of occurrences in a particular interval.

Q: Does λ have to be an integer?

A: Not at all. λ can be any non-negative number. It can't be negative as it's the mean number of occurrences in an interval, and it doesn't make sense to have a negative number of occurrences.

Q: What's that "e" in the formula all about?

A: *e* is a constant in mathematics that stands for the number 2.718. So you can substitute in 2.718 for e in the formula for calculating Poisson probabilities.

The constant *e* is used frequently in calculus, and it also has many other applications in everything from calculating compound interest to advanced probability theory. Further discussion of e is outside the scope of this book, though.

Q: I keep getting the wrong answer when I try to calculate probabilities using the Poisson distribution. Where am I going wrong?

A: There are two main areas where it's easy to trip up. The first thing is to make sure you're using the right formula. It's easy to get the r and the λ mixed up, so make sure you've got them the right way round.

The second thing is to make sure you're using the e^x function correctly on your calculator. One way of doing this is to leave the $e^{-\lambda}$ calculation until the end. Calculate everything else first, then multiply by $e^{-\lambda}$.

Where's my drink? I want a drink to go with my popcorn. Give me my drink now!

The Statsville Cinema has another problem.

It's not just the popcorn machine that keeps breaking down, now the drinks machine has begun malfunctioning too. The mean number of breakdowns per week of the drinks machine is 2.3.

The cinema manager can't afford for *anything* to go wrong next week when the promotion is on. What's the probability that there will be no breakdowns next week, either with the popcorn machine nor the drinks machine?

BRAIN POWER

What's the probability distribution of the drinks machine? How can we find the probability that neither the popcorn machine nor the drinks machine go wrong next week?

So what's the probability distribution?

Let's take a closer look at this situation.

We have two machines, a popcorn machine and a drinks machine, and we know the mean number of breakdowns of each machine in a week. We want to find the probability that there will be no breakdowns next week.

Here are the distributions of the two machines:

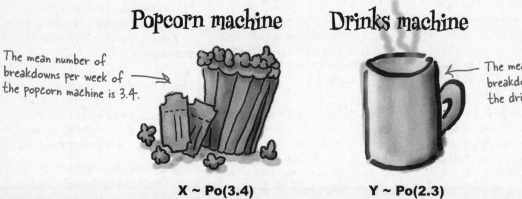

Popcorn machine

The mean number of breakdowns per week of the popcorn machine is 3.4.

X ~ Po(3.4)

Drinks machine

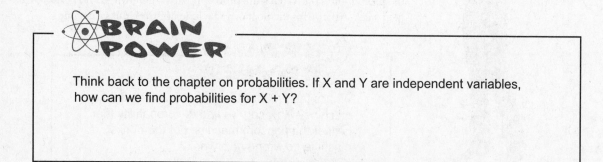

The mean number of breakdowns per week of the drinks machine is 2.3.

Y ~ Po(2.3)

If X represents the number of breakdowns of the popcorn machine and Y represents the number of breakdowns of the drinks machine, then both X and Y follow Poisson distributions. What's more, X and Y are independent. In other words, the popcorn machine breaking down has no impact on the probability that the drinks machine will malfunction, and the drinks machine breaking down has no impact on the probability that the popcorn machine will malfunction.

We need to find the probability that the total number of malfunctions next week is 0. In other words, we need to find

$$P(X + Y = 0)$$

BRAIN POWER

Think back to the chapter on probabilities. If X and Y are independent variables, how can we find probabilities for X + Y?

Combine Poisson variables

You saw in previous chapters that if X and Y are independent random variables, then

$$P(X + Y) = P(X) + P(Y)$$

$$E(X + Y) = E(X) + E(Y)$$

This means that if $X \sim Po(\lambda_x)$ and $Y \sim Po(\lambda_y)$,

$$\mathbf{X + Y \sim Po(\lambda_x + \lambda_y)}$$

This means that if X and Y both follow Poisson distributions, then so does X + Y. In other words, we can use our knowledge of the way both X and Y are distributed to find probabilities for X + Y.

Sharpen your pencil

If X is the number of times the popcorn machine malfunctions and Y is the number of times the drinks machine malfunctions, then $X \sim Po(3.4)$ and $Y \sim Po(2.3)$.

1. What's the distribution of X + Y?

2. Once you've found how X + Y is distributed, you can use it to find probabilities. What's $P(X + Y = 0)$?

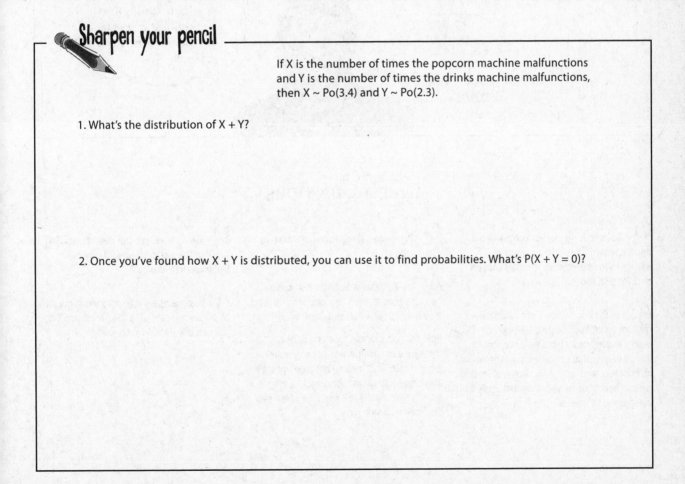

Sharpen your pencil Solution

If X is the number of times the popcorn machine malfunctions and Y is the number of times the drinks machine malfunctions, then $X \sim Po(3.4)$ and $Y \sim Po(2.3)$.

1. What's the distribution of X + Y?

$$\lambda_x + \lambda_y = 3.4 + 2.3$$
$$= 5.7$$
$$X + Y \sim Po(5.7)$$

2. Once you've found how X + Y is distributed, you can use it to find probabilities. What's P(X + Y = 0)?

$$P(X + Y = 0) = \frac{e^{-\lambda} \lambda^r}{r!}$$
$$= \frac{e^{-5.7} \times 5.7^0}{0!}$$
$$= \frac{e^{-5.7} \times 1}{1}$$
$$= 0.003$$

Only a .003 chance of no breakdowns next week? Guess we better get some new machines after all.

there are no Dumb Questions

Q: Does that mean that the probability and expectation shortcuts we saw earlier in the book work for the Poisson distribution too?

A: Yes they do. X and Y are independent random variables, because the popcorn machine malfunctioning does not affect the probability that the drinks machine will malfunction, and vice versa. This means that we can use all of the shortcuts that apply to independent variables.

Q: Why does X + Y follow a Poisson distribution?

A: X + Y follows a Poisson distribution because both X and Y are independent, and they both follow a Poisson distribution.

Both the popcorn machine and drinks machine each malfunction at random but at a mean rate. This means that together they also breakdown at random and at a mean rate. Together, they still meet the criteria for the Poisson distribution.

Q: So can we use the distribution of X + Y in the same we would any other Poisson distribution?

A: Yes, we use it in exactly the same way, so once you know what the parameter λ is, you can use it to find probabilities.

The Case of the Broken Cookies

Kate works at the Statsville cookie factory, and her job is to make sure that boxes of cookies meet the factory's strict rules on quality control.

Kate know that the probability that a cookie is broken is 0.1, and her boss has asked her to find the probability that there will be 15 broken cookies in a box of 100 cookies. "It's easy," he says. "Just use the binomial distribution where n is 100, and p is 0.1."

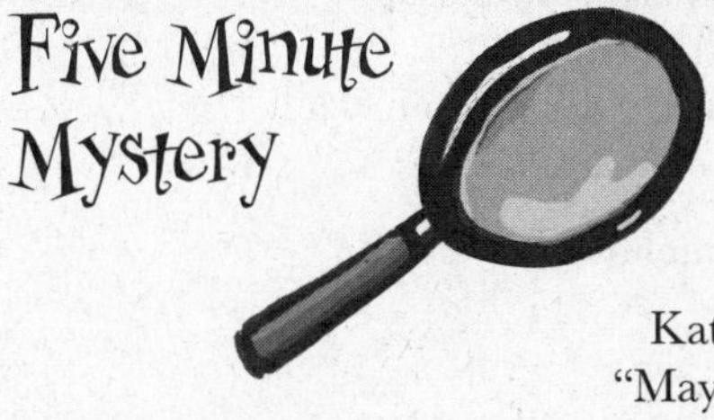

Kate picks up her calculator, but when she tries to calculate 100!, her calculator displays an error because the number is too big. "Well," says her boss, "you'll just have to calculate it manually. But I'm going home now, so have a nice night."

Kate stares at her calculator, wondering what to do. Then she smiles. "Maybe I can leave early tonight, after all."

Within a minute, Kate's calculated the probability. She's managed to find the probability and has managed to avoid calculating 100! altogether. She picks up her coat and walks out the door.

How did Kate find the probability so quickly, and avoid the error on her calculator?

The Poisson in disguise

The Poisson distribution has another use too. Under certain circumstances it can be used to approximate the binomial distribution.

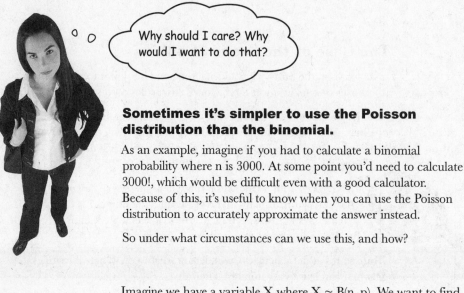

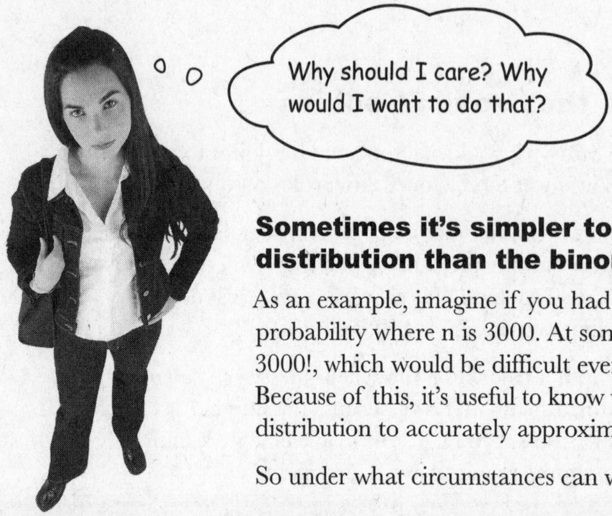

Why should I care? Why would I want to do that?

Sometimes it's simpler to use the Poisson distribution than the binomial.

As an example, imagine if you had to calculate a binomial probability where n is 3000. At some point you'd need to calculate 3000!, which would be difficult even with a good calculator. Because of this, it's useful to know when you can use the Poisson distribution to accurately approximate the answer instead.

So under what circumstances can we use this, and how?

Imagine we have a variable X where X ~ B(n, p). We want to find a set of circumstances where B(n, p) is similar to Po(λ).

Let's start off by looking at the expectation and variance of the two distributions. We want to find the circumstances in which the expectation and variance of the Poisson distribution are like those of the Binomial distribution. In other words, we want

Expectation → **λ** to be like **np**

Variance → **λ** to be like **npq**

→ **np** to be like **npq**

np and npq are close to each other if q is close to 1 and n is large. In other words:

X ~ B(n, p) can be approximated by X ~ Po(np) if n is large and p is small

The approximation is typically very close if n is larger than 50, and p is less than 0.1.

Exercise

A student needs to take an exam, but hasn't done any revision for it. He needs to guess the answer to each question, and the probability of getting a question right is 0.05. There are 50 questions on the exam paper. What's the probability he'll get 5 questions right? Use the Poisson approximation to the binomial distribution to find out.

there are no Dumb Questions

Q: Why would I ever want to use the Poisson distribution to approximate the binomial distribution?

A: When n is very large, it can be difficult to calculate ${}^nC_r$. Some calculators run out of memory, and the results can be so large they're just unwieldy. Using the Poisson distribution in this way is a way round this sort of problem.

Q: So when can I use this approximation?

A: You can use it when n is large (say over 50) and p is small (say less than 0.1). When this is the case, the binomial distribution and the Poisson distribution are approximately the same.

Q: Why do we use np as the parameter for the Poisson distribution?

A: The Poisson distribution takes one parameter, λ, and E(X) = λ. This means that if we have use the Poisson approximation of the binomial distribution, we can substitute in the expectation of the binomial distribution, np.

Exercise Solution

A student needs to take an exam, but hasn't done any revision for it. He needs to guess the answer to each question, and the probability of getting a question right is 0.05. There are 50 questions on the exam paper. What's the probability he'll get 5 questions right? Use the Poisson approximation to the binomial distribution to find out.

Let's use X to represent the number of questions the student gets right. In this problem, $n = 50$ and $p = 0.05$, $np = 2.5$. This means we can use $X \sim Po(2.5)$ to approximate the probability.

$$P(X = 5) = \frac{e^{-\lambda}\lambda^r}{r!}$$

$$= \frac{e^{-2.5} \times 2.5^5}{5!}$$

$$= \frac{e^{-2.5} \times 97.65625}{120}$$

$$= e^{-2.5} \times 0.8138$$

$$= 0.067$$

Solved: The Case of the Broken Cookies

How did Kate find the probability so quickly, and avoid the Out of Memory error on her calculator?

Five Minute Mystery Solved

Kate spotted that even though she needed to use the binomial distribution, her values of n and p were such that she could approximate the probability using the Poisson distribution instead.

A lot of calculators can't cope with high factorials, and this can sometimes make the binomial distribution unwieldy. Knowing how to approximate it with the Poisson distribution can sometimes save you quite a bit of time.

Anyone for popcorn?

You've covered a lot of ground in this chapter. You've built on your existing knowledge of probability and statistics by tackling three of the most important discrete probability distributions. Moreover, you've gained a deeper understanding of how probability distributions work and the sort of shortcuts you can make to save yourself time and produce reliable results, skills that will come in useful in the rest of the book.

So sit back and enjoy the popcorn — you've earned it.

Your quick guide to the Poisson distribution

Here's a quick summary of everything you could possibly need to know about the Poisson distribution

When do I use it?

Use the Poisson distribution if you have independent events such as malfunctions occurring in a given interval, and you know λ, the mean number of occurrences in a given interval. You're interested in the number of occurrences in one particular interval.

How do I calculate probabilities, and the expectation and variance?

Use

$$P(X = r) = \frac{e^{-\lambda} \lambda^r}{r!} \qquad E(X) = \lambda \qquad Var(X) = \lambda$$

How do I combine independent random variables?

If $X \sim Po(\lambda_x)$ and $Y \sim Po(\lambda_y)$, then

$$X + Y \sim Po(\lambda_x + \lambda_y)$$

What connection does it have to the binomial distribution?

If $X \sim B(n, p)$, where n is large and p is small, then X can be approximated using

$$X \sim Po(np)$$

Long Exercise

Here are some scenarios. Your job is to say which distribution each of them follows, say what the expectation and variance are, and find any required probabilities.

1. A man is bowling. The probability of him knocking all the pins over is 0.3. If he has 10 shots, what's the probability he'll knock all the pins over less than three times?

2. On average, 1 bus stops at a certain point every 15 minutes. What's the probability that no buses will turn up in a single 15 minute interval?

3. 20% of cereal packets contain a free toy. What's the probability you'll need to open fewer than 4 cereal packets before finding your first toy?

Long Exercise Solution

Here are some scenarios. Your job is to say which distribution each of them follows, say what the expectation and variance are, and find any required probabilities.

1. A man is bowling. The probability of him knocking all the pins over is 0.3. If he has 10 shots, what's the probability he'll knock all the pins down less than three times?

If X is the number of times the man knocks all the pins over, then $X \sim B(10, 0.3)$

$$E(X) = np = 10 \times 0.3 = 3$$

$$Var(X) = npq = 10 \times 0.3 \times 0.7 = 2.1$$

For a general probability, $P(X = r) = {}^nC_r \times p^r \times q^{n-r}$

$$\begin{aligned} P(X = 0) &= {}^{10}C_0 \times 0.3^0 \times 0.7^{10} \\ &= 1 \times 1 \times 0.028 \\ &= 0.028 \end{aligned}$$

$$\begin{aligned} P(X = 1) &= {}^{10}C_1 \times 0.3^1 \times 0.7^9 \\ &= 10 \times 0.3 \times 0.04035 \\ &= 0.121 \end{aligned}$$

$$\begin{aligned} P(X = 2) &= {}^{10}C_2 \times 0.3^2 \times 0.7^8 \\ &= 45 \times 0.09 \times 0.0576 \\ &= 0.233 \end{aligned}$$

$$\begin{aligned} P(X < 3) &= P(X = 0) + P(X = 1) + P(X = 2) \\ &= 0.028 + 0.121 + 0.233 \\ &= 0.382 \end{aligned}$$

2. On average, 1 bus stops at a certain point every 15 minutes. What's the probability that no buses will turn up in a single 15 minute interval?

If X is the number of buses that stop in a 15 minute interval, then $X \sim Po(1)$

$E(X) = \lambda$
$= 1$

$Var(X) = \lambda$
$= 1$

For a general probability, $P(X = r) = \frac{e^{-\lambda} \lambda^r}{r!}$

$$P(X = 0) = \frac{e^{-1} \times 1^0}{0!}$$
$$= \frac{e^{-1} \times 1}{1}$$
$$= 0.368$$

3. 20% of cereal packets contain a free toy. What's the probability you'll need to open fewer than 4 cereal packets before finding your first toy?

If X is the number of cereal packets that need to be opened in order to find your first toy, then $X \sim Geo(0.2)$

$E(X) = 1/p$
$= 1/0.2$
$= 5$

$Var(X) = q/p^2$
$= 0.8/0.2^2$
$= 0.8/0.04$
$= 20$

For a general probability, $P(X \leq r) = 1 - q^r$

$$P(X \leq 3) = 1 - q^r$$
$$= 1 - 0.8^3$$
$$= 1 - 0.512$$
$$= 0.488$$

BULLET POINTS

- The **geometric distribution** applies when you run a series of independent trials, there can be either a success or failure for each trial, the probability of success is the same for each trial, and the main thing you're interested in is how many trials are needed in order to get your first success.
- If the conditions are met for the geometric distribution, X is the number of trials needed to get the first successful outcome, and p is the probability of success in a trial, then

  $X \sim Geo(p)$

- The following probabilities apply if $X \sim Geo(p)$:

  $P(X = r) = pq^{r-1}$
  $P(X > r) = q^r$
  $P(X \leq r) = 1 - q^r$

- If $X \sim Geo(p)$ then

  $E(X) = 1/p$
  $Var(X) = q/p^2$

- The **binomial distribution** applies when you run a series of finite independent trials, there can be either a success or failure for each trial, the probability of success is the same for each trial, and the main thing you're interested in is the number of successes in the n independent trials.
- If the conditions are met for the binomial distribution, X is the number of successful outcomes out of n trials, and p is the probability of success in a trial, then

  $X \sim B(n, p)$

- If $X \sim B(n, p)$, you can calculate probabilities using

  $P(X = r) = {}^nC_r \, p^r q^{n-r}$

  where

  $${}^nC_r = \frac{n!}{r!\,(n-r)!}$$

- If $X \sim B(n, p)$, then

  $E(X) = np$
  $Var(X) = npq$

- The **Poisson distribution** applies when individual events occur at random and independently in a given interval, you know the mean number of occurrences in the interval or the rate of occurrences and this is finite, and you want to know the number of occurrences in a given interval.
- If the conditions are met for the Poisson distribution, X is the number of occurrences in a particular interval, and λ is the rate of occurrences, then

  $X \sim Po(\lambda)$

- If $X \sim Po(\lambda)$ then

  $$P(X = r) = \frac{e^{-\lambda} \lambda^r}{r!}$$

  $E(X) = \lambda$
  $Var(X) = \lambda$

- If $X \sim Po(\lambda_x)$, $Y \sim Po(\lambda_y)$ and X and Y are independent,

  $X + Y \sim Po(\lambda_x + \lambda_y)$

- If $X \sim B(n, p)$ where n is large and p is small, you can approximate it with $X \sim Po(np)$.

8 using the normal distribution

Being Normal

Whoever told you that's
normal was kidding.

Discrete probability distributions can't handle every situation.

So far we've looked at probability distributions where we've been able to specify exact values, but this isn't the case for every set of data. Some types of data just **don't fit** the probability distributions we've encountered so far. In this chapter, we'll take a look at how **continuous probability distributions** work, and introduce you to one of the most important probability distributions in town—the **normal distribution**.

Discrete data takes exact values...

So far we've looked at probability distributions where the data is ***discrete***. By this we mean the data is composed of distinct numeric values, and we're been able to calculate the probability of each of these values. As an example, when we looked at the probability distribution for the winnings on a slot machine, the possible amounts we could win on each game were very precise. We knew exactly what amounts of money we could win, and we knew we'd win one of them.

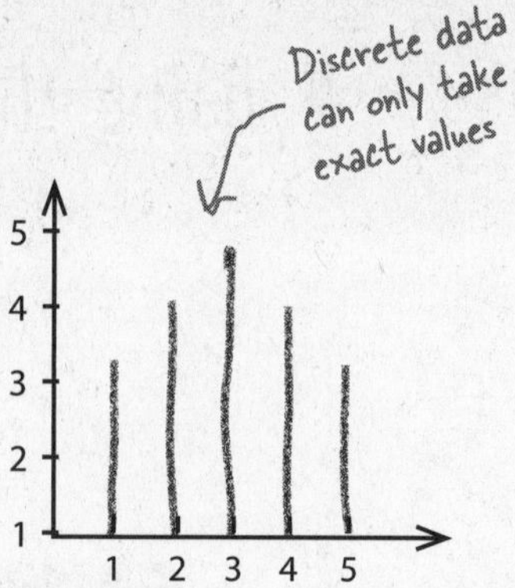

If data is discrete, it's numeric and can take only exact values. It's often data that can be *counted* in some way, such as the number of gumballs in a gumball machine, the number of questions answered correctly in a game show, or the number of breakdowns in a particular period.

You can think of discrete data as being like a series of stepping stones. You can step from value to value, and there are definite breaks between each value.

...but not all numeric data is discrete

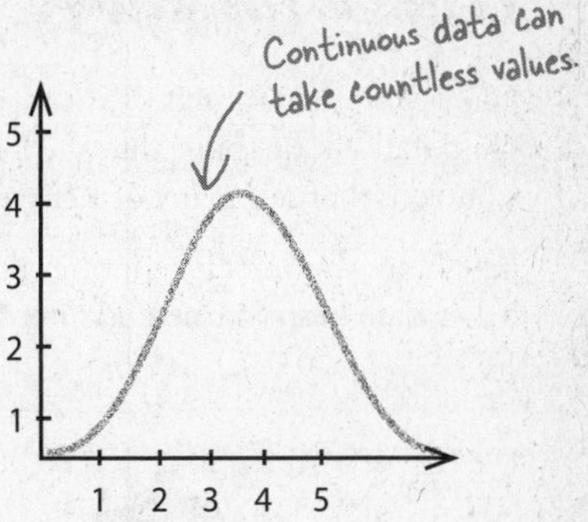

It's not always possible to say what all the values should be in a set of data. Sometimes data covers a range, where any value within that range is possible. As an example, suppose you were asked to accurately measure pieces of string that are between 10 inches and 11 inches long. You could have measurements of 10 inches, 10.1 inches, 10.01 inches, and so on, as the length could be anything within that range.

Numeric data like this is called ***continuous***. It's frequently data that is *measured* in some way rather than counted, and a lot depends on the degree of precision you need to measure to.

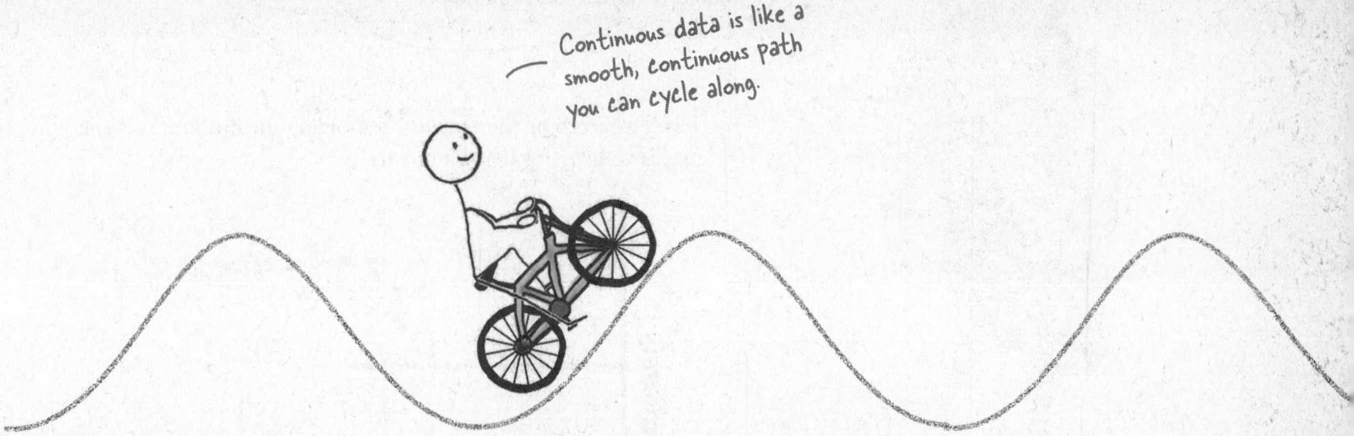

Continuous data is like a smooth, continuous path you can cycle along.

But why should I care about continuous data?

The type of data you have affects how you find probabilities.

So far we've only looked at probability distributions that deal with discrete data. Using these probability distributions, we've been able to find the probabilities of exact discrete values.

The problem is that a lot of real-world problems involve continuous data, and discrete probability distributions just don't work with this sort of data. To find probabilities for continuous data, you need to know about continuous data and continuous probability distributions.

Meanwhile, someone has a problem...

What's the delay?

Julie is a student, and her best friend keeps trying to get her fixed up on blind dates in the hope that she'll find that special someone. The only trouble is that not many of her dates are punctual—or indeed turn up.

Julie hates waiting alone for her date to arrive, so she's made herself a rule: if her date hasn't turned up after 20 minutes, then she leaves.

Meet Julie, a girl on a mission to find the perfect partner for herself.

Here's a sketch of the frequency showing the amount of time Julie spends waiting for her date to arrive:

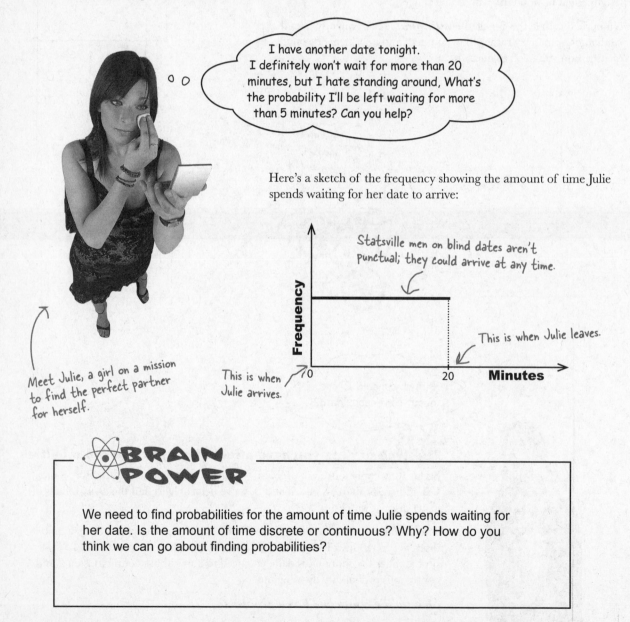

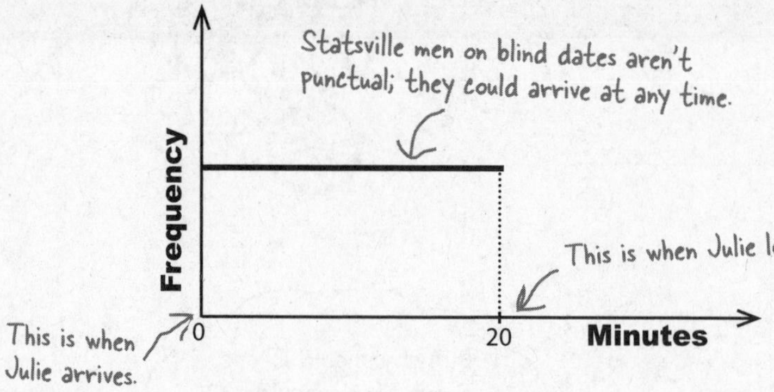

Statsville men on blind dates aren't punctual; they could arrive at any time.
Frequency
This is when Julie leaves.
0
20
Minutes
This is when Julie arrives.

BRAIN POWER

We need to find probabilities for the amount of time Julie spends waiting for her date. Is the amount of time discrete or continuous? Why? How do you think we can go about finding probabilities?

We need a probability distribution for continuous data

We need to find the probability that Julie will have to wait for more than 5 minutes for her date to turn up. The trouble is, the amount of time Julie has to wait is continuous data, which means the probability distributions we've learned thus far don't apply.

When we were dealing with discrete data, we were able to produce a specific probability distribution. We could do this by either showing the probability of each value in a table, or by specifying whether it followed a defined probability distribution, such as the binomial or Poisson distribution. By doing this, we were able to specify the probability of each possible value. As an example, when we found the probability distribution for the winnings per game for one of Fat Dan's slot machines, we knew all of the possible values for the winnings and could calculate the probability of each one..

With discrete data, we could give the probability of each value.

x	-1	4	9	14	19
P(X = x)	0.977	0.008	0.008	0.006	0.001

For continuous data, it's a different matter. We can no longer give the probability of each value because it's impossible to say what each of these precise values is. As an example, Julie's date might turn up after 4 minutes, 4 minutes 10 seconds, or 4 minutes 10.5 seconds. Counting the number of possible options would be impossible. Instead, we need to focus on a particular level of accuracy and the probability of getting a **range** of values.

I get it. For discrete probability distributions, we look at the probability of getting a particular value; for continuous probability distributions, we look at the probability of getting a particular range.

Probability density functions can be used for continuous data

We can describe the probability distribution of a continuous random variable using a ***probability density function***.

A probability density function f(x) is a function that you can use to find the probabilities of a continuous variable across a range of values. It tells us what the shape of the probability distribution is.

Here's a sketch of the probability density function for the amount of time Julie spends waiting for her date to turn up:

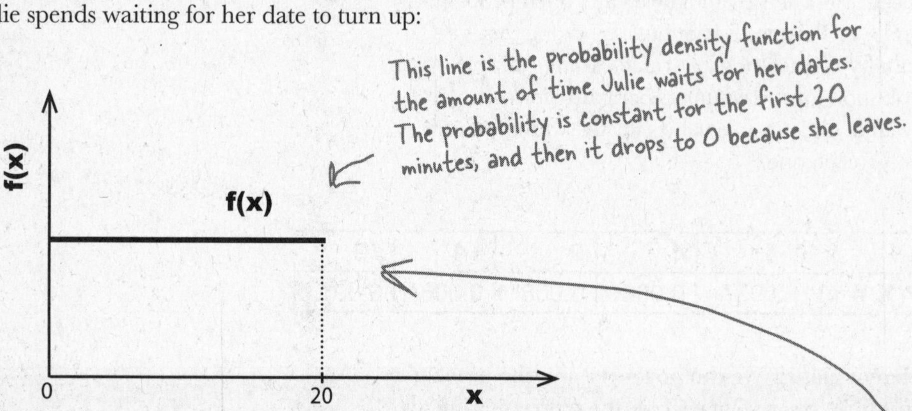

Can you see how it matches the shape of the frequency? This isn't just a coincidence.

Probability is all about how likely things are to happen, and the frequency tells you how often values occur. The higher the relative frequency, the higher the probability of that value occurring. As the frequency for the amount of time Julie has to wait is constant across the 20 minute period, this means that the probability density function is constant too.

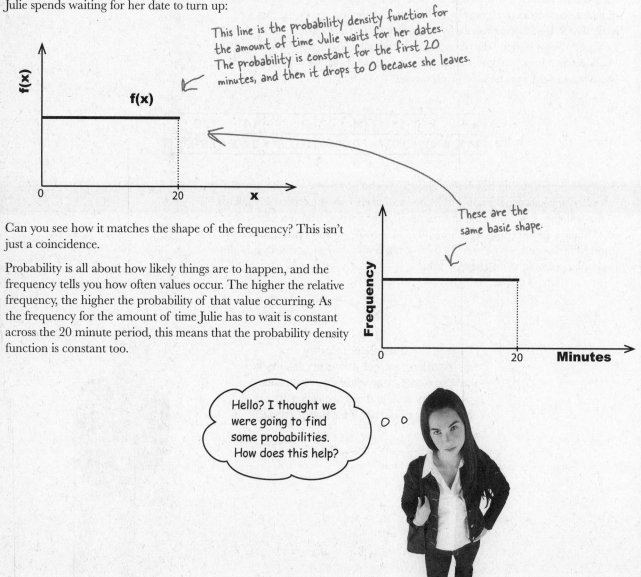

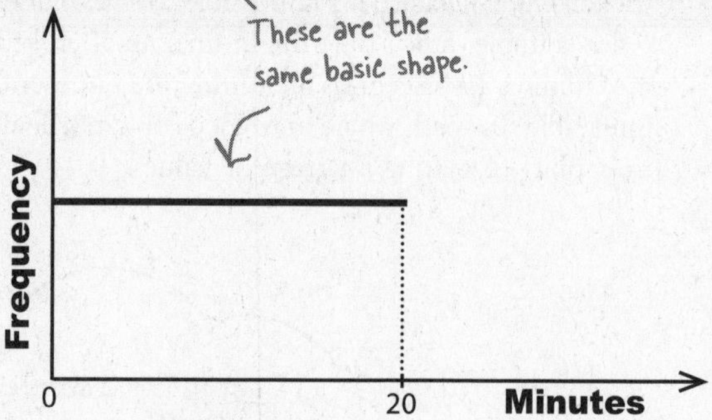

These are the same basic shape.
Frequency
0
20
Minutes

Hello? I thought we were going to find some probabilities. How does this help?

Probability = area

For continuous random variables, probabilities are given by area. To find the probability of getting a particular range of values, we start off by sketching the probability density function. The probability of getting a particular range of values is given by the area under the line between those values.

As an example, we want to find the probability that Julie has to wait for between 5 and 20 minutes for her date to turn up. We can find this probability by sketching the probability density function, and then working out the area under it where x is between 5 and 20.

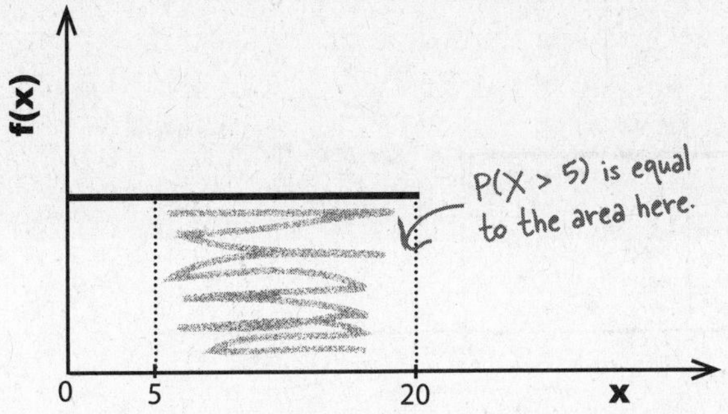
f(x)
P(X > 5) is equal to the area here.
0
5
20
x

The total area under the line must be equal to 1, as the total area represents the total probability. This is because for any probability distribution, the total probability must be equal to 1, and, therefore, the area must be too.

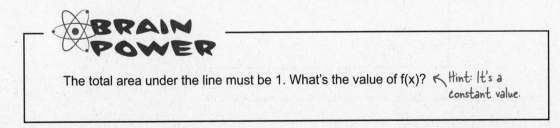

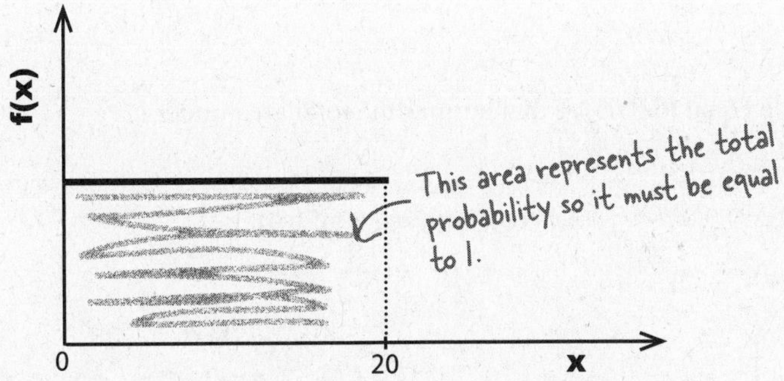
f(x)
This area represents the total probability so it must be equal to 1.
0
20
x

Let's use this to help us find the probability that Julie will need to wait for over 5 minutes for her date to arrive.

BRAIN POWER

The total area under the line must be 1. What's the value of f(x)? ← Hint: It's a constant value.

To calculate probability, start by finding f(x)...

Before we can find probabilities for Julie, we need to find f(x), the probability density function.

So far, we know that f(x) is a constant value, and we know that the total area under it must be equal to 1. If you look at the sketch of f(x), the area under it forms a rectangle where the width of the base is 20. If we can find the height of the rectangle, we'll have the value of f(x).

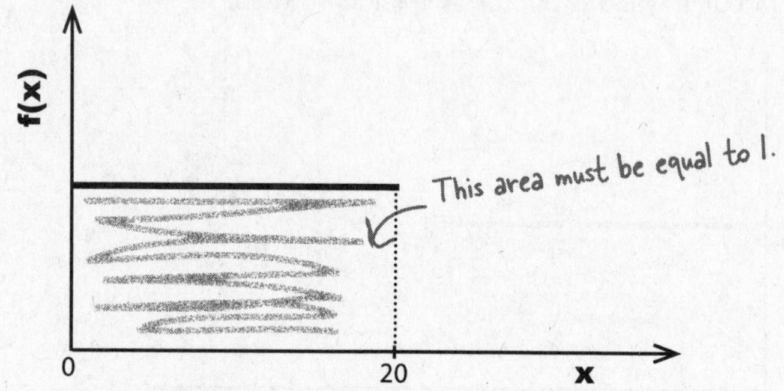

We find the area of a rectangle by multiplying its width and height together. This means that

$$1 = 20 \times \text{height}$$

$$\text{height} = 1/20$$

$$= 0.05$$

This means that f(x) must be equal to 0.05, as that ensures the total area under it will be 1. In other words,

$$f(x) = 0.05 \qquad \text{where x between 0 and 20}$$

Here's a sketch:

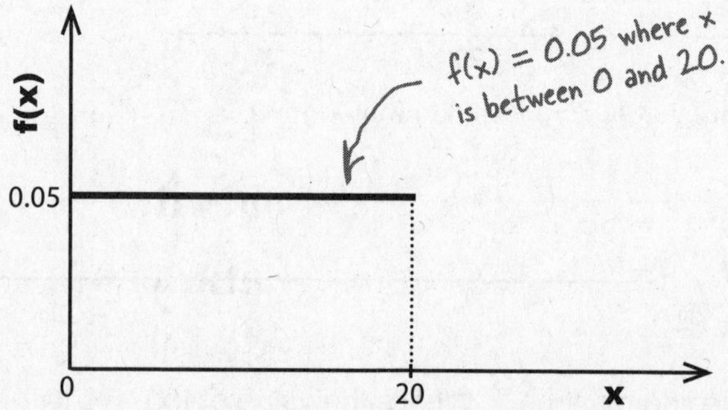

Now that we've found the probability density function, we can find P(X > 5).

...then find probability by finding the area

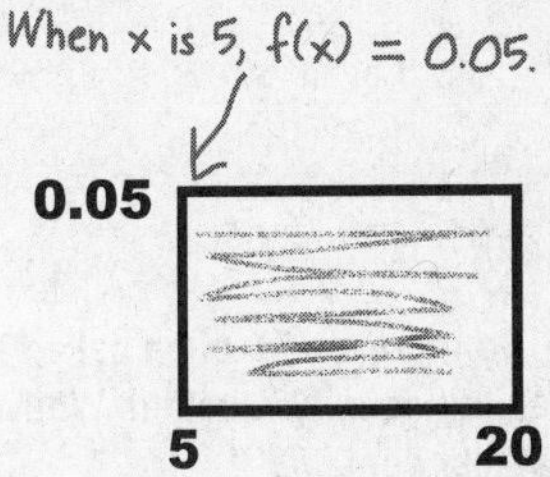

The area under the probability density line between 5 and 20 is a rectangle. This means that calculating the area of this rectangle will give us the probability P(X > 5).

$$P(X > 5) = (20 - 5) \times 0.05$$
$$= 0.75$$

← Area of rectangle = base x height.

So the probability that Julie will have to wait for more than 5 minutes is 0.75.

Do I **have** to use area to find probability? Can't I just pick all the exact values in that range and add their probabilities together? That's what we did for discrete probabilities.

That doesn't work for continuous probabilities.

For continuous probabilities, we *have* to find the probability by calculating the area under the probability density line.

We can't add together the probability of getting each value within the range as there are an infinite number of values. It would take forever.

The only way we can find the probability for continuous probability distributions is to work out the area underneath the curve formed by the probability density function.

When dealing with continuous data, you calculate probabilities for a range of values.

there are no Dumb Questions

Q: So there's a function called the probability density function. What's probability density?

A: Probability density tells you how high probabilities are across ranges, and it's described by the probability density function. It's very similar to frequency density, which we encountered back in Chapter 1. Probability density uses area to tell you about probabilities, and frequency density uses area to tell you about frequencies.

Q: So aren't probability density and probability the same thing?

A: Probability density gives you a means of *finding* probability, but it's not the probability itself. The probability density function is the line on the graph, and the probability is given by the area underneath it for a specific range of values.

Q: I see, so if you have a chart showing a probability density function, you find the probability by looking at area, instead of reading it directly off the chart.

A: Exactly. For continuous data, you need to find probability by calculating area. Reading probabilities directly off a chart only works for discrete probabilities.

Q: Doesn't finding the probability get complicated if you have to calculate areas? I mean, what if the probability density function is a curve and not a straight line?

A: It's still possible to do it, but you need to use calculus, which is why we're not expecting you to do that in this book. The key thing is that you see where the probabilities come from and how to interpret them.

If you're *really* interested in working out probabilities using calculus, by all means, give it a go. We don't want to hold you back.

Q: You've talked a lot about probability ranges. How do I find the probability of a precise value?

A: When you're dealing with continuous data, you're really talking about acceptable degrees of accuracy, and you form a range based on these values. Let's look at an example:

Suppose you wanted a piece of string that's 10 inches long to the nearest inch. It would be tempting to say that you need a piece of string that's exactly 10 inches long, but that's not entirely accurate. What you're *really* after is a piece of string that's between 9.5 inches and 10.5 inches, as you want string that 10 inches in length *to the nearest inch*. In other words, you want to find the probability of the length being in the range 9.5 inches to 10.5 inches.

Q: But what if I want to find the probability of a precise single value?

A: This may not sound intuitive at first, but it's actually 0. What you're really talking about is the probability that you have a precise value *to an infinite number of decimal places*.

If we go back to the string length example, what would happen if you needed a piece of string exactly 10 inches long? You would need to have a length of string measuring 10 inches long to the nearest atom and examined under a powerful microscope.

The probability of the string being precisely 10 inches long is virtually impossible.

Q: But I'm sure that degree of accuracy isn't needed. Surely it would be enough to measure it to the nearest hundredth of an inch?

A: Ah, but that brings us back to the degree of accuracy you need in order for the length to pass as 10 inches, rather than finding the probability of a value to an infinite degree of precision. You use your degree of accuracy to construct your range of acceptable measurements so that you can work out the probability.

BE the probability density function

A bunch of probability density functions have lost track of their probabilities. Your job is to play like you're the probability density function and work out the probability between the specified ranges. Draw a sketch if you think that will help.

1. $f(x) = 0.05$ where $0 < x < 20$

   Find $P(X < 5)$

2. $f(x) = 1$ where $0 < x < 1$

   Find $P(X < 0.5)$

3. $f(x) = 1$ where $0 < x < 1$

   Find $P(X > 2)$

4. $f(x) = 0.1 - 0.005x$ where $0 < x < 20$

   Find $P(X > 5)$

BE the probability density function solution

A bunch of probability density functions have lost track of their probabilities. Your job is to play like you're the probability density function and work out the probability between the specified ranges. Draw a sketch if you think that will help.

1. $f(x) = 0.05$ where $0 < x < 20$

Find $P(X < 5)$

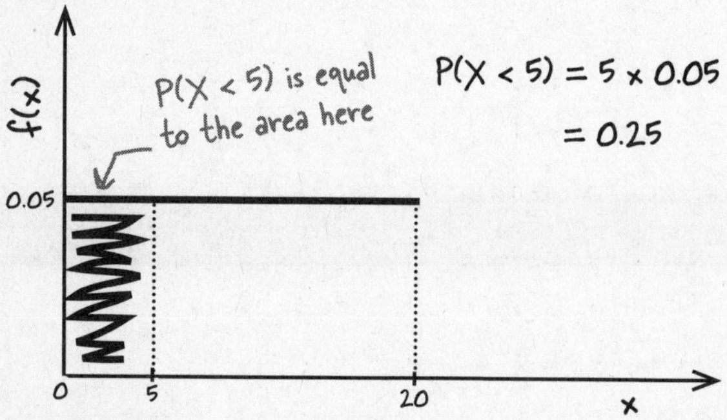

$P(X < 5) = 5 \times 0.05$

$= 0.25$

2. $f(x) = 1$ where $0 < x < 1$

Find $P(X < 0.5)$

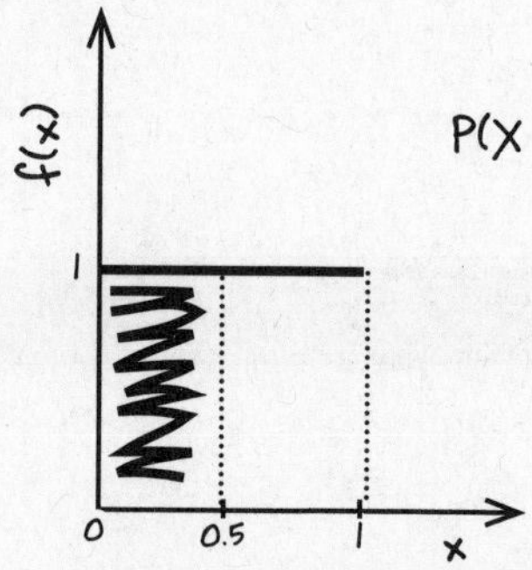

$P(X < 0.5) = 1 \times 0.5$

$= 0.5$

3. $f(x) = 1$ where $0 < x < 1$

Find $P(X > 2)$

The upper limit of x for this probability density function is 1, which means that it's 0 above this.

$P(X > 2) = 0$

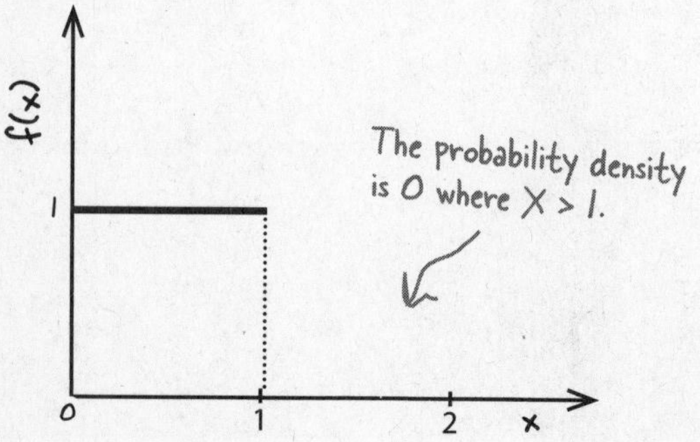

4. $f(x) = 0.1 - 0.005x$ where $0 < x < 20$

Find $P(X > 5)$

When $x = 5$, $f(x) = 0.075$. This means we have to find the area of a right-angled triangle with height 0.075 and width 15.

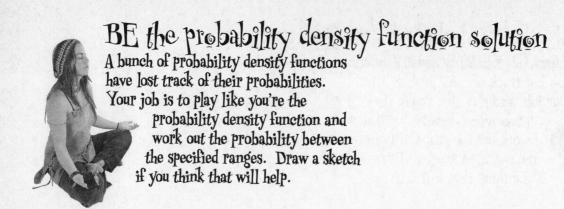

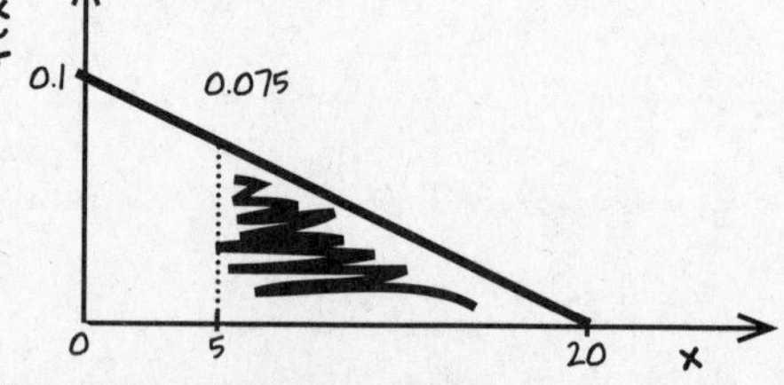

$P(X > 5) = (0.075 \times 15)/2$

$= 1.125/2$

$= 0.5625$

The area of a triangle is 1/2 the base multiplied by the height.

BULLET POINTS

- **Discrete data** is composed of distinct numeric values.
- **Continuous data** covers a range, where any value within that range is possible. It's frequently data that is measured in some way, rather than counted.
- Continuous probability distributions can be described with a probability density function.
- You find the probability for a range of values by calculating the area under the probability density function between those values. So to find $P(a < X < b)$, you need to calculate the area under the probability density function between a and b.
- The total area under the probability density function must equal 1.

We've found the probability

So far, we've looked at how you can use probability density functions to find probabilities for continuous data. We've found that the probability that Julie will have to wait for more than 5 minutes for her date to turn up is 0.75.

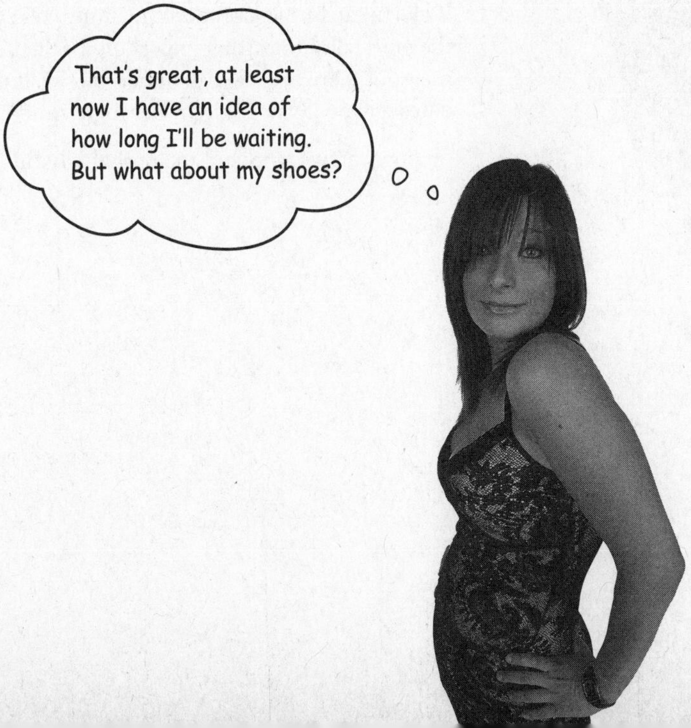

That's great, at least now I have an idea of how long I'll be waiting. But what about my shoes?

Searching for a ~~soul~~ sole mate

As well as preferring men who are punctual, Julie has preconceived ideas about what the love of her like should be like.

I need a man who'll be taller than me when I wear my highest heels. Shoes definitely come first.

Julie loves wearing high-heeled shoes, and the higher the heel, the happier she is. The only problem is that she insists that her dates should be taller than her when she's wearing her most extreme set of heels, and she's running out of suitable men.

Unfortunately, the last couple of times Julie was sent on a blind date, the guys fell short of her expectations. She's wondering how many men out there are taller than her and what the probability is that her dates will be tall enough for her high standards.

So how can we work out the probability this time?

Male modelling

So far we've looked at very simple continuous distributions, but it's unlikely these will model the heights of the men Julie might be dating. It's likely we'll have several men who are quite a bit shorter than average, a few really tall ones, and a lot of men somewhere in between. We can expect most of the men to be average height.

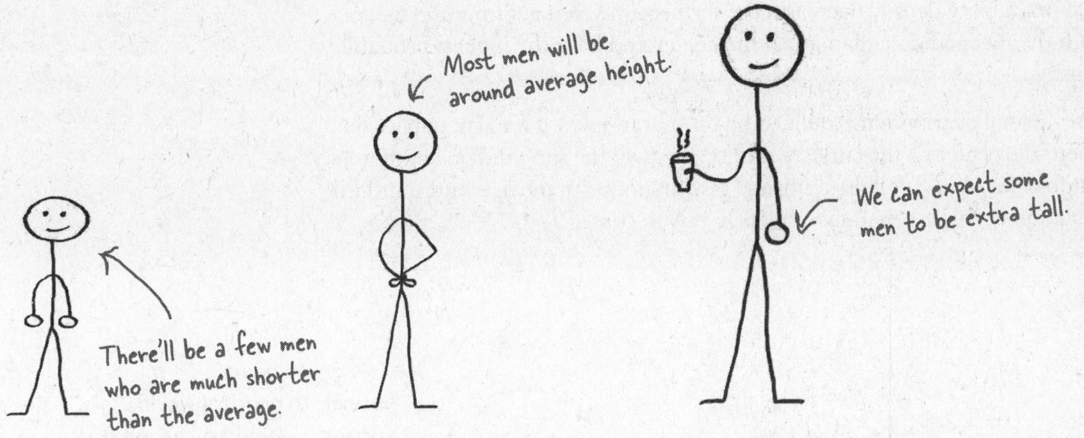

Most men will be around average height.
We can expect some men to be extra tall.
There'll be a few men who are much shorter than the average.

Given this pattern, the probability density of the height of the men is likely to look something like this.

There are fewer shorter guys, so the probability density is low.
Most men will be average height.
There'll be a smaller number of tall guys.

This shape of distribution is actually fairly common and can be applied to lots of situations. It's called the **normal distribution**.

The normal distribution is an "ideal" model for continuous data

The normal distribution is called normal because it's seen as an ideal. It's what you'd "normally" expect to see in real life for a lot of continuous data such as measurements.

The normal distribution is in the shape of a bell curve. The curve is symmetrical, with the highest probability density in the center of the curve. The probability density decreases the further away you get from the mean. Both the mean and median are at the center and have the highest probability density.

The normal distribution is defined by two parameters, μ and σ^2. μ tells you where the center of the curve is, and σ gives you the spread. If a continuous random variable X follows a normal distribution with mean μ and standard deviation σ, this is generally written $X \sim N(\mu, \sigma^2)$.

$$X \sim N(\mu, \sigma^2)$$

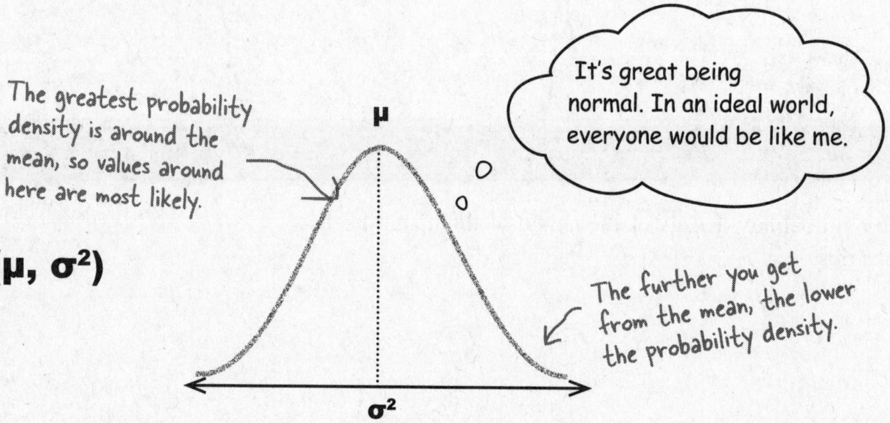

The greatest probability density is around the mean, so values around here are most likely.
μ
It's great being normal. In an ideal world, everyone would be like me.
The further you get from the mean, the lower the probability density.
σ²

So what effect do μ and σ really have on the shape of the normal distribution?

We said that μ tells you where the center of the curve is, and σ^2 indicates the spread of values. In practice, this means that as σ^2 gets larger, the flatter and wider the normal curve becomes.

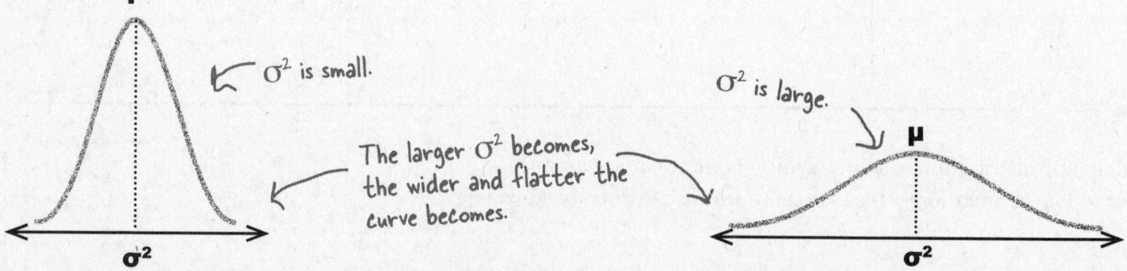

The larger σ² becomes, the wider and flatter the curve becomes.

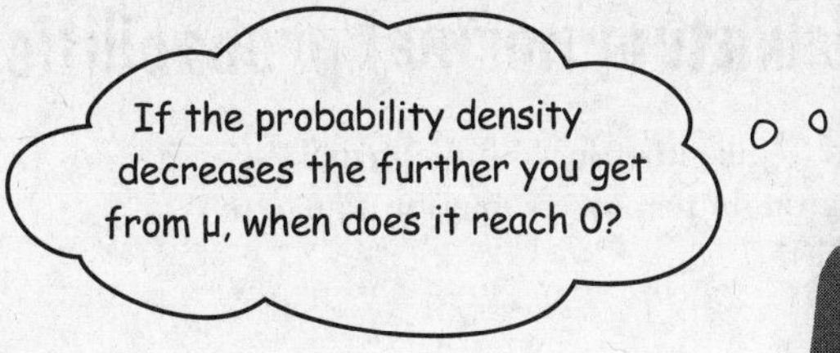
If the probability density decreases the further you get from μ, when does it reach 0?

No matter how far you go out on the graph, the probability density never equals 0.

The probability density gets closer and closer to 0, but never quite reaches it. If you looked at the probability density curve a very long way from μ, you'd find that the curve just skims above 0.

Another way of looking at this is that events become more and more unlikely to occur, but there's always a tiny chance they might.

So how do we find normal probabilities?

As with any other continuous probability distribution, you find probabilities by calculating the area under the curve of the distribution. The curve gives the probability density, and the probability is given by the area between particular ranges. If, for instance, you wanted to find the probability that a variable X lies between a and b, you'd need to find the area under the curve between points a and b.

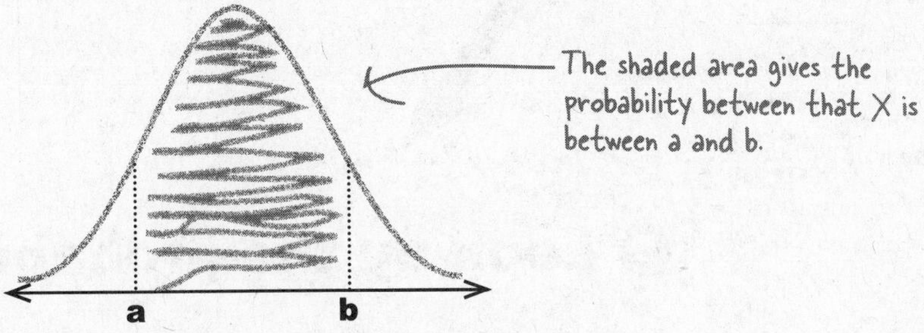
The shaded area gives the probability between that X is between a and b.
a
b

Sound complicated? Don't worry, it's easier than you might think.

Working out the area under the normal curve would be difficult if you had to do it all by yourself, but fortunately you have a helping hand in the form of probability tables. All you need to do is work out the range of the area you want to find, and then look up the corresponding probability in the table.

Three steps to calculating normal probabilities

There are a few steps you need to take in order to find normal probabilities. We'll guide you through the process, but for now here's a roadmap of where we're headed.

If the normal distribution applies to your situation, see if you can find what the mean and standard deviation are. You'll need these before you can find your probabilities. You also need to figure out what area you need to find.

1 Grab your distribution and range

Don't worry about this for now; we'll show you how to do this really soon.

2 Standardize it

Once you've transformed your normal curve, you can look up probabilities using handy probability tables. Job done!

3 Look up the probabilities

Step 1: Determine your distribution

The first thing we need to do is determine the distribution of the data.

Julie has been given the mean and standard deviation of the heights of eligible men in Statsville. The mean is 71 inches, and the variance is 20.25 inches. This means that if X represents the heights of the men, X ~ N(71, 20.25).

This is shorthand for "The variable X follows a normal distribution, and has a mean of 71 and a variance of 20.25."

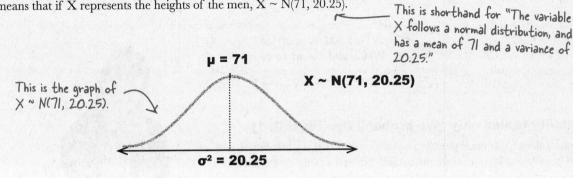

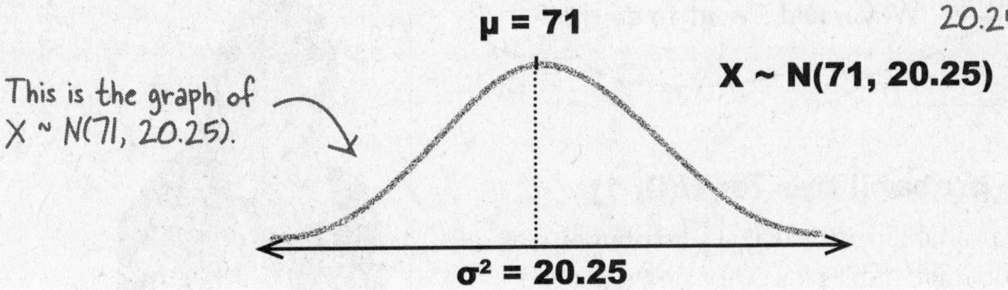

This is the graph of X ~ N(71, 20.25).

We also need to know which range of values will give us the right probability area. In this case, we need to find the probability that Julie's blind date will be sufficiently tall.

That's easy. Julie wants her date to be taller than her, so we can work out probabilities based on her height.

Julie is 64 inches tall, so we'll find the probability that her date is taller. Here's a sketch:

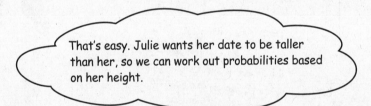

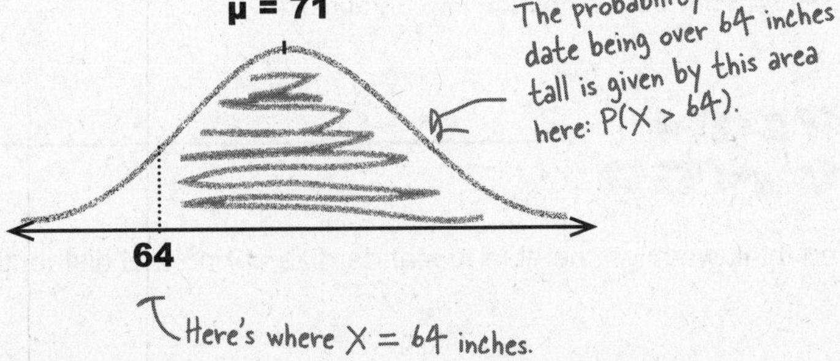

Here's where X = 64 inches.

Step 2: Standardize to N(0, 1)

The next step is to standardize our variable X so that the mean becomes 0 and the standard deviation 1. This gives us a standardized normal variable Z where $Z \sim N(0, 1)$.

Is that your idea of fun?
Why would I want to do that?

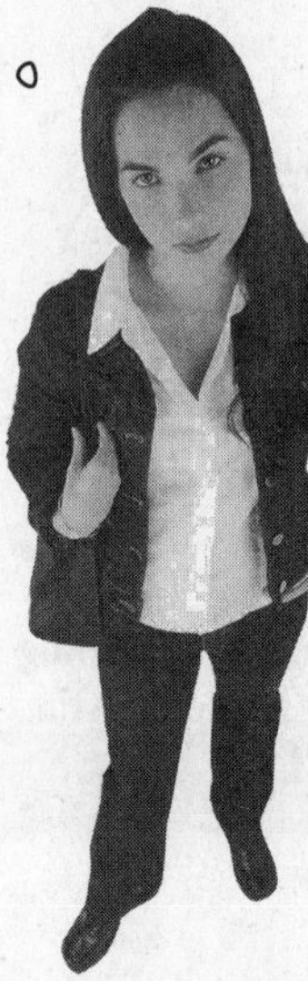

Probability tables only give probabilities for N(0, 1).

Probability tables focus on giving the probabilities for N(0, 1) distributions, as it would be impossible to produce probability tables for every single normal distribution curve. There are an infinite number of possible values for μ and σ^2, and as the normal curve uses these as parameters to indicate the center and spread of the curve, there are also an infinite number of possible normal distribution curves.

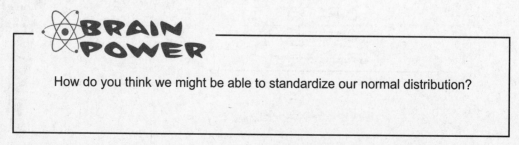

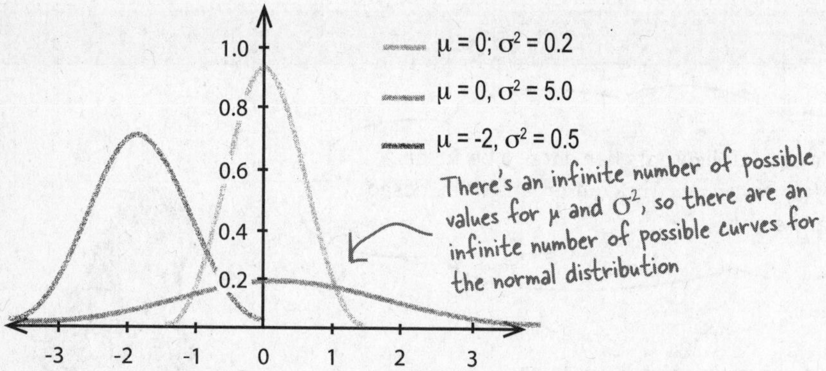

Being able to use a standard normal distribution means that we can use the same set of probability tables for all possible values of μ and σ^2. There's just one question—how do we convert out normal distribution into a standard form?

BRAIN POWER

How do you think we might be able to standardize our normal distribution?

To standardize, first move the mean...

Let's start off by transforming our normal distribution so that the mean becomes 0 rather than 71. To do this, we move the curve to the left by 71.

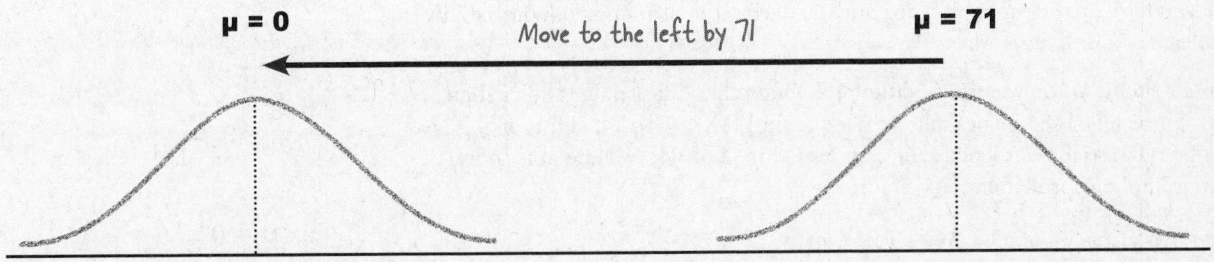
μ = 0
Move to the left by 71
μ = 71

This gives us a new distribution of

$$X - 71 \sim N(0, 20.25)$$

...then squash the width

We also need to adjust the variance. To do this, we "squash" our distribution by dividing by the standard deviation. We know the variance is 20.25, so the standard deviation is 4.5. ← Recall that the standard deviation is the square root of the variance.

Doing this gives us $\dfrac{X - 71}{4.5} \sim N(0, 1)$

or $Z \sim N(0, 1)$ where

$$Z = \frac{X - 71}{4.5}$$

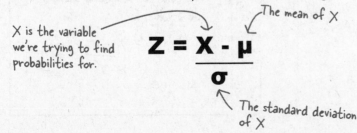

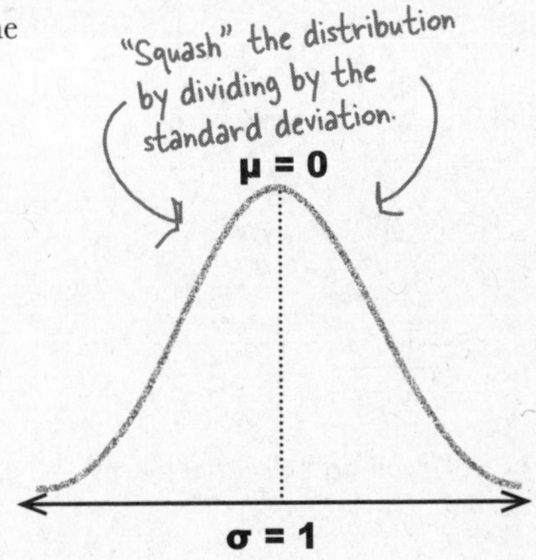
"Squash" the distribution by dividing by the standard deviation.
μ = 0
σ = 1

Look familiar? This is the ***standard score*** we encountered when we first looked at the standard deviation in Chapter 3. In general, you can find the standard score for any normal variable X using

$$Z = \frac{X - \mu}{\sigma}$$

X is the variable we're trying to find probabilities for.

The mean of X

The standard deviation of X

Now find Z for the specific value you want to find probability for

So far we've looked at how our probability distribution can be standardized to get from $X \sim N(\mu, \sigma^2)$ to $Z \sim N(0, 1)$. What we're most interested in is actual probabilities. What we need to do is take the range of values we want to find probabilities for, and find the standard score of the limit of this range. Then we can look up the probability for our standard score using normal probability tables.

In our situation, we want to find the probability that Julie's date is taller than her. Since Julie is 64 inches tall, we need to find $P(X > 64)$. The limit of this range is 64, so if we calculate the standard score z of 64, we'll be able to use this to find our probability.

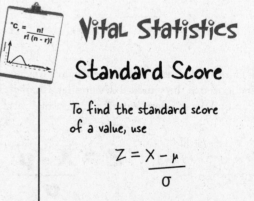

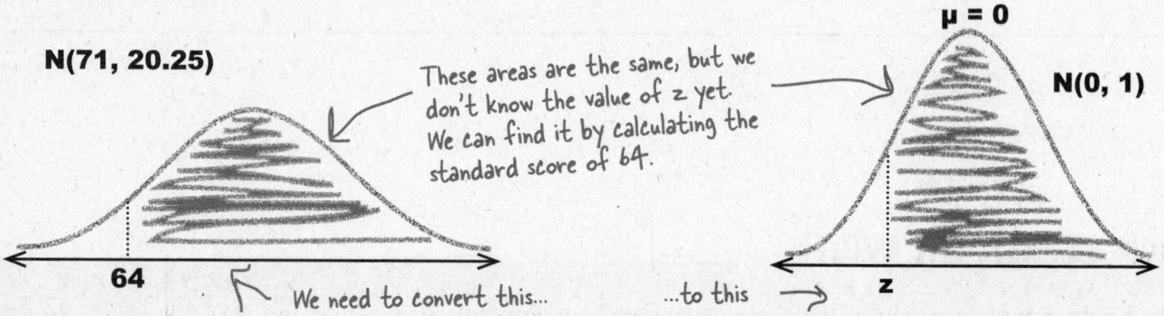

N(71, 20.25)
These areas are the same, but we don't know the value of z yet. We can find it by calculating the standard score of 64.
μ = 0
N(0, 1)
64
We need to convert this...
...to this
z

Let's find the standard score of 64.

$$z = \frac{x - \mu}{\sigma}$$

$$= \frac{64 - 71}{4.5}$$

$$= -1.56 \text{ (to 2 decimal places)}$$

So -1.56 is the standard score of 64, using the mean and standard deviation of the men's heights in Statsville.

Now that we have this, we can move onto the final step, using tables to look up the probability.

Vital Statistics

Standard Score

To find the standard score of a value, use

$$Z = \frac{X - \mu}{\sigma}$$

there are no Dumb Questions

Q: Is this the same standard score that we saw before?

A: Yes it is. It has more uses than just the normal distribution, but it's particularly useful here as it allows us to use standard normal probability tables.

Q: Is the probability for my standardized range really the same as for my original distribution? How does that work?

A: The probabilities work out the same, but using the probability tables is a lot more convenient.

When we standardize our original normal distribution, everything keeps the same proportion. The overall area doesn't grow or shrink, and as it's area that gives the probability, the probability stays the same too.

Sharpen your pencil

It's time to standardize. We'll give you a distribution and value, and you have to tell us what the standard score is.

1. N(10, 4), value 6

2. N(6.3, 9), value 0.3

3. N(2, 4). If the standard score is 0.5, what's the value?

4. The standard score of value 20 is 2. If the variance is 16, what's the mean?

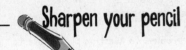

Sharpen your pencil Solution

It's time to standardize. We'll give you a distribution and value, and you have to tell us what the standard score is.

1. N(10, 4), value 6

$$z = \frac{x - \mu}{\sigma}$$

$$= \frac{6 - 10}{2}$$

$$= -2$$

2. N(6.3, 9), value 0.3

$$z = \frac{x - \mu}{\sigma}$$

$$= \frac{0.3 - 6.3}{3}$$

$$= -2$$

3. N(2, 4). If the standard score is 0.5, what's the value?

This is the reverse of previous problems. We're given the standard score, and we have to find the original value. We can do this by substituting in the values we know, and finding x.

$$z = \frac{x - \mu}{\sigma}$$

$$0.5 = \frac{x - 2}{2}$$

$$0.5 \times 2 = x - 2$$

$$x = 1 + 2$$

$$= 3$$

4. The standard score of value 20 is 2. If the variance is 16, what's the mean?

This is a similar problem to question 3. We have to substitute in the values we know to find μ.

$$z = \frac{x - \mu}{\sigma}$$

$$2 = \frac{20 - \mu}{4}$$

$$2 \times 4 = 20 - \mu$$

$$\mu = 20 - 8$$

$$= 12$$

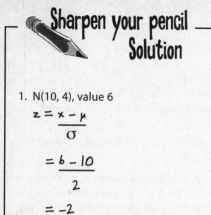

So we've found our distribution, standardized it, and found Z. Now can we find the probability of my blind date being taller than me?

Step 3: Look up the probability in your handy table

Now that we have a standard score, we can use probability tables to find our probability. Standard normal probability tables allow you to look up any value z, and then read off the corresponding probability P(Z < z).

We've put all the probability tables you need in Appendix ii of the book.

Just flip to pages 658-659 for the normal distribution tables you need to find probabilities in this chapter.

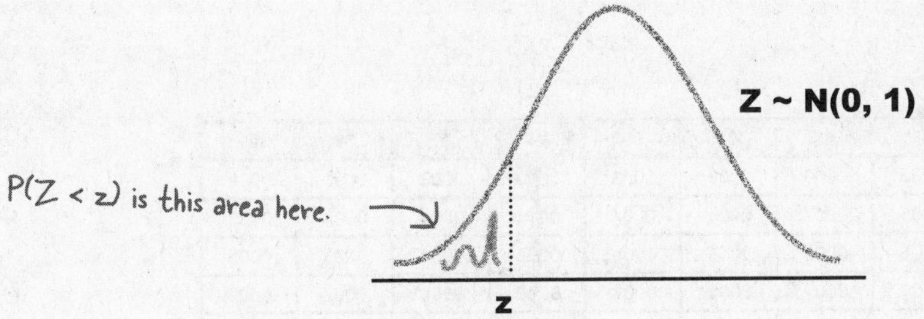

So how do you use probability tables?

Start off by calculating z to 2 decimal places. This is the value that you will need to look up in the table.

To look up the probability, you need to use the first column and the top row to find your value of z. The first column gives the value of z to 1 decimal place (without rounding), and the top row gives the second decimal place. The probability is where the two intersect.

As an example, if you wanted to find P(Z < –3.27), you'd find –3.2 in the first column, .07 in the top row, and read off a probability of 0.0005.

Here's the column for .07, the second decimal place for z

Here's the row for z = –3.2x, where x is some number.

This is where –3.2 and .07 meet. It gives the value of P(Z < z).

z	.00	.01	.02	.03	.04	.05	.06	.07	.08	.09
–3.4	.0003	.0003	.0003	.0003	.0003	.0003	.0003	.0003	.0003	.0002
–3.3	.0005	.0005	.0005	.0004	.0004	.0004	.0004	.0004	.0004	.0003
–3.2	.0007	.0007	.0006	.0006	.0006	.0006	.0006	.0005	.0005	.0005
–3.1	.0010	.0009	.0009	.0009	.0008	.0008	.0008	.0008	.0007	.0007
–3.0	.0013	.0013	.0013	.0012	.0012	.0011	.0011	.0011	.0010	.0010
–2.9	.0019	.0018	.0018	.0017	.0016	.0016	.0015	.0015	.0014	.0014
–2.8	.0026	.0025	.0024	.0023	.0023	.0022	.0021	.0021	.0020	.0019
–2.7	.0035	.0034	.0033	.0032	.0031	.0030	.0029	.0028	.0027	.0026
–2.6	.0047	.0045	.0044	.0043	.0041	.0040	.0039	.0038	.0037	.0036
–2.5	.0062	.0060	.0059	.0057	.0055	.0054	.0052	.0051	.0049	.0048
–2.4	.0082	.0080	.0078	.0075	.0073	.0071	.0069	.0068	.0066	.0064

Julie's probability is in the table

Let's go back to our problem with Julie. We want to find P(Z > -1.56), so let's look up -1.56 in the probability table and see what this gives us.

You can find normal probability tables in the appendix at the back of the book.

Here's the column for .06, the second decimal place for z

z	.00	.01	.02	.03	.04	.05	.06	.07	.08	.09
-3.4	.0003	.0003	.0003	.0003	.0003	.0003	.0003	.0003	.0003	.0002
-3.3	.0005	.0005	.0005	.0004	.0004	.0004	.0004	.0004	.0004	.0003
-3.2	.0007	.0007	.0006	.0006	.0006	.0006	.0006	.0005	.0005	.0005
-3.1	.0010	.0009	.0009	.0009	.0008	.0008	.0008	.0008	.0007	.0007
-3.0	.0013	.0013	.0013	.0012	.0012	.0011	.0011	.0011	.0010	.0010
-2.9	.0019	.0018	.0018	.0017	.0016	.0016	.0015	.0015	.0014	.0014
-2.8	.0026	.0025	.0024	.0023	.0023	.0022	.0021	.0021	.0020	.0019
-2.7	.0035	.0034	.0033	.0032	.0031	.0030	.0029	.0028	.0027	.0026
-2.6	.0047	.0045	.0044	.0043	.0041	.0040	.0039	.0038	.0037	.0036
-2.5	.0062	.0060	.0059	.0057	.0055	.0054	.0052	.0051	.0049	.0048
-2.4	.0082	.0080	.0078	.0075	.0073	.0071	.0069	.0068	.0066	.0064
-2.3	.0107	.0104	.0102	.0099	.0096	.0094	.0091	.0089	.0087	.0084
-2.2	.0139	.0136	.0132	.0129	.0125	.0122	.0119	.0116	.0113	.0110
-2.1	.0179	.0174	.0170	.0166	.0162	.0158	.0154	.0150	.0146	.0143
-2.0	.0228	.0222	.0217	.0212	.0207	.0202	.0197	.0192	.0188	.0183
-1.9	.0287	.0281	.0274	.0268	.0262	.0256	.0250	.0244	.0239	.0233
-1.8	.0359	.0351	.0344	.0336	.0329	.0322	.0314	.0307	.0301	.0294
-1.7	.0446	.0436	.0427	.0418	.0409	.0401	.0392	.0384	.0375	.0367
-1.6	.0548	.0537	.0526	.0516	.0505	.0495	[illegible]	.0475	[illegible]	.0455
-1.5	.0668	.0655	.0643	.0630	.0618	.0606	.0594	.0582	.0571	.0559
-1.4	.0808	.0793	.0778	.0764	.0749	.0735	.0721	.0708	.0694	.0681
-1.3	.0968	.0951	.0934	.0918	.0901	.0885	.0869	.0853	.0838	.0823
-1.2	.1151	.1131	.1112	.1093	.1075	.1056	.1038	.1020	.1003	.0985
-1.1	.1357	.1335	.1314	.1292	.1271	.1251	.1230	.1210	.1190	.1170

Here's the row for z = -1.5x, where x is some number.

This is where -1.5 and .06 meet. It gives the value of P(Z < z).

So, looking up the value of –1.56 in the probability table gives us a probability of 0.0594. In other words, P(Z < –1.56) = 0.0594. This means that

$$P(Z > -1.56) = 1 - P(Z < -1.56)$$

← The total probability is 1, so the total area under the curve is 1.

$$= 1 - 0.0594$$

$$= 0.9406$$

In other words, the probability that Julie's date is taller than her is 0.9406.

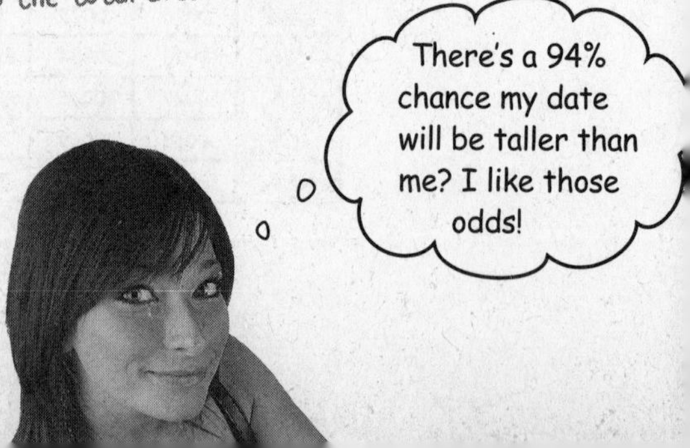

There's a 94% chance my date will be taller than me? I like those odds!

Probability Tables Up Close

Probability tables allow you to look up the probability $P(Z < z)$ where z is some value. The problem is you don't always want to find this sort of probability; sometimes you want to find the probability that a continuous random variable is greater than z, or between two values. How can you use probability tables to find the probability you need?

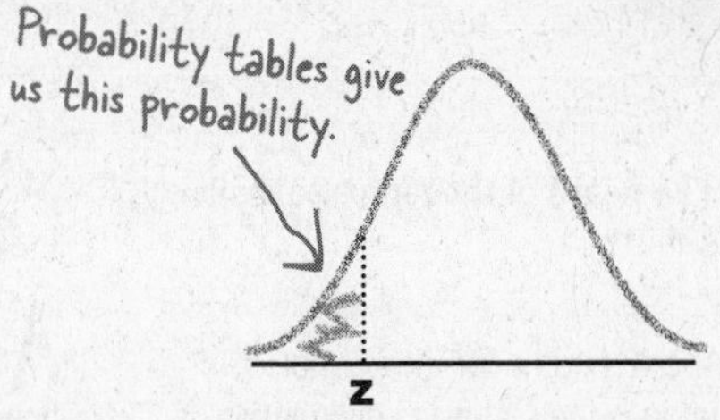

The big trick is to find a way of using the probability tables to get to what you want, usually by finding a whole area and then subtracting what you don't need.

Finding P(Z > z)

We can find probabilities of the form $P(Z > z)$ using

$$P(Z > z) = 1 - P(Z < z)$$

We've already used this to find the probability that Julie is taller than her date.

In other words, take the area where $Z < z$ away from the total probability.

Finding P(a < Z < b)

Finding this sort of probability is slightly more complicated to calculate, but it's still possible. You can calculate this sort of probability using

$$P(a < Z < b) = P(Z < b) - P(Z < a)$$

You could use this to find the probability that the height of Julie's date is within a particular range.

In other words, calculate $P(Z < b)$, and take away the area for $P(Z < a)$.

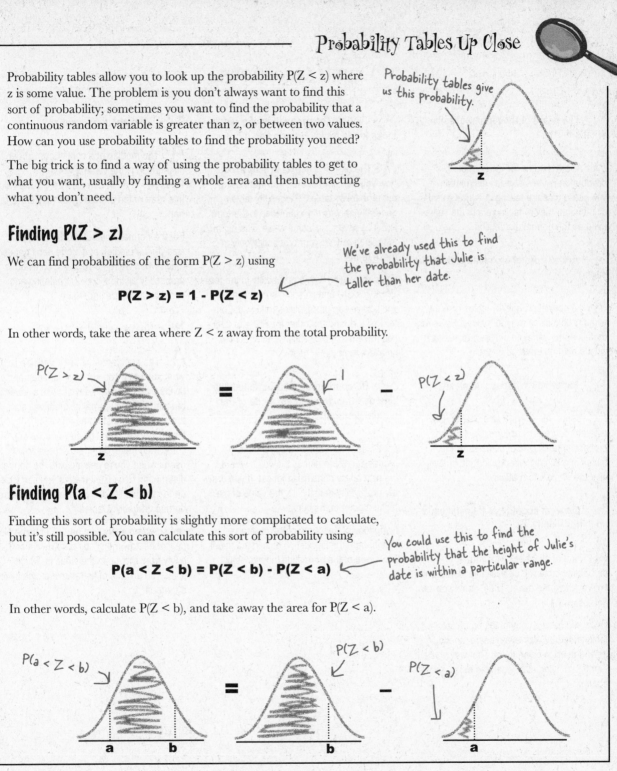

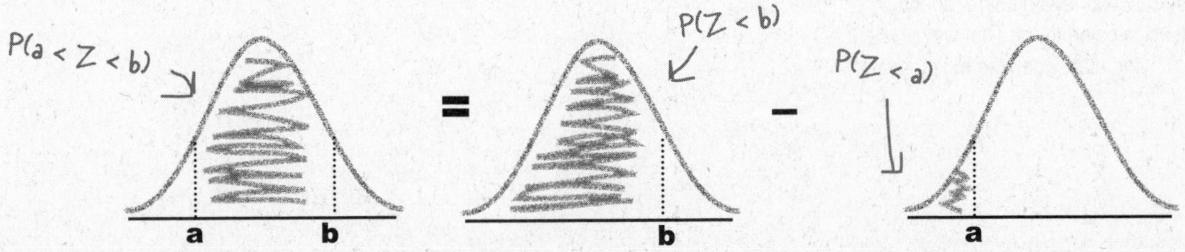

there are no Dumb Questions

Q: I've heard of the term "Gaussian." What's that?

A: Another name for the normal distribution is the Gaussian distribution. If you hear someone talking about a Gaussian distribution, they're talking about the same thing as the normal distribution.

Q: Are all normal probability tables the same?

A: All normal probability tables give the same probabilities for your values. However, there's some variation between tables as to what's actually covered by them.

Q: Variation? What do you mean?

A: Some tables and exam boards use different degrees of accuracy in their probability tables. Also, some show the tables in a slightly different format, but still give the same information.

Q: So what should I do if I'm taking a statistics exam?

A: First of all, check what format of probability table will be available to you while you're sitting the exam. Then, see if you can get a copy.

Once you have a copy of the probability tables used by your exam board, spend time getting used to using them. That way you'll be off to a flying start when the exam comes around.

Q: Finding the probability of a range looks kinda tricky. How do I do it?

A: The big thing here is to think about how you can get the area you want using the probability tables. Probability tables generally only give probabilities in the form $P(Z < z)$ where z is some value. The big trick, then, is to rewrite your probability only in these terms.

If you're dealing with a probability in the form $P(a < Z < b)$—that is, some sort of range—you'll have two probabilities to look up, one for $P(Z < a)$ and the other for $P(Z < b)$. Once you have these probabilities, subtract the smallest from the largest.

Q: Do continuous distributions have a mode? Can you find the mode of the normal distribution?

A: Yes. The mode of a continuous probability distribution is the value where the probability density is highest. If you draw the probability density, it's the value of the highest point of the curve.

If you look at the curve of the normal distribution, the highest point is in the middle. The mode of the normal distribution is μ.

Q: What about the median?

A: The median of a continuous probability distribution is the value *a* where $P(X < a) = 0.5$. In other words, it's the value that area of the probability density curve in half.

For the normal distribution, the median is also μ. The median and mode don't get used much when we're dealing with continuous probability distributions. Expectation and variance are more important.

Q: What's a standard score?

A: The standard score of a variable is what you get if you subtract its mean and divide by its standard deviation. It's a way of standardizing normal distributions so that they are transformed into a N(0, 1) distribution, and that gives you a way of comparing them. Standard scores are useful when you're dealing with the normal distribution because it means you can look up the probability of a range using standard normal probability tables.

The standard score of a particular value also describes how many standard deviations away from the mean the value is, which gives you an idea of its relative proximity to the mean.

Sharpen your pencil

It's time to put your probability table skills to the test. See if you can solve the following probability problems.

1. $P(Z < 1.42)$

2. $P(-0.15 < Z < 0.5)$

3. $P(Z > z) = 0.1423$. What's z?

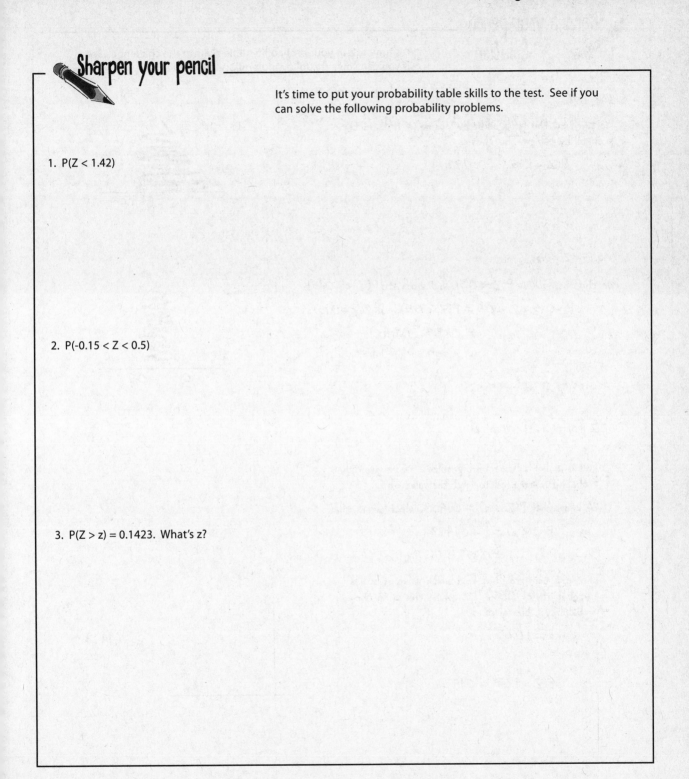

Sharpen your pencil Solution

It's time to put your probability table skills to the challenge. See if you can solve the following probability problems.

1. $P(Z < 1.42)$

We can find this probability by looking up 1.42 in the probability tables. This gives us

$$P(Z < 1.42) = 0.9222$$

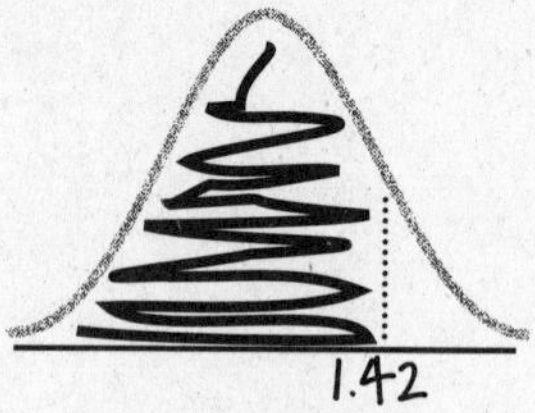

2. $P(-0.15 < Z < 0.5)$

For this one, look up $P(Z < 0.5)$, and subtract $P(Z < -0.15)$

$$P(-0.15 < Z < 0.5 = P(Z < 0.5) - P(Z < -0.15)$$
$$= 0.6915 - 0.4404$$
$$= 0.2511$$

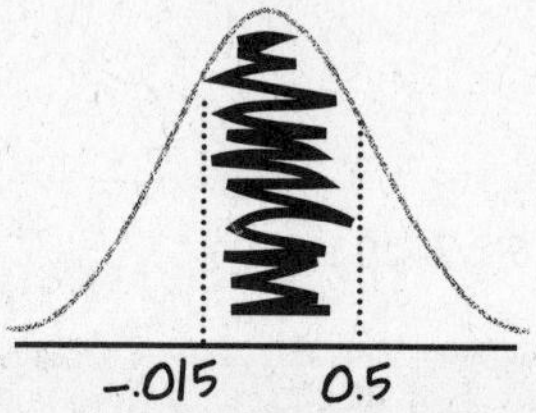

3. $P(Z > z) = 0.1423$. What's z?

This is a slightly different problem. We're given the probability, and need to find the value of z

We know that $P(Z > z) = 0.1423$, which means that

$$P(Z < z) = 1 - 0.1423$$
$$= 0.8577$$

The next thing to do is find which value of z has a probability of 0.8577. Looking this up in the probability tables gives us

$$z = 1.07$$

so

$$P(Z > 1.07) = 0.1423$$

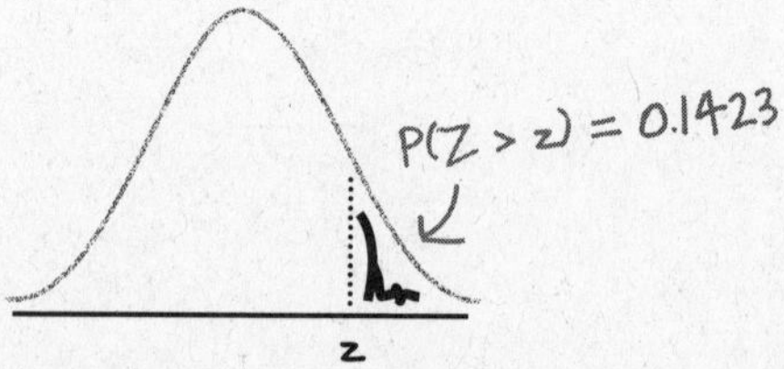

Exercise

Wait a sec, if I wear my
5-inch heels, I'm much taller.
Won't that affect the probability
that my date is taller than me?

Julie has a problem. When we calculated the probability of her date being taller than her, we failed to take her high heels into account. See if you can find the probability of Julie's date being taller than her while she's wearing shoes with 5 inch heels.

As a reminder, Julie is 64 inches tall and $X \sim N(71, 20.25)$ where X is the height of men in Statsville.

Exercise Solution

Julie has a problem. When we calculated the probability of her date being taller than her, we failed to take her high heels into account. See if you can find the probability of Julie's date being taller than her while she's wearing shoes with 5 inch heels.

As a reminder, Julie is 64 inches tall and X ~ N(71, 20.25) where X is the height of men in Statsville.

When Julie is wearing 5 inch high heels, her height is 69 inches. We need to find $P(X > 69)$.

We need to start by finding the standard score of 178 so that we can use probability tables to look up the probabilities.

$$Z = \frac{X - \mu}{\sigma}$$

$$= \frac{69 - 71}{4.5}$$

← The variance is 20.25, so the standard deviation is the square root, 4.5.

$$= \frac{-2}{4.5}$$

$$= -0.44 \text{ (to 2 decimal places)}$$

Now we've found z, we need to find $P(Z > z)$ i.e. $P(Z > -0.44)$

$$P(Z > -0.44) = 1 - P(Z < -0.44)$$

$$= 1 - 0.3300$$

$$= 0.67$$

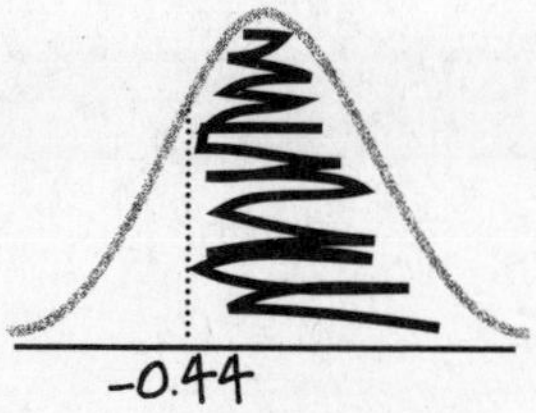

So the probability that Julie's date is taler than her when she's wearing shoes with a 5 inch heel is 0.67.

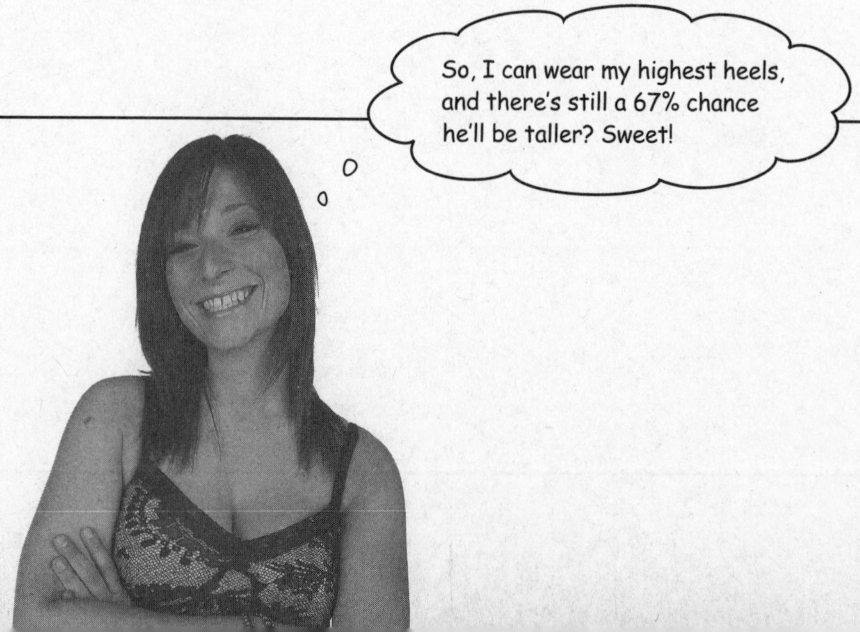

So, I can wear my highest heels, and there's still a 67% chance he'll be taller? Sweet!

The Case of the Missing Parameters

Will at Manic Mango Games has a problem. He needs to give his boss the mean and standard deviation of the number of minutes people take to complete level one of their new game. This shouldn't be difficult, but unfortunately a ferocious terrier has eaten the piece of paper he wrote them on.

Will only has three clues to help him.

First of all, Will knows that the number of minutes people spend playing level one follows a normal distribution.

Secondly, he knows that the probability of a player playing for less than 5 minutes is 0.0045.

Finally, the probability of someone taking less than 15 minutes to complete level one is 0.9641.

How can Will find the mean and standard deviation?

The Case of the Missing Parameters: Solved

How can Will find the mean and standard deviation?

Will can use probability tables and standard scores to get expressions for the mean and standard deviation that he can then solve.

First of all, we know that $P(X < 5) = 0.0045$. From probability tables, $P(X < z_1)$ where $z_1 = -2.61$, which means that the standard score of 5 is -2.61. If we put this into the standard score formula, we get

$$-2.61 = \frac{5 - \mu}{\sigma}$$

Similarly, $P(X < 15) = 0.9641$, which means that the standard score of 15 is 1.8. This gives us

$$1.8 = \frac{15 - \mu}{\sigma}$$

This gives us two equations we can solve to find μ and σ.

$$\left.\begin{aligned} -2.61\sigma &= 5 - \mu \\ 1.8\sigma &= 15 - \mu \end{aligned}\right\}$$

This is a pair of equations we can now solve.

If we subtract the first equation from the second, we get

$$1.8\sigma + 2.61\sigma = 15 - \mu - 5 + \mu$$

$$4.41\sigma = 10$$

$$\sigma = 2.27$$

If we then substitute this into the second equation, we get

$$1.8 \times 2.27 = 15 - \mu$$

$$\mu = 15 - 4.086$$

$$= 10.914$$

In other words,

$$\left.\begin{aligned} \mu &= 10.914 \\ \sigma &= 2.27 \end{aligned}\right\}$$

These are the values of μ and σ.

And they all lived happily ever after

Just as the odds predicted, Julie's latest blind date was a success! Julie had to make sure her intended soulmate was compatible with her shoes, so she made sure she wore her highest heels to put him to the test. What's more, he was already at the venue when she arrived, so she didn't have to wait around.

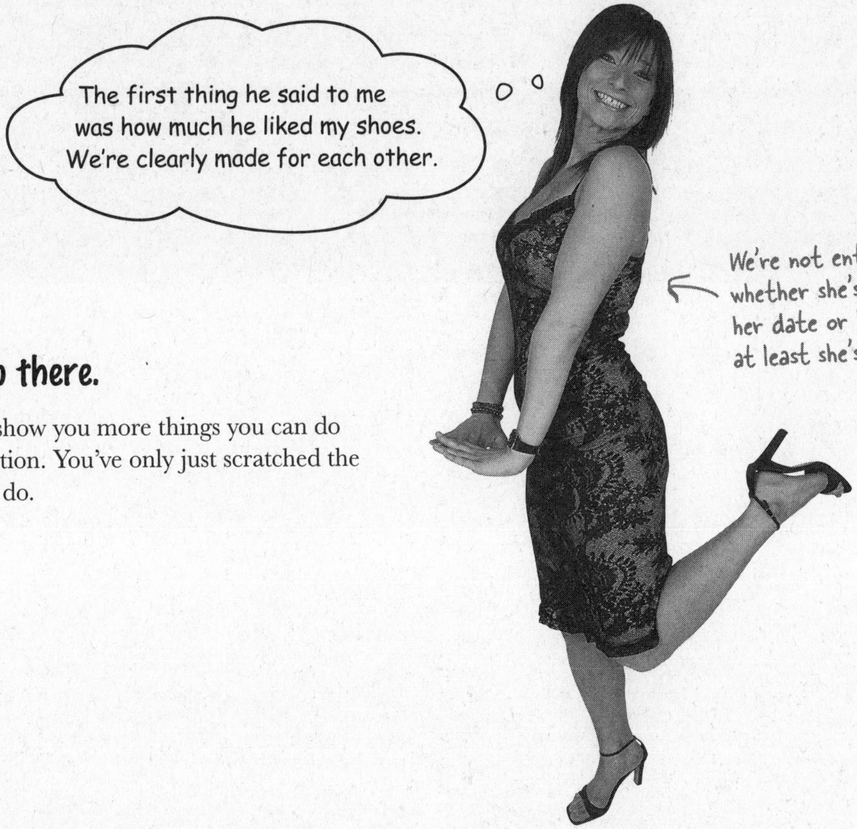

The first thing he said to me was how much he liked my shoes. We're clearly made for each other.

We're not entirely sure whether she's referring to her date or her shoes, but at least she's happy.

But it doesn't stop there.

Keep reading and we'll show you more things you can do with the normal distribution. You've only just scratched the surface of what you can do.

BULLET POINTS

- The normal distribution forms the shape of a symmetrical bell curve. It's defined using $N(\mu, \sigma^2)$.
- To find normal probabilities, start by identifying the probability range you need. Then find the standard score for the limit of this range using

$$Z = \frac{X - \mu}{\sigma} \quad \text{where } Z \sim N(0, 1).$$

- You find normal probabilities by looking up your standard score in probability tables. Probability tables give you the probability of getting this value or lower.

9 using the normal distribution ii

Beyond Normal

If only all probability distributions were normal.

Life can be so much ***simpler*** with the normal distribution. Why spend all your time working out individual probablities when you can look up entire ranges in one swoop, and still leave time for game play? In this chapter, you'll see how to **solve more complex problems** in the blink of an eye, and you'll also find out how to bring some of that normal goodness to **other probability distributions**.

Love is a roller coaster

The wedding market is big business nowadays, and Dexter has an idea for making that special day truly memorable. Why get married on the ground when you can get married on a roller coaster?

Dexter's convinced there's a lot of money to be made from his innovative Love Train ride, if only it passes the health and safety regulations.

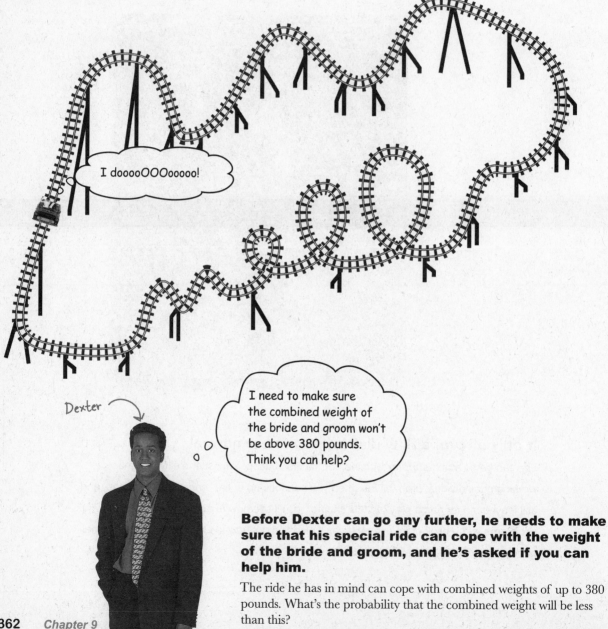

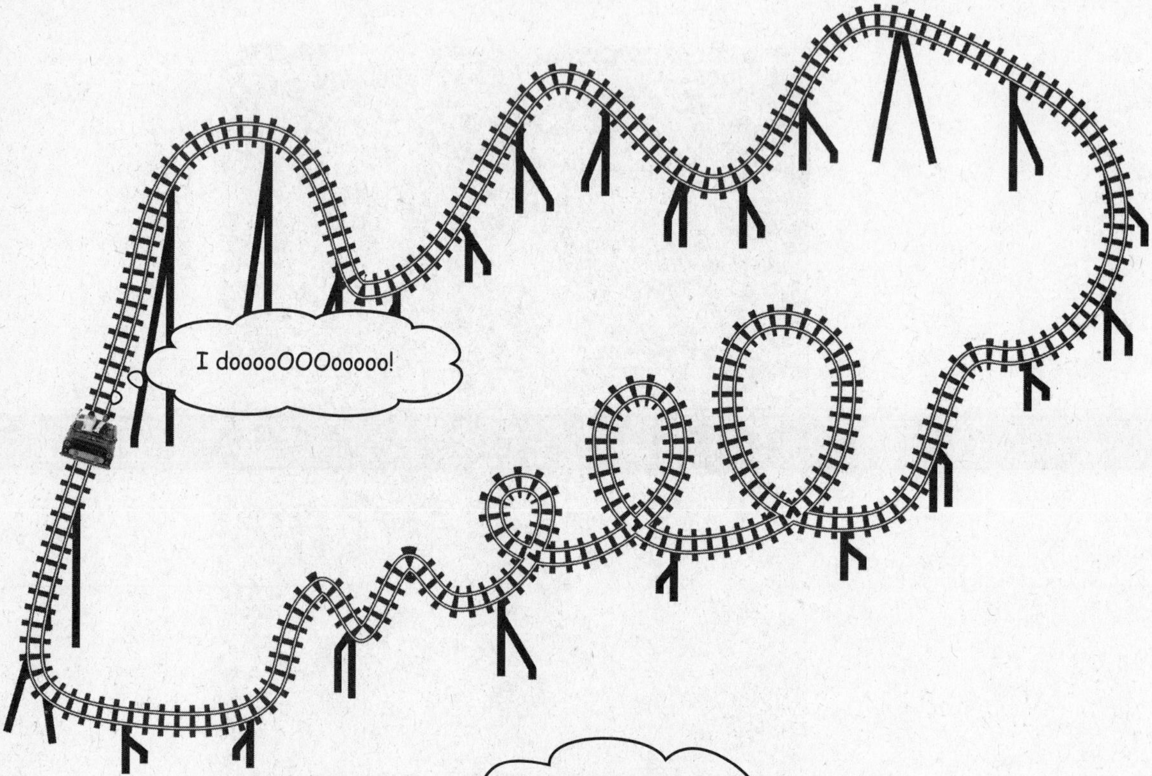
I doooooOOOooooo!

Dexter
I need to make sure the combined weight of the bride and groom won't be above 380 pounds. Think you can help?

Before Dexter can go any further, he needs to make sure that his special ride can cope with the weight of the bride and groom, and he's asked if you can help him.

The ride he has in mind can cope with combined weights of up to 380 pounds. What's the probability that the combined weight will be less than this?

All aboard the Love Train

Before we start, we need to know how the weights of brides and grooms in Statsville are distributed, taking into account the weight of all their wedding clothes. Both follow a normal distribution, with the bride weight distributed as N(150, 400) and the groom weight as N(190, 500). Their weights are measured in pounds.

Bride ~ N(150, 400)

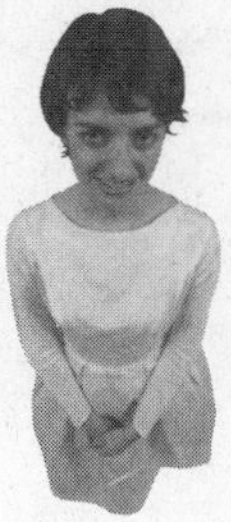

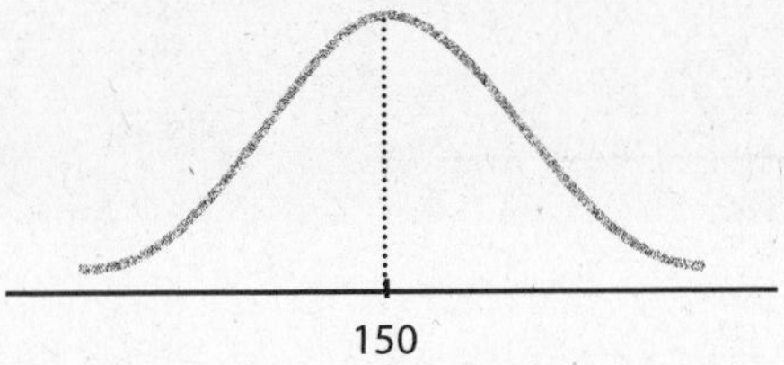

Groom ~ N(190, 500)

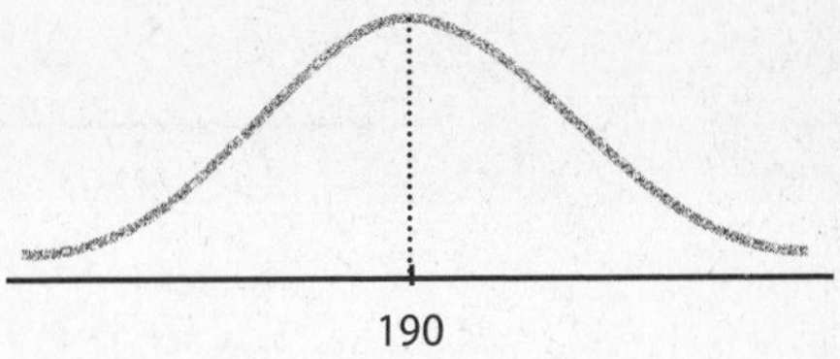

We need to use these two probability distributions to somehow work out the probability that the weight of a bride and groom will be less than the maximum weight allowance on the ride. If the probability is sufficiently high, we can be confident the ride is feasible.

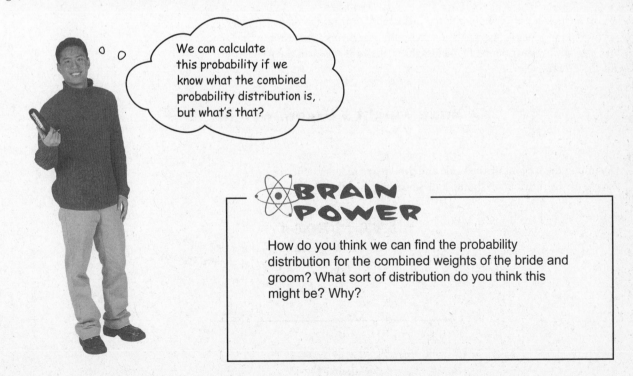

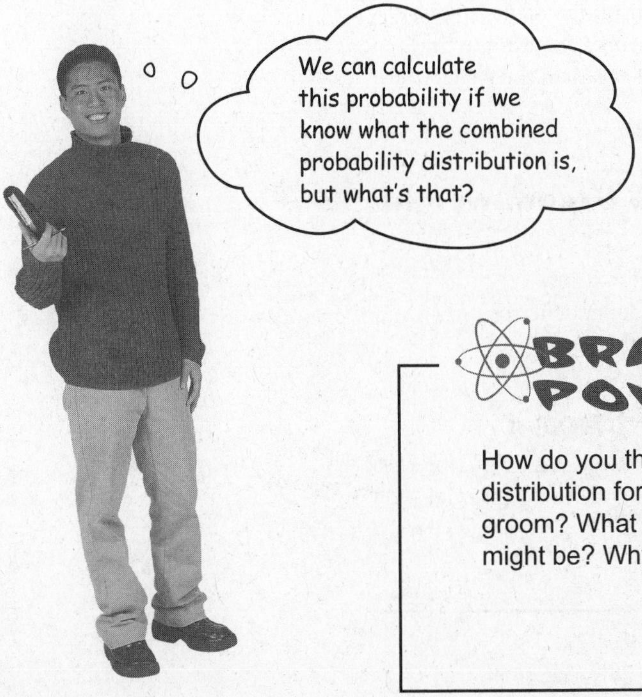

We can calculate this probability if we know what the combined probability distribution is, but what's that?

BRAIN POWER

How do you think we can find the probability distribution for the combined weights of the bride and groom? What sort of distribution do you think this might be? Why?

Normal bride + normal groom

Let's start by taking a closer look at how the weights of the bride and groom are distriuted.

As you know, the weights follow normal distributions like this:

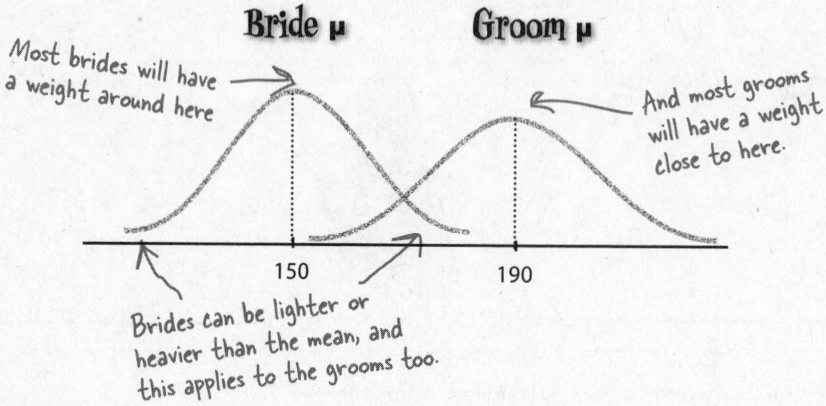

What we're *really* after, though, is the probability distribution of the combined weight of the bride and groom. In other words, we want to find the probability distribution of the weight of the bride *added* to the weight of the groom.

Bride weight + Groom weight ~ ?

Assuming the weights of the bride and groom are independent, the shape of the distribution should look something like this:

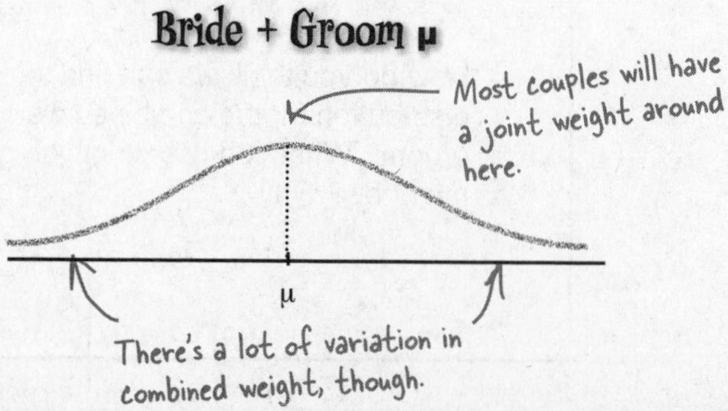

It's still just weight

Can you remember when we first looked at continuous data and looked at how data such as height and weight tend to be distributed? We found that data such as height and weight are continuous, and they also tend to follow a normal distribution.

This time we're looking at the combined weight of the happy couple. Even though it's *combined* weight, it's still just weight, and we already know how weight tends to be distributed. The combined weight is still **continuous**. What's more, the combined weight is still **distributed normally**. In other words, the combined weight of the bride and groom follows a normal distribution.

Knowing that the combined weight of the bride and groom follows a normal distribution helps us a lot. It means that we'll be able to use probability tables just like we did before to look up probabilities, which means we'll be able to look up the probability that the combined weight is less than 380 pounds—just what we need for the ride.

There's only one problem—before we can go any further, we need to know the mean and variance of the combined weight of the bride and groom. How can we find this?

The combined weight of the bride and groom follows a normal distribution, but what's the mean and variance?

Bride weight + Groom weight ~ N(?, ?)

Sharpen your pencil

It's time for a trip down memory lane. Can you remember the discrete shortcuts for the following formulas? Assume X and Y are independent.

1. E(X + Y)

2. Var(X + Y)

3. E(X - Y)

4. Var(X - Y)

Sharpen your pencil Solution

It's time for a trip down memory lane. Can you remember the discrete shortcuts for the following formulas? Assume X and Y are independent.

1. $E(X + Y)$

   $E(X + Y) = E(X) + E(Y)$

2. $Var(X + Y)$

   $Var(X + Y) = Var(X) + Var(Y)$

3. $E(X - Y)$

   $E(X - Y) = E(X) - E(Y)$

4. $Var(X - Y)$

   $Var(X + Y) = Var(X) + Var(Y)$

Remember that we ADD the variances, even though it's for $X - Y$.

I don't see how these shortcuts help us. They're for discrete data, and we're dealing with continuous now.

The shortcuts apply to continuous data too.

When we originally encountered these shortcuts, we were dealing with discrete data. Fortunately, the same rules and shortcuts also apply to continuous data.

BRAIN POWER

How do you think we can use these shortcuts to find the probability distribution of the weight of the bride + the weight of the groom?

How's the combined weight distributed?

So far, we've found that the combined weight of the bride and groom are normally distributed, and this means we can use probability tables to look up the probability of the combined weight being less than a certain amount.

Let's try rewriting the bride and groom weight distributions in terms of X and Y. If X represents the weight of the bride and Y the weight of the groom, and X and Y are independent, then we want to find μ and σ where

$$X + Y \sim N(\mu, \sigma^2)$$

X + Y means "the weight of the bride + the weight of the groom." But how do we know what the mean and variance are?

In other words, before we go any further we need to find the mean and variance of X + Y. But how?

Take a look at the answers to the last exercise. When we were working with discrete probability distributions, we saw that as long as X and Y are independent we could work out E(X + Y) and Var(X + Y) by using

$$E(X + Y) = E(X) + E(Y) \quad \text{and} \quad Var(X + Y) = Var(X) + Var(Y)$$

So if we know what the expectation and variance of X and Y are, we can use these to work out the expectation and variance of X + Y.

That means that if we know the distribution of X and Y, we can figure out the distribution of X + Y too.

We can use what we already know to figure out what we don't.

Because we know how the weight of the bride and the weight of the groom are distributed, we can find the distribution of the combined weight of the bride and groom.

Let's look at this in more detail.

X + Y Distribution Up Close

Being able to find the distribution of X + Y is useful if you're working with combinations of normal variables. If independent random variables X and Y are normally distributed, then X + Y is normal too. What's more, you can use the mean and variance of X and Y to calculate the distribution of X + Y.

← Remember, two variables are independent if they have no impact on each other's probabilities.

To find the mean and variance of X + Y, you can use the same formulae that we used for discrete probability distributions. In other words, if

$$X \sim N(\mu_x, \sigma_x^2) \quad \text{and} \quad Y \sim N(\mu_y, \sigma_y^2)$$

then

$$X + Y \sim N(\mu, \sigma^2)$$

If you add the means of X and Y together, you get the mean of X + Y. Similarly, summing the variances of X and Y gives you the variance of X + Y

where

$$\mu = \mu_x + \mu_y \qquad \qquad \sigma^2 = \sigma_x^2 + \sigma_y^2$$

We can use these shortcuts if X and Y are independent, which makes life very easy indeed

In other words, the mean of X + Y is equal to the mean of X plus the mean of Y, and the variance of X + Y is equal to the variance of X plus the variance of Y.

Let's look at a sketch of this. What do you notice about the variance of X + Y?

$$X \sim N(\mu_x, \sigma_x^2) \qquad Y \sim N(\mu_y, \sigma_y^2) \qquad X + Y \sim N(\mu_x + \mu_y, \sigma_x^2 + \sigma_y^2)$$

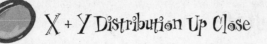

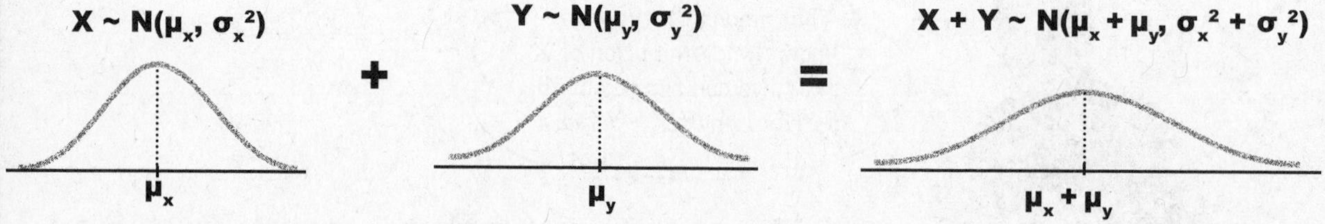

The variance of X + Y is greater than the variance of X and also greater than the variance of Y, which means that the curve of X + Y is more elongated than either. This is true for any normal X and Y. By adding the two variables together, you are in effect increasing the amount of variability, and this elongates the shape of the distribution. This in turn means that the shape of the distribution gets flatter so that the total area under the curve is still 1.

X - Y Distribution Up Close

Sometimes X + Y just won't give you the sorts of probabilities you're after. If you need to find probabilities involving the *difference* between two variables, you'll need to use X - Y instead.

X - Y follows a normal distribution if X and Y are independent random variables and are both normally distributed. This is exactly the same criteria as for X + Y.

To find the mean and variance, we again use the same shortcuts that we used for discrete probability distributions. If

$$X \sim N(\mu_x, \sigma_x^2) \quad \text{and} \quad Y \sim N(\mu_y, \sigma_y^2)$$

then

$$\mathbf{X - Y \sim N(\mu, \sigma^2)}$$

We ADD the variances together, just like we did for discrete probability distributions.

where

$$\boldsymbol{\mu = \mu_x - \mu_y} \qquad \boldsymbol{\sigma^2 = \sigma_x^2 + \sigma_y^2}$$

In other words, the mean of X – Y is equal to the mean of Y *subtracted* from the mean of X, and you find the variance of X – Y by *adding* the X and Y variances together.

Subtract the mean, add the variance.

$$X \sim N(\mu_x, \sigma_x^2) \quad - \quad Y \sim N(\mu_y, \sigma_y^2) \quad = \quad X - Y \sim N(\mu_x - \mu_y, \sigma_x^2 + \sigma_y^2)$$

μ_x μ_y $\mu_x - \mu_y$

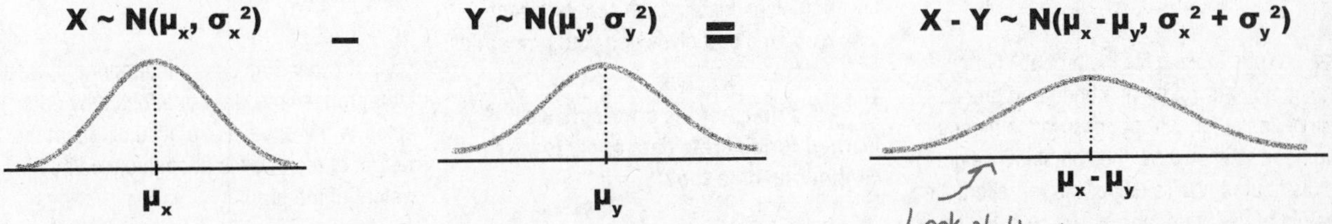

Look at the shape. It's the same shape as for X + Y but with the center of the curve in a different place. They have the same shape because they have the same variance.

Adding the variances together may not make intuitive sense at first, but it's exactly the same as when we worked with discrete probability distributions. Even though we're subtracting Y from X, we're actually still *increasing* the amount of variability. Adding the variances together reflects this. As with the X + Y distribution, this leads to a flatter, more elongated shape than either X or Y

If you look at the actual shape of the X - Y distribution, it's the same shape curve as for X + Y distribution, except that the center has moved. The two distributions have the same variances, but different means.

Finding probabilities

Now that we know how to calculate the distribution of X + Y, we can look at how to use it to calculate probabilities. Here are the steps you need to go through.

We know we need to use X + Y, and we have a way of working out the mean and variance.

1. **Work out the distribution and range**

Once we know the distribution and the range, we standardize it.

2. **Standardize it**

We can then look up the probability in standard normal probability tables.

3. **Look up the probabilities**

Sound familiar? These are exactly the same steps that we went through in the previous chapter for the normal distribution.

there are no Dumb Questions

Q: Remind me, why did we need to find the distribution of X + Y?

A: We're looking for the probability that the combined weight of a bride and groom will be less than 380 pounds, which means we need to know how the combined weight is distributed. We're using X to represent the weight of the bride, and Y to represent the weight of the groom, which means we need to use the distribution of X + Y.

Q: You say we can look up probabilities for X + Y using probability tables. How?

A: In exactly the same way as we did before. We take our probability distribution, calculate the standard score, and then look this value up in probablity tables.

Looking up probabilities for X + Y is no different from looking up probabilities for anything else. Just find the standard score, look it up, and that gives you your probability.

Q: So do all of the shortcuts we learned for discrete data apply to continuous data too?

A: Yes, they do. This means we have an easy way of combining random variables and finding out how they're distributed, which in turn means we can solve more complex problems.

The key thing to remember is that these shortcuts apply as long as the random variables are independent.

Q: Can you remind me what independent means?

A: If two variables are independent, then their probabilities are not affected by each other. In our case, we're assuming that the weight of the bride is not influenced by the weight of the groom.

Q: What if X and Y aren't independent? What then?

A: If X and Y aren't independent, then we can't use these shortcuts. We'd need to do a lot more work to find out how X + Y is distributed because you'd have to find out what the relationship is between X and Y.

Sharpen your pencil

Find the probability that the combined weight of the bride and groom is less than 380 pounds using the following three steps.

1. X is the weight of the bride and Y is the weight of the groom, where $X \sim N(150, 400)$ and $Y \sim N(190, 500)$. With this information, find the probability distribution for the combined weight of the bride and groom.

2. Then, using this distribution, find the standard score of 380 pounds.

3. Finally, use the standard score to find $P(X + Y < 380)$.

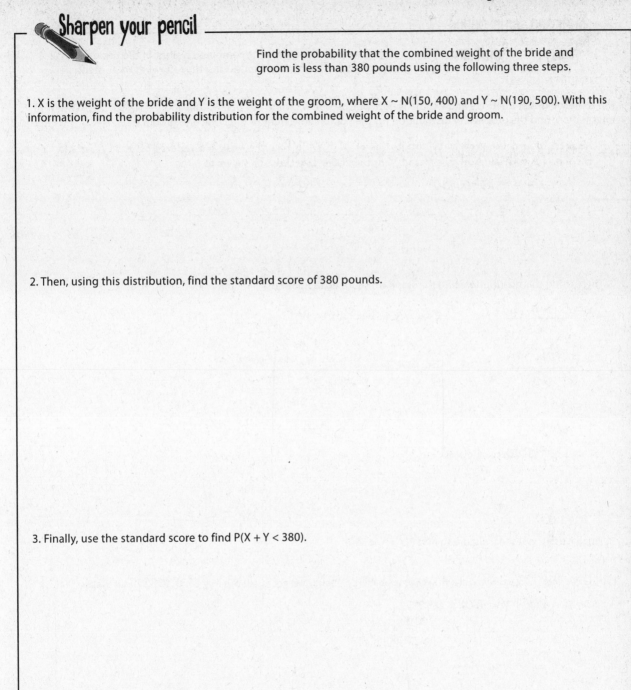

Sharpen your pencil Solution

Find the probability that the combined weight of the bride and groom is less than 380 pounds using the following three steps.

1. X is the weight of the bride and Y is the weight of the groom, where X ~ N(150, 400) and Y ~ N(190, 500). With this information, find the probability distribution for the combined weight of the bride and groom.

We need to find the probability distribution of X + Y. To find the mean and variance of X + Y, we add the means and variances of the X and Y distributions together. This gives us

$$X + Y \sim N(340, 900)$$

2. Then, using this distribution, find the standard score of 380 pounds.

$$z = \frac{(x + y) - \mu}{\sigma}$$

$$= \frac{380 - 340}{30}$$

$$= \frac{40}{30}$$

$= 1.33$ (to 2 decimal places)

Remember how before we used $z = \frac{x - \mu}{\sigma}$?

This time around we're using the distribution of X + Y, so we use $z = \frac{(x + y) - \mu}{\sigma}$

3. Finally, use the standard score to find P(X + Y < 380)

If we look 1.33 up in standard normal probability tables, we get a probability of 0.9082. This means that

$$P(X + Y < 380) = 0.9082$$

Exercise

Julie's matchmaker is at it again. What's the probability that a man will be at least 5 inches taller than a woman?

In Statsville, the height of men in inches is distributed as N(71, 20.25), and the height of women in inches is distributed as N(64, 16).

Exercise Solution

Julie's matchmaker is at it again. What's the probability that a man will be at least 5 inches taller than a woman?

In Statsville, the height of men in inches is distributed as N(71, 20.25), and the height of women in inches is distributed as N(64, 16).

Let's use X to represent the height of the men and Y to represent the height of the women. This means that $X \sim N(71, 20.25)$ and $Y \sim N(64, 16)$.

We need to find the probability that a man is at least 5 inshes taller than a woman. This means we need to find

$$P(X > Y + 5)$$

or

$$P(X - Y > 5)$$

To find the mean and variance of X – Y, we take the mean of Y from the mean of X, and add the variances together. This gives us

$$X - Y \sim N(7, 36.25)$$

We need to find the standard score of 5 inches

$$z = \frac{(x - y) - \mu}{\sigma}$$

$$= \frac{5 - 7}{6.02}$$

$$= -0.33 \text{ (to 2 decimal places)}$$

We can use this to find $P(X - Y > 5)$.

$$P(X - Y > 5) = 1 - P(X - Y < 5)$$

$$= 1 - 0.3707$$

$$= 0.6293$$

More people want the Love Train

It looks like there's a good chance that the combined weight of the happy couple will be less than the maximum the ride can take. But why restrict the ride to the bride and groom?

Customers are demanding that we allow more members of the wedding party to join the ride, and they'll pay good money. That's great, but will the Love Train be able to handle the extra load?

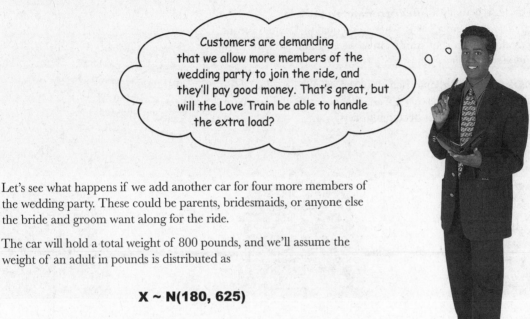

Let's see what happens if we add another car for four more members of the wedding party. These could be parents, bridesmaids, or anyone else the bride and groom want along for the ride.

The car will hold a total weight of 800 pounds, and we'll assume the weight of an adult in pounds is distributed as

$$X \sim N(180, 625)$$

where X represents the weight of an adult. But how can we work out the probability that the combined weight of four adults will be less than 800 pounds?

BRAIN POWER

Think back to the shortcuts you can use when you calculate expectation and variance. What's the difference between independent observations and linear transformations? What effect does each have on the expectation and variance? Which is more appropriate for this problem?

Linear transforms describe underlying changes in values...

Let's start off by looking at the probability distribution of 4X, where X is the weight of one adult. Is 4X appropriate for describing the probability distribution for the weight of 4 people?

The distribution of 4X is actually a ***linear transform*** of X. It's a transformation of X in the form aX + b, where a is equal to 4, and b is equal to 0. This is exactly the same sort of transform as we encountered earlier with discrete probability distributions.

Linear transforms describe underlying changes to the size of the values in the probability distribution. This means that 4X actually describes the weight of an individual adult whose weight has been multiplied by 4.

The 4X probability distribution describes adults whose weights have been multiplied by 4. The weight is changed, not the number of adults.

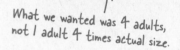

What we wanted was 4 adults, not 1 adult 4 times actual size.

So what's the distribution of a linear transform?

Suppose you have a linear transform of X in the form aX + b, where $X \sim N(\mu, \sigma^2)$. As X is distributed normally, this means that aX + b is distributed normally too. But what's the expectation and variance?

Let's start with the expectation. When we looked at discrete probability distributions, we found that $E(aX + b) = aE(X) + b$. Now, X follows a normal distribution where $E(X) = \mu$, so this gives us $E(aX + b) = a\mu + b$.

We can take a similar approach with the variance. When we looked at discrete probability distributions, we found that $Var(aX + b) = a^2 Var(X)$. We know that Var(X) in this case is given by $Var(X) = \sigma^2$, so this means that $Var(aX + b) = a^2\sigma^2$.

Putting both of these together gives us

$$\mathbf{aX + b \sim N(a\mu + b,\ a^2\sigma^2)}$$

The new variance is the SQUARE of a multiplied by the original variance.

In other words, the new mean becomes $a\mu + b$, and the new variance becomes $a^2\sigma^2$.

So what about independent observations?

...and independent observations describe how many values you have

Rather than transforming the weight of each adult, what we *really* need to figure out is the probability distribution for the combined weight of four separate adults. In other words, we need to work out the probability distribution of four ***independent observations*** of X.

The weight of each adult is an observation of X, so this means that the weight of each adult is described by the probability distribution of X. We need to find the probability distribution of four independent observations of X, so this means we need to find the probability distribution of

$$X_1 + X_2 + X_3 + X_4$$

where X_1, X_2, X_3 and X_4 are independent observations of X.

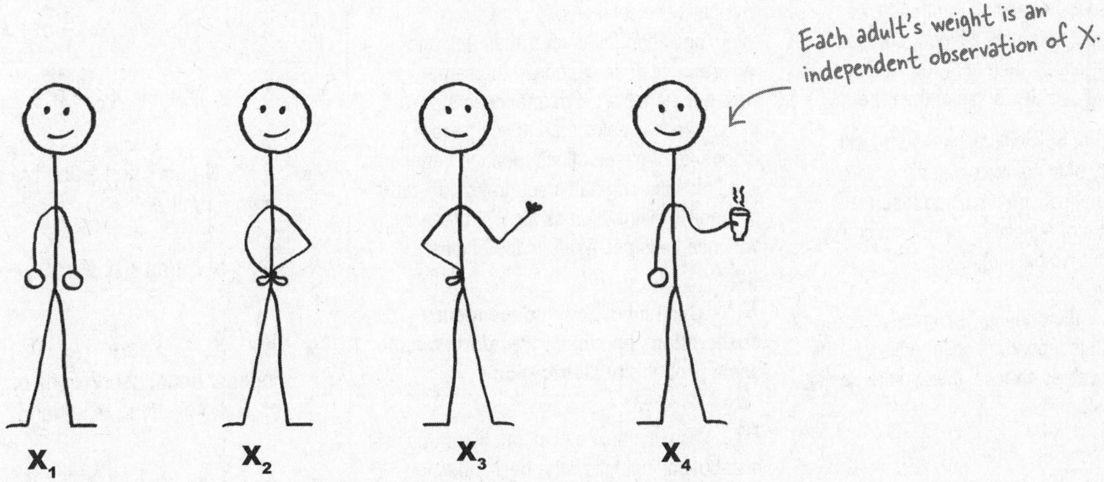

Expectation and variance for independent observations

When we looked at the expectation and variance of independent observations of discrete random variables, we found that

$$E(X_1 + X_2 + \dots X_n) = nE(X)$$

and

$$Var(X_1 + X_2 + \dots + X_n) = nVar(X)$$

As you'd expect, these same calculations work for *continuous* random variables too. This means that if $X \sim N(\mu, \sigma^2)$, then

$$\mathbf{X_1 + X_2 + \dots + X_n \sim N(n\mu, n\sigma^2)}$$

there are no Dumb Questions

Q: So what's the difference between linear transforms and independent observations?

A: Linear transforms affect the underlying values in your probability distribution. As an example, if you have a length of rope of a particular length, then applying a linear transform affects the length of the rope.

Independent observations have to do with the quantity of things you're dealing with. As an example, if you have *n* independent observations of a piece of rope, then you're talking about *n* pieces of rope.

In general, if the quantity changes, you're dealing with independent observations. If the underlying values change, then you're dealing with a transform.

Q: Do I really have to know which is which? What difference does it make?

A: You have to know which is which because it make a difference in your probability calculations. You calculate the expectation for linear transforms and independent observations in the same way, but there's a big difference in the way the variance is calculated. If you have *n* independent observations then the variance is *n* times the original. If you transform your probability distribution as aX + b, then your variance becomes a^2 times the original.

Q: Can I have both independent observations and linear transforms in the same probability distribution?

A: Yes you can. To work out the probability distribution, just follow the basic rules for calculating expectation and variance. You use the same rules for both discrete and continuous probability distributions.

BULLET POINTS

- If $X \sim N(\mu_x, \sigma^2_x)$ and $Y \sim N(\mu_y, \sigma^2_y)$, and X and Y are independent, then

  $X + Y \sim N(\mu_x + \mu_y, \sigma^2_x + \sigma^2_y)$

  $X - Y \sim N(\mu_x - \mu_y, \sigma^2_x + \sigma^2_y)$

- If $X \sim N(\mu, \sigma^2)$ and a and b are numbers, then

  $aX + b \sim N(a\mu + b, a^2\sigma^2)$

- If $X_1, X_2, \dots, X_n$ are independent observations of X where $X \sim N(\mu, \sigma^2)$, then

  $X_1 + X_2 + \dots + X_n \sim N(n\mu, n\sigma^2)$

Exercise

Let's solve Dexter's Love Train dilemma. What's the probability that the combined weight of 4 adults will be less than 800 pounds? Assume the weight of an sdult is distributed as N(180, 625).

Exercise Solution

Let's solve Dexter's Love Train dilemma. What's the probability that the combined weight of 4 adults will be less than 800 pounds? Assume the weight of an sdult is distributed as N(180, 625).

If we represent the weight of an adult as X, then $X \sim N(180, 625)$. We need to start by finding how the weight of 4 adults is distributed. To find the mean and variance of this new distribution, we multiply the mean and variance of X by 4. This gives us

$$X_1 + X_2 + X_3 + X_4 \sim N(720, 2500)$$

To find $P(X_1 + X_2 + X_3 + X_4 < 800)$, we start by finding the standard score.

$$z = \frac{x - \mu}{\sigma}$$

$$= \frac{800 - 720}{50}$$

$$= \frac{80}{50}$$

$$= 1.6$$

Looking this value up in standard normal probability tables gives us a value of 0.9452. This means that

$$P(X_1 + X_2 + X_3 + X_4 < 800) = 0.9452$$

We interrupt this chapter to bring you...

WHO WANTS TO WIN A
SWIVEL CHAIR
Hello, and welcome back to Who Wants To Win A Swivel Chair, Statsville's favorite quiz show. We've got some more fiendishly difficult questions on tonight's show.

We've got some more great questions lined up for you today, so let's get on with the show. In this round I'm going to ask you forty questions, and you need to get thirty or more right to get through to the next round. Or you can walk away and take a consolation prize. For each question there are four possible answers. The title of this round is "Even More About Me." Good luck!

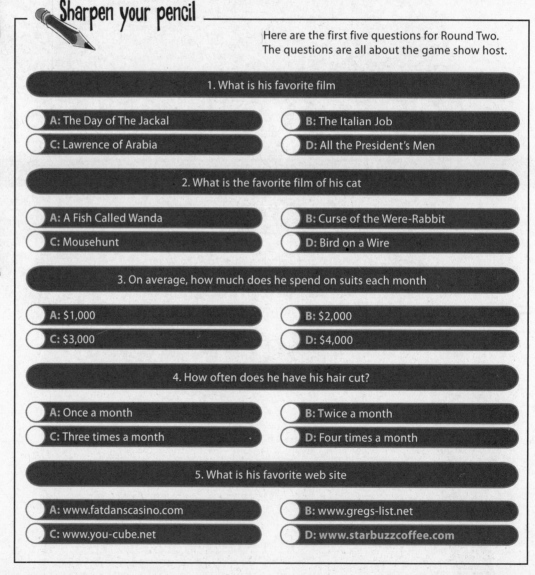

Sharpen your pencil

Here are the first five questions for Round Two. The questions are all about the game show host.

1. What is his favorite film

- **A:** The Day of The Jackal
- **B:** The Italian Job
- **C:** Lawrence of Arabia
- **D:** All the President's Men

2. What is the favorite film of his cat

- **A:** A Fish Called Wanda
- **B:** Curse of the Were-Rabbit
- **C:** Mousehunt
- **D:** Bird on a Wire

3. On average, how much does he spend on suits each month

- **A:** $1,000
- **B:** $2,000
- **C:** $3,000
- **D:** $4,000

4. How often does he have his hair cut?

- **A:** Once a month
- **B:** Twice a month
- **C:** Three times a month
- **D:** Four times a month

5. What is his favorite web site

- **A:** www.fatdanscasino.com
- **B:** www.gregs-list.net
- **C:** www.you-cube.net
- **D:** www.starbuzzcoffee.com

Should we play, or walk away?

As before, it's unlikely you'll know the game show host well enough to answer questions about him. It looks like you'll need to give random answers to the questions again.

So what's the probability of getting 30 or more questions right out of 40? That will help us determine whether to keep playing, or walk away.

Sharpen your pencil

How would you find the probability of getting at least 30 out of 40 questions correct? What steps would you need to go through to get the right answer? How would you find the mean and variance?

We're not asking you to find the probability—just say how you'd go about finding it.

Sharpen your pencil Solution

How would you find the probability of getting at least 30 out of 40 questions correct? What steps would you need to go through to get the right answer? How would you find the mean and variance?

We're not asking you to find the probability—just say how you'd go about finding it.

There are 40 questions, which means there are 40 trials. The outcome of each trial can be a success or failure, and we want to find the probability of getting a certain number of successes. In order to do this, we need to use the binomial distribution. We use $n = 40$, and as each question has four possible answers, p is 1/4 or 0.25..

If X is the number of questions we get right, then we want to find $P(X > 30)$. This means we have to calculate and add together the probabilities for $P(X = 30)$ up to $P(X = 40)$.

We can find the mean and variance using n, p and q, where $q = 1 - p$. The mean is equal to np, and the variance is equal to npq. This gives us a mean of $40 \times 0.25 = 10$, and a variance of $40 \times 0.25 \times 0.75 = 7.5$.

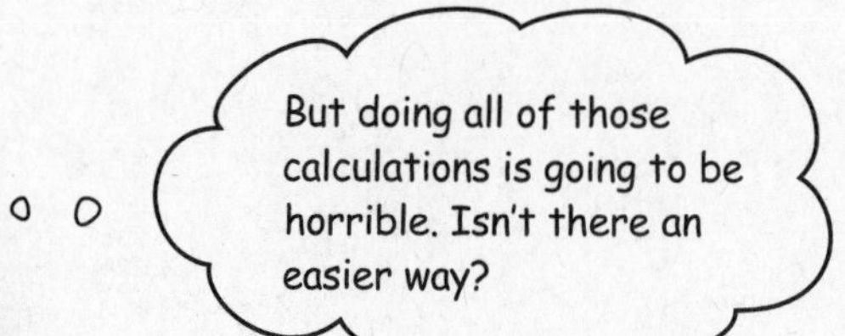

But doing all of those calculations is going to be horrible. Isn't there an easier way?

Using the binomial distribution can be a lot of work.

In order to find the probability that we answer 30 or more questions correctly, we need to add together 11 individual probabilities. Each of these probabilities is tricky to find, and it would be very easy to make a mistake somewhere along the way.

What we really need is an easier way of calculating binomial probabilities.

Wouldn't it be dreamy if there was a way of making other distributions as easy to work with as the normal? But I know it's just a fantasy...

Normal distribution to the rescue

We've seen that life with the binomial distribution can be tough at times. Some of the calculations can be tricky and repetitive, which in turn means that it's easy to make mistakes and spend a lot of time only to come up with the wrong answer.

Sound hopeless? Don't worry, there's an easy way out.

In certain circumstances, you can use the normal distribution to approximate the binomial distribution.

You're saying the normal distribution can approximate the binomial? I thought the Poisson did that. What gives?

The Poisson distribution can approximate the binomial in some situations, but the normal can in others.

Knowing how to approximate the binomial distribution with other distributions is useful because it can cut down on all sorts of complexities, and in some situations the Poisson distribution can help us work out some tricky binomial probabilities.

In certain *other* circumstances, we can use the normal distribution to approximate the binomial instead. There are some huge advantages with this, as it means that instead of performing calculations, we can use normal probability tables to simply look up the probabilities we need.

All we need to do is figure out the circumstances under which this works.

BRAIN POWER

It's been a while since we looked at how we could use the Poisson distribution to approximate the binomial. Under what circumstances is it appropriate?

$B(n, p)$ can be approximated by the Poisson when $n > 50$ and $p < 0.1$.

BE the Distribution

Below you'll see some binomial distributions for different values of n and p. Your job is to play like you're the distribution and say which one you think can best be approximated by the normal. Take a good look at the shape of each distribution and say which one is most normal.

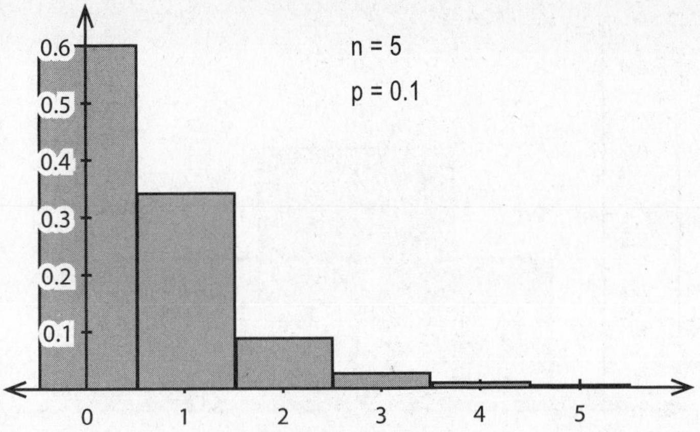

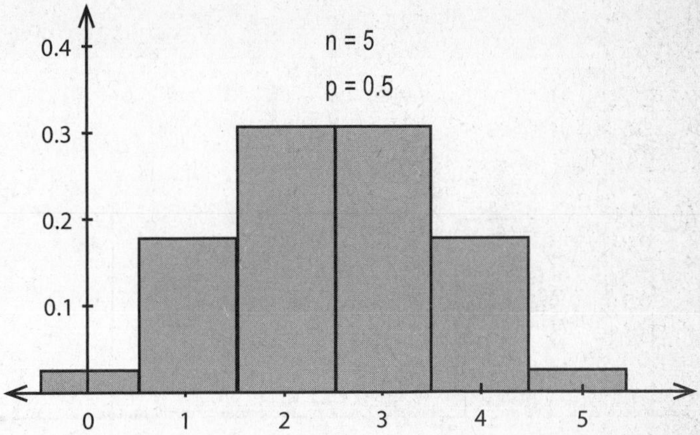

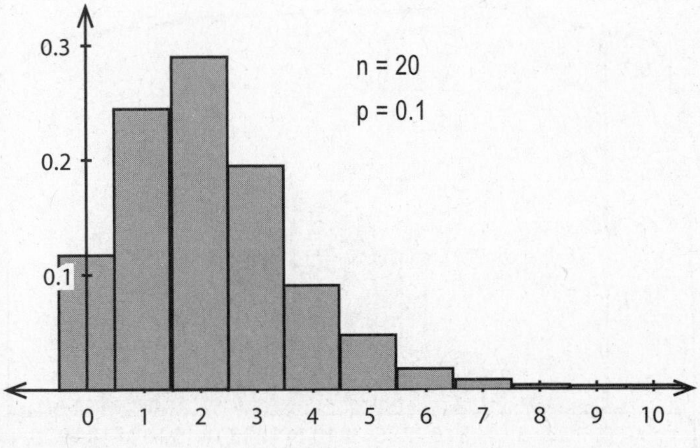

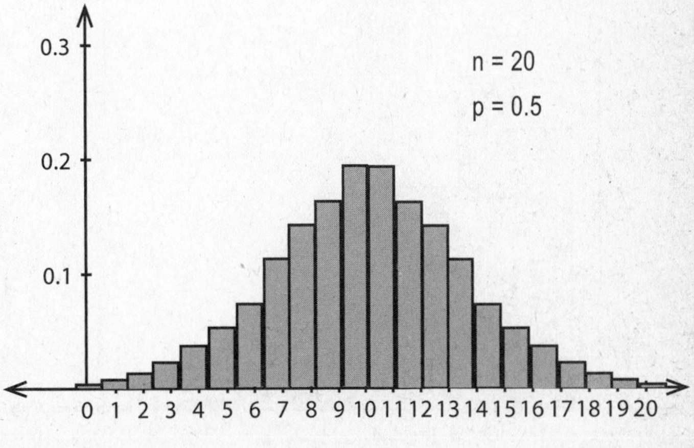

BE the Distribution Solution

Below you'll see some binomial distributions for different values of n and p. Your job is to play like you're the distribution and say which one you think can best be approximated by the normal. Take a good look at the shape of each distribution and say which one is most normal.

This distribution is symmetrical, but it's difficult to pick out a smooth bell-shaped curve. It's unlikely that using the normal distribution to approximate this would be accurate enough.

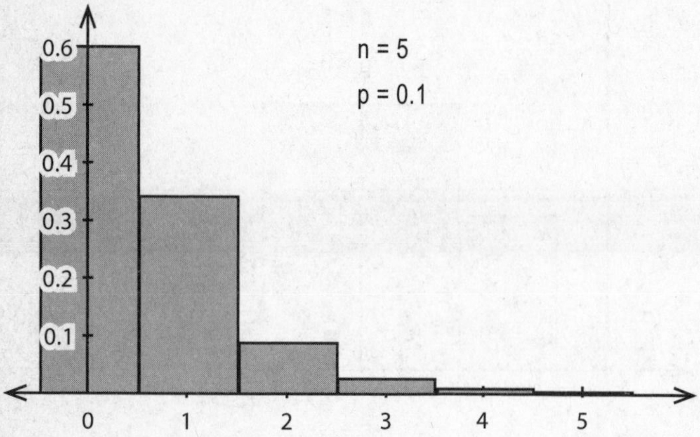

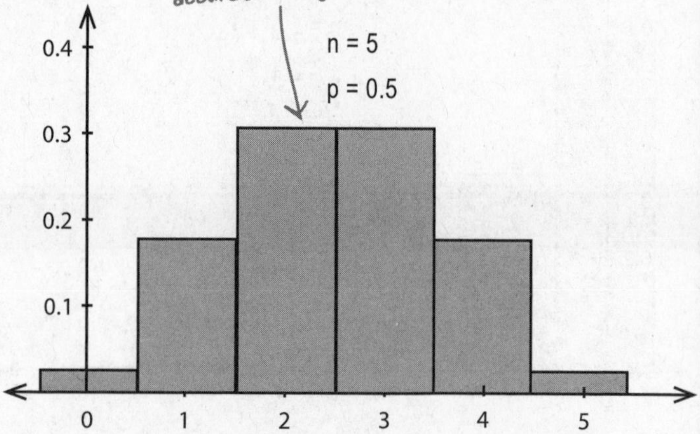

Out of all these distributions, this is the one that can best be approximated by the normal distribution. When n = 20 and p = 0.5, the shape of the distribution is very similar to that of the normal.

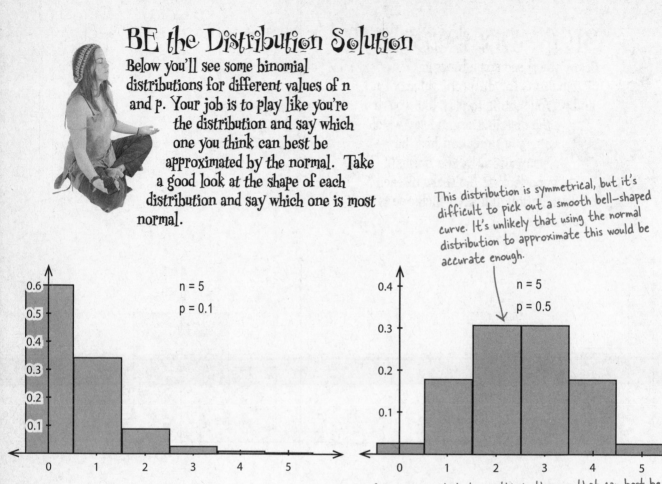

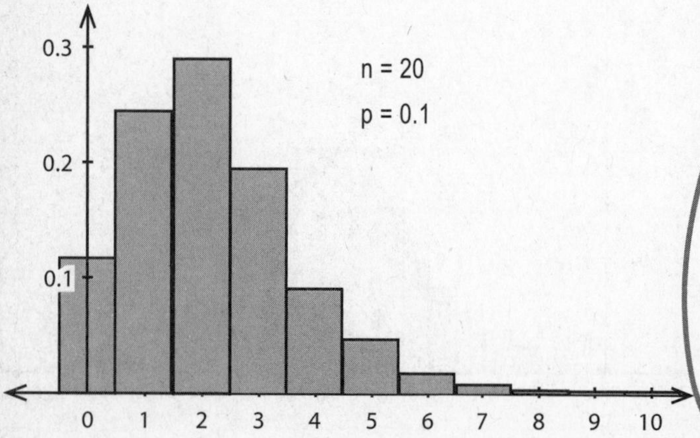

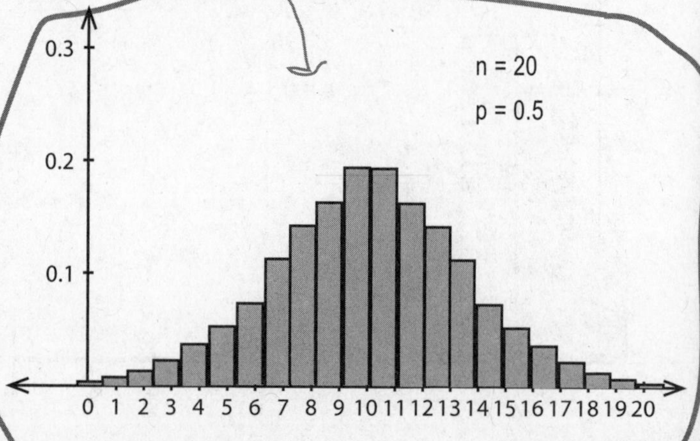

When to approximate the binomial distribution with the normal

Under certain circumstances, the shape of the binomial distribution looks very similar to the normal distribution. In these situations, we can use the normal distribution *in place* of the binomial to give a close approximation of its probabilities. Instead of calculating lots of individual probabilities, we can look up whole ranges in standard normal probability tables.

So under what circumstances can we do this?

We saw in the last exercise that the binomial distribution looks very similar to the normal distribution where p is around 0.5, and n is around 20. As a general rule, you can use the normal distribution to approximate the binomial when np and nq are both greater than 5.

n is the number of values, p is the probability of success, and q is $1 - p$.

Finding the mean and variance

Before we can use normal probability tables to look up probabilities, we need to know what the mean and variance is so that we can calculate the standard score. We can take these directly from the binomial distribution. When we originally looked at the binomial distribution, we found that:

$\mu = np$ and $\sigma^2 = npq$

We can use these as parameters for our normal approximation.

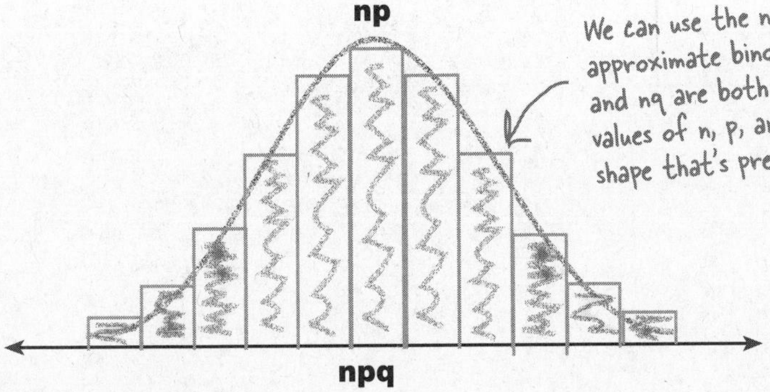

We can use the normal distribution to approximate binomial probabilities when np and nq are both greater than 5. These values of n, p, and q give us a nice, smooth shape that's pretty close to the normal.

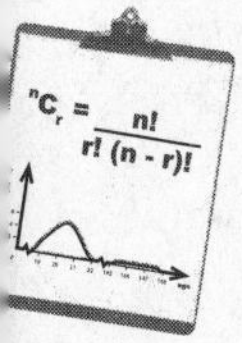

Vital Statistics

Approximating the Binomial Distribution

If $X \sim B(n, p)$ and $np > 5$ and $nq > 5$, you can use $X \sim N(np, npq)$ to approximate it.

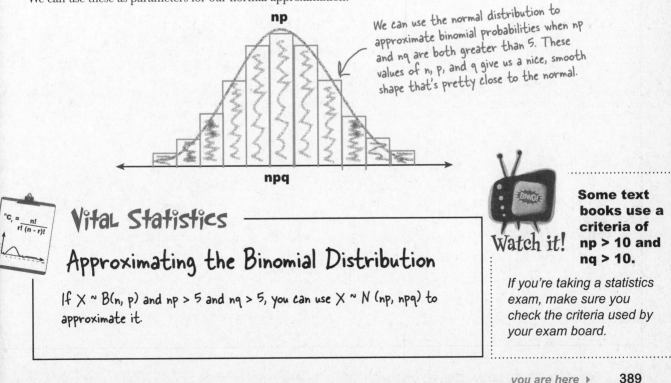

Watch it!

Some text books use a criteria of np > 10 and nq > 10.

If you're taking a statistics exam, make sure you check the criteria used by your exam board.

Long Exercise

Before we use the normal distribution for the full 40 questions for Who Wants To Win A Swivel Chair, let's tackle a simpler problem to make sure it works. Let's try finding the probability that we get 5 or fewer questions correct out of 12, where there are only two possible choices for each question.

Let's start off by working this out using the binomial distribution. Use the binomial distribution to find $P(X < 6)$ where $X \sim B(12, 0.5)$.

Now let's try using the normal approximation to the binomial and check that we get the same result. First of all, if X ~ B(12, 0.5), what normal distribution can we use to approximate this? Once you've found that, what's P(X < 6)?

Long Exercise

Before we use the normal distribution for the full 40 questions for Who Wants To Win A Swivel Chair, let's tackle a simpler problem to make sure it works. Let's try finding the probability that we get 5 or fewer questions correct out of 12, where there are only two possible choices for each question.

Let's start off by working this out using the binomial distribution. Use the binomial distribution to find $P(X < 6)$ where $X \sim B(12, 0.5)$.

To find individual probabilities, we use the formula

$$P(X = r) = {}^nC_r p^r q^{n-r} \quad \text{where} \quad {}^nC_r = \frac{n!}{r!(n-r)!}$$

We need to find $P(X < 6)$ where $X \sim B(12, 0.5)$. To do this, we need to find $P(X = 0)$ through $P(X = 5)$, and then add all the probabilities together.

The individual probabilities are

$$P(X = 0) = {}^{12}C_0 \times 0.5^{12} = 0.5^{12}$$

$$P(X = 1) = {}^{12}C_1 \times 0.5 \times 0.5^{11} = 12 \times 0.5^{12}$$

$$P(X = 2) = {}^{12}C_2 \times 0.5^2 \times 0.5^{10} = 66 \times 0.5^{12}$$

$$P(X = 3) = {}^{12}C_3 \times 0.5^3 \times 0.5^9 = 220 \times 0.5^{12}$$

$$P(X = 4) = {}^{12}C_4 \times 0.5^4 \times 0.5^8 = 495 \times 0.5^{12}$$

$$P(X = 5) = {}^{12}C_5 \times 0.5^5 \times 0.5^7 = 792 \times 0.5^{12}$$

Adding these together gives us an overall probability of

$$P(X < 6) = (1 + 12 + 66 + 220 + 495 + 792) \times 0.5^{12}$$

$$= 1586 \times 0.5^{12}$$

$$= 0.387 \text{ (to 3 decimal places)}$$

Now let's try using the normal approximation to the binomial and check we get the same result. First of all, if X ~ B(12, 0.5), what normal distribution can we use to approximate this? Once you've found that, what's P(X < 6)?

X ~ B(12, 0.5), which means that n = 12, p = 0.5 and q = 0.5. A good approximation to this is X ~ N(np, npq), or X ~ N(6, 3).

We want to find P(X < 6), so we start by calculating the standard score.

$$z = \frac{x - \mu}{\sigma}$$

$$= \frac{6 - 6}{\sqrt{3}}$$

$$= 0$$

Looking this up in probability tables gives us

P(X < 6) = 0.5

Did I miss something? In what way was that a good approximation?

The two methods of calculating the probability have given quite different results.

Using the binomial distribution, P(X < 6) comes to 0.387, but using the normal distribution it comes to 0.5. We should have been able to use the normal distribution in place of the binomial, but the results aren't close enough.

BRAIN POWER

What do you think could have gone wrong? How do you think we could fix it?

Revisiting the normal approximation

So what went wrong? Let's take a closer look at the problem and see if we can figure out what happened and also what we can do about it.

First off, here's the probability distribution for X ~ B(12, 0.5). We wanted to find the probability of getting fewer than 6 questions correct, and we achieved this by calculating P(X < 6).

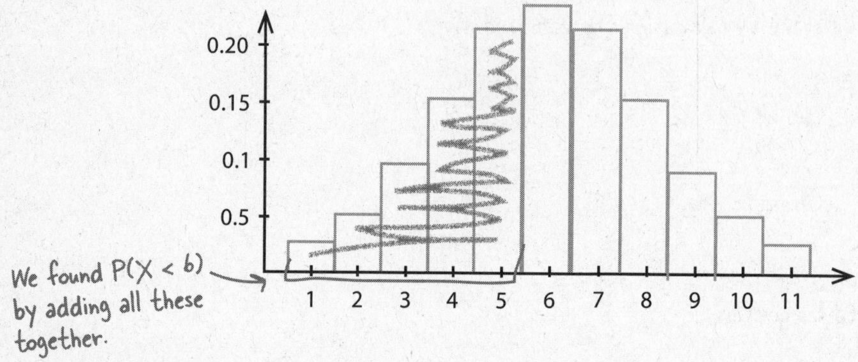

We then approximated the distribution by using X ~ N(6, 3), and as needed to find P(X < 6) for the binomial distribution, we calculated P(X < 6) using the normal distribution:

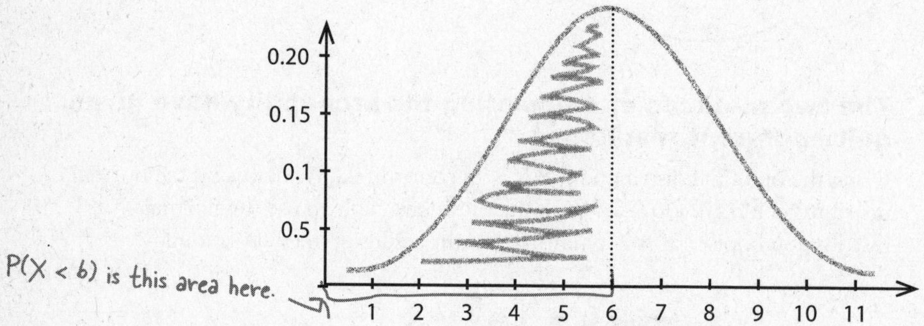

Take a really close look at the two probability distributions. It's tricky to spot, but there's a crucial difference between the two—the ranges we used to calculate the two probabilities are slightly different. We actually used a slightly larger range when we used the normal distribution, and this accounts for the larger probability.

We'll look at this in more detail on the next page.

The binomial is discrete, but the normal is continuous

There's one thing we overlooked when we calculated the two probabilities—we didn't make allowances for one distribution being discrete (the binomial), and the other being continuous (the normal). This is important, as the probability range we use can make a big difference to the resulting probabilities.

Here are the probability distributions for X ~ B(12, 0.5) and N(6, 3), both shown on the same chart. We've highlighted where the probability range we used with the normal distribution extends beyond the range we used for the binomial distribution.

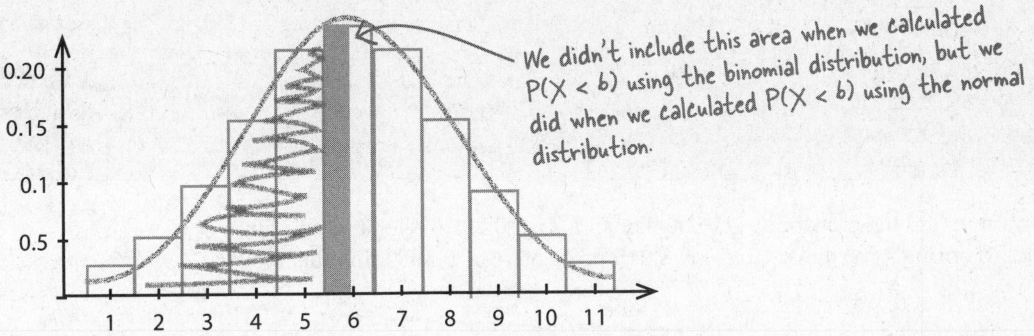

Can you see where the problem lies?

When we take integers from a discrete probability distribution and translate them onto a continuous scale, we don't just look at those precise values in isolation. Instead, we look at the range of numbers that round to each of the values.

Let's take the discrete value 6 as an example. When we translate the number 6 to a continuous scale, we need to consider all of the numbers that round to it—in other words, the entire range of numbers from 5.5 to 6.5.

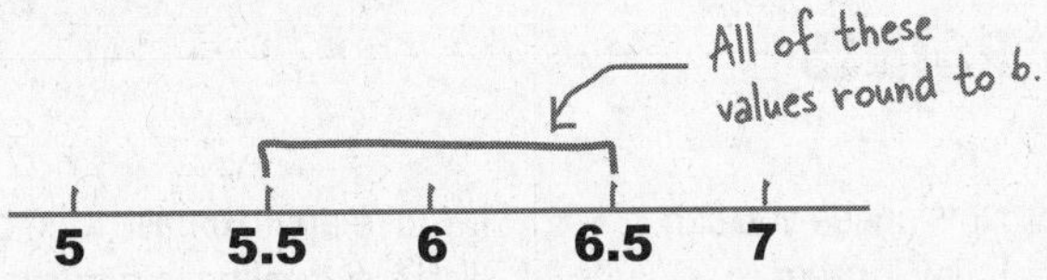

So how does this apply to our probability problem?

When we tried using the normal distribution to approximate the probability of getting fewer than 6 questions correct, we didn't look at how the discrete value 6 translates onto a continuous scale. The discrete value 6 actually covers a range from 5.5 to 6.5, so instead of using the normal distribution to find $P(X < 6)$, we should have tried calculating $P(X < 5.5)$ instead.

This adjustment is called a ***continuity correction***. A continuity correction is the small adjustment that needs to be made when you translate discrete values onto a continuous scale.

Apply a continuity correction before calculating the approximation

Let's try finding $P(X < 5.5)$ where $X \sim N(6, 3)$, and see how good an approximation this is for the probability of getting five or fewer questions correct. Using the binomial distribution we found that the probability we're aiming for is around 0.387.

Let's see how close an approximation the normal distribution gives us.

We want to find $P(X < 5.5)$ where $X \sim (6, 3)$, so let's start by calculating the standard score.

$$z = \frac{x - \mu}{\sigma}$$

$$= \frac{5.5 - 6}{\sqrt{3}}$$

$$= -0.29 \text{ (to 2 decimal places)}$$

We want to find the probability given by the area $Z < -0.29$, and looking this up in standard normal probability tables gives us a probability of 0.3859. In other words,

$$P(X < 5.5) = 0.3859$$

Look at these two probabilities. They're really close, so it looks like the continuity correction did the trick.

This is really close to the probability we came up with using the binomial distribution. The binomial distribution gave us a probability of 0.387, so the normal distribution gives us a pretty close approximation.

BULLET POINTS

- In particular circumstances you can **use the normal distribution to approximate the binomial**. If $X \sim B(n, p)$ and $np > 5$ and $nq > 5$ then you can approximate X using $X \sim N(np, npq)$
- If you're approximating the binomial distribution with the normal distribution, then you need to **apply a continuity correction** to make sure your results are accurate.

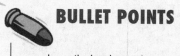

Continuity Corrections Up Close

The big trick with using the normal distribution to approximate binomial probabilities is to make sure you apply the right continuity correction. As you've seen, small changes in the probability range you choose can lead to significant errors in the actual probabilities. This might not sound like too big a deal, but using the wrong probability could lead to you making the wrong decisions.

Let's take a look at the kinds of continuity corrections you need to make for different types of probability problems.

Finding ≤ probabilities

When you work with probabilities of the form $P(X \leq a)$, the key thing you need to make sure of is that you choose your range so that it includes the discrete value a. On a continuous scale, the discrete value a goes up to $(a + 0.5)$. This means that if you're using the normal distribution to find $P(X \leq a)$, you actually need to calculate $P(X < a + 0.5)$ to come up with a good approximation. In other words, you add an extra 0.5.

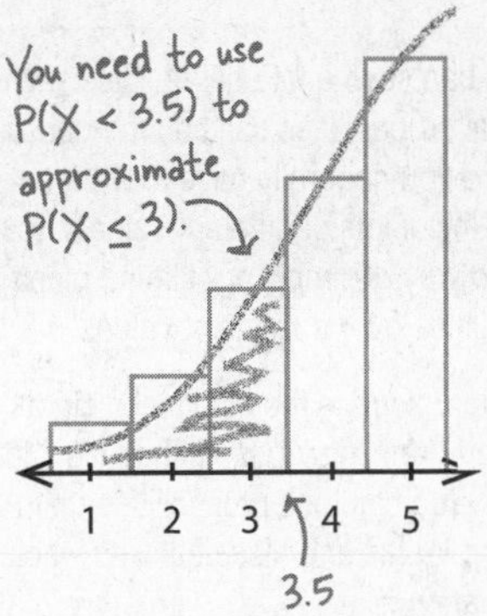

Finding ≥ probabilities

If you need to find probabilities of the form $P(X \geq b)$, you need to make absolutely sure that your range includes the discrete value b. The value b extends down to $(b - 0.5)$ on a continuous scale so you need to use a range of $P(X > b - 0.5)$ to make sure that you include it. In other words, you need to subtract an extra 0.5.

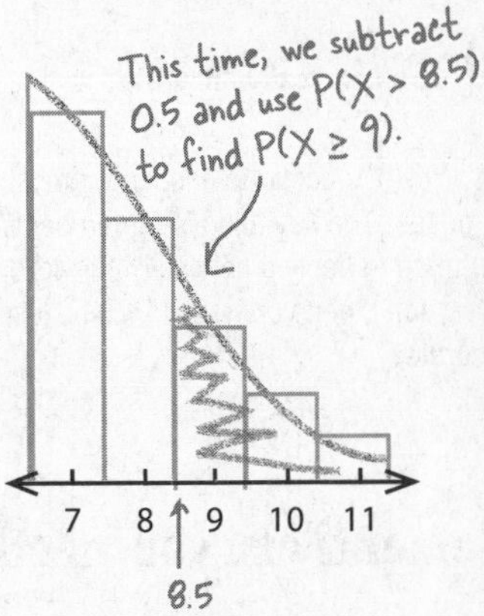

Finding "between" probabilities

Probabilities of the form $P(a \leq X \leq b)$ need continuity corrections to make sure that both a and b are included. To do this, we need to extend the range out by 0.5 either side. To approximate this probability using the normal distribution, we need to find $P(a - 0.5 < X < b + 0.5)$. This is really just a combination of the two types above.

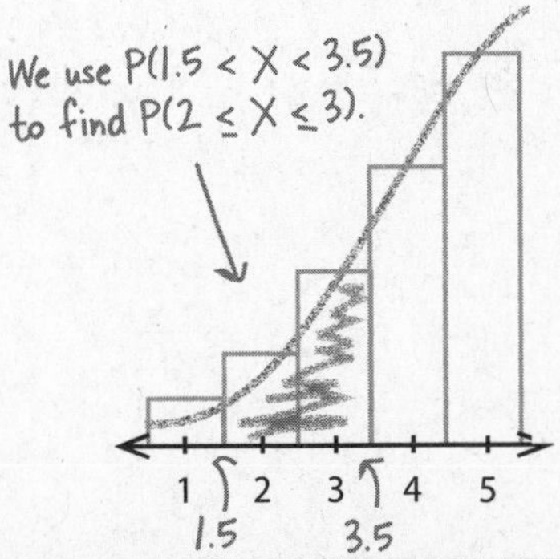

there are no Dumb Questions

Q: Does it really save time to approximate the binomial distribution with the normal?

A: It can save a lot of time. Calculating binomial probabilities can be time-consuming because you generally have to work out the probability of lots of different values. You have no way of simply calculating binomial probabilities over a range of values.

If you approximate the binomial distribution with the normal distribution, then it's a lot quicker. You can look probabilities up in standard tables and also deal with whole ranges at once.

Q: So is it really accurate?

A: Yes, It's accurate enough for most purposes. The key thing to remember is that you need to apply a continuity correction. If you don't then your results will be less accurate.

Q: What about continuity corrections for < and >? Do I treat those the same way as the ones for ≤ and ≥?

A: There's a difference, and it all comes down to which values you want to include and exclude.

When you're working out probabilities using ≤ and ≥, you need to make sure that you include the value in the inequality in your probability range. So if, say, you need to work out $P(X \leq 10)$, you need to make sure your probability includes the value 10. This means you need to consider $P(X < 10.5)$.

When you're working out probabilities using < or >, you need to make sure that you exclude the value in the inequality from your probability range. This means that if you need to work out $P(X < 10)$, you need to make sure that your probability excludes 10. You need to consider $P(X < 9.5)$.

Q: You can approximate the binomial distribution with both the normal and Poisson distributions. Which should I use?

A: It all depends on your circumstances. If $X \sim B(n, p)$, then you can use the normal distribution to approximate the binomial distribution if $np > 5$ and $nq > 5$.

You can use the Poisson distribution to approximate the binomial distribution if $n > 50$ and $p < 0.1$

Remember, you need to apply a continuity correction when you approximate the binomial distribution with the normal distribution.

Pool Puzzle

Your **job** is to take snippets from the pool and place them into the blank lines so that you get the right continuity correction for each dscrete probability range. You **may** use the same snippet more than once, and you won't need to use all the snippets.

$X < 3$ → ____________

$X > 3$ → ____________

$X \leq 3$ → ____________

$X \geq 3$ → ____________

$3 \leq X < 10$ → ____________

$X = 0$ → ____________

$3 \leq X \leq 10$ → ____________

$3 < X \leq 10$ → ____________

$X > 0$ → ____________

$3 < X < 10$ → ____________

Note: each thing from the pool can be used more than once!

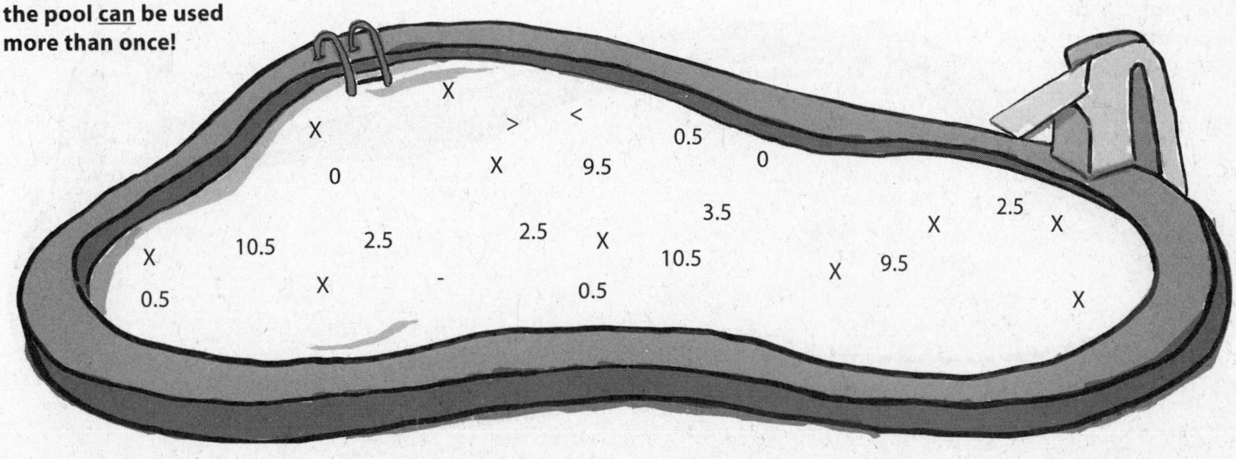

Pool Puzzle

Your **job** is to take snippets from the pool and place them into the blank lines so that you get the right continuity correction for each dscrete probability range. You **may** use the same snippet more than once, and you won't need to use all the snippets.

Here, we're looking for values less than 3. 2.5 rounds to 3, so we only want to include values less than 2.5 in our range.

All the numbers from −0.5 to 0.5 round to 0, so they must be included in the range.

Here, we're looking for values less than or equal to 3. All the numbers between 2.5 and 3.5 round to 3, so we need to include values less than 3.5 in our range.

$X < 3 \rightarrow X < 2.5$

$X > 3 \rightarrow X > 3.5$

$X \leq 3 \rightarrow X < 3.5$

$X \geq 3 \rightarrow X > 2.5$

$3 \leq X < 10 \rightarrow 2.5 < X < 9.5$

$X = 0 \rightarrow -0.5 < X < 0.5$

$3 \leq X \leq 10 \rightarrow 2.5 < X < 10.5$

$3 < X \leq 10 \rightarrow 3.5 < X < 10.5$

$X > 0 \rightarrow X > 0.5$

$3 < X < 10 \rightarrow 3.5 < X < 9.5$

Note: each thing from the pool can be used more than once!

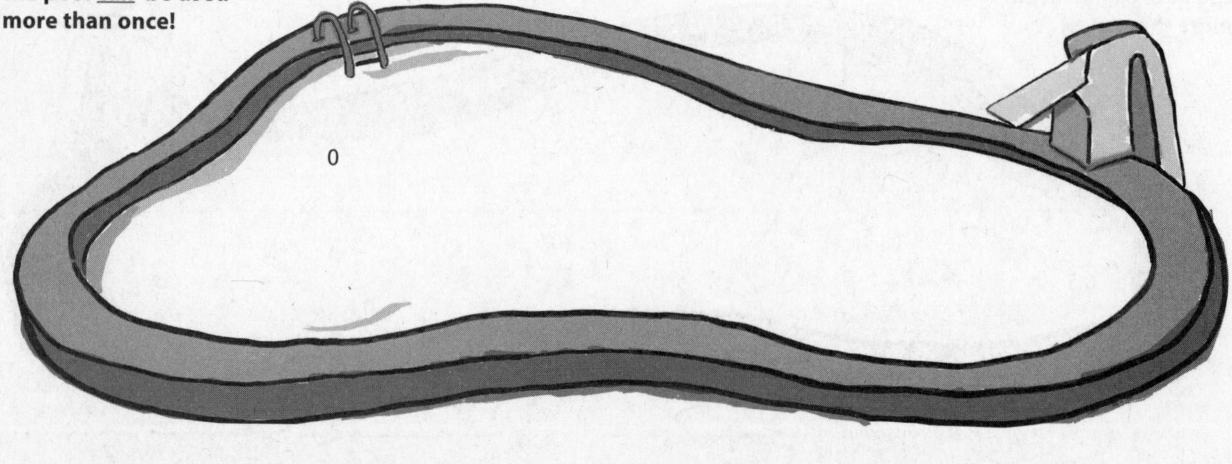

Exercise

What's the probability of you winning the jackpot on today's edition of Who Wants to Win a Swivel Chair? See if you can find the probability of getting at least 30 questions correct out of 40, where each question has a choice of 4 possible answers.

Exercise Solution

What's the probability of you winning the jackpot on today's edition of Who Wants to Win a Swivel Chair? See if you can find the probability of getting at least 30 questions correct out of 40, where each question has a choice of 4 possible answers.

If X is the number of questions we get right, then we want to find $P(X \geq 30)$ where $X \sim B(40, 0.25)$.

As np and nq are both greater than 5, it's appropriate for us to use the normal distribution to approximate this probability. $np = 10$ and $npq = 30$, which means we need to find $P(X > 29.5)$ where $X \sim N(10, 30)$.

Let's start by finding the standard score.

$$z = \frac{x - \mu}{\sigma}$$

$$= \frac{29.5 - 10}{30}$$

$$= \frac{19.5}{30}$$

$$= 0.65$$

Looking up 0.65 in probability tables gives us a probability of 0.7422. This means that

$$P(X > 29.5) = 1 - 0.7422$$

$$= 0.2578$$

So, looks like you've only got about a 26% chance at that swivel chair. If you lose, you'll miss out on our great consolation prize. Why don't you take the prize and run?

Sorry to see you go. It's been great having you back as a contestant on the show, but we've just had an urgent email from someone called Dexter...

Sharpen your pencil Solution

Here are the first five questions for Round Two. The questions are all about the game show host.

1. What is his favorite film

- ✓ **A:** The Day of The Jackal
- **B:** The Italian Job
- **C:** Lawrence of Arabia
- **D:** All the President's Men

2. What is the favorite film of his cat

- **A:** A Fish Called Wanda
- **B:** Curse of the Were-Rabbit
- ✓ **C:** Mousehunt
- **D:** Bird on a Wire

3. On average, how much does he spend on suits each month

- **A:** $1,000
- **B:** $2,000
- **C:** $3,000
- ✓ **D:** $4,000

4. How often does he have his hair cut?

- **A:** Once a month
- ✓ **B:** Twice a month
- **C:** Three times a month
- **D:** Four times a month

5. What is his favorite web site

- ✓ **A:** www.fatdanscasino.com
- **B:** www.gregs-list.net
- **C:** www.you-cube.net
- **D:** www.starbuzzcoffee.com

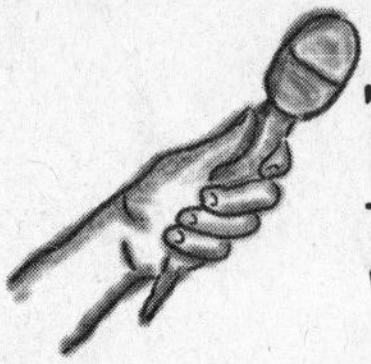

The Normal Distribution Exposed

This week's interview:
Why Being Normal Isn't Dull

Head First: Hey, Normal, glad you could make it on the show.

Normal: Thanks for inviting me, Head First.

Head First: Now, my first question is about your name. Why are you called Normal?

Normal: It's really because I'm so representative of a lot of types of data. They have a probability distribution that has a distinctive shape and a smooth, bell-curved shape, and that's me. I'm something of an ideal.

Head First: Can you give me an example?

Normal: Sure. Imagine you have a baker's shop that sells loaves of bread. Now, each loaf of a particular sort of bread should theoretically weigh about the same, but in practice, the actual weight of each loaf of bread will vary.

Head First: But surely they'll all weigh about the same?

Normal: More or less, but with variation. I model that variation.

Head First: So why's that so important?

Normal: Well, it means that you can use me to work out probabilities. Say you want to find the probability of a randomly chosen loaf of bread being below a particular weight. That sounds like something that could be quite difficult, but with me, it's easy.

Head First: Easy? How do you mean?

Normal: With a lot of the other probability distributions, there can be lots of complicated calculations involved. With Binomial you have factorials, and with Poisson you have to work with exponentials. With me there's none of that. Just look me up in a table and away you go.

Head First: Surely it's not quite as simple as that?

Normal: Well, you do have to convert me to a standard score first, but that's nothing, not in the grand scheme of things.

Head First: So tell me, do you think you're better than the other probability distributions?

Normal: I wouldn't say that I'm better as such, but I'm a lot more flexible, and I'm useful in lots of situations. I'm also a lot more robust. When the numbers get high for Poisson and Binomial distributions, they run into trouble. Mind you, I do what I can to help out.

Head First: You do? How?

Normal: Well under certain circumstances both Binomial and Poisson look like me. It's uncanny; they're often stopped at parties by people asking them if they're Normal. I tell them to take it as a compliment.

Head First: So how does that help?

Normal: Well, because they look like me, it means that you can actually use my probability tables to work out their probabilities. How cool is that? No more late nights slaving over a calculator; just look it up.

Head First: I'm afraid that's all we've got time for tonight. Normal, thanks for coming along, it's been a pleasure.

Normal: You're welcome, Head First.

All aboard the Love Train

Remember Dexter's Love Train? He's started running trials of the ride, and everyone who's given it a trial run thinks it's great. There's just one problem: sometimes the ride breaks down and causes delays, and delays cost money.

Dexter's found some statistics on the Internet about the model of roller coaster he's been trying out, and according to one site, you can expect the ride to break down 40 times a year.

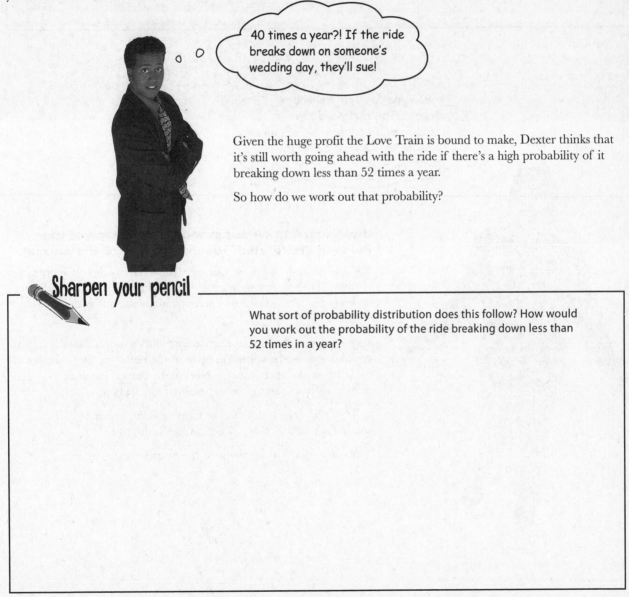

40 times a year?! If the ride breaks down on someone's wedding day, they'll sue!

Given the huge profit the Love Train is bound to make, Dexter thinks that it's still worth going ahead with the ride if there's a high probability of it breaking down less than 52 times a year.

So how do we work out that probability?

Sharpen your pencil

What sort of probability distribution does this follow? How would you work out the probability of the ride breaking down less than 52 times in a year?

Sharpen your pencil Solution

What sort of probability distribution does this follow? How would you work out the probability of the ride breaking down less than 52 times in a year?

Situations where you're dealing with things breaking down at a mean rate follow a Poisson distribution, taking a parameter of the mean. If X represents the number of breakdowns in a year, then $X \sim Po(40)$.

We need to find $P(X < 52)$. To find this, we'd need to find each individual probability for all values of X up to 52.

Working out that probability is gonna be tricky and time-consuming. I wonder if we can take a shortcut like we did with the binomial.

Under certain circumstances, the shape of the Poisson distribution resembles that of the normal.

The advantage of this is we can use standard normal probability tables to work out whole ranges of probabilities. This means that we don't have to calculate lots of individual probabilities in order to find what we need.

Approximating the Poisson distribution with the normal is very similar to when you use the normal in place of the binomial. Once you have the right set of circumstances, you take the Poisson mean and variance, and use them as parameters in a normal distribution.

If $X \sim Po(\lambda)$, this means that the corresponding normal approximation is $X \sim N(\lambda, \lambda)$. But when is this true?

It all comes down to the shape of the distribution.

When to approximate the binomial distribution with the normal

We can use the normal distribution to approximate the Poisson whenever the Poisson distribution adopts a shape that's like the normal, but when does this happen? Let's take a look.

When λ is small...

When λ is small, the shape of the Poisson distribution is different from that of the normal distribution. The shape isn't symmetrical, and the curve looks as though it's "pulled" over to the right.

As the Poisson distribution doesn't resemble the normal for small values of λ, the normal distribution isn't a suitable approximation for the poisson distribution where λ is small.

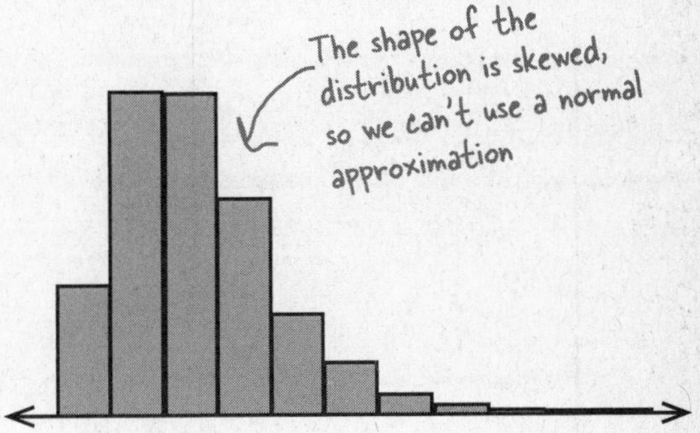
The shape of the distribution is skewed, so we can't use a normal approximation

When λ is large...

As λ gets larger, the shape of the Poisson distribution looks increasingly like that of the normal distribution. The main part of the shape is reasonably symmetrical, and it forms a smooth curve that's just like the one for the normal.

This means that as λ gets larger, the normal distribution can be used to give a better and better approximation of it.

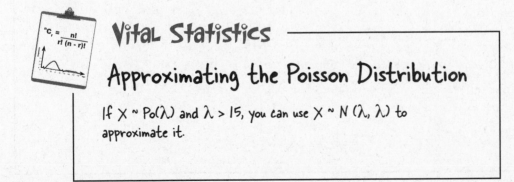

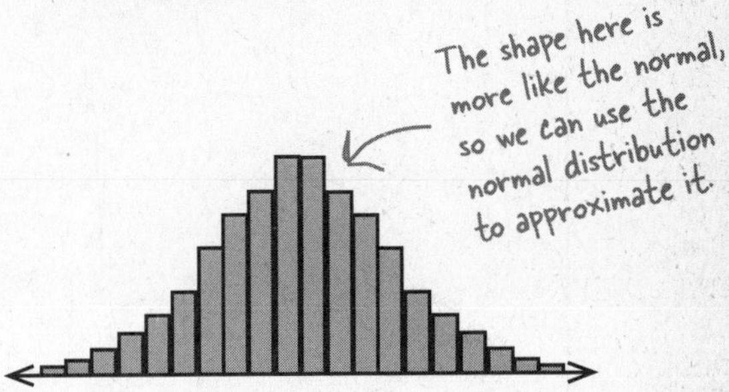
The shape here is more like the normal, so we can use the normal distribution to approximate it.

So how large is large enough?

We've seen that the Poisson resembles the normal distribution when λ is large, but how big does it have to be before we can use the normal?

λ actually gets sufficiently large when λ is greater than 15. This means that if $X \sim Po(\lambda)$ and $\lambda > 15$, we can approximate this using $X \sim N(\lambda, \lambda)$.

Vital Statistics

Approximating the Poisson Distribution

If $X \sim Po(\lambda)$ and $\lambda > 15$, you can use $X \sim N(\lambda, \lambda)$ to approximate it.

The number of breakdowns on Dexter's Love Train follows a Poisson distribution where λ = 40. What's the probability that there will be fewer than 52 breakdowns in the first year?

Hint: Use a normal approximation, and remember your continuity corrections.

It's time to test your statistical knowledge. Complete the table below, saying what normal distribution suits each situation, and what conditions there are.

Situation	Distribution	Condition
$X + Y$ $X \sim N(\mu_x, \sigma^2_x), Y \sim (\mu_y, \sigma^2_y)$	$X + Y \sim N(\mu_x + \mu_y, \sigma^2_x + \sigma^2_y)$	X, Y are independent
$X - Y$ $X \sim N(\mu_x, \sigma^2_x), Y \sim (\mu_y, \sigma^2_y)$		
$aX + b$ $X \sim N(\mu, \sigma^2)$		
$X_1 + X_2 + \dots + X_n$ $X \sim N(\mu, \sigma^2)$		
Normal approximation of X $X \sim B(n, p)$		
Normal approximation of X $X \sim Po(\lambda)$		

Exercise Solution

The number of breakdowns on Dexter's Love Train follows a Poisson distribution where $\lambda = 40$. What's the probability that there will be fewer than 52 breakdowns in the first year?

If we use X to represent the number of breakdowns in a year, then $X \sim Po(40)$

As λ is large, we can use the normal distribution to approximate this. In other words, we use $X \sim N(40, 40)$

We need to find the probability that there are fewer than 52 breakdowns. As we're approximating a discrete probability distribution with a continuous one, we have to apply a continuity correction. We don't want to include 52, so we need to find $P(X \leq 51.5)$.

Before we can find the probability using standard normal tables, we need to calculate the standard score.

$$z = \frac{x - \mu}{\sigma}$$

$$= \frac{51.5 - 40}{6.32}$$

$$= 1.82 \text{ (to 2 decimal places)}$$

Looking this up in probability tables gives us 0.9656. This means that the probability of there being fewer than 52 breakdowns in a year is 0.9656.

It's time to test your statistical knowledge. Complete the table below, saying what normal distribution suits each situation, and what conditions there are.

Situation	Distribution	Condition
$X + Y$ **$X \sim N(\mu_x, \sigma^2_x)$, $Y \sim (\mu_y, \sigma^2_y)$**	$X + Y \sim N(\mu_x + \mu_y, \sigma^2_x + \sigma^2_y)$	X, Y are independent
$X - Y$ **$X \sim N(\mu_x, \sigma^2_x)$, $Y \sim (\mu_y, \sigma^2_y)$**	$X - Y \sim N(\mu_x - \mu_y, \sigma^2_x + \sigma^2_y)$	X, Y are independent
$aX + b$ **$X \sim N(\mu, \sigma^2)$**	$aX + b \sim N(a\mu + b, a^2\sigma^2)$	a, b are constant values
$X_1 + X_2 + \dots + X_n$ **$X \sim N(\mu, \sigma^2)$**	$X_1 + X_2 + \dots + X_n \sim N(n\mu, n\sigma^2)$	$X_1, X_2, \dots, X_n$ are independent observations of X
Normal approximation of X **$X \sim B(n, p)$**	$X \sim N(np, npq)$	$np > 5$, $npq > 5$ Continuity correction required
Normal approximation of X **$X \sim Po(\lambda)$**	$X \sim N(\lambda, \lambda)$	$\lambda > 15$ Continuity correction required

BULLET POINTS

- In particular circumstances you can use the normal distribution to approximate the Poisson.
- If $X \sim Po(\lambda)$ and $\lambda > 15$ then you can approximate X using $X \sim N(\lambda, \lambda)$
- If you're approximating the Poisson distribution with the normal distribution, then you need to apply a continuity correction to make sure your results are accurate.

there are no Dumb Questions

Q: You can approximate the binomial and Poisson distributions with the normal, but what about the geometric disribution? Can the normal distribution ever approximate that?

A: We were able to use the normal distribution in place of the binomial and Poisson distributions because under particular circumstances, these distributions adopt the same shape as the normal.

The geometric distribution, on the other hand, never looks like the normal, so the normal can never effectively approximate it.

Q: Do I have to use a continuity correction if I approximate the Poisson distribution with the normal distribution?

A: Yes. This is because you're approximating a discrete probability distribution with a continuous one. This means that you need to apply a continuity correction, just as you would for the binomial distribution.

Q: What's the advantage of approximating the binomial or poisson distribution with the normal? Won't my results be more accurate if I just use the original distribution?

A: Your results will be more accurate if you use the original distribution, but using them can be time consuming. If you wanted to find the probability of a range of values using the binomial or poisson distribution, you'd need to find the probability of every single value within that range. Using the normal distribution, on the other hand, you can look up probabilities for whole ranges, and so they're a lot easier to find.

Use a continuity correction if you approximate the Poisson distribution with the normal distribution.

A runaway success!

Thanks to your savvy statistical analysis,the Love Train is open for business, and demand has outstripped Dexter's highest expectations. Here are some of Dexter's happy customers:

10 using statistical sampling

Taking Samples

Stay nice and relaxed, and this won't hurt a bit.

Statistics deal with data, but where does it come from?

Some of the time, data's easy to collect, such as the ages of people attending a health club or the sales figures for a games company. But what about the times when data isn't so easy to collect? Sometimes the number of things we want to collect data about are so huge that it's difficult to know where to start. In this chapter, we'll take a look at how you can **effectively gather data** in the real world, in a way that's efficient, accurate, and can also save you time and money to boot. Welcome to the world of sampling.

The Mighty Gumball taste test

Mighty Gumball is the leading vendor of a wide variety of candies and chocolates. Their signature product is their super-long-lasting gumball. It comes in all sorts of colors to suit all tastes.

Mighty Gumball plans to run a series of television commercials to attract even more customers, and as part of this, they want to advertise just how long the flavor of their gumballs lasts for. The problem is, how do they get the data?

They've decided to implement a taste test, and they've hired a bunch of tasters to help with the tests. There are just two problems: the tasters are using up all of the gumballs, and their dental plans are costing the company a fortune.

Well, gumball #1,466 ran out of flavor after 55 minutes, but gumball #1,467 is still going strong after an hour...
Please, no more gumballs! I'm running out of teeth

They're running out of gumballs

The fatal flaw with the Mighty Gumball taste test is that the tasters are trying out **all** of the gumballs. Not only is this having a bad effect on the tasters' teeth, it also means that there are no gumballs left to sell. After all, they can hardly reuse their gumballs once the tasters have finished with them.

The whole point of the taste test is for Mighty Gumball to figure out how long the flavor lasts for. But does this really mean that the tasters have to try out *every single gumball?*

BRAIN POWER

What would you do to establish how long the gumball flavor lasts for? What do you need to consider? Write your answer below in as much detail as possible.

Test a gumball sample, not the whole gumball population

Mighty Gumball is running into problems because they're tasting *every single gumball* as part of their taste test. It's costing them time, money, and teeth, and they have no gumballs left to actually sell to their customers.

So what should Mighty Gumball do differently? Let's start by looking at the difference between populations and samples.

Gumball populations

At the moment, Mighty Gumball is carrying out their taste test using every single gumball that they have available. In statistical terms, they are conducting their test using an entire **population**.

A statistical ***population*** refers to the entire group of things that you're trying to measure, study, or analyze. It can refer to anything from humans to scores to gumballs. The key thing is that a population refers to all of them.

A ***census*** is a study or survey involving the entire population, so in the case of Mighty Gumball, they're conducting a census of their gumball population by tasting every single one of them. A census can provide you with accurate information about your population, but it's not always practical. When populations are large or infinite, it's just not possible to include every member.

Gumball samples

You don't have to taste every gumball to get an idea of how long the flavor lasts for. Instead of testing the entire population, you can test a **sample** instead.

A statistical ***sample*** is a selection of items taken from a population. You choose your sample so that it's fairly representative of the population as a whole; it's a representative subset of the population. For Mighty Gumball, a sample of gumballs means just a **small selection** of gumballs rather than every single one of them.

A study or survey involving just a sample of the population is called a ***sample survey***. A lot of the time, conducting a survey is more practical than a census. It's usually less time-consuming and expensive, as you don't have to deal with the entire population. And because you don't use the whole population, taking a sample survey of the gumballs means that there'll be plenty left over when you're done.

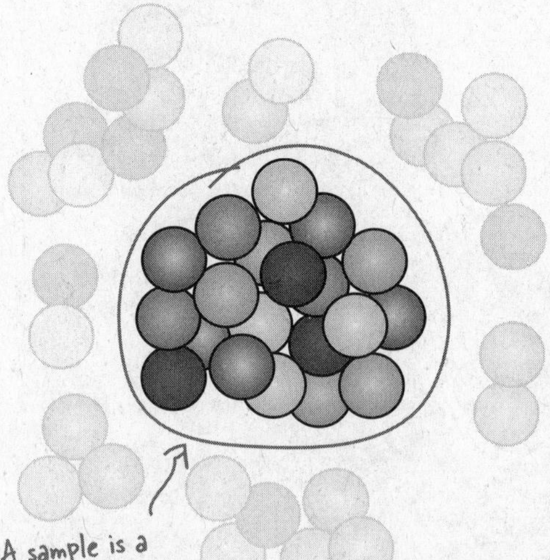

A sample is a subset of the population, so just some of the gumballs.

So how can you use samples to find out about a population? Let's see.

How sampling works

The key to creating a good sample is to choose one that is as close a match to your population as possible. If your sample is representative, this means it has similar characteristics to the population. And this, in turn, means that you can use your sample to predict what characteristics the population will have.

Suppose you use a representative sample of gumballs to test how long the flavor of each gumball lasts for. The distribution of the results might look something like this:

Even though you've only tried a small sample of gumballs, you still have an impression of the shape of the distribution, and the more gumballs you try, the clearer the shape is. As an example, you can get a rough impression of where the center of the population distribution is by looking at the shape of the sample distribution.

Let's compare this with the actual population:

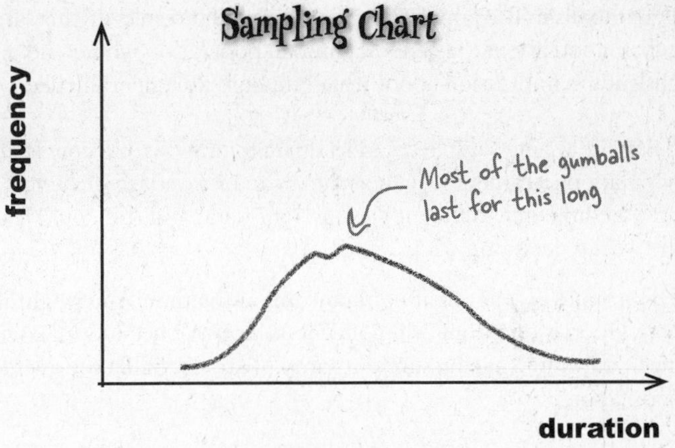

Here's the chart for the population. Can you see how closely the sample and population distributions agree?

If you compare the two charts, the overall shape is very similar, even though one is for *all* of the gumballs and the other is for just *some* of them. They share key characteristics such as where the center of the data is, and this means you can use the sample data to make predictions about the population.

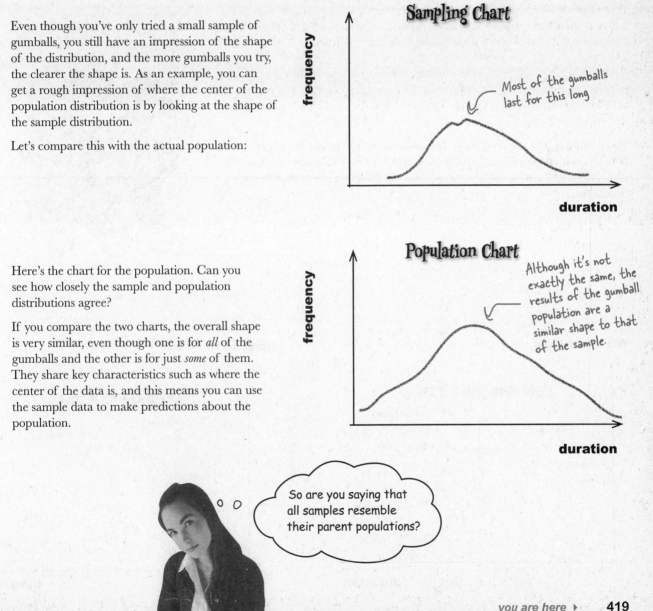

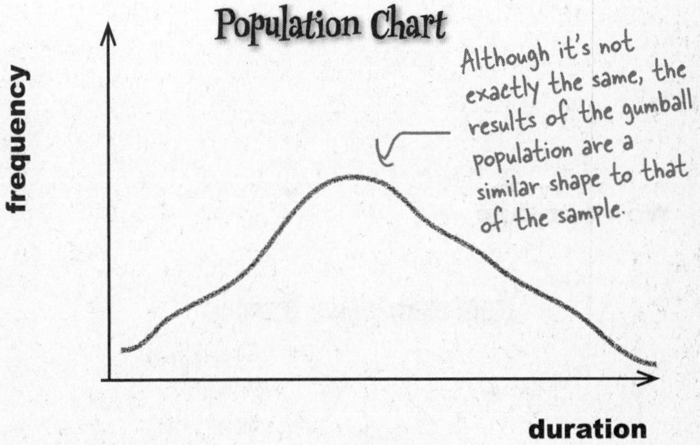

Population Chart
frequency
Although it's not exactly the same, the results of the gumball population are a similar shape to that of the sample.
duration

So are you saying that all samples resemble their parent populations?

When sampling goes wrong

If only we could guarantee that every sample was a close match to the population it comes from. Unfortunately, not every sample closely resembles its population. This may not sound like a big deal, but using a misleading sample can actually lead you to draw the wrong conclusions about your population.

As an example, imagine if you took a sample of gumballs to find out how long flavor typically lasts for, but your sample only contained red gumballs. Your sample might be representative of *red* gumballs, but not so representative of *all* gumballs in the population. If you used the results of this sample to gather information about the general gumball population, you could end up with a misleading impression about what gumballs are generally like.

Using the wrong sample could lead you to draw wrong conclusions about population parameters, such as the mean or standard deviation. You might be left with a completely different view of your data, and this could lead you to make the wrong decisions.

The trouble is, you might not know this at the time. You might think your population is one thing when in fact it's not. We need to make sure we have some mechanism for making sure our samples are a reliable representation of the population.

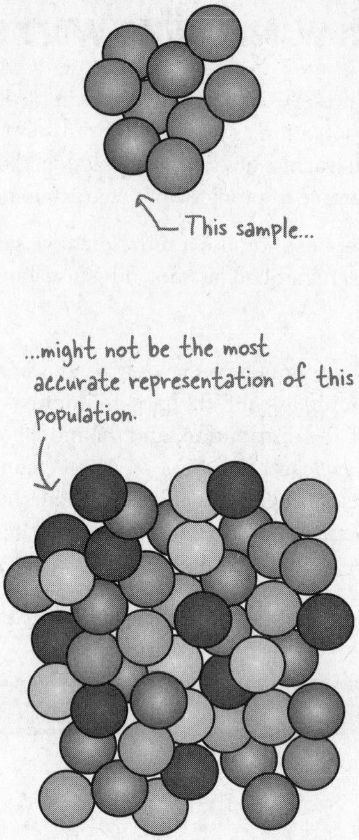

This sample...
...might not be the most accurate representation of this population.

We want this:

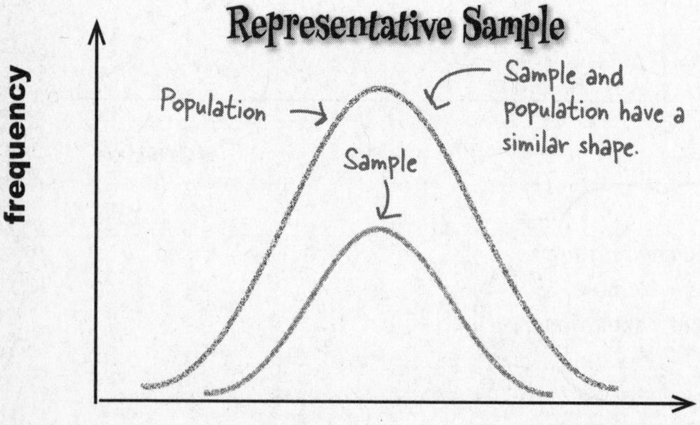

Instead of this:

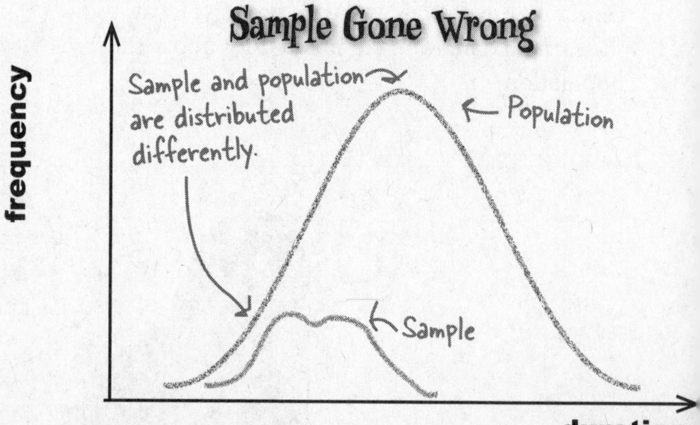

The Case of the Lost Coffee Sales

The Starbuzz CEO has an idea for a brand-new coffee he wants to sell in his coffee shops, but he's not sure how popular it's going to be with his customers. He asks his new intern to conduct a survey to help predict the customers' opinions. The intern will ask customers to taste the new brew, and tell him what they think.

The intern is really happy to be given such a great opportunity. First off, he's been told that if he does the job well, he stands to get a bonus at the end of the month. Secondly, he gets to give out free coffee to friendly Starbuzz customers and hear lots of positive things. Thirdly, he's been looking for an excuse to talk to one particular girl who's a regular visitor to his local coffee shop, and this could be just the break he needs.

After the intern conducts his survey, he's delighted to tell the CEO that everyone loves the new coffee, and it's bound to be a huge success. "That's great," says the CEO. "We'll launch it next season."

When the new coffee is finally launched, sales are poor, and the CEO has to cancel the range. What do you think went wrong?

Why didn't the new coffee sell well?

How to design a sample

You use samples to make inferences about the population in general, and to make sure you get accurate results, you need to choose your sample wisely. Let's start off by pinning down what your population really is, so you can get as representative a sample as possible.

Define your target population

The first thing to be clear about is what your **target population** is so that you know where you're collecting your sample from. By ***target population***, we mean the group that you're reseraching and want to collect results for. The target population you choose depends, to a large extent, on the purpose of your study. For example, do you want to gather data about all the gumballs in the world, one particular brand, or one particular type?

Try to be as precise as possible, as that way it's easier to make your sample as representative of your population as possible.

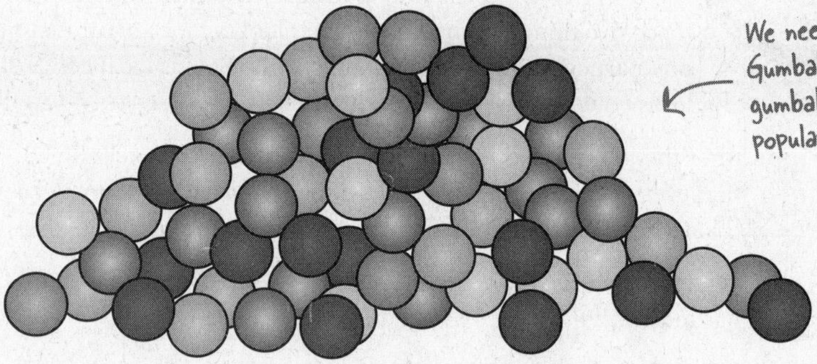

Define your sampling units

Once you've defined your target population, you need to decide what sort of object you're going to sample. Normally these will be the sorts of things you described when you defined your target population. As an example, this could be a single gumball or maybe a packet of gumballs.

The sampling unit in the taste test is a single Mighty Gumball super-long-lasting gumball.

Define your sampling frame

Finally, you need a list of all the sampling units within your target population, preferably with each sampling unit either named or numbered. This is called the ***sampling frame***. It's basically a list from which you can choose your sample.

Sometimes it's not possible to come up with a list that covers the entire target population. As an example, if you want to collect the views of people living within a certain area, people moving in or out of an area can affect who you have on your list of names. If you're dealing with similar objects such as gumballs, it might not be possible or practical to name or number each one.

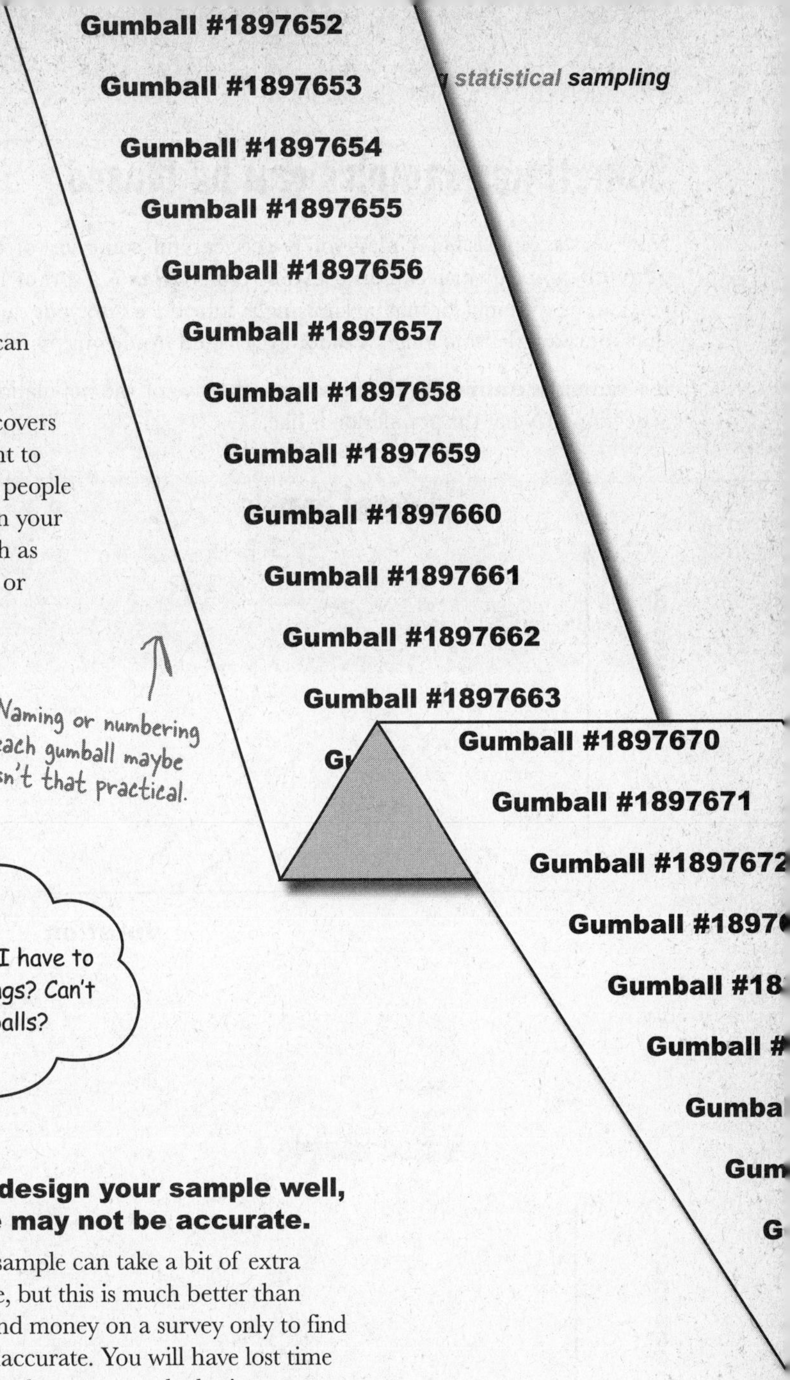

Naming or numbering each gumball maybe isn't that practical.

This seems like a waste of time. Do I have to do all of these things? Can't I just sample gumballs?

If you don't design your sample well, your sample may not be accurate.

Designing your sample can take a bit of extra preparation time, but this is much better than spending time and money on a survey only to find the results are inaccurate. You will have lost time and money doing the survey, and what's more, someone might make wrong decisions based on it.

A poorly designed sample can introduce ***bias***. Let's look at this in more detail.

Sometimes samples can be biased

Not every sample is fair. Unless you're very careful, some sort of bias can creep in to the sample, which can distort your results. ***Bias*** is a sort of favoritism that you can unwittingly (or maybe knowingly) introduce into your sample, meaning that your sample is no longer randomly selected from your population

If a sample is **unbiased**, then it's representative of the population. It's a fair reflection of what the population is like.

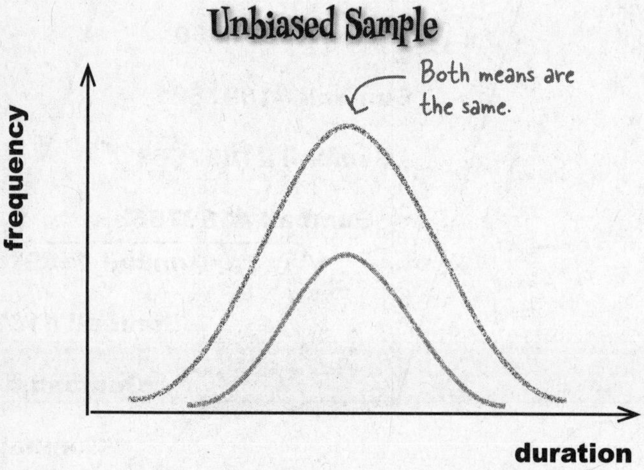

Unbiased samples

An ***unbiased sample*** is representative of the target population. This means that it has similar characteristics to the population, and we can use these to make inferences about the population itself.

The shape of the distribution of an unbiased sample is similar to the shape of the population it comes from. If we know the shape of the sample distribution, we can use it to predict that of the population to a reasonable level of confidence.

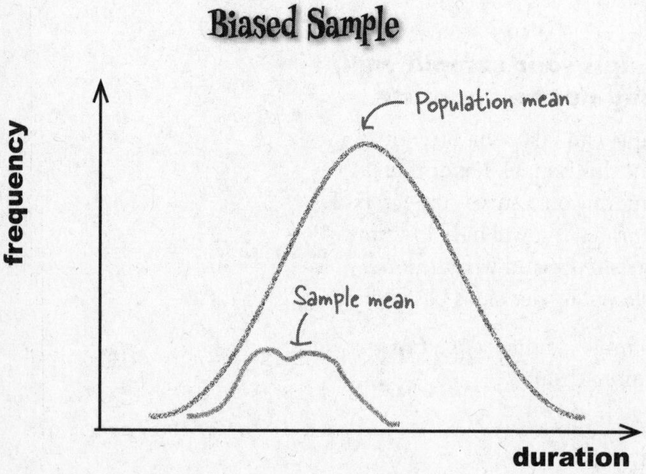

Biased samples

A ***biased sample*** is not representative of the target population. We can't use it to make inferences about the population because the sample and population have different characteristics. If we try to predict the shape of the population distribution from that of the sample, we'd end up with the wrong result.

This sounds hopeless. How can I be certain I avoid bias? Where does it come from anyway?

Sources of bias

So how does bias creep into samples? Through any of the following and more:

- **A sampling frame where items have been left off**, such that not everything in the target population is included. If it's not in your sampling frame, it won't be in your sample.

- **An incorrect sampling unit.** Instead of individual gumballs, maybe the sampling unit should have been boxes of gumballs instead.

- **Individual sampling units you chose for your sample weren't included in your actual sample.** As an example, you might send out a questionnaire that not everybody responds to.

- **Poorly designed questions in a questionnaire.** Design your questions so that they're neutral and everyone can answer them. An example of a biased question is "Mighty Gumball candy is tastier than any other brand, do you agree?" It would be better to ask the person being surveyed for the name of their favorite brand of confectionary.

- **Samples that aren't random.** As an example, if you're conducting a survey on the street, you may avoid questioning anyone that looks too busy to stop, or too aggressive. This means that you exclude aggressive or busy-looking people from your survey.

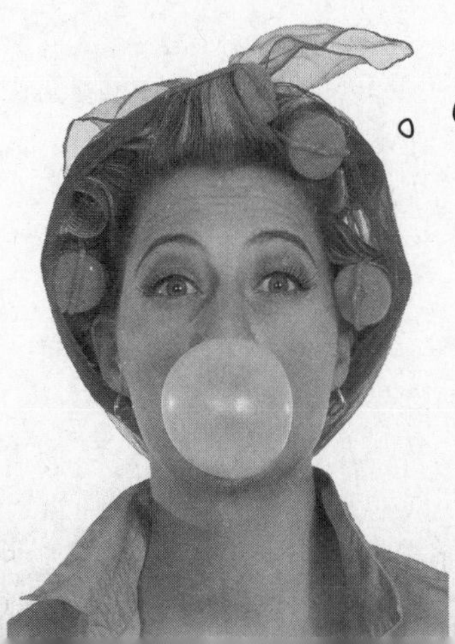

You mean I can't just try the pink ones???

As you can see, there are lots of sources of bias, and a lot of it comes down to how you choose your sample.

We need to take a look at ways in which you can choose your sample to minimize the chances of introducing bias.

there are no Dumb Questions

Q: So is the sampling frame a list of everything that we're sampling?

A: The sampling frame lists all the individual units in the population, and it's used as a basis for the sample. It's not the sample itself, as we don't sample everything on it.

Q: How do I put together the sampling frame?

A: How you do it and what you use depends on your target population. As an example, if your target population is all car owners, then you can use a list of registered car owners. If your target population is all the students attending a particular college, you can use the college registrar.

Q: How about things like telephone listings? Can I use those for my sampling frame?

A: It all depends on your target population. Telephone listings exclude households without a telephone, and there may also be households who have elected not to be listed. If your target population is households with a listed telephone number, then using telephone lists is a good idea. If your target population is all households with a telephone or even all telephones, then your sampling frame won't be entirely accurate—and that can introduce bias.

Q: Can I always compile a sampling frame?

A: Not always. Imagine if you had to survey all the fish in the sea. It would be impossible to name and number every individual fish.

Q: Will I always have to have a target population?

A: Yes. You need to know what your target population is so that you can make sure your sample is representative of it. Thinking carefully about what your target population is can help you avoid bias.

If you're sampling for someone else, get as much detail as possible about who the target population should be. Make sure you know exactly what is included and what is excluded.

Q: Why is bias so bad?

A: Bias is bad because it can mislead you into drawing wrong conclusions about your target population, which in turn can lead you into making wrong decisions. If, for example, you only sampled pink gumballs, your survey results might be accurate for all pink gumballs, but not for all gumballs in general. There may be significant differences between the different color gumballs.

Q: How can the questions in a questionnaire cause bias?

A: Bias often creeps in through the phrasing of questions.

First off, if you present a series of statements and ask respondents to agree or disagree, it's more likely that people will agree unless they have strong negative feelings. This means that the results of your survey will be biased towards people agreeing.

Bias can also occur if you give a set of possible answers that don't cover all eventualities. As an example, imagine you need to ask people how often they exercise in a typical week. You would introduce bias if you give answers such as "more than 5 times a week," "3–5 times a week," "1–2 times a week," and "I don't value my health, so I don't exercise." Someone may not exercise, but disagree with the statement that they don't value their health. This would mean that they wouldn't be able to answer the question.

Sharpen your pencil

Look at the following scenarios. What would you choose as a target population? What's the sampling unit? How would you develop a sampling frame? What other things might you need to consider when forming your sample?

1. Choc-O-Holic Inc. manufactures chocolates, and they have just finished a limited-edition run of chocolates for the holiday season. They want to check the quality of those chocolates.

2. The Statsville Health Club wants to conduct a survey to see what their customers think of their facilities.

Sharpen your pencil Solution

Look at the following scenarios. What would you choose as a target population? What's the sampling unit? How would you develop a sampling frame? What other things might you need to consider when forming your sample?

1. Choc-O-Holic Inc. manufactures chocolates, and they have just finished a limited-edition run of chocolates for the holiday season. They want to check the quality of those chocolates.

The target population is all the chocolates in the limited edition run.

The sampling unit is one chocolate.

The sampling frame needs to cover all of the chocolates; as it's a limited-edition run, it's possible that Choc-O-Holic has records of how many chocolates are in the run, including numbers of each type of chocolate.

When forming the sample, you need to make sure that it is representative of the target population and unbiased. If there are different types of chocolate in the run, you'd need to make sure that you included each sort of chocolate.

2. The Statsville Health Club wants to conduct a survey to see what their customers think of their facilities.

The target population is all the customers of the Statsville Health Club.

The sampling unit is one customer.

The sampling frame needs to cover all of the customers. It's likely that the health club has a list of registered customers, so you could use this as the sampling frame.

As before, you need to make sure that your sample is representative of the population and unbiased. You'd need to make sure that each of the classes is fairly represented by customer gender, customer age group, and so on.

Solved: The Case of the Lost Coffee Sales

Why didn't the coffee sell well?

We don't know for certain, but there's a very good chance that the sample of people surveyed by the intern wasn't representative of the target population.

First of all, the intern was looking forward to giving away free coffee to friendly Starbuzz customers and hearing positive things. Does this mean he only spoke to customers who looked friendly to him? Did he get their *real* opinions about the coffee, or did he only ask them whether they agreed it tasted nice?

The intern also hoped to use the job as an opportunity to speak to a girl at his local coffee shop. Did he spend most of his time in this particular coffee shop? Did the girl influence his sample choice?

Finally, the CEO launched the new coffee in a different season from the one in which the survey took place, and this may have affected sales too.

Any or all of these factors could have lead to the sample being misleading, which in turn led to the wrong decision being made.

How to choose your sample

We've looked at how to design your sample and explored types of bias that need to be avoided. Now we need to select our actual sample from the sample frame. But how should we go about this?

Simple random sampling

One option is to choose the sample at random. Imagine you have a population of N sampling units, and you need to pick a sample of *n* sampling units. ***Simple random sampling*** is where you choose a sample of *n* using some random process, and all possible samples of size *n* are equally likely to be selected.

With simple random sampling, you have two options. You can either sample **with replacement** or **without replacement**.

Sampling with replacement

Sampling with replacement means that when you've selected each unit and recorded relevant information about it, you put it back into the population. By doing this, there's a chance that a sampling unit might be chosen more than once. You'd be sampling with replacement if you decided to question people on the street at random without checking if you had already questioned them before. If you stop a person for questioning and then let them go once you've finished asking them questions, you are in effect releasing them back into the population. It means that you may question them more than once.

Sampling without replacement

Sampling without replacement means that the sampling unit isn't replaced back into the population. An example of this is the gumball taste test; you wouldn't want to put gumballs that have been tasted back into the population.

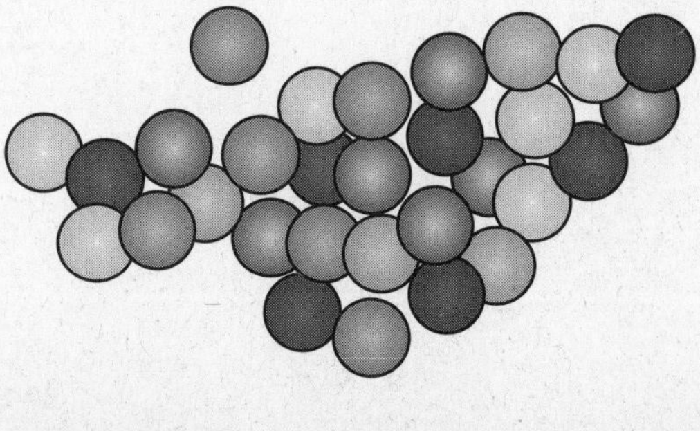

You wouldn't want to replace gumballs once you've tasted them, so this would be simple random samping without replacement.

How to choose a simple random sample

There are two main ways of using simple random sampling: by drawing lots or using random numbers.

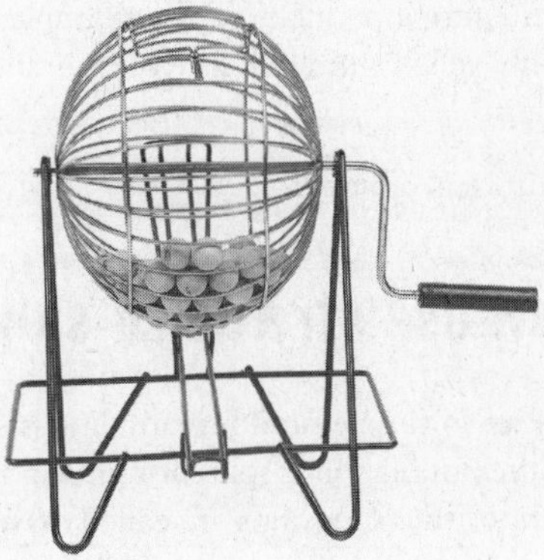

Drawing lots

Drawing lots is just like pulling names out of a hat. You write the name or number of each member of the sampling frame on a piece of paper or ball, and then place them all into a container. You then draw out *n* names or numbers at random so that you have enough for your sample.

Random number generators

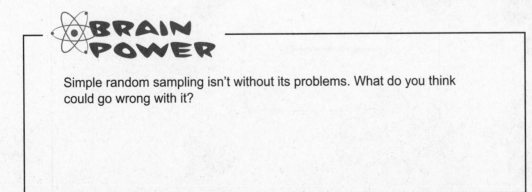

If you have a large sampling frame, drawing lots might not be practical, so an alternative is to use a random number generator, or random number tables. For this, you give each member of the sampling frame a number, generate a set of *n* random numbers, and then pick the members of the set whose assigned numbers correspond to the random numbers that were generated.

It's important to make sure that each number has an equal chance of occuring so that there's no bias.

BRAIN POWER

Simple random sampling isn't without its problems. What do you think could go wrong with it?

There are other types of sampling

Even simple random sampling has its problems.

With simple random sampling, there's still a chance that your sample will not represent the target population. For example, you might end up randomly drawing only yellow gumballs for your sample, and the other colors would be left out.

So how can we avoid this?

We can use stratified sampling...

An alternative to simple random sampling is ***stratified sampling***. With this type of sampling, the population is split into similar groups that share similar characteristics. These charateristics or groups are called ***strata***, and each individual group is called a ***stratum***. As an example, we could split up the gumballs into the different colors, yellow, green, red, and pink, so that each color forms a different stratum.

Once you've done this, you can perform simple random sampling on each stratum to ensure that each group is represented in your overall sample. To do this, look at the proportions of each stratum within the overall population and take a proportionate number from each. As an example, if 50% of the gumballs that Mighty Gumball produce are red, half of your sample should consist of red gumballs.

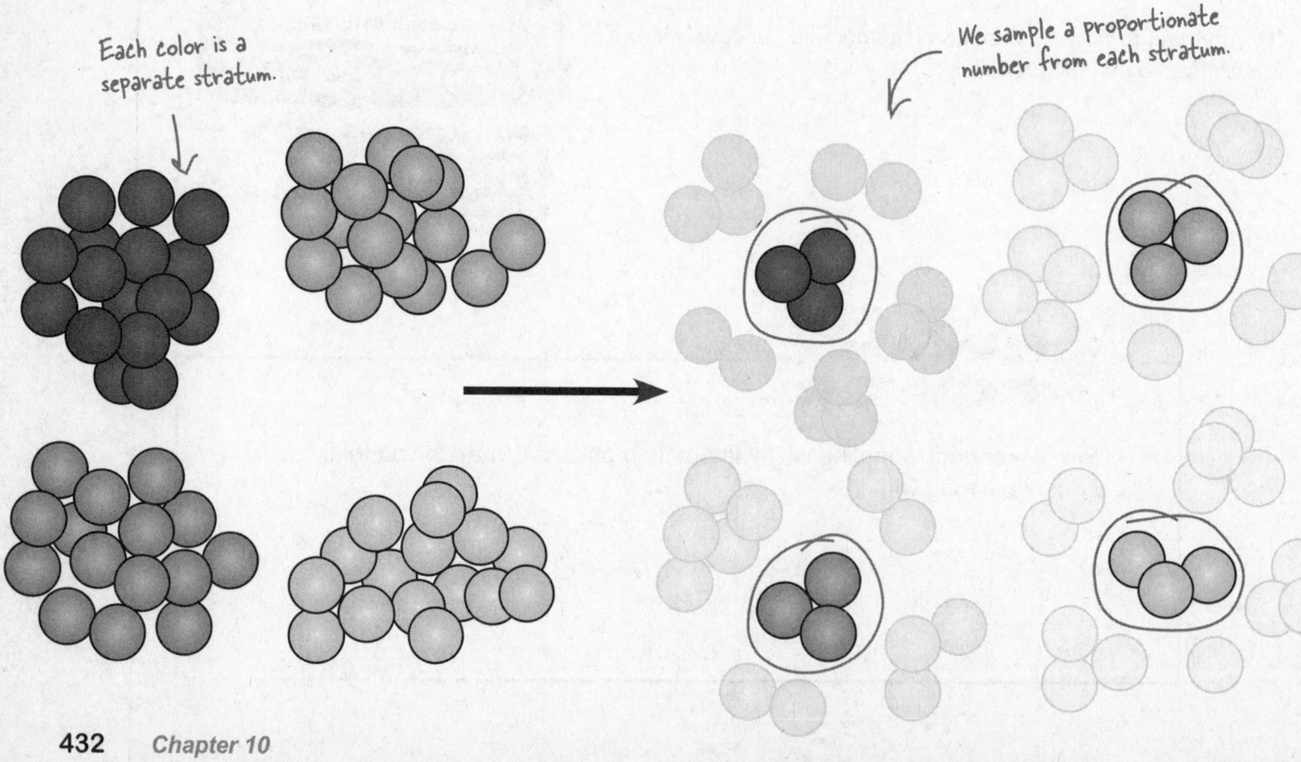
Each color is a separate stratum.
We sample a proportionate number from each stratum.

...or we can use cluster sampling...

Cluster sampling is useful if the population has a number of similar groups or clusters. As an example, gumballs might be sold in packets, with each packet containing a similar number of gumballs with similar colors. Each packet would form a ***cluster***.

With cluster sampling, instead of taking a simple random sample of units, you **draw a simple random sample of clusters**, and then survey everything within each of these clusters. As an example, you could take a simple random sample of *packets* of gumballs, and then taste all the gumballs in these packets.

Cluster sampling works because each cluster is similar to the others, and an added advantage is that you don't need a sampling frame of the whole population in order to achieve it. As an example, if you were surveying trees and used particular forests as your cluster, you would only need to know about each tree within only the forests you'd selected.

The problem with cluster sampling is that it might not be entirely random. As an example, it's likely that all of the gumballs in a packet will have been produced by the same factory. If there are differences between the factories, you may not pick these up.

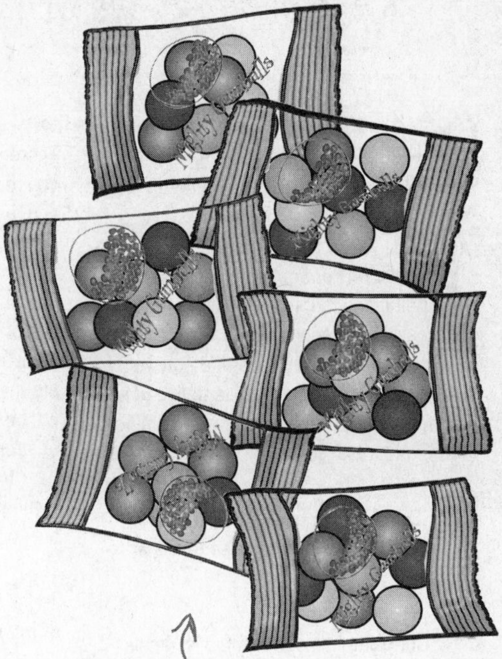

Each packet forms a cluster.

...or even systematic sampling

With systematic sampling, you list the population in some sort of order, and then survey every *k*th item, where *k* is some number. As an example, you could choose to sample every 10th gumball.

Systematic sampling is relatively quick and easy, but there's one key disadvantage. If there's some sort of cyclic pattern in the population, your sample will be biased. As an example, if gumballs are produced such that every 10th gumball is red, you will end up only sampling red gumballs, and this could lead to you drawing misleading conclusions about your population.

You can pick every 10th gumball to get a systematic sample.

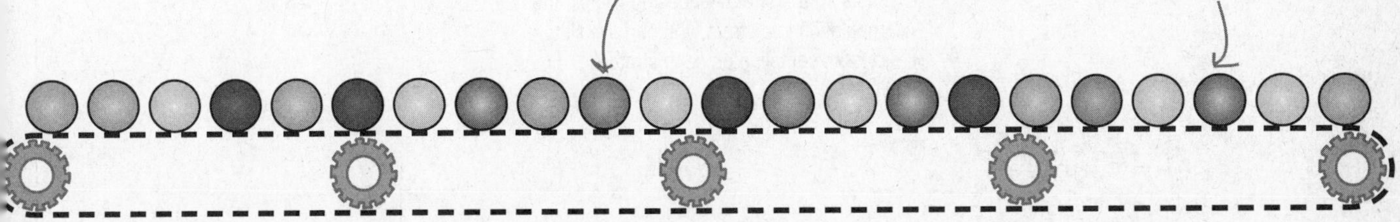

there are no Dumb Questions

Q: Does using one of these methods of sampling guarantee that the sample won't be biased?

A: They don't *guarantee* that the sample won't be biased, but they do minimize the chances of this happening. By really thinking about your target population and how you can make your sample representative of it, you stand a much better chance of coming up with an unbiased, representative sample.

Q: Do I have to use any of these methods? Can't I just choose items at random.

A: Choosing items at random is simple random sampling. Yes, this is one approach you can take, but one thing to be aware of is that there is a chance your sample will not be representative of the population at large.

Q: But why? Surely if I choose items at random, then my sample is bound to be representative of the target population.

A: Not necessarily. You see, if you choose sampling units at random, then there's a chance that purely at random, you could choose a sample that doesn't effectively represent the target population. As an example, if you choose customers of the Statsville Health Club completely at random, there's a chance that you might choose only attendees of one particular class, or of one particular gender.

There might also be a case where you think you're sampling at random, when really you're not. As an example, if you conduct a survey to find out customer satisfaction, but leave it up to customers whether or not they respond, you may well end up with a biased sample as customers have to be sufficiently motivated to respond. The customers who are most motivated to take part in the survey will be those who are either strongly satisfied or strongly dissatisfied. You are less likely to hear from those customers without strong feelings, yet those people may make up the bulk of the population.

Q: How about if I just increase the size of my sample? Will that get around bias?

A: The larger your sample, the less chance there is of your sample being biased, and this is one way of minimizing the chances of getting a biased sample using simple random sampling. The trouble is, the larger your sample, the more cumbersome and time-consuming it can be to gather data.

Q: What's the difference between stratified sampling and clustered sampling?

A: With stratified sampling, you divide the population into different groups or strata, where all the units within a stratum are as similar to each other as possible. In other words, you take some characteristic or property such as gender, and use this as the basis for the strata. Once you've split the population into strata, you perform simple random sampling on each stratum.

With clustered sampling, your aim is to divide the population into clusters, trying to make the clusters as alike as possible. You then use simple random sampling to choose clusters, and then sample everything in those clusters.

Q: I see. So with stratified sampling, you make each stratum as different as possible, and with clustered sampling, you make each cluster as similar as possible.

A: Exactly.

Q: So what about systematic sampling?

A: With systematic sampling, you choose a number, *k*, and then choose every *k*th item for your sample. This way of sampling is fairly quick and easy, but it doesn't mean that your sample will be representative of the population. In fact, this sort of sampling can only be used effectively if there are no repetitive patterns or organization in the sampling frame

Q: Drawing lots sounds antiquated. Do people still do that?

A: It's not as common as it used to be, but it's still a way of sampling.

You've been given 10 boxes of chocolates and been asked to sample the chocolates in them. There are whilte, milk, and dark chocolates in the boxes. Your target population is all of the chocolates, and the sampling unit is one chocolate.

1. How could you apply simple random sampling to this problem?

2. How could you apply stratified sampling?

3. What about cluster sampling?

Exercise Solution

You've been given 10 boxes of chocolates and been asked to sample the chocolates in them. There are whilte, milk, and dark chocolates in the boxes. Your target population is all of the chocolates, and the sampling unit is one chocolate.

1. How could you apply simple random sampling to this problem?

You could apply simple random sampling by choosing chocolates at random, either through drawing lots or using random numbers. That way, each chocolate stands an equal chance of being sampled.

2. How could you apply stratified sampling?

For stratified smpling, you divide the chocolates into strata and apply simple random sampling to each one. Each strata comprises of a group of chocolates with similar characteristics, so you could use the different types of chocolate. One stratum could be white chocolates, another one could be milk chocolates, and the final one could be dark chocolates.

3. What about cluster sampling?

For cluster sampling, you divide the chocolates into groups, but this time each group needs to be similar. Assuming each box of chocolates is similar, you could take one of the boxes, and sample all of the chocolates in it.

Exercise

How would you go about conducting a sample survey of Mighty Gumball's super-long-lasting gumballs? The gumballs come in four different colors, and they're all made in the same factory. Assume you have to start your sample from scratch.

Exercise Solution

How would you go about conducting a sample survey of Mighty Gumball's super-long-lasting gumballs? The gumballs come in four different colors, and they're all made in the same factory. Assume you have to start your sample from scratch.

The target population is all of Mighty Gumball's super-long-lasting gumballs, and the sampling unit is an individual gumball. For the sampling frame, we ideally need some sort of numbered list of the gumballs, but it's likely that this isn't practical. Instead, we'll settle for a list showing how many gumballs there are in the population for each color.

The type of sampling you use is subjective, but we'd choose to use stratified sampling, as this may be the best way of coming up with an unbiased sample. We'd divide the gumballs into their different colors and then use simple random sampling to choose a proportionate number of each of the four colors. We would then use these for our sample.

Don't worry if you got a different answer. The key thing is to think through how you can best make your survey representative of the population, and you may have different ideas.

BULLET POINTS

- A **population** is the entire collection of things you are studying.
- A **sample** is a relatively small selection taken from the population that you can use to draw conclusions about the population itself.
- To take a sample, start off by defining your target population, the population you want to study. Then decide on your sampling units, the sorts of things you need to sample. Once you've done that, draw up a sampling frame, a list of all the sampling units in your target population.
- A sample is **biased** if it isn't representative of your target population.
- **Simple random sampling** is where you choose sampling units at random to form your sample. This can be with or without replacement. You can perform simple random sampling by drawing lots or using random number generators.
- **Stratified sampling** is where you divide the population into groups of similar units or strata. Each stratum is as different from the others as possible. Once you've done this, you perform simple random sampling within each stratum.
- **Cluster sampling** is where you divide the population into clusters where each cluster is as similar to the others as possible. You use simple random sampling to choose a selection of clusters. You then sample every unit in these clusters.
- **Systematic sampling** is where you choose a number, *k*, and sample every *k*th unit.

Mighty Gumball has a sample

With your help, Mighty Gumball has gathered a sample of their super-long-lasting gumballs. This means that rather than perform taste tests on the entire gumball population, they can use their sample instead.

That's great! It means we'll save time, money, and teeth.

So what's next?

We've looked at how we can put together a representative sample, but what we *haven't* looked at is how we can use it. We know that an unbiased sample shares the same characteristics as its parent population, but what's the best way of analyzing this?

Keep reading, and we'll show you how in the next chapter.

11 estimating populations and samples

Making Predictions

...I mean, men! They're all the same. Once you've met one, you've met them all!

Wouldn't it be great if you could tell what a population was like, just by taking one sample?

Before you can claim **full sample mastery**, you need to know how to use your samples to best effect once you've collected them. This means using them to **accurately predict** what the population will be like and coming up with a way of saying how **reliable** your predictions are. In this chapter, we'll show you how knowing your sample helps you **get to know your population**, and vice versa.

So how long does flavor really last for?

With your help, Mighty Gumball has pulled together an unbiased sample of super-long-lasting gumballs. They've tested each of the gumballs in the sample and collected lots of data about how long gumball flavor within the sample lasts.

There's just one problem...

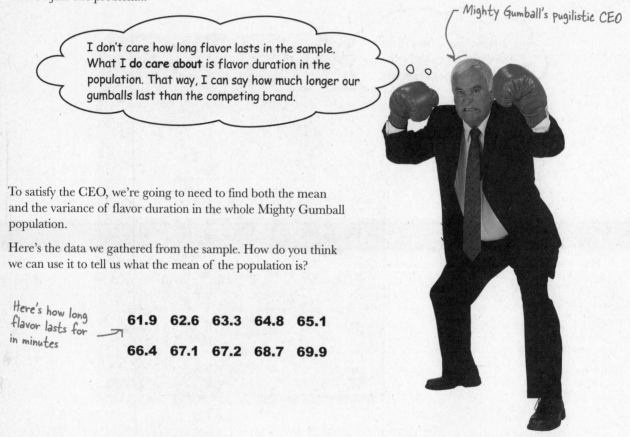

Mighty Gumball's pugilistic CEO
I don't care how long flavor lasts in the sample. What I **do care about** is flavor duration in the population. That way, I can say how much longer our gumballs last than the competing brand.

To satisfy the CEO, we're going to need to find both the mean and the variance of flavor duration in the whole Mighty Gumball population.

Here's the data we gathered from the sample. How do you think we can use it to tell us what the mean of the population is?

Here's how long flavor lasts for in minutes

61.9	**62.6**	**63.3**	**64.8**	**65.1**
66.4	**67.1**	**67.2**	**68.7**	**69.9**

BRAIN POWER

Take a look at the data. How would you use this data to estimate the mean and variance of the population? How reliable do you think your estimate will be? Why?

Let's start by estimating the population mean

So how can we use the results of the sample taste test to tell us the mean amount of time gumball flavor lasts for in the general gumball population?

The answer is actually pretty intuitive. We assume that the mean flavor duration of the gumballs in the sample matches that of the population. In other words, we find the mean of the sample and use it as the mean for the population too.

Here's a sketch showing the distribution of the sample, and what you'd expect the distribution of the population to look like based on the sample. You'd expect the distribution of the population to be a similar shape to that of the sample, so you can assume that the mean of the sample and population have about the same value.

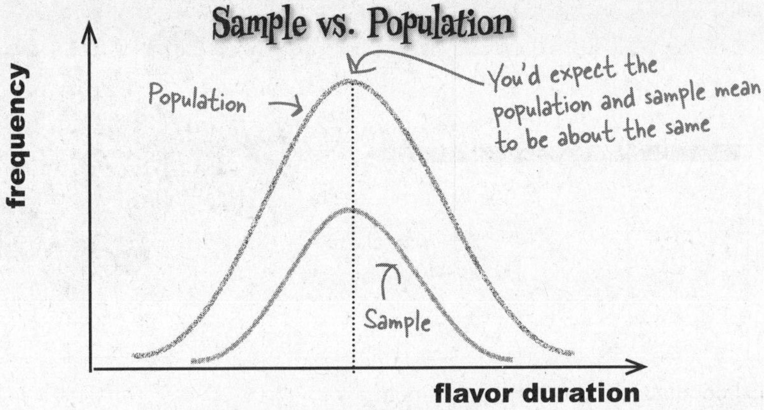

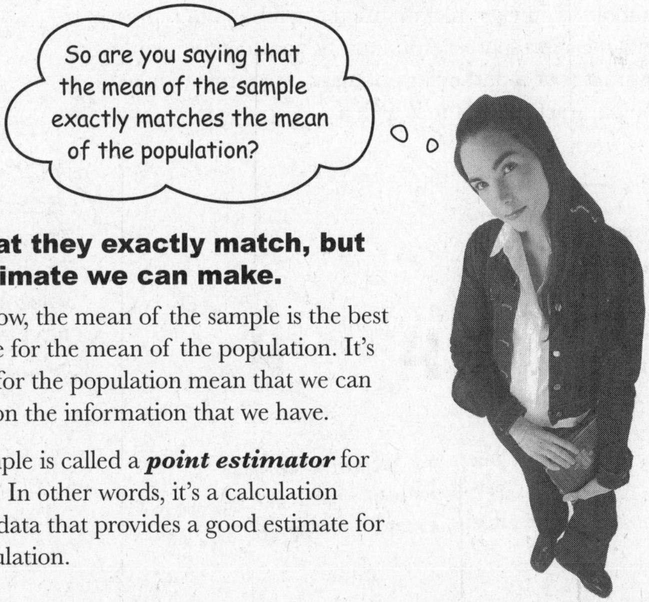

We can't say that they exactly match, but it's the best estimate we can make.

Based on what we know, the mean of the sample is the best estimate we can make for the mean of the population. It's the most likely value for the population mean that we can come up with based on the information that we have.

The mean of the sample is called a ***point estimator*** for the population mean. In other words, it's a calculation based on the sample data that provides a good estimate for the mean of the population.

Point estimators can approximate population parameters

Up until now, we've been dealing with actual values of population parameters such as the mean, μ, or the variance, σ^2. We've either been able to calculate these for ourselves, or we've been told what they are.

This time around we don't know the exact value of the population parameters. Instead of **calculating** them using the population, we **estimate** them using the sample data instead. To do this, we use point estimators to come up with a best guess of the population parameters.

A ***point estimator*** of a population parameter is some function or calculation that can be used to estimate the value of the population parameter. As an example, the point estimator of the population mean is the mean of the sample, as we can use the sample mean to estimate the population mean.

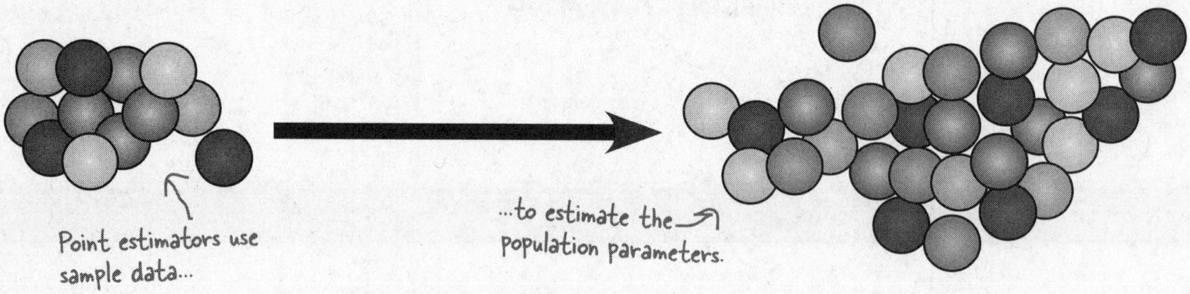

Point estimators use sample data...
...to estimate the population parameters.

We differentiate between an actual population parameter and its point estimator using the ^ symbol. As an example, we use the symbol μ to represent the population mean, and $\hat{\mu}$ to represent its estimator. So to show that you're dealing with the point estimator of a particular population parameter, take the symbol of the population parameter, and top it with a ^.

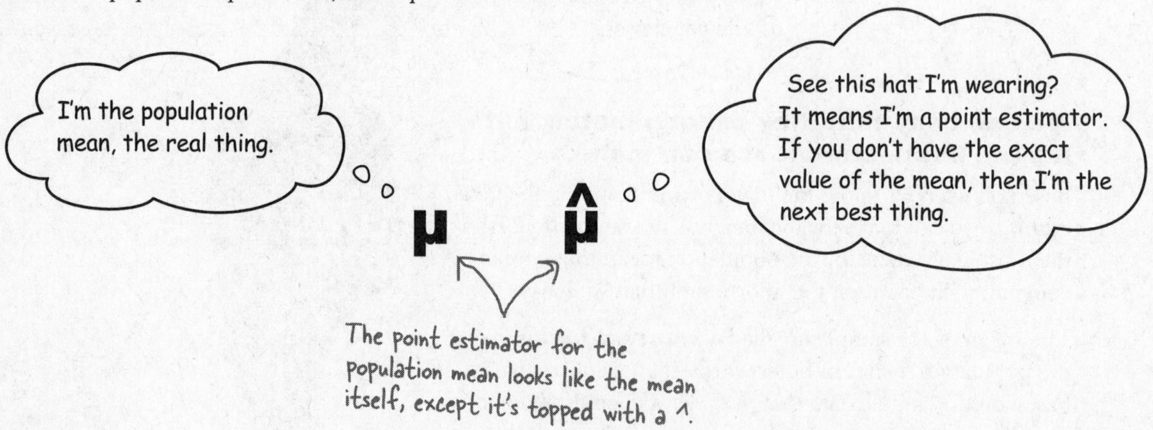

I'm the population mean, the real thing.
μ
μ̂
See this hat I'm wearing? It means I'm a point estimator. If you don't have the exact value of the mean, then I'm the next best thing.
The point estimator for the population mean looks like the mean itself, except it's topped with a ^.

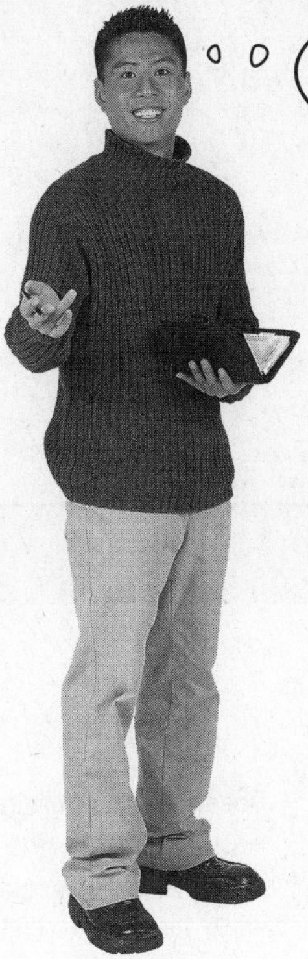

It occurs to me that we have a symbol for the population mean and one for its point estimator. Is there a symbol for the sample mean too?

There's a shorthand way of writing the sample mean.

The symbol μ has a very precise meaning. It's the mean of the population. We have a different way to represent the mean of the sample so that we don't get confused about which mean we're talking about. To represent the sample mean, we use the symbol $\bar{x}$ (pronounced "x bar"). That way, we know that if someone refers to μ, they're referring to the population mean, and if they refer to $\bar{x}$, they're referring to the sample mean.

$\bar{x}$ is the sample equivalent of μ, and you calculate it in the same way you would the population mean. You add together all the data in your sample, and then divide by however many items there are. In other words, if your sample size is *n*,

$\bar{x}$ is the mean of the sample.

$$\bar{x} = \frac{\Sigma x}{n}$$

Add together the numbers in the sample, and divide by how many there are.

We can use this to write a shorthand expression for the point estimator for the population. Since we can estimate the population mean using the mean of the sample, this means that

We estimate the mean of the population...

$$\hat{\mu} = \bar{x}$$

...using the mean of the sample.

Sharpen your pencil

Use the sample data to estimate the value of the population mean. Here's a reminder of the data:

61.9 62.6 63.3 64.8 65.1 66.4 67.1 67.2 68.7 69.9

Sharpen your pencil Solution

Use the sample data to estimate the value of the population mean. Here's a reminder of the data:

61.9 62.6 63.3 64.8 65.1 66.4 67.1 67.2 68.7 69.9

We can estimate the population mean by calculating the mean of the sample.

$$\hat{\mu} = \bar{x} = \frac{61.9 + 62.6 + 63.3 + 64.8 + 65.1 + 66.4 + 67.1 + 67.2 + 68.7 + 69.9}{10}$$

$$= 657/10$$

$$= 65.7$$

there are no Dumb Questions

Q: Surely the mean is just the mean. Why are there so many different symbols for it?

A: There are three different concepts at work. There's the mean of the population, the mean of the sample, and the point estimator for the population mean.

The **population mean** is represented by μ. This is the sort of mean that we've encountered throughout the book so far, and you find it by adding together all the data in the population and dividing by the size of the population.

The **sample mean** is represented by $\bar{x}$. You find it in the same way that you find μ, except that this time your data comes from a sample. To calculate $\bar{x}$, you add together the data in your sample, and divide by the size of it.

The **point estimator** for μ is represented by $\hat{\mu}$. It's effectively a best guess for what you think the population mean is, based on the sample data.

Q: So does that mean that we can find μ by just taking the mean of a sample?

A: We can't find the exact value of μ using a sample, but if the sample is unbiased, it gives us a very good estimate. In other words, we can use the sample data to find $\hat{\mu}$, not the true value of μ itself.

Q: But what about if the sample is biased? How do we come up with an estimate for μ then?

A: This is where it's important to make your sample as unbiased as possible. If all the data you have comes from your sample, then that's what you need to use as the basis for your estimate. If your sample is biased, then this means that your estimate for μ is likely to be inaccurate, and it may lead you into making wrong decisions.

Q: Does the size of the sample matter?

A: In general, the larger the size of your sample, the more accurate your point estimator is likely to be.

μ is the mean of the population, $\bar{x}$ is the mean of the sample, and $\hat{\mu}$ is the point estimator for μ.

BULLET POINTS

- A **point estimator** is an estimate for the value of a population parameter, derived from sample data.
- The ^ symbol is added to the population parameter when you're talking about its point estimator. As an example, the point estimator for μ is $\hat{\mu}$.
- The mean of a sample is represented as $\bar{x}$. To find the mean of the sample, use the formula

$$\bar{x} = \frac{\Sigma x}{n}$$

where *x* represents the values in the sample, and *n* is the sample size.

- The point estimator for the population mean is found by calculating $\bar{x}$. In other words,

$$\hat{\mu} = \bar{x}$$

This means that if you want a good estimate for the true value of the population mean, you can use the mean of the sample.

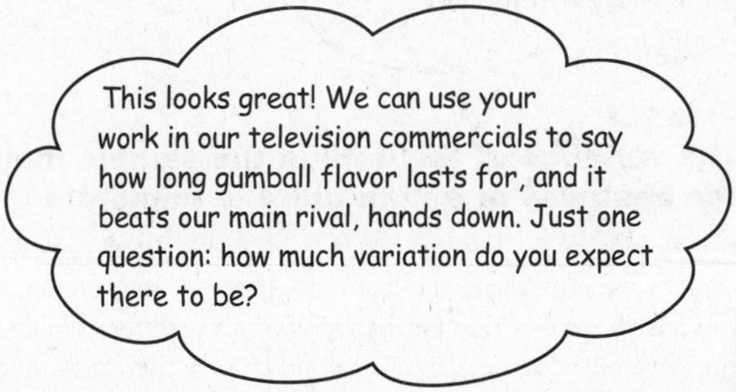

This looks great! We can use your work in our television commercials to say how long gumball flavor lasts for, and it beats our main rival, hands down. Just one question: how much variation do you expect there to be?

You've come up with a good estimate for the population mean, but what about the variance?

If we can come up with a good estimate for the population variance, then the CEO will be able to tell how much variation in flavor duration there's likely to be in the gumball population, based on the results of the sample data.

Let's estimate the population variance

So far we've seen how we can use the sample mean to estimate the mean of the population. This means that we have a way of estimating what the mean flavor duration is for the super-long-lasting gumball population.

To satisfy the Mighty Gumball CEO, we also need to come up with a good estimate for the population variance.

So what can we use as a point estimator for the population variance? In other words, how can we use the sample data to find $\hat{\sigma}^2$?

That's easy. The variance of the sample is bound to be the same as that of the population. We can use the sample variance to estimate the population variance.

The variance of the data in the sample may not be the best way of estimating the population variance.

You already know that the variance of a set of data measures the way in which values are dispersed from the mean. When you choose a sample, you have a smaller number of values than with the population, and since you have fewer values, there's a good chance they're more clustered around the mean than they would be in the population. More extreme values are less likely to be in your sample, as there are generally fewer of them.

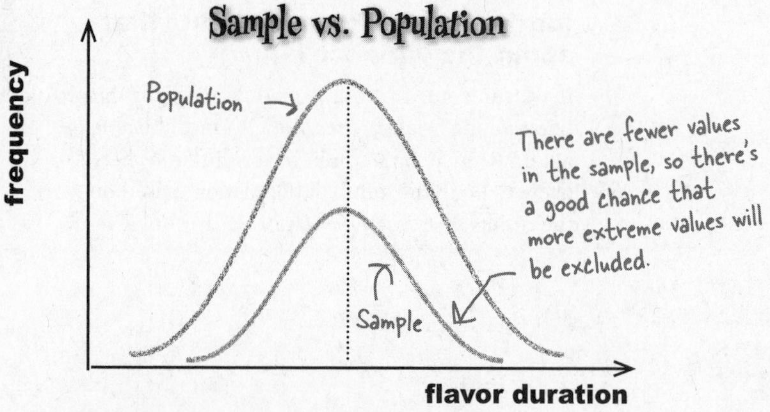
Sample vs. Population
frequency
Population
Sample
There are fewer values in the sample, so there's a good chance that more extreme values will be excluded.
flavor duration

So what would be a better estimate of the population variance?

We need a different point estimator than sample variance

The problem with using the sample variance to estimate that of the population is that it tends to be slightly too low. The sample variance tends to be slightly less than the variance of the population, and the degree to which this holds depends on the number of values in the sample. If the number in the sample is small, there's likely to be a bigger difference between the sample and population variances than if the size of the sample is large.

What we need is a better way of estimating the variance of the population, some function of the sample data that gives a slightly higher result than the variance of all the values in the sample.

So what <u>is</u> the estimator?

Rather than take the variance of all the data in the sample to estimate the population variance, there's something else we can use instead. If the size of the sample is *n*, we can estimate the population variance using

$$\hat{\sigma}^2 = \frac{\Sigma(x - \bar{x})^2}{n - 1}$$

Estimator for the population variance → $\hat{\sigma}^2$

Take each item in the sample, subtract the sample mean, square the result, then add the lot together. → $\Sigma(x - \bar{x})^2$

Divide by the number in the sample minus 1. → $n - 1$

In other words, we take each item in the sample, subtract the sample mean, and then square the result. We then add all of the results together, and divide by the number of items in the sample minus 1. This is just like finding the variance of the values in the sample, but dividing by $n - 1$ instead of n.

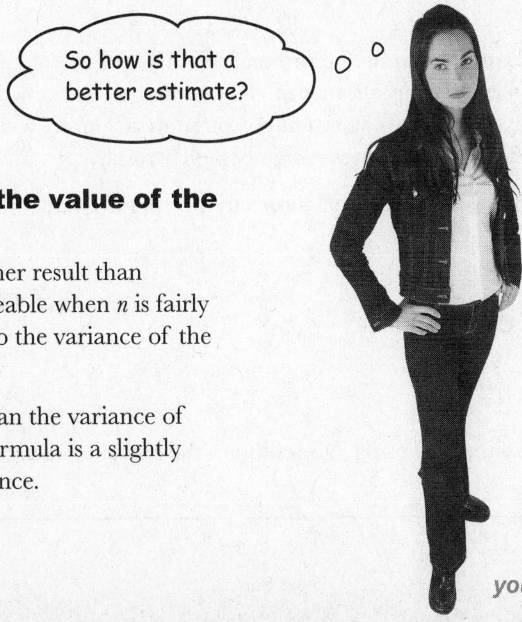
So how is that a better estimate?

This formula is a closer match to the value of the population variance.

Dividing a set of numbers by $n - 1$ gives a higher result than dividing by n, and this difference is most noticeable when n is fairly small. This means that the formula is similar to the variance of the sample data, but gives a slightly higher result.

The population variance tends to be higher than the variance of the data in the sample. This means that this formula is a slightly better point estimator for the population variance.

Variance Up Close

Knowing what formula you should use to find the variance can be confusing. There's one formula for population variance σ^2, and a slightly different one for its point estimator $\hat{\sigma}^2$. So which formula should you use when?

Population variance

If you want to find the exact variance of a population and you have data for the whole population, use

$$\sigma^2 = \frac{\Sigma(x - \mu)^2}{n}$$

Population variance → σ^2; Population mean → μ; Size of the population → n

In this situation, you have all the data for your population. You know what the mean is for your population, and you want to find the variance of all of these values. This is the calculation that you've seen throughout this book so far.

Estimating the population variance

If you need to *estimate* the variance of a population using sample data, use

$$\hat{\sigma}^2 = \frac{\Sigma(x - \bar{x})^2}{n - 1}$$

Point estimator for the population variance, based on your sample. → $\hat{\sigma}^2$

Sample mean → $\bar{x}$

n – 1, not n, where n is the size of the sample. This time it's an estimate..

Instead of calculating the variance of an actual population of *n* values, you have to *estimate* the variance of the population, based on the sample of data you have. To make you estimate a bit more accurate, you divide by *n* - 1 instead of *n*, as this gives a slightly higher result.

The formula for the population variance point estimator is usually written s^2, so

$$\hat{\sigma}^2 = s^2 \quad \text{where} \quad s^2 = \frac{\Sigma(x - \bar{x})^2}{n - 1}$$

Point estimator for the population variance → $\hat{\sigma}^2$

s^2 gives the formula based on the sample data

This is similar to using $\bar{x}$ to represent the sample mean.

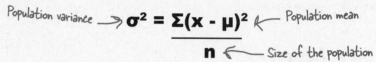

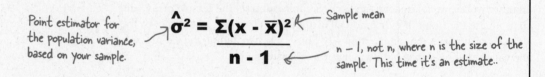

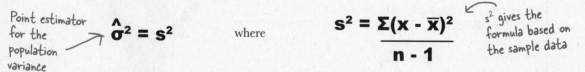

Which formula's which?

Sometimes it can be tricky deciding whether you should divide by n for the variance, or whether you should divide by $n - 1$. The golden rule to remember is that dividing by n gives you the **actual variance for the set of data that you have**.

If you have the data for the entire population, then dividing by n gives you the actual variance of the population. You need to use the formula for σ^2 and divide by n.

If you have a sample of data from the population, then chances are you'll want to use this to estimate the variance of the population. This means that you need to use the formula for s^2 and divide by $n - 1$.

Watch it!

Some books tell you to divide by *n* – 1 for a sample, and some tell you to divide by *n*.

This is because different books make different assumptions about what you're using your sample for. If you're using the sample to estimate the population variance, then you need to divide by n – 1. You only need to divide by n if you want to calculate the variance of that exact set of values.

If you're taking a statistics exam, check the approach that your exam board takes.

Sharpen your pencil

Here's a reminder of the data from the Mighty Gumball sample. What do you estimate the population variance to be?

61.9 62.6 63.3 64.8 65.1 66.4 67.1 67.2 68.7 69.9

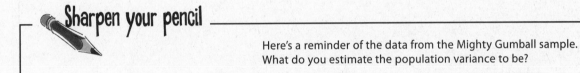

Sharpen your pencil Solution

Here's a reminder of the data from the Mighty Gumball sample. What do you estimate the population variance to be?

61.9 62.6 63.3 64.8 65.1 66.4 67.1 67.2 68.7 69.9

We can estimate the population variance by calculating s^2.

$$\hat{\sigma}^2 = s^2$$

$$= \frac{\sum(x - \bar{x})^2}{n - 1}$$

$$= \frac{(-3.8)^2 + (-3.1)^2 + (-2.4)^2 + (-0.9)^2 + (-0.6)^2 + (0.7)^2 + (1.4)^2 + (1.5)^2 + (3)^2 + (4.2)^2}{9}$$

$$= \frac{14.44 + 9.61 + 5.76 + 0.81 + 0.36 + 0.49 + 1.96 + 2.25 + 9 + 17.64}{9}$$

$$= 62.32/9$$

$$= 6.92 \text{ (to 2 decimal places)}$$

there are no Dumb Questions

Q: Why do I divide by *n* – 1 for the sample variance? Why can't I divide by *n*?

A: You divide by *n* – 1 for a sample because most of the time, you use your sample data to estimate the variance of the population. Dividing by *n* – 1 gives you a slightly more accurate result than dividing by *n*. This is because the variance of values in the sample is likely to be slightly lower than the population variance.

Q: Is there some mathematical basis for this?

A: Yes there is. It's something that we're going to touch upon at the end of the chapter, but hold onto that thought; it's a good one.

Q: How do I remember which symbols are used for the population, and which are used for the sample?

A: In general, Greek letters are used for the population, and normal Roman letters are used for the mean and variance for the sample.

Q: Is there a point estimator for the standard deviation in the same way that there is for the variance? How do I find it?

A: If you need to estimate the standard deviation, start by calculating the estimator for the variance. The estimator for the standard deviation is the square root of this.

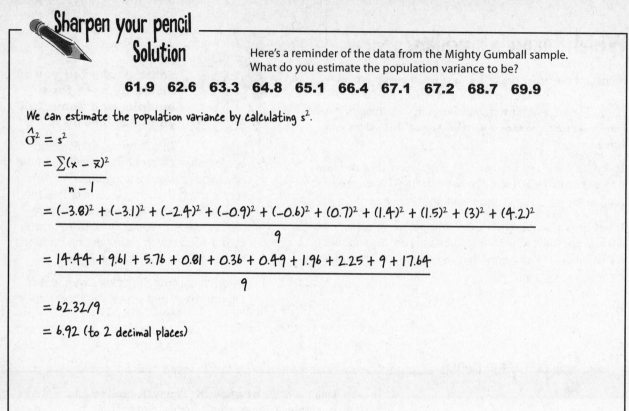

Mighty Gumball has done more sampling

The Mighty Gumball CEO is so inspired by the results of the taste test that he's asked for another sampling exercise that he can use for his television advertisements. This time, the CEO wants to be able to say how popular Mighty Gumball's candy is compared with that of their main rival.

The Mighty Gumball staff have asked a random sample of people whether they prefer gumballs produced by Mighty Gumball or whether they prefer those of their main rival. They're hoping they can use the results to predict what proportion of the population is likely to prefer Mighty Gumball.

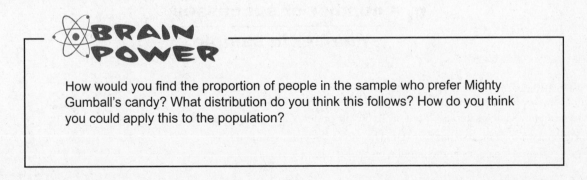

I'll choose Mighty Gumball over any other manufacturer.

Ewww! These gumballs are gross.

Mighty Gumball has found that in a sample of 40 people, 32 of them prefer their gumballs. The other 8 prefer those of their rival.

BRAIN POWER

How would you find the proportion of people in the sample who prefer Mighty Gumball's candy? What distribution do you think this follows? How do you think you could apply this to the population?

It's a question of proportion

For the latest Mighty Gumball sample, the thing the CEO is interested in is whether or not each person prefers Mighty Gumball confectionery to that of their chief rival. In other words, every person who prefers Mighty Gumball candy can be classified as a success.

So how do we use the sample data to predict the proportion of successes in the population?

Predicting population proportion

If we use X to represent the number of successes in the population, then X follows a binomial distribution with parameters n and p. n is the number of people in the population, and p is the proportion of successes.

In the same way that our best estimate of the population mean is the mean of the sample, our best guess for the proportion of successes in the population has to be the proportion of successes in the sample. This means that if we can find the proportion of people in the sample who prefer Mighty Gumball's treats, we'll have a good estimate for the proportion of people who prefer Mighty Gumball in the general population.

We can find the proportion of successes in the sample by taking the total number of people who prefer Mighty Gumball, and then dividing by the total number of people in the sample. If we use p_s to represent the proportion of successes in the sample, then we can estimate the proportion of successes in the population using

Point estimator for the proportion of successes in the population →

$$\hat{p} = p_s$$

← Proportion of successes in the sample

where

$$p_s = \frac{\text{number of successes}}{\text{number in sample}}$$

In other words, we can use the proportion of successes in the sample as a point estimator for the proportion of successes in the population. In the case of the company's latest sample, 32 out of 40 people prefer Mighty Gumball confectionery, which means that $p_s = 0.8$. Therefore, the point estimator for the proportion of successes in the population is also 0.8.

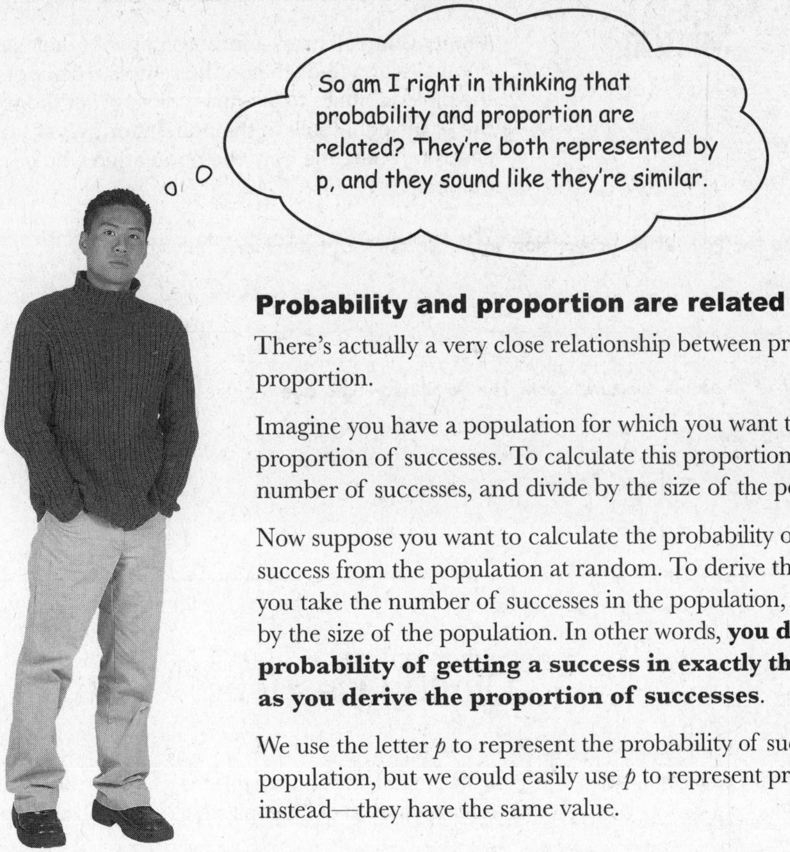
So am I right in thinking that probability and proportion are related? They're both represented by p, and they sound like they're similar.

Probability and proportion are related

There's actually a very close relationship between probability and proportion.

Imagine you have a population for which you want to find the proportion of successes. To calculate this proportion, you take the number of successes, and divide by the size of the population.

Now suppose you want to calculate the probability of choosing a success from the population at random. To derive this probability, you take the number of successes in the population, and divide by the size of the population. In other words, **you derive the probability of getting a success in exactly the same way as you derive the proportion of successes**.

We use the letter *p* to represent the probability of success in the population, but we could easily use *p* to represent proportion instead—they have the same value.

p = probability = proportion

Sharpen your pencil

Mighty Gumball takes another sample of their super-long-lasting gumballs, and finds that in the sample, 10 out of 40 people prefer the pink gumballs to all other colors. What proportion of people prefer pink gumballs in the population? What's the probability of choosing someone from the population who *doesn't* prefer pink gumballs?

Sharpen your pencil Solution

Mighty Gumball takes another sample of their super-long-lasting gumballs, and finds that in the sample, 10 out of 40 people prefer the pink gumballs to all other colors. What proportion of people prefer pink gumballs in the population? What's the probability of choosing someone from the population who *doesn't* prefer pink gumballs?

We can estimate the population proportion with the sample proportion. This gives us

$$\hat{p} = p_s = 10/40$$
$$= 0.25$$

The probability of choosing someone from the population who doesn't prefer pink gumballs is

$$P(\text{Preference not Pink}) = 1 - \hat{p}$$
$$= 1 - 0.25$$
$$= 0.75$$

there are no Dumb Questions

Q: So is proportion the same thing as probability?

A: The proportion is the number of successes in your population, divided by the size of your population. This is the same calculation you would use to calculate probability for a binomial distribution.

Q: Does proportion just apply to the binomial distribution? What about other probability distributions?

A: Out of all the probability distributions we've covered, the only one which has any bearing on proportion is the binomial distribution. It's specific to the sorts of problems you have with this distribution.

Q: Is the proportion of the sample the same as the proportion of the population?

A: The proportion of the sample can be used as a point estimator for the proportion of the population. It's effectively a best guess as to what the value of the population proportion is.

Q: Is that still the case if the sample is biased in some way? How do I estimate proportion from a biased sample?

A: The key here is to make sure that your sample is unbiased, as this is what you base your estimate on. If your sample is biased, this means that you will come with an inaccurate estimate for the population proportion. This is the case with other point estimators too.

Q: So how do I make sure my sample is unbiased?

A: Going through the points we raised in the previous chapter is a good way of making sure your sample is as representative as possible. The hard work you put in to preparing your sample is worth it because it means that your point estimators are a more accurate reflection of the population itself.

This is awesome! We have a lot of impressive statistics we can use in our advertising.
10/40
people prefer pink!

BULLET POINTS

- The point estimator for the population variance is given by

$$\hat{\sigma}^2 = s^2$$

where s^2 is given by

$$\frac{\Sigma(x - \bar{x})^2}{n - 1}$$

- The population proportion is represented using p. It's the proportion of successes within the population.

- The point estimator for p is given by p_s, where p_s is the proportion of successes in the sample.

$$\hat{p} = p_s$$

- You calculate p_s by dividing the number of successes in the sample by the size of the sample.

$$p_s = \frac{\text{number of successes}}{\text{number in sample}}$$

Buy your gumballs here!

Remember the Statsville Cinema? They're recently been authorized to sell Mighty Gumball products to film-goers, and it's a move that's proving popular with most of their customers.

The trouble is, not everybody's happy.

I really like red gumballs, and I'd rather not eat the other colors. How many red gumballs come in the box?

Introducing new jumbo boxes

The cinema sells mixed boxes of gumballs, and this weekend, they're putting on a film marathon of classic films.

The event looks like it's going to be popular, and tickets are selling well. The trouble is, some people get cranky if they don't get their fix of red gumballs.

A jumbo box of gumballs is meant for sharing, and each box contains 100 gumballs. 25% of gumballs in the entire gumball population are red.

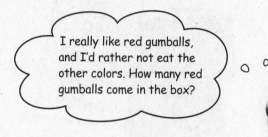

I need 40 red gumballs to make it through the movie. Is that likely? If there aren't enough red gumballs in the box, I'll get another snack instead.

We need to find the probability that in one particular jumbo box, 40 or more of the gumballs will be red.

Since there are 100 gumballs per box, that means we need to find the probability that 40% of the gumballs in this box are red, given that 25% of the gumball population is red.

So how does this relate to sampling?

So far, we've looked at how to put together an unbiased sample, and how to use samples to find point estimators for population parameters.

This time around, the situation's different. Here, we're told what the population parameters are, and we have to work out probabilities for one particular jumbo box of gumballs. In other words, instead of working out probabilities for the population, we need to work out probabilities for the sample proportion.

Isn't that the sort of thing that we were doing before? What's the big deal?

This time, we're looking for probabilities for a sample, not a population.

Rather than work out the probability of getting particular frequencies or values in a probability distribution, this time around we need to find probabilities **for the sample proportion itself**. We need to figure out the probability of getting this particular result in this particular box of gumballs.

Before we can work out probabilities for this, we need to figure out the probability distribution for the sample proportion. Here's what we need to do:

1. **Look at all possible samples the same size as the one we're considering.**
   If we have a sample of size *n*, we need to consider all possible samples of size *n*. There are 100 gumballs in the box, so in this case *n* is 100.

2. **Look at the distribution formed by all the samples, and find the expectation and variance for the proportion.**
   Every sample is different, so the proportion of red gumballs in each box of gumballs will probably vary.

3. **Once we know how the proportion is distributed, use it to find probabilities.**
   Knowing how the proportion of successes in a sample is distributed means we can use it to find probabilities for the proportion in a random sample—in this case, a jumbo box of gumballs.

Let's take a look at how to do this.

The sampling distribution of proportions

So how do we find the distribution of the sample proportions?

Let's start with the gumball population. We've been told what the proportion of red gumballs is in the population, and we can represent this as *p*. In other words, p = 0.25.

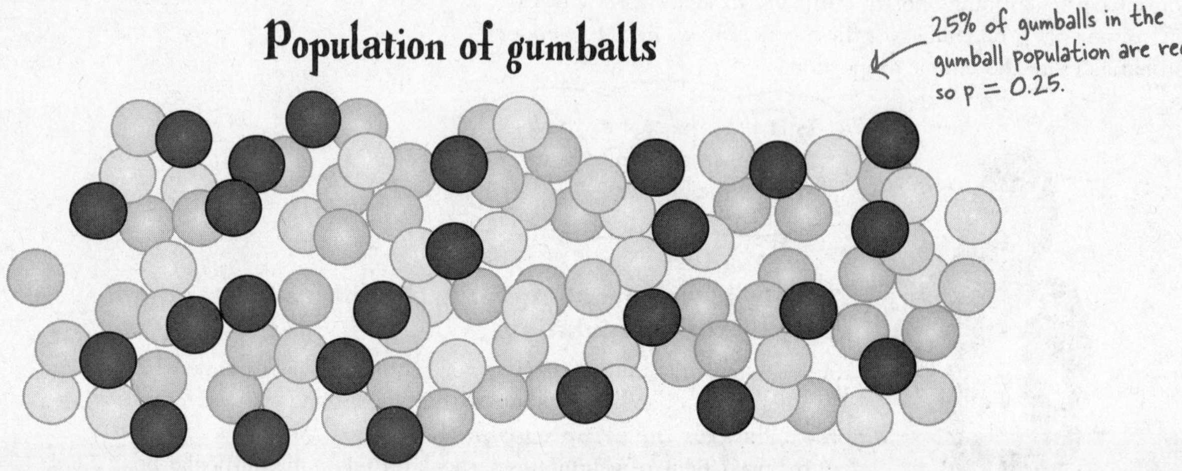

Each jumbo box of gumballs is effectively a sample of gumballs taken from the population. Each box contains 100 gumballs, so the sample size is 100. Let's represent this with *n*.

If we use the random variable X to represent the number of red gumballs in the sample, then X ~ B(n, p), where n = 100 and p = 0.25.

The proportion of red gumballs in the sample depends on X, the number of red gumballs in the sample. This means that the proportion itself is a random variable. We can write this as P_s, where $P_s = X/n$

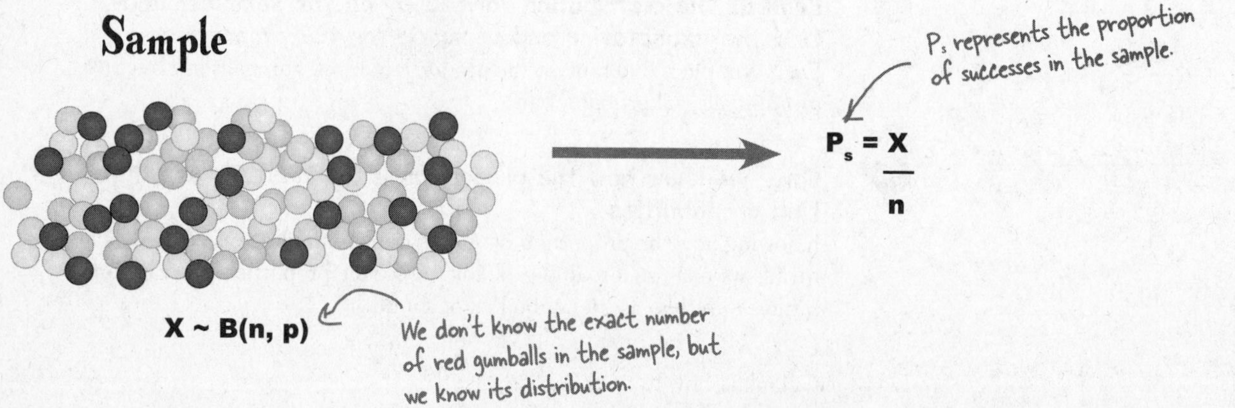

Sample
X ~ B(n, p)
We don't know the exact number of red gumballs in the sample, but we know its distribution.
P_s = X/n
P_s represents the proportion of successes in the sample.

There are *many* possible samples we could have taken of size *n*. Each possible sample would comprise *n* gumballs, and the number of red gumballs in each would follow the same distribution. For each sample, the number of red gumballs is distributed as B(n, p), and the proportion of successes is given by X/n.

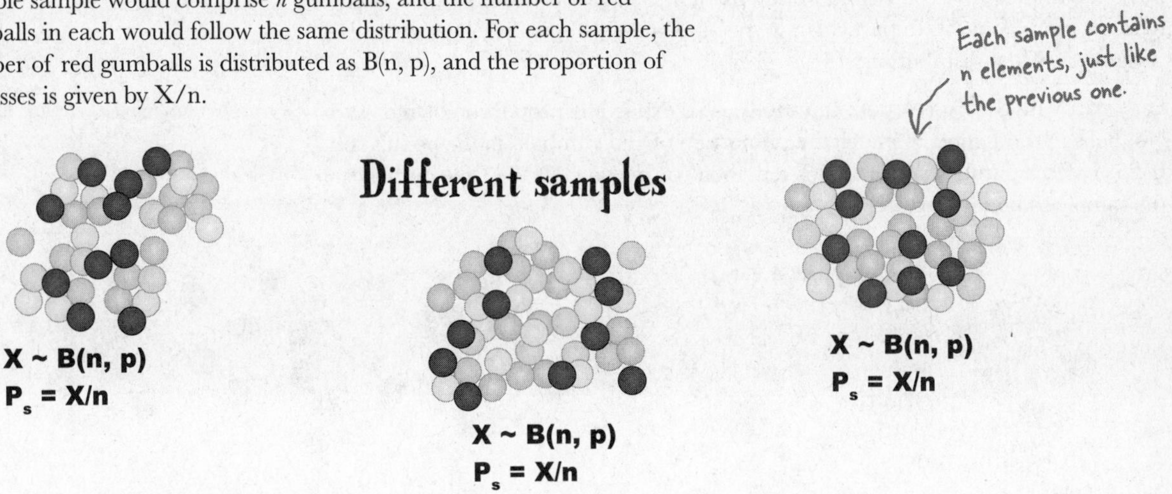
Each sample contains n elements, just like the previous one.
Different samples
X ~ B(n, p)
P_s = X/n
X ~ B(n, p)
P_s = X/n
X ~ B(n, p)
P_s = X/n

We can form a distribution out of all the sample proportions using all of the possible samples. This is called the ***sampling distribution of proportions***, or the distribution of P_s.

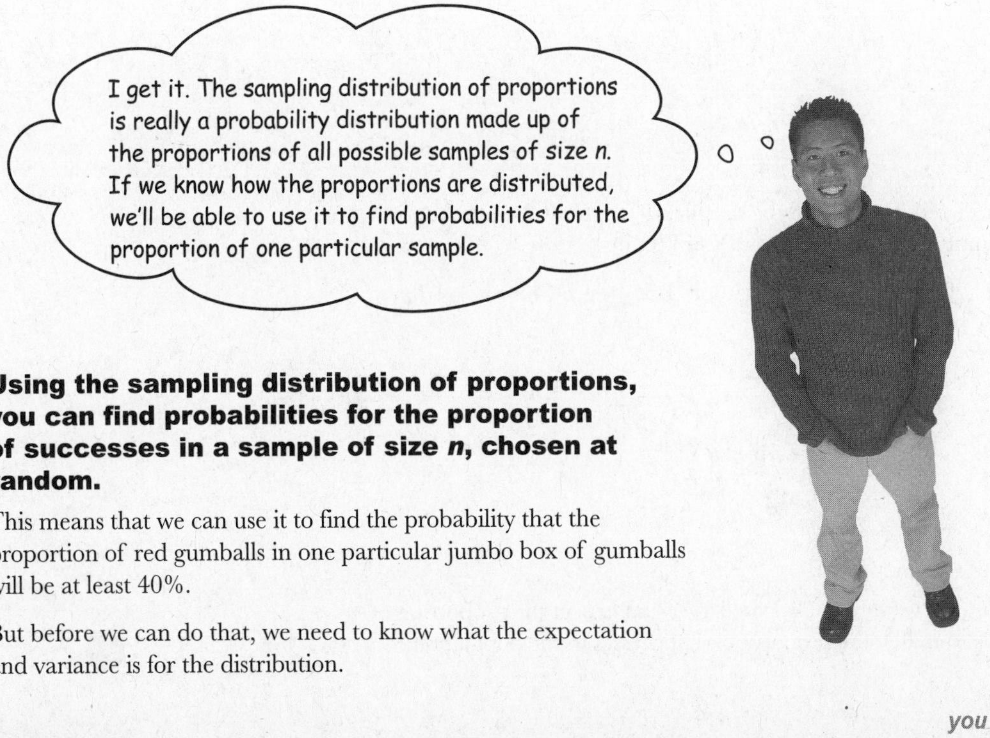
I get it. The sampling distribution of proportions is really a probability distribution made up of the proportions of all possible samples of size n. If we know how the proportions are distributed, we'll be able to use it to find probabilities for the proportion of one particular sample.

Using the sampling distribution of proportions, you can find probabilities for the proportion of successes in a sample of size *n*, chosen at random.

This means that we can use it to find the probability that the proportion of red gumballs in one particular jumbo box of gumballs will be at least 40%.

But before we can do that, we need to know what the expectation and variance is for the distribution.

So what's the expectation of P_s?

So far we've seen how we can form a distribution from the proportions of all possible samples of size *n*. Before we can use it to calculate probabilities, we need to know more about it. In particular, we need to know what the expectation and variance is of the distribution.

Let's start with the expectation. Intuitively, we'd expect the proportion of red gumballs in the sample to match the proportion of red gumballs in the population. If 25% of the gumball population is red, then you'd expect 25% of the gumballs in the sample to be red also.

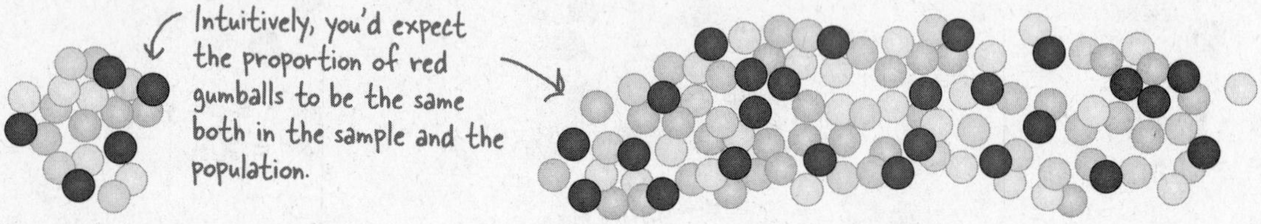
Intuitively, you'd expect the proportion of red gumballs to be the same both in the sample and the population.

So what's the expectation of P_s?

We want to find $E(P_s)$, where $P_s = X/n$. In other words, we want to find the expected value of the sample proportion, where the sample proportion is equal to the number of red gumballs divided by the total number of gumballs in the sample. This gives us

$$E(P_s) = E\left(\frac{X}{n}\right)$$

$$= \frac{E(X)}{n}$$

Now X is the number of red gumballs in the sample. If we count the number of red gumballs as the number of successes, then $X \sim B(n, p)$.

You've already seen that for a binomial distribution, $E(X) = np$. This means that

$$E(P_s) = \frac{E(X)}{n}$$

$$= \frac{\cancel{n}p}{\cancel{n}} \quad \leftarrow E(X) = np$$

$$= p$$

This result ties in with what we intuitively expect. We can expect the proportion of successes in the sample to match the proportion of successes in the population.

And what's the variance of P_s?

Before we can find out any probabilities for the sample proportion, we also need to know what the variance is for P_s. We can find the variance in a similar way to how we find the expectation.

So what's $Var(P_s)$? Let's start as we did before by using $P_s = X/n$.

$$Var(P_s) = Var\left(\frac{X}{n}\right)$$

$$= \frac{Var(X)}{n^2}$$

This comes from $Var(aX) = a^2Var(X)$. In this case, $a = 1/n$.

As we've said before, X is the number of number of red gumballs in the sample. If we count the number of red gumballs as the number of successes, then $X \sim B(n, p)$. This means that $Var(X) = npq$, as this is the variance for the binomial distribution. This gives us

$$Var(P_s) = \frac{Var(X)}{n^2}$$

$$= \frac{npq}{n^2}$$

← $Var(X) = npq$

$$= \frac{pq}{n}$$

Taking the square root of the variance gives us the standard deviation of P_s, and this tells us how far away from *p* the sample proportion is likely to be. It's sometimes called the ***standard error of proportion***, as it tells you what the error for the proportion is likely to be in the sample.

Standard error of proportion = $\sqrt{\frac{pq}{n}}$

The standard error of proportion gets smaller as *n* gets larger. This means that the more items there are in your sample, the more reliable your sample proportion becomes as an estimator for *p*.

So how can we use the expectation and variance values we found to calculate probabilities for the proportion? Let's take a look.

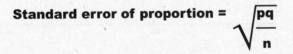

Find the distribution of P_s

So far we've found the expectation and variance for P_s, the sampling distribution of proportions. We've found that if we form a distribution from all the sample proportions, then

$$E(P_s) = p \qquad Var(P_s) = \frac{pq}{n}$$

We can use this to help us find the probability that the proportion of red gumballs in a sample of 100 is at least 40%.

But how? Don't we need to know what the distribution of P_s is first?

Right, the distribution of P_s actually depends on the size of the samples.

Here's a sketch of the distribution for P_s when *n* is large.

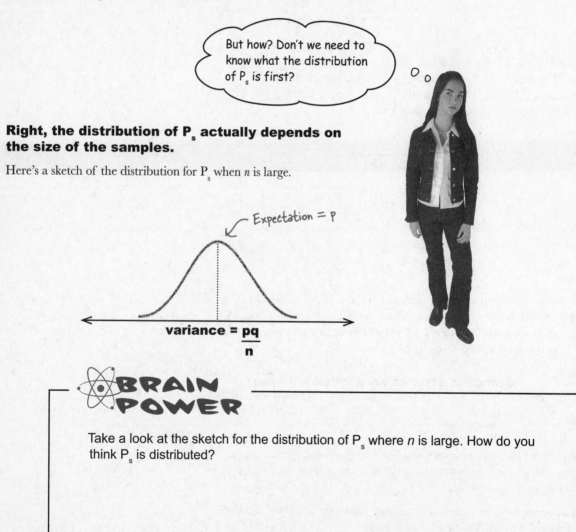

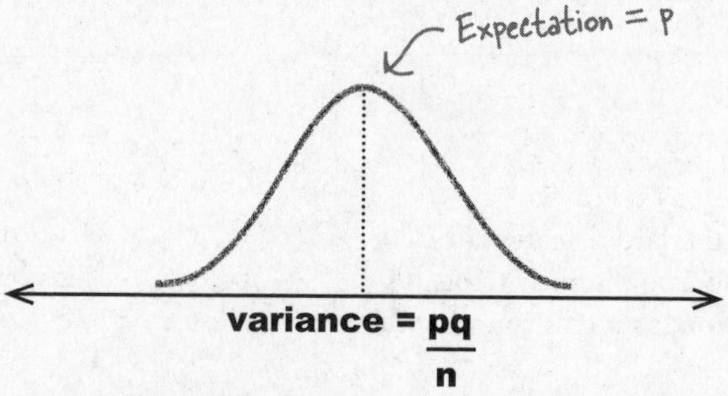

BRAIN POWER

Take a look at the sketch for the distribution of P_s where *n* is large. How do you think P_s is distributed?

P_s follows a normal distribution

When n is large, the distribution of P_s becomes approximately normal. By large, we mean greater than 30. The larger *n* gets, the closer the distribution of P_s gets to the normal distribution.

We've already found the expectation and variance of P_s, so this means that if n is large,

$$P_s \sim N\left(p, \frac{pq}{n}\right)$$

As P_s follows a normal distribution for n > 30, this means that we can use the normal distribution to solve our gumball problem. We can use the normal distribution to calculate the probability that the proportion of red gumballs in a jumbo box of gumballs will be at least 40%.

There's just one thing to remember: the sampling distribution needs a continuity correction.

Sometimes statisticians disagree about how large n needs to be.

If you're taking a statistics exam, make sure that you check how your exam board defines this.

P_s—continuity correction required

The number of successes in each of the samples is discrete, and as it's used to calculate proportion, you need to apply a continuity correction when you use the normal distribution to find probabilities.

We've seen before that if X represents the number of successes in the sample, then $P_s = X/n$. The normal continuity correction for X is ±(1/2).

If we substitute this in place of X in the formula $P_s = X/n$, this means that the continuity correction for P_s is given by

$$\text{Continuity correction} = \frac{\pm(1/2)}{n}$$

$$= \frac{\pm 1}{2n}$$

In other words, if you use the normal distribution to approximate probabilities for P_s, make sure you apply a continuity correction of ±1/2n. The exact continuity correction depends on the value of *n*.

If n is very large, the continuity correction can be left out.

As *n* gets larger, the continuity correction becomes very small, and this means that it makes very little difference to the overall probability. Some textbooks omit the continuity correction altogether.

there are no Dumb Questions

Q: What's a sampling distribution?

A: A sampling distribution is what you get if you take lots of different samples from a single population, all of the same size and taken in the same way, and then form a distribution out of some key characteristic of each sample. This means that the sampling distribution of proportions is what you get if you form a sampling distribution out of the proportions for each of the samples.

Q: Do we actually have to gather all possible samples?

A: No, we don't have to physically form all of the samples. Instead we imagine that we do, and then come up with expressions for the expectation and variance.

Q: So a sampling distribution has an expectation and variance? Why?

A: A sampling distribution is a probability distribution in the same way as any other probability distribution, so it has an expectation and variance.

The expectation of the sampling population of proportions is like the average value of a sample proportion; it's what you expect the proportion of a sample taken from a particular population to be.

Q: Why isn't the variance of P_s the same as the population variance σ^2?

A: The variance for the sampling distribution of proportions describes how the sample proportions vary. It doesn't describe how the values themselves vary. The variance has a different value because it describes a different concept.

Q: So what use does the sampling distribution of proportions have?

A: It allows you to work out probabilities for the proportion of a sample taken from a known population. It gives you an idea of what you can expect a sample to be like.

Q: What does the standard error of proportion really mean?

A: The standard error is the square root of the variance for the sampling distribution. In effect, it tells you how far away you can expect the sample proportion to be from the true value of the population proportion. This means it tells you what sort of error you can expect to have.

BULLET POINTS

- The **sampling distribution of proportions** is what you get if you consider all possible samples of size *n* taken from the same population and form a distribution out of their proportions. We use P_s to represent the sample proportion random variable.
- The expectation and variance of P_s are defined as

$$E(P_s) = p$$

$$Var(P_s) = pq/n$$

where *p* is the population proportion.

- The **standard error of proportion** is the standard deviation of this distribution. It's given by

$$\sqrt{Var(P_s)}$$

- If $n > 30$, then P_s follows a normal distribution, so

$$P_s \sim N(p, pq/n)$$

for large *n*. When working with this, you need to apply a continuity correction of

$$\pm \frac{1}{2n}$$

Exercise

25% of the gumball population are red. What's the probability that in a box of 100 gumballs, at least 40% will be red? We'll guide you through the steps.

1. If P_s is the proportion of red gumballs in the box, how is P_s distributed?

2. What's the value of $P(P_s \geq 0.4)$?

Hint: Remember to apply a continuity correction.

25% of the gumball population are red. What's the probability that in a box of 100 gumballs, at least 40% will be red? We'll guide you through the steps.

1. If P_s is the proportion of red gumballs in the box, how is P_s distributed?

Let's use p to represent the probability that a gumball is red. In other words, $p = 0.25$.

Let's use P_s to represent the proportion of gumballs in the box that are red.

$P_s \sim N(p, pq/n)$, where $p = 0.25$, $q = 0.75$, and $n = 100$. As pq/n is equal to $0.25 \times 0.75 / 100 = 0.001875$, this gives us

$P_s \sim N(0.25, 0.001875)$

2. What's the value of $P(P_s \geq 0.4)$? Hint: Remember that you need to apply a continuity correction.

$$\begin{aligned} P(P_s \geq 0.4) &= P(P_s > 0.4 - 1/(2 \times 100)) \\ &= P(P_s > 0.395) \end{aligned}$$

As $P_s \sim N(0.25, 0.001875)$, we need to find the standard score of 0.395 so that we can look up the result in probability tables. This gives us

$$\begin{aligned} z &= \frac{0.395 - 0.25}{\sqrt{0.001875}} \\ &= 3.35 \end{aligned}$$

$$\begin{aligned} P(Z > z) &= 1 - P(Z < 3.35) \\ &= 1 - 0.9996 \\ &= 0.0004 \end{aligned}$$

In other words, the probability that in a box of 100 gumballs, at least 40% will be red, is 0.0004.

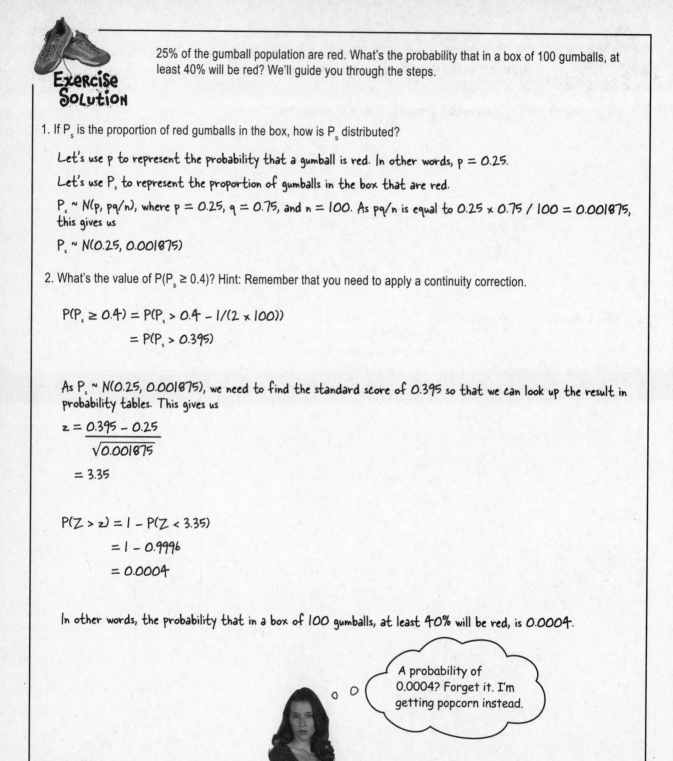

A probability of 0.0004? Forget it. I'm getting popcorn instead.

Sampling Distribution of Proportions Up Close

The sampling distribution of proportions is the distribution formed by taking the proportions of all possible samples of size *n*. The proportion of successes in a sample is represented by P_s, and it is is distributed as

$$E(P_s) = p$$

$$Var(P_s) = \frac{pq}{n}$$

When *n* is large, say bigger than 30, the distribution of P_s becomes approximately normal, so

$$P_s \sim N\left(p, \frac{pq}{n}\right)$$

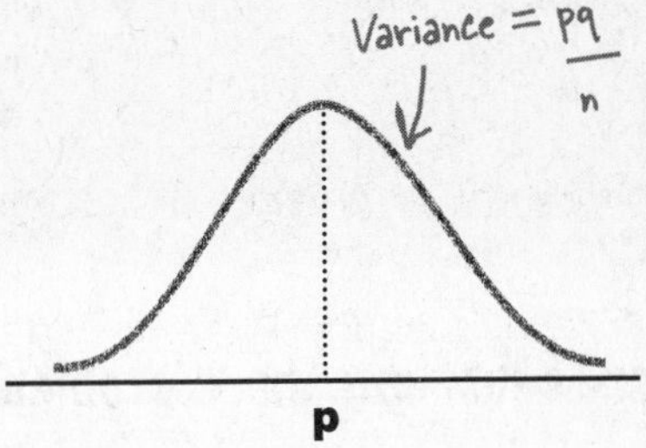

Knowing the probability distribution of P_s is useful because it means that given a particular population, we can calculate probabilities for the proportion of successes in the sample. We can approximate this with the normal distribution, and the larger the size of the sample, the more accurate the approximation.

The sampling distribution continuity correction

When you use the normal distribution in this way, it's important to apply a ***continuity correction***. This is because the number of successes in the sample is discrete, and it's used in the calculation of proportion.

If X represents the number of successes in the sample, then $P_s = X/n$. The continuity correction for X is $\pm(1/2)$, so this means the continuity correction is given by

$$\text{Continuity correction} = \pm\frac{1}{2n}$$

In other words, if you use the normal distribution to approximate probabilities for the sampling proportion, make sure you apply a continuity correction of $\pm 1/2n$.

How many gumballs?

Using the sampling distribution of proportions, you've successfully managed to find the probability of getting a certain proportion of successes in one particular sample. This means that you can now use samples to predict what the population will be like, and also use your knowledge of the population to make predictions about samples.

I'm impressed. Really impressed. Now, there's just one more thing that needs sorting out...

There's just one more problem...

The Mighty Gumball CEO has one more problem for you to work on. In addition to selling jumbo boxes, gumballs are also sold in handy pocket-sized packets that you can carry with you wherever you go.

According to Mighty Gumball's statistics for the population, the mean number of gumballs in each packet is 10, and the variance is 1. The trouble is they've had a complaint. One of their most faithful customers bought 30 packets of gumballs, and he found that the average number of gumballs per packet in his sample is only 8.5.

The CEO is concerned that he will lose one of his best customers, and he wants to offer him some form of compensation. The trouble is, he doesn't want to compensate all of his customers. He needs to know what the probability is of this happening again.

BRAIN POWER

What do you need to know in order to solve this sort of problem?

We need probabilities for the sample mean

This is a slightly different problem from last time. We're told what the population mean and variance are for the packets of gumballs, and we have a sample of packets we need to figure out probabilities for. Instead of working out probabilities for the sample proportion, this time we need to work out probabilities for the sample mean.

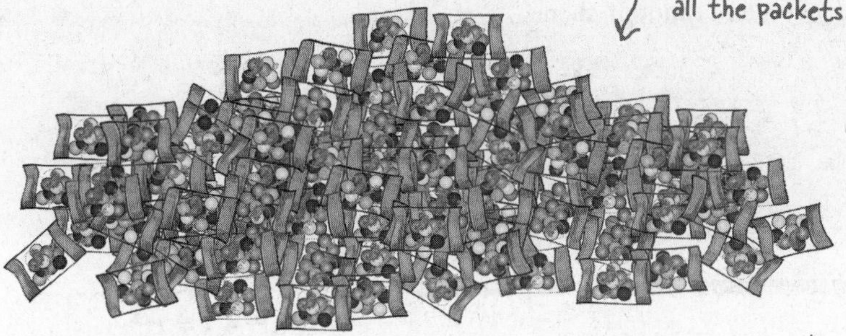
The population in this case is all the packets of gumballs.

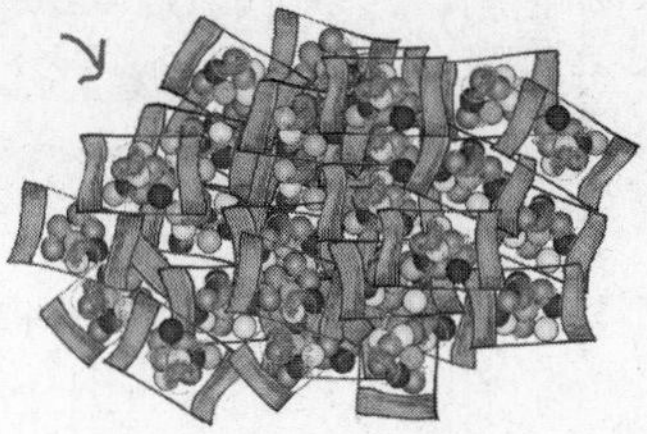
The sample comprises 30 such packets.

Before we can work out probabilities for the sample mean, we need to figure out its probability distribution. Here's what we need to do:

1. **Look at all possible samples the same size as the one we're considering.**
   If we have a sample of size *n*, we need to consider all possible samples of size *n*. There are 30 packets of gumballs, so in this case, *n* is 30.

2. **Look at the distribution formed by all the samples, and find the expectation and variance for the sample mean.**
   Every sample is different, and the number of gumballs in each packet varies.

3. **Once we know how the sample mean is distributed, use it to find probabilities.**
   If we know how the means of all possible samples are distributed, we can use it to find probabilities for the mean in a random sample, in this case the, packets of gumballs.

Let's take a look at how to tackle this.

The sampling distribution of the mean

So how do we find the distribution of the sample means?

Let's start with the population of gumball packets. We've been told what the mean and variance is for the population, and we'll represent these with μ and σ^2. We can represent the number of gumballs in a packet with X.

Each packet chosen at random is an **independent observation** of X, so each gumball packet follows the same distribution. In other words, if X_i represents the number of gumballs in a packet chosen at random, then each X_i has an expectation of μ and variance σ^2.

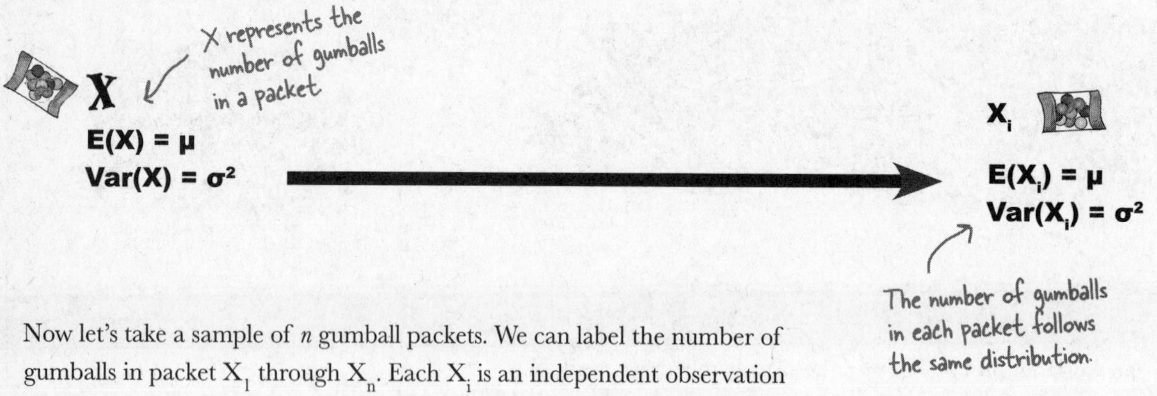
X represents the number of gumballs in a packet
X
E(X) = μ
Var(X) = σ²
X_i
E(X_i) = μ
Var(X_i) = σ²
The number of gumballs in each packet follows the same distribution.

Now let's take a sample of n gumball packets. We can label the number of gumballs in packet X_1 through X_n. Each X_i is an independent observation of X, which means that they follow the same distribution. Each X_i has an expectation of μ and variance σ^2.

We can represent the mean of gumballs in these n packets of gumballs with $\overline{X}$. The value of $\overline{X}$ depends on how many gumballs are in each packet of the n pockets, and to calculate it, you add up the total number of gumballs and divide by n.

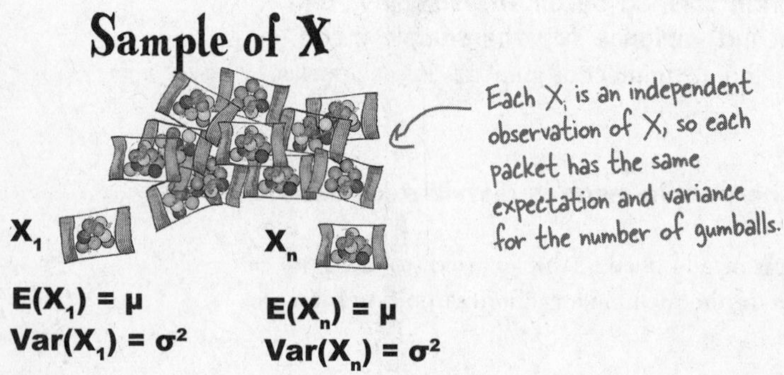
Sample of X
Each X_i is an independent observation of X, so each packet has the same expectation and variance for the number of gumballs.
X_1
E(X_1) = μ
Var(X_1) = σ²
X_n
E(X_n) = μ
Var(X_n) = σ²

This is the mean of the sample, the mean number of gumballs in the packets.

$$\overline{X} = \frac{X_1 + X_2 + \ldots + X_n}{n}$$

There are *many* possible samples we could have taken of size *n*. Each possible sample comprises *n* packets, which means that each sample comprises *n* independent observations of X. The number of gumballs in each randomly chosen packet follows the same distribution as all the others, and we calculate the mean number of gumballs for each sample in the same way.

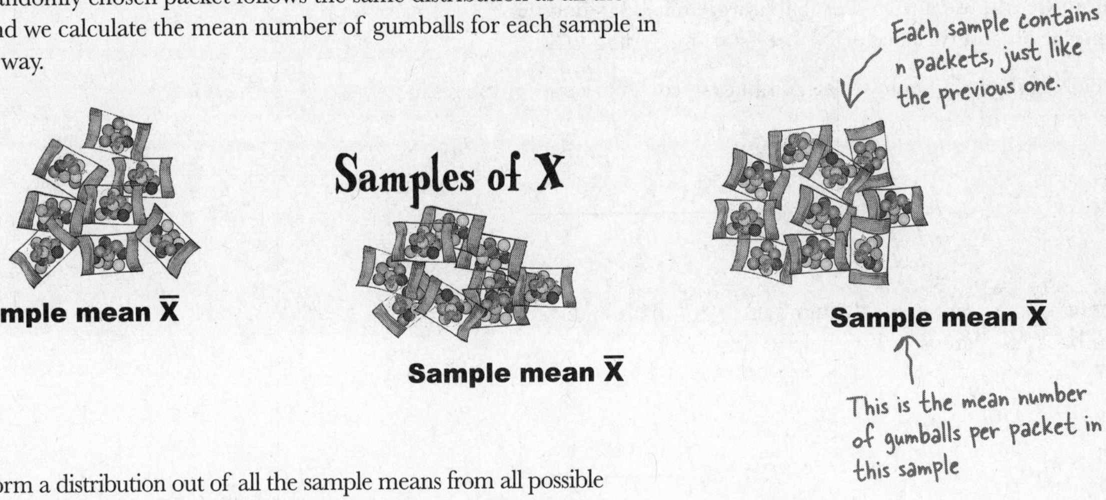

Each sample contains n packets, just like the previous one.

We can form a distribution out of all the sample means from all possible samples. This is called the ***sampling distribution of means***, or the distribution of $\overline{X}$.

So does this really help us? What does it give us?

The sampling distribution of means gives us a way of calculating probabilities for the mean of a sample.

Before you can work out the probability of any variable, you need to know about its probability distribution, and this means that if you want to calculate probabilities for the sample mean, you need to know how the sample means are distributed. In our particular case, we want to know what the probability is of there being a mean of 8.5 gumballs or fewer in a sample of 30 packets of gumballs.

Just as with the sampling distribution of proportions, before we can start calculating probabilities, we need to know what the expectation and variance are of the distribution.

Find the expectation for $\overline{X}$

So far, we've looked at how to construct the sampling distribution of means. In other words, we consider all possible samples of size *n* and form a distribution out of their means.

Before we can use it to find probabilities, we need to find the expectation and variance of $\overline{X}$. Let's start by finding $E(\overline{X})$.

Now $\overline{X}$ is the mean number of gumballs in each packet of gumballs in our sample. In other words,

$$\overline{X} = \frac{X_1 + X_2 + \dots + X_n}{n}$$

where each X_i represents the number of gumballs in the i'th packet of gumballs. We can use this to help us find $E(\overline{X})$.

$$E(\overline{X}) = E\left(\frac{X_1 + X_2 + \dots + X_n}{n}\right)$$

These two expressions are the same, just written in a different way.

$$= E\left(\frac{1}{n}X_1 + \frac{1}{n}X_2 + \dots + \frac{1}{n}X_n\right)$$

$$= E\left(\frac{1}{n}X_1\right) + E\left(\frac{1}{n}X_2\right) + \dots + E\left(\frac{1}{n}X_n\right)$$

We can split this into n separate expectations because $E(X + Y) = E(X) + E(Y)$.

Every expectation includes 1/n, so we can take it out of the expression. This comes from $E(aX) = aE(X)$.

$$= \frac{1}{n}\left(E(X_1) + E(X_2) + \dots + E(X_n)\right)$$

This means that if we know what the expectation is for each X_i, we'll have an expression for $E(\overline{X})$.

Now each X_i is an independent observation of X, and we already know that $E(X) = \mu$. This means that we can substitute μ for each $E(X_i)$ in the above expression.

So where does this get us?

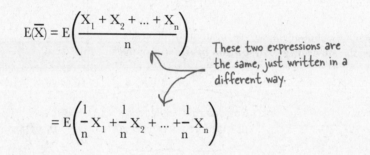

Let's replace each $E(X_i)$ with μ.

$$E(\overline{X}) = \frac{1}{n}(\mu + \mu + \dots + \mu)$$

← The expectation of X is μ. $E(X_i) = \mu$ for every i.

$$= \frac{1}{n}(n\mu)$$

← There are n of these

$$= \mu$$

This means that $E(\overline{X}) = \mu$. In other words, the average of all the possible sample means of size n is the mean of the population they're taken from. You're, in effect, finding the mean of all possible means.

This is actually quite intuitive. It means that overall, you'd expect the average number of gumballs per packet in a sample to be the same as the average number of gumballs per packet in the population. In our situation, the mean number of gumballs in each packet in the population is 10, so this is what we'd expect for the sample too.

If the population mean is 10 gumballs per packet, you can expect the sample mean to be 10 gumballs per packet too.

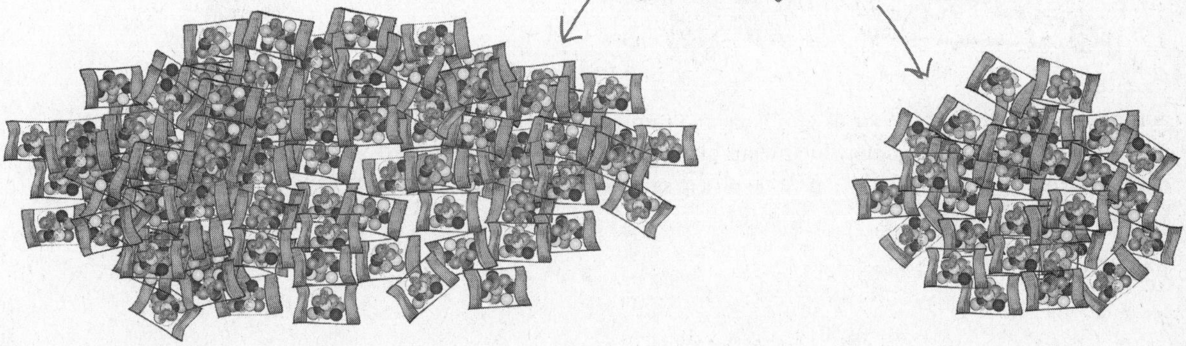

BRAIN POWER

What else do we need to know in order to find probabilities for the sample mean? How do you think we can find this?

What about the the variance of $\overline{X}$?

So far we know what $E(\overline{X})$ is, but before we can figure out any probabilities for the sample mean, we need to know what $Var(\overline{X})$ is. This will bring us one step closer to finding out what the distribution of $\overline{X}$ is like.

Why do we need to find Var($\overline{X}$)? Isn't it the same as Var(X)? Isn't it just σ^2?

The distribution of $\overline{X}$ is different from the distribution of X.

X represents the number of gumballs in a packet. We've been told what the mean number of gumballs in a packet is, and we've also been given the variance.

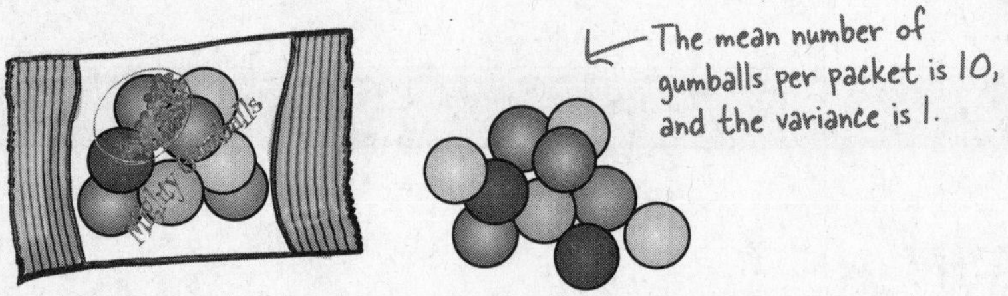
The mean number of gumballs per packet is 10, and the variance is 1.

$\overline{X}$ represents the mean number of gumballs in a sample, so the distribution of $\overline{X}$ represents how the means of all possible samples are distributed. $E(\overline{X})$ refers to the mean value of the sample means, and $Var(\overline{X})$ refers to how they vary.

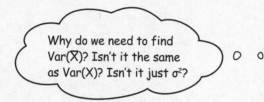

Finding $Var(\overline{X})$ is actually a very similar process to finding $E(\overline{X})$.

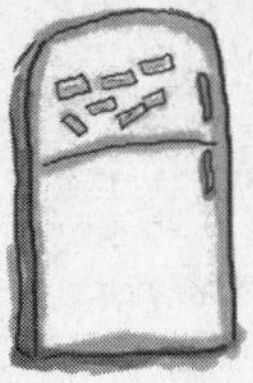

Statistics Magnets

Here are some equations for finding an expression for the variance of the sample mean. Unfortunately, some parts of the equations have fallen off. Your task is to fill in the blanks below by putting the magnets back in the right positions and derive the variance of the sample mean.

Hint: Look back at how we found $E(\overline{X})$. It might help you.

$$Var(\overline{X}) = Var\left(\frac{X_1 + X_2 + \dots + X_n}{n}\right)$$

$$= Var\left(\_\_\_\_\_\_\_\_\_\_\right)$$

$$= Var\left(\_\_\_\_\_\_\_\_\right) + Var\left(\_\_\_\_\_\_\_\_\right) + \dots + Var\left(\_\_\_\_\_\_\_\_\right)$$

$$= \_\_\_\_\_\_\_\_\_\_ \;(Var(X_1) + Var(X_2) + \dots + Var(X_n))$$

$$= \frac{1}{n^2}\;(\_\_\_\_\_\_\_\_\_\_)$$

$$= n \times 1\,\sigma^2 \over n^2$$

$$= \_\_\_\_\_\_\_\_\_\_$$

$\sigma^2 + \sigma^2 + \dots + \sigma^2$

$\frac{1}{n}X_1$

$\frac{\sigma^2}{n}$

$\frac{1}{n}X_2$

$\left(\frac{1}{n}\right)^2$

$\frac{1}{n}X_n$

$\frac{1}{n}X_1 + \frac{1}{n}X_2 + \dots + \frac{1}{n}X_n$

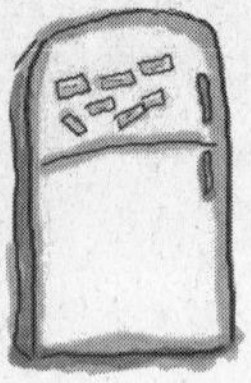

Statistics Magnets Solution

Here are some equations for finding the variance of the sample mean. Unfortunately, some parts of the equations have fallen off. Your task is to put the magnets back in the right positions and derive the variance of the sample mean.

$$Var(\overline{X}) = Var\left(\frac{X_1 + X_2 + \ldots + X_n}{n}\right)$$

$$= Var\left(\frac{1}{n}X_1 + \frac{1}{n}X_2 + \ldots + \frac{1}{n}X_n\right)$$

$$= Var\left(\frac{1}{n}X_1\right) + Var\left(\frac{1}{n}X_2\right) + \ldots + Var\left(\frac{1}{n}X_n\right)$$

$$= \left(\frac{1}{n}\right)^2 (Var(X_1) + Var(X_2) + \ldots + Var(X_n))$$

$$= \frac{1}{n^2}(\sigma^2 + \sigma^2 + \ldots + \sigma^2$$

$$= n \times \frac{1}{n^2}\sigma^2$$

$$= \frac{\sigma^2}{n}$$

Well done if you got this far. The derivation was a bit tricky, but we've found out what the variance of $\overline{X}$ is. We know how much the sample mean might vary.

Relax

Don't worry if you didn't complete this exercise; it's hard.

Most exam boards won't ask you to derive this, and in real life, you'll just need to remember the result. We're just showing you where it comes from.

Sampling Distribution of the Means Up Close

Let's take a closer look at the sampling distribution of means.

We started off by looking at the distribution of a population X. The mean of X is given by μ, and the variance by σ^2, so $E(X) = \mu$ and $Var(X) = \sigma^2$.

We then took all possible samples of size *n* taken from the population X and formed a distribution out of all the sample means, the distribution of $\overline{X}$. The mean and variance of this distribution are given by:

$$E(\overline{X}) = \mu$$

$$Var(\overline{X}) = \frac{\sigma^2}{n}$$

The standard deviation of $\overline{X}$ is the square root of the variance. The standard deviation tells you how far away from μ the sample mean is likely to be, so it's known as the ***standard error of the mean***.

$$\text{Standard error of the mean} = \frac{\sigma}{\sqrt{n}}$$

The standard error of the mean gets smaller as *n* gets larger. This means that the more items there are in your sample, the more reliable your sample mean becomes as an estimator for the population mean.

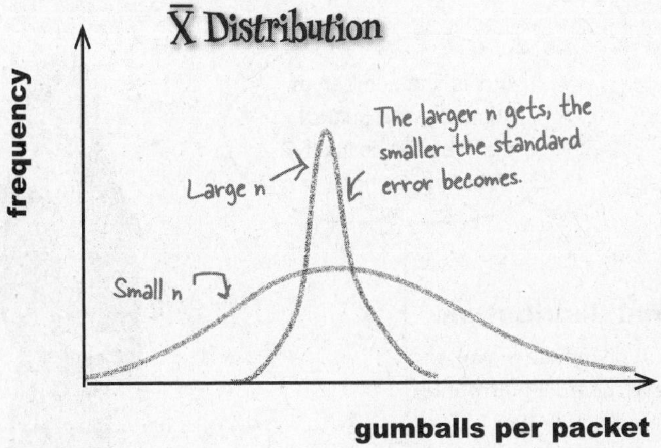

So how is $\overline{X}$ distributed?

So far we've found what the expectation and variance is for $\overline{X}$. Before we can find probabilities, though, we need to know exactly how $\overline{X}$ is distributed.

Let's start by looking at the distribution of $\overline{X}$ if X is normal.

Here's a sketch of the distribution for $\overline{X}$ for different values of μ, σ^2, and n, where X is normally distributed. What do you notice?

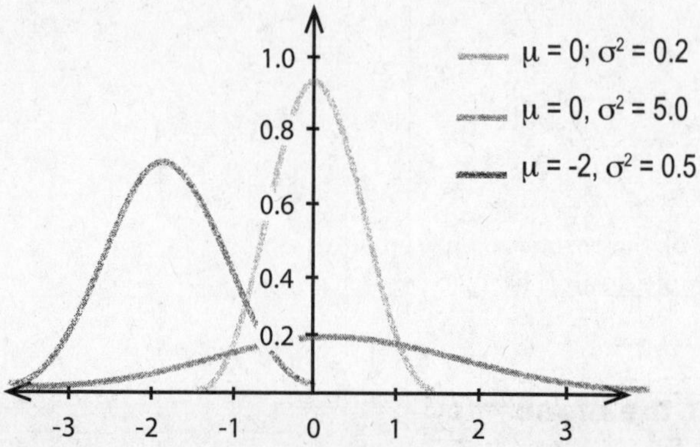

For each of these combinations, the distribution of $\overline{X}$ is normal. In other words

If $X \sim N(\mu, \sigma^2)$, then $\overline{X} \sim N(\mu, \sigma^2/n)$

These are the mean and variance of $\overline{X}$ that we found earlier.

But is the number of gumballs in a packet distributed normally? What if it's not?

X might not follow a normal distribution.

We need to know how $\overline{X}$ is distributed so that we can work out probabilities for the sample mean. The trouble is, we don't know how X is distributed.

We need to know what distribution $\overline{X}$ follows if X isn't normal.

If n is large, $\overline{X}$ can still be approximated by the normal distribution

As n gets larger, $\overline{X}$ gets closer and closer to a normal distribution. We've already seen that $\overline{X}$ is normal if X is normal. If X *isn't* normal, then we can still use the normal distribution to approximate the distribution of $\overline{X}$ if n is sufficiently large.

In our current situation, we know what the mean and variance are for the population, but we don't know what its distribution is. However, since our sample size is 30, this doesn't matter. We can still use the normal distribution to find probabilities for $\overline{X}$.

This is called the ***central limit theorem***.

Introducing the Central Limit Theorem

The central limit theorem says that if you take a sample from a non-normal population X, and if the size of the sample is large, then the distribution of $\overline{X}$ is approximately normal. If the mean and variance of the population are μ and σ^2, and n is , say, over 30, then

$$\overline{X} \sim N(\mu, \sigma^2/n)$$

This is the mean and variance of $\overline{X}$.

Does this look familiar? It's the same distribution that we had when X followed a normal distribution. The only difference is that if X is normal, it doesn't matter what size sample you use.

By the central limit theorem, if your sample of X is large, then $\overline{X}$'s distribution is approximately normal.

Using the central limit theorem

So how does the central limit theorem work in practice? Let's take a look.

The binomial distribution

Imagine you have a population represented by $X \sim B(n, p)$ where *n* is greater than 30. As we've seen before, $\mu = np$, and $\sigma^2 = npq$.

The central limit theorem tells us that in this situation, $\overline{X} \sim N(\mu, \sigma^2/n)$. To find the distribution of $\overline{X}$, we substitute in the values for the population. This means that if we substitute in values for $\mu = np$ and $\sigma^2 = npq$, we get

$$\overline{X} \sim N(np, pq)$$

For the binomial distribution, the mean of the population is np, and the variance is npq. If we substitute this into the sampling distribution, we get $\overline{X} \sim N(np, pq)$.

The Poisson distribution

Now suppose you have a population that follows a Poisson distribution of $X \sim Po(\lambda)$, again where *n* is greater than 30. For the Poisson distribution, $\mu = \sigma^2 = \lambda$.

As before, we can use the normal distribution to help us find probabilities for $\overline{X}$. If we substitute population parameters into $\overline{X} \sim N(\mu, \sigma^2/n)$, we get:

$$\overline{X} \sim N(\lambda, \lambda/n)$$

For the Poisson distribution, the mean and variance are both λ. If we substitute this into the sampling distribution, we get $\overline{X} \sim N(\lambda, \lambda/n)$.

In general, you take the distribution $\overline{X} \sim N(\mu, \sigma^2/n)$ and substitute in values for μ and σ^2.

Finding probabilities

Since $\overline{X}$ follows a normal distribution, this means that you can use standard normal probability tables to look up probabilities. In other words, you can find probabilities in exactly the same way you would for any other normal distribution.

Let's apply this to Mighty Gumball's problem.

The mean number of gumballs per packet is 10, and the variance is 1. If you take a sample of 30 packets, what's the probability that the sample mean is 8.5 gumballs per packet or fewer? We'll guide you through the steps.

1. What's the distribution of $\overline{X}$?

2. What's the value of $P(\overline{X} < 8.5)$?

Let's apply this to Mighty Gumball's problem.

The mean number of gumballs per packet is 10, and the variance is 1. If you take a sample of 30 packets, what's the probability that the sample mean is 8.5 gumballs per packet or fewer? We'll guide you through the steps.

1. What's the distribution of $\bar{X}$?

We know that $\bar{X} \sim N(\mu, \sigma^2/n)$, $\mu = 10$, $\sigma^2 = 1$, and $n = 30$, and $1/30 = 0.0333$. So this gives us

$\bar{X} \sim N(10, 0.0333)$

2. What's the value of $P(\bar{X} < 8.5)$?

As $\bar{X} \sim N(10, 0.0333)$, we need to find the standard score of 8.5 so that we can look up the result in probability tables. This gives us

$$z = \frac{8.5 - 10}{\sqrt{0.0333}}$$

$$= -8.22 \text{ (to 2 decimal places)}$$

$$P(Z < z) = P(Z < -8.22)$$

This probability is so small that it doesn't appear on the probability tables. We can assume that an event with a probability this small will hardly ever happen.

there are no Dumb Questions

Q: Do I need to use any continuity corrections with the central limit theorem?

A: Good question, but no you don't. You use the central limit theorem to find probabilities associated with the sample mean rather than the values in the sample, which means you don't need to make any sort of continuity correction.

Q: Is there a relationship between point estimators and sampling distributions?

A: Yes, there is.

Let's start with the mean. The point estimator for the population mean is $\bar{x}$, which means that $\hat{\mu} = \bar{x}$. Now, if we look at the expectation for the sampling distribution of means, we get $E(\bar{X}) = \mu$. The expectation of all the sample means is given by μ, and we can estimate μ with the sample mean.

Similarly, the point estimator for the population proportion is p_s, the sample proportion, which means that $p = p_s$. If we take the expectation of all the sample proportions, we get $E(P_s) = p$. The expectation of all the sample proportions is given by p, and we can estimate p with the sample proportion.

We're not going to prove it, but we get a similar result for the variance. We have $\sigma^2 = s^2$, and $E(S^2) = \sigma^2$.

Q: So is that a coincidence?

A: No, it's not. The estimators are chosen so that the expectation of a large number of samples, all of size *n* and taken in the same way, is equal to the true value of the population parameter. We call these estimators ***unbiased*** if this holds true.

An unbiased estimator is likely to be accurate because on average across all possible samples, it's expected to be the value of the true population parameter.

Q: How does standard error come into this?

A: The best unbiased estimator for a population parameter is generally one with the smallest variance. In other words, it's the one with the smallest standard error.

BULLET POINTS

- The **sampling distribution of means** is what you get if you consider all possible samples of size *n* taken from the same population and form a distribution out of their means. We use $\bar{X}$ to represent the sample mean random variable.
- The **expectation and variance of $\bar{X}$** are defined as

$$E(\bar{X}) = \mu$$

$$Var(\bar{X}) = \sigma^2/n$$

where μ and σ^2 are the mean and variance of the population.

- The **standard error of the mean** is the standard deviation of this distribution. It's given by

$$\sqrt{Var(\bar{X})}$$

- If $X \sim N(\mu, \sigma^2)$, then $\bar{X} \sim N(\mu, \sigma^2/n)$.
- The **central limit theorem** says that if *n* is large and X doesn't follow a normal distribution, then

$$\bar{X} \sim N(\mu, \sigma^2/n)$$

Sampling saves the day!

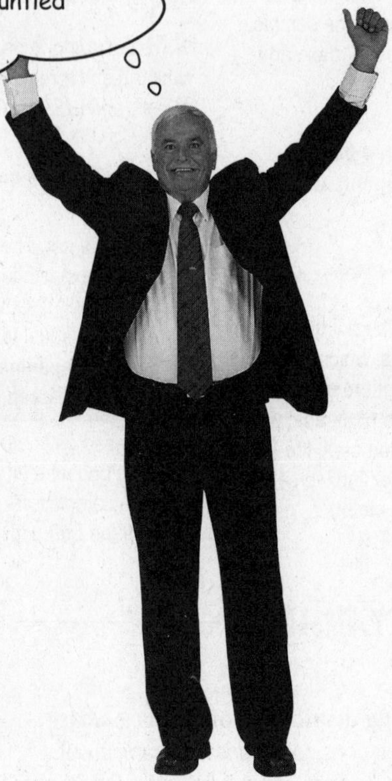

The work you've done is awesome! My top customer found an average of 8.5 gumballs in a sample of 30 packets, and you've told me the probability of getting that result is extremely unlikely. That means I don't have to worry about compensating disgruntled customers, which means more money for me!

You've made a lot of progress

Not only have you been able to come up with point estimators for population parameters based on a single sample, you've also been able to use population to calculate probabilities in the sample. That's pretty powerful.

12 constructing confidence intervals

Guessing with Confidence

I put this in the oven for 2.5 hours, but if you bake yours for 1—5 hours, you should be fine.

Sometimes samples don't give quite the right result.

You've seen how you can use point estimators to estimate the **precise value** of the population mean, variance, or proportion, but the trouble is, how can you be certain that your estimate is completely accurate? After all, your assumptions about the population rely on just one sample, and what if your sample's off? In this chapter, you'll see **another way of estimating population statistics**, one that **allows for uncertainty**. Pick up your probability tables, and we'll show you the ins and outs of **confidence intervals**.

Mighty Gumball is in trouble

The Mighty Gumball CEO has gone ahead with a range of television advertisements, and he's proudly announced exactly how long the flavor of the super-long-lasting gumballs lasts for, right down to the last second.

Unfortunately...

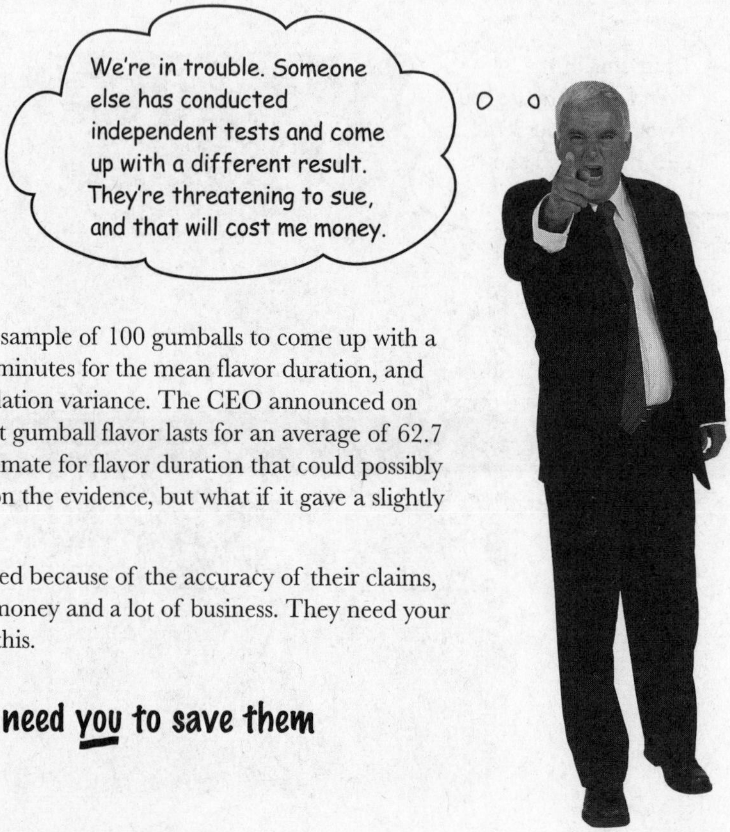

We're in trouble. Someone else has conducted independent tests and come up with a different result. They're threatening to sue, and that will cost me money.

Mighty Gumball used a sample of 100 gumballs to come up with a point estimator of 62.7 minutes for the mean flavor duration, and 25 minutes for the population variance. The CEO announced on primetime television that gumball flavor lasts for an average of 62.7 minutes. It's the best estimate for flavor duration that could possibly have been made based on the evidence, but what if it gave a slightly wrong result?

If Mighty Gumball is sued because of the accuracy of their claims, they could lose a lot of money and a lot of business. They need your help to get them out of this.

They need <u>you</u> to save them

BRAIN POWER

What do you think could have gone wrong? Should Mighty Gumball have used the precise value of the point estimator in their advertising? Why? Why not?

The problem with precision

As you saw in the last chapter, point estimators are the best estimate we can possibly give for population statistics. You take a representative sample of data and use this to estimate key statistics of the population such as the mean, variance, and proportion. This means that the point estimator for the mean flavor duration of super-long-lasting gumballs was the best possible estimate we could possibly give.

The problem with deriving point estimators is that we rely on the results of a single sample to give us a very precise estimate. We've looked at ways of making the sample as representative as possible by making sure the sample is unbiased, but we don't know with absolute certainty that it's 100% representative, purely because we're dealing with a sample.

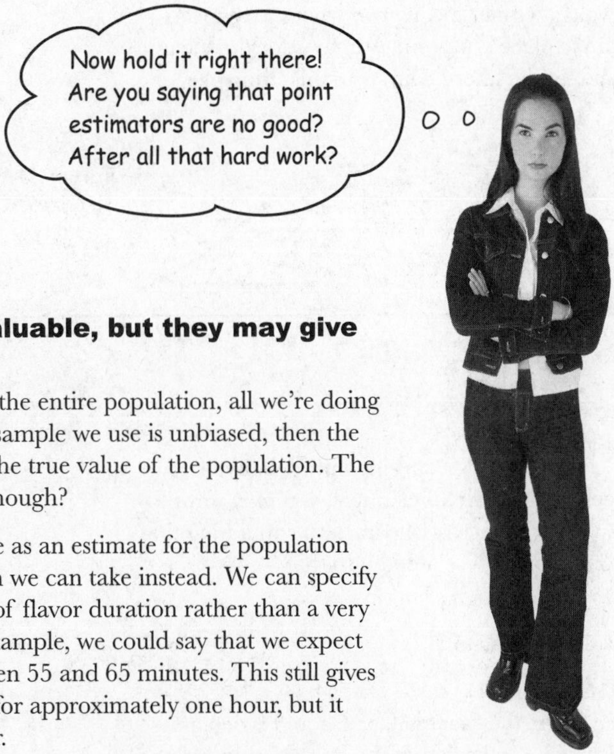

Now hold it right there!
Are you saying that point
estimators are no good?
After all that hard work?

Point estimators are valuable, but they may give slight errors.

Because we're not dealing with the entire population, all we're doing is giving a best estimate. If the sample we use is unbiased, then the estimate is likely to be close to the true value of the population. The question is, how close is close enough?

Rather than give a precise value as an estimate for the population mean, there's another approach we can take instead. We can specify some interval as an estimation of flavor duration rather than a very precise length of time. As an example, we could say that we expect gumball flavor to last for between 55 and 65 minutes. This still gives the impression that flavor lasts for approximately one hour, but it allows for some margin of error.

The question is, how do we come up with the interval? It all depends how confident you want to be in the results...

Introducing confidence intervals

Up until now, we've estimated the mean amount of time that gumball flavor lasts for by using a point estimator, based upon a sample of data. Using the point estimator, we've been able to give a very precise estimate for the mean duration of the flavor. Here's a sketch showing the distribution of flavor duration in the sample of gumballs.

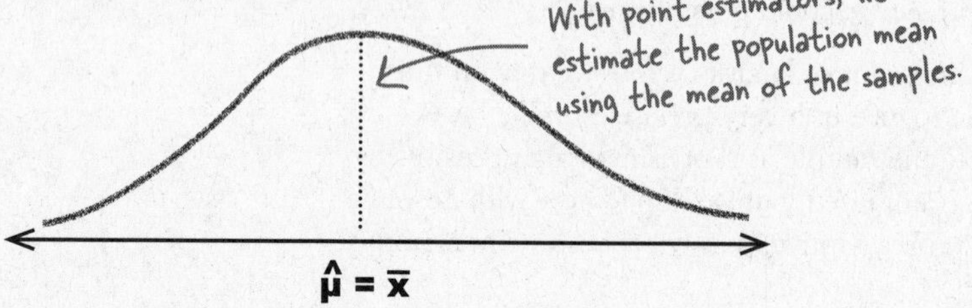
With point estimators, we estimate the population mean using the mean of the samples.
$\hat{\mu} = \bar{x}$

So what happens if we specify an interval for the population mean instead? Rather than specify an exact value, we can specify two values we expect flavor duration to lie between. We place our point estimator for the mean in the center of the interval and set the interval limits to this value plus or minus some margin of error.

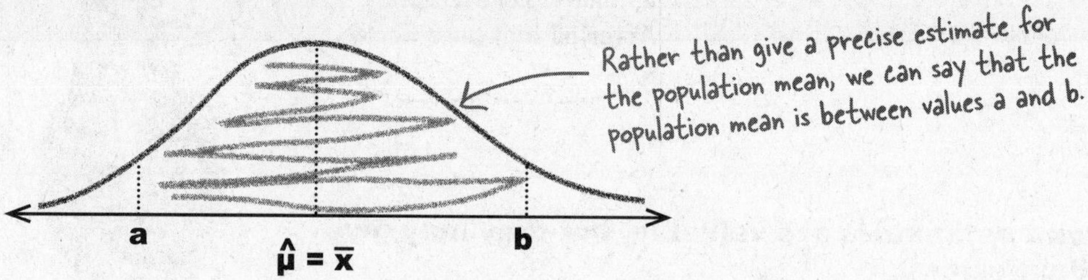
Rather than give a precise estimate for the population mean, we can say that the population mean is between values a and b.
a
b
$\hat{\mu} = \bar{x}$

The interval limits are chosen so that there's a specified probability of the population mean being between a and b. As an example, you may want to choose a and b so that there's a 95% chance of the interval containing the population mean. In other words, you choose a and b so that

$$P(a < \mu < b) = 0.95$$

We represent this interval as (a, b). As the exact value of a and b depends on the degree of confidence you want to have that the interval contains the population mean, (a, b) is called a ***confidence interval***.

So how do we find the confidence interval for the population mean?

Four steps for finding confidence intervals

Here are the broad steps involved in finding confidence intervals. Don't worry if you don't get what each step is about just yet, we'll go through them in more detail soon.

1. **Choose your population statistic** ← This is the population statistic you want to construct a confidence interval for.
2. **Find its sampling distribution** ← ...ou encountered ...mpling distributions in ...e last chapter.
3. **Decide on the level of confidence** ← The probability that your interval contains the statistic
4. **Find the confidence limits** ← To find the confidence limits, we need to know the level of confidence and the sampling distribution.

Let's see if we can construct a confidence interval for the Mighty Gumball CEO that he can use in his television commercials. Let's find a confidence interval for the mean amount of time gumball flavor lasts for.

there are no Dumb Questions

Q: So can you construct a confidence interval for any population statistic?

A: Broadly speaking, you can construct a confidence interval for any population statistic where you know what the sampling distribution is like. We've looked at sampling distributions for the mean and proportion, so we can construct confidence intervals for both of these.

Q: What about the variance? Can we construct a confidence interval for that?

A: Theoretically, yes, but we haven't covered the sampling distribution for the variance, and we're not going to. It's more common to construct confidence intervals for the mean and proportion, and these are what tend to be covered by statistics exams.

Q: Do these steps relate to the confidence interval for the mean or the confidence interval for the proportion?

A: They're general steps that apply to either. You can use them for the population mean and for the population proportion.

Q: Does it matter how the population is distributed?

A: The key thing is the sampling distribution of the statistic you're trying to construct a confidence interval for. If you want a confidence interval for the mean, you need to know the sampling distribution of means, and if you want a confidence interval for the proportion, you need to know the sampling distribution of proportions.

The main impact the population distribution has on the confidence interval is the effect it has on the sampling distribution. We'll see how later on.

Step 1: Choose your population statistic

The first step is to pick the statistic you want to construct a confidence interval for. This all depends on the problem you want to solve.

We want to construct a confidence interval for the mean amount of time that gumball flavor lasts for, so in this case, we want to construct a confidence interval for the population mean, μ.

Now that we've chosen the population statistic, we can move onto the next step.

Step 2: Find its sampling distribution

To find a confidence interval for the population mean, we need to know what the sampling distribution is for the mean. In other words, we need to know what the expectation and variance of $\overline{X}$ is, and also what distribution it follows.

Let's start with the expectation and variance. If we go back to the work we did in the last chapter, then the sampling distribution of means has the following expectation and variance:

$$E(\overline{X}) = \mu \qquad Var(\overline{X}) = \frac{\sigma^2}{n}$$

In order to use this to find the confidence interval for μ, we substitute in values for the population variance, σ^2, and the sample size, n.

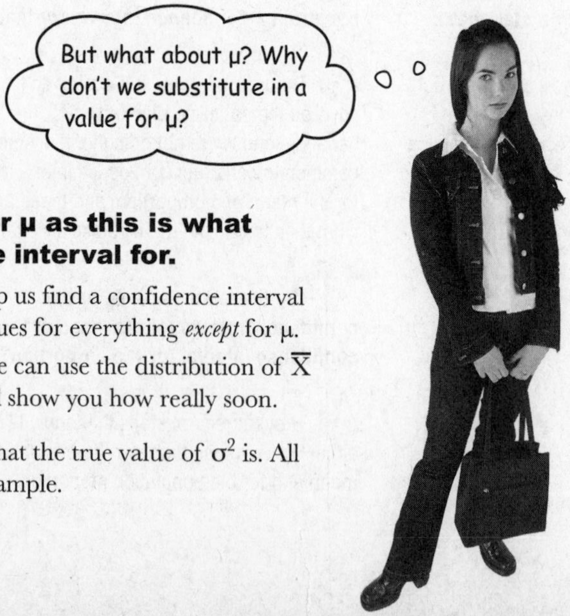
But what about μ? Why don't we substitute in a value for μ?

We don't substitute in a value for μ as this is what we're trying to find a confidence interval for.

We're using the sampling distribution to help us find a confidence interval for μ, so this means that we substitute in values for everything *except* for μ. By substituting in the values for σ^2 and n, we can use the distribution of $\overline{X}$ to help us find the confidence interval. We'll show you how really soon.

There's just one problem. We don't know what the true value of σ^2 is. All we have to go on is estimates based on the sample.

Point estimators to the rescue

So what can we use as the value for σ^2?

Even though we don't know what the true value is for the population variance, σ^2, we can estimate its value using its point estimator. Rather than use σ^2, we can use $\hat{\sigma}^2$ in its place, or s^2.

This means that the expectation and variance for the sampling distribution of means is

$$E(\overline{X}) = \mu \qquad \mathrm{Var}(\overline{X}) = \frac{s^2}{n}$$

This is the point estimator for the variance. We don't know what the true value of the population variance is, so we use the sample variance to estimate it instead..

Mighty Gumball used a sample of 100 gumballs to come up with their estimates, and they have calculated that $s^2 = 25$. This means that

$$\mathrm{Var}(\overline{X}) = \frac{s^2}{n} = \frac{25}{100} = 0.25$$

There's still one thing we have left to do. Before we can find the confidence interval for μ, we need to know exactly how $\overline{X}$ is distributed.

Sharpen your pencil

Assume that $X \sim N(\mu, \sigma^2)$ and that the number in the sample is large. What distribution does $\overline{X}$ follow? Use $E(\overline{X})$ and $\mathrm{Var}(\overline{X})$ above to help you.

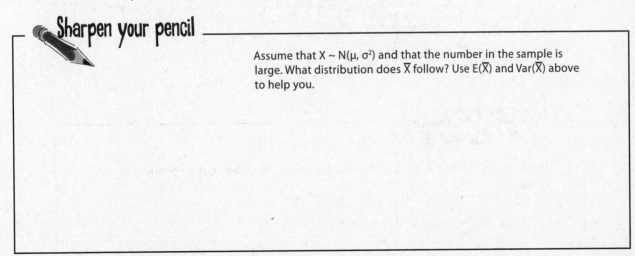

Sharpen your pencil Solution

Assume that $X \sim N(\mu, \sigma^2)$ and that the number in the sample is large. What distribution does $\overline{X}$ follow? Use $E(\overline{X})$ and $Var(\overline{X})$ above to help you.

If X follows a normal distribution, then $\overline{X}$ does too. Substituting in the point estimator for σ^2, we get:

$$\overline{X} \sim N(\mu, s^2/n)$$

or:

$$\overline{X} \sim N(\mu, 0.25)$$

We've found the distribution for $\overline{X}$

Now that we know how $\overline{X}$ is distributed, we have enough information to move onto the next step.

Step 3: Decide on the level of confidence

The level of confidence lets you say how sure you want to be that the confidence interval contains your population statistic. As an example, suppose we want a confidence level of 95% for the population mean. This means that the probability of the population mean being inside the confidence interval is 0.95.

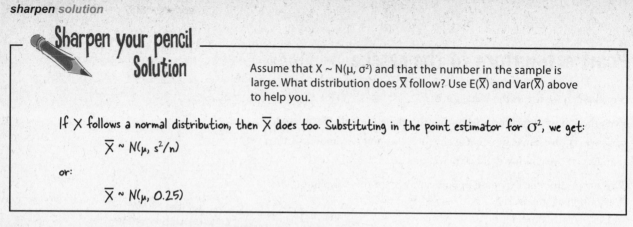

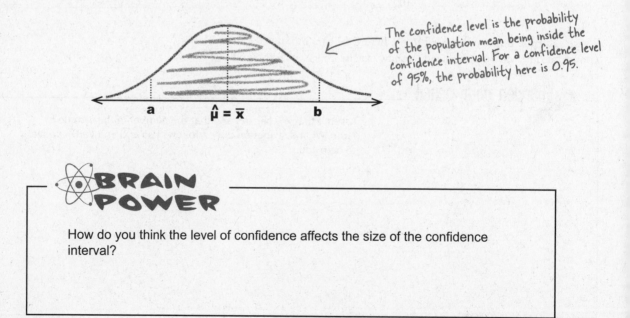

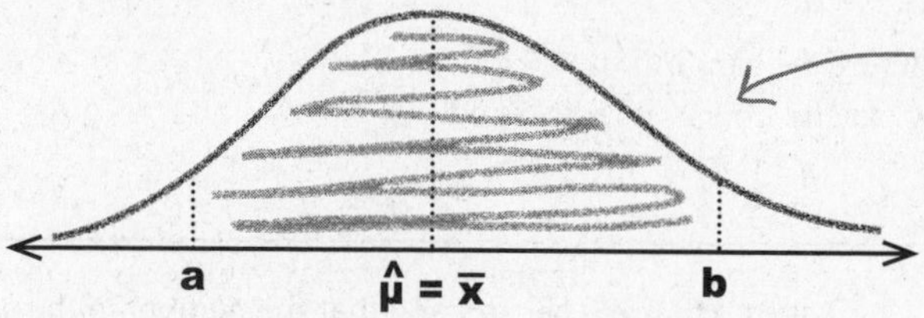

The confidence level is the probability of the population mean being inside the confidence interval. For a confidence level of 95%, the probability here is 0.95.

BRAIN POWER

How do you think the level of confidence affects the size of the confidence interval?

How to select an appropriate confidence level

So who decides what the level of confidence should be? What's the right level of confidence?

The answer to this really depends on your situation and how confident you need to be that your interval contains the population statistic. A 95% confidence level is common, but sometimes you might want a different one, such as 90% or 99%. As an example, the Mighty Gumball CEO might want to have a higher degree of confidence that the population mean falls inside the confidence interval, as he intends to use it in his television advertisements.

The key thing to remember is that the higher the confidence level is, the wider the interval becomes, and the more chance there is of the confidence interval containing the population statistic.

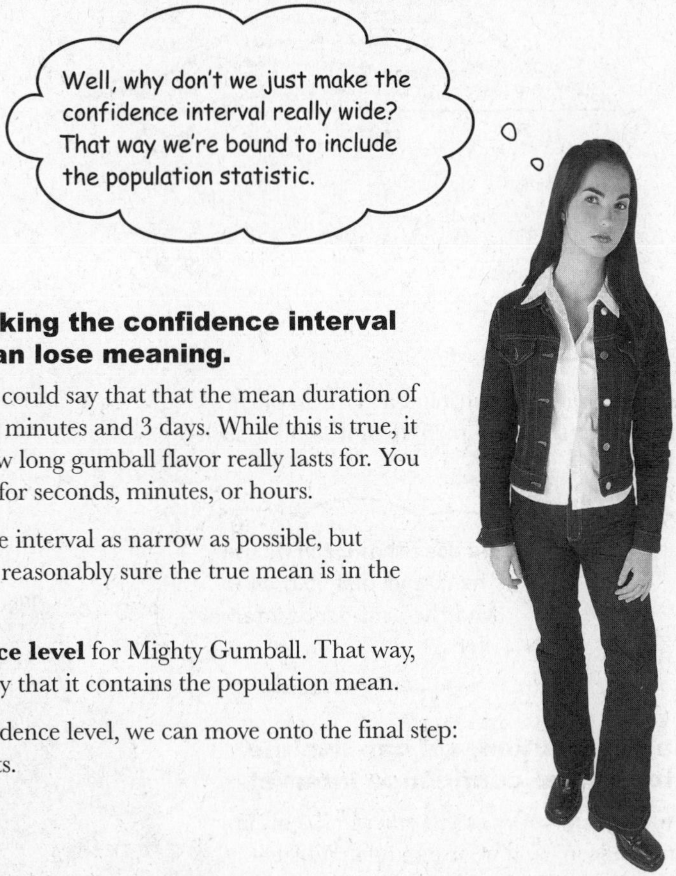

Well, why don't we just make the confidence interval really wide? That way we're bound to include the population statistic.

The trouble with making the confidence interval too wide is that it can lose meaning.

As an extreme example, we could say that that the mean duration of gumball flavor is between 0 minutes and 3 days. While this is true, it doesn't give you an idea how long gumball flavor really lasts for. You don't know whether it lasts for seconds, minutes, or hours.

The key thing is to make the interval as narrow as possible, but wide enough so you can be reasonably sure the true mean is in the interval.

Let's use a **95% confidence level** for Mighty Gumball. That way, there'll be a high probability that it contains the population mean.

Now that we have the confidence level, we can move onto the final step: finding the confidence limits.

Step 4: Find the confidence limits

The final step is to find a and b, the ***limits*** of the confidence interval, which indicate the left and right borders of the range in which there's a 95% probability of the mean falling. The exact value of a and b depends on the sampling distribution we need to use and the level of confidence that we need to have.

For this problem, we need to find the 95% confidence level for the mean duration of gumball flavor. Meaning, there must be a .95 probability that μ lies between the a and b that we find. We also know that $\overline{X}$ follows a normal distribution, where $\overline{X} \sim N(\mu, 0.25)$.

Here's a sketch of what we need:

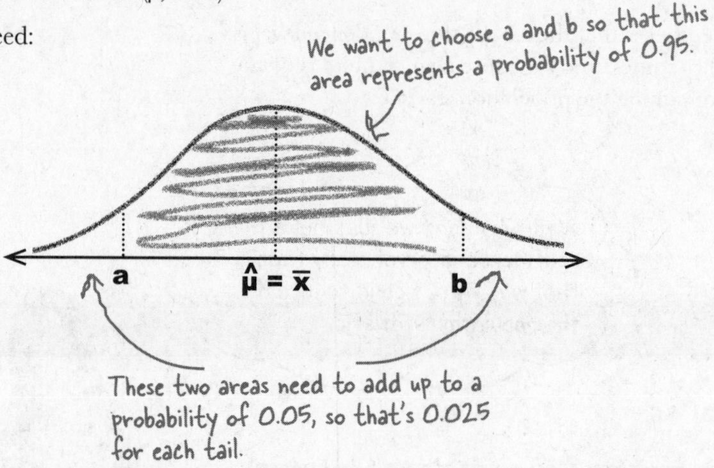
We want to choose a and b so that this area represents a probability of 0.95.
a
$\hat{\mu} = \overline{x}$
b
These two areas need to add up to a probability of 0.05, so that's 0.025 for each tail.

We can find the values of a and b using the distribution of $\overline{X}$. In other words, we can use the distribution of $\overline{X} \sim N(\mu, 0.25)$ to find a and b, such that $P(\overline{X} < a) = 0.025$ and $P(\overline{X} > b) = 0.025$.

So does that mean we use the normal distribution to find the confidence interval for μ?

As $\overline{X}$ follows a normal distribution, we can use the normal distribution to find the confidence interval.

We can do this in a similar way to how we've solved other problems in the past. We calculate a standard score, and we use standard normal probability tables to help us get the result we need.

Start by finding Z

Before we can use normal probability tables, we need to standardize $\overline{X}$. We know that $\overline{X} \sim N(\mu, 0.25)$, so this means that if we standardize, we get

$$Z = \frac{\overline{X} - \mu}{\sqrt{0.25}} \quad \text{where} \quad Z \sim N(0, 1)$$

Here's a sketch of the standardized version of the confidence interval.

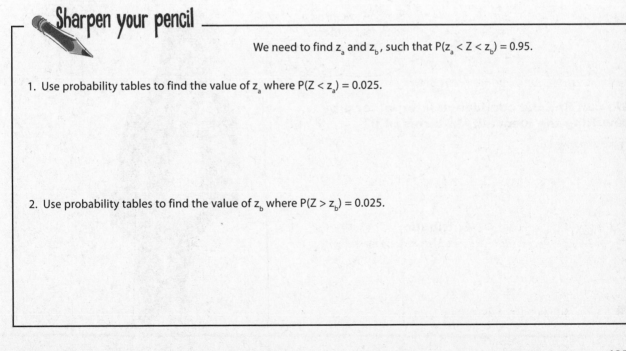

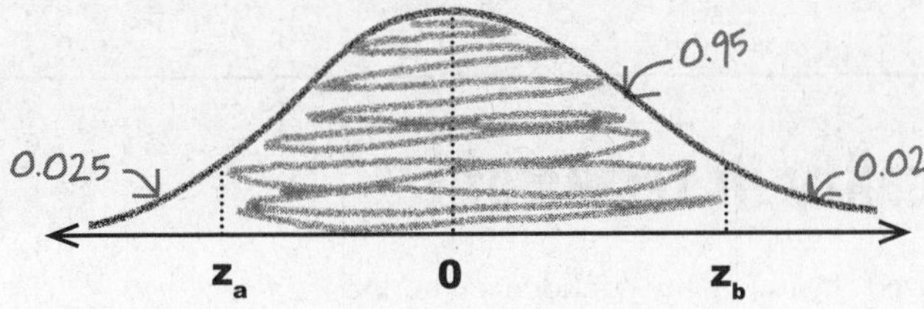

We need to find z_a and z_b where $P(z_a < Z < z_b) = 0.95$. In other words, the standardized confidence limits are given by z_a and z_b where $P(Z < z_a) = 0.025$ and $P(Z > z_b) = 0.025$. We can find the values of z_a and z_b using probability tables.

Sharpen your pencil

We need to find z_a and z_b, such that $P(z_a < Z < z_b) = 0.95$.

1. Use probability tables to find the value of z_a where $P(Z < z_a) = 0.025$.

2. Use probability tables to find the value of z_b where $P(Z > z_b) = 0.025$.

Sharpen your pencil Solution

We need to find z_a and z_b such that $P(z_a < Z < z_b) = 0.95$.

1. Use probability tables to find the value of z_a where $P(Z < z_a) = 0.025$.

   If we look up the probability 0.025 in standard normal probability tables, this gives us $z_a = -1.96$

2. Use probability tables to find the value of z_b where $P(Z > z_b) = 0.025$.

   To find z_b, we need to look up a value of 0.975. This gives us $z_b = 1.96$.

Rewrite the inequality in terms of μ

So far, we've found a standardized version of the confidence interval. We've found that P(-1.96 < Z < 1.96) = 0.95. In other words,

$$P\left(-1.96 < \frac{\overline{X} - \mu}{0.5} < 1.96\right) = 0.95$$

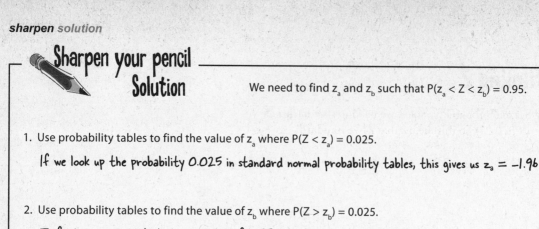

But don't we need a confidence interval for μ? How do find it?

We can find the confidence interval for μ by rewriting the inequality in terms of μ.

If we can rewrite

$$-1.96 < \frac{\overline{X} - \mu}{0.5} < 1.96$$

in the form

$$a < \mu < b$$

This will give us a confidence interval for μ.

We'll have our confidence limits for μ.

Pool Puzzle

Your job is to rewrite $-1.96 < (\overline{X} - \mu)/0.5 < 1.96$ and come up with a confidence interval for μ. Take snippets from the pool and place them into the blank lines. You may **not** use the same snippet more than once.

This gives you the lefthand side of the inequality → $-1.96 < \frac{\overline{X} - \mu}{0.5} < 1.96$ ← This gives you the righthand side

Lefthand side	Righthand side
$-1.96 < \frac{\overline{X} - \mu}{0.5}$ (.........)	$\frac{\overline{X} - \mu}{0.5} < 1.96$ (.........)
$-1.96 \times$ .... $< \overline{X} - \mu$	$\overline{X} - \mu <$ .... $\times 0.5$
.... $+ \mu < \overline{X}$	$\overline{X} <$ .... $+ \mu$
$\mu <$	.... $< \mu$

$\overline{X} - 0.98 < \mu < \overline{X} + 0.98$ ← This is what you get if you put the two sides of the inequality together again

Note: Each thing from the pool can only be used once!

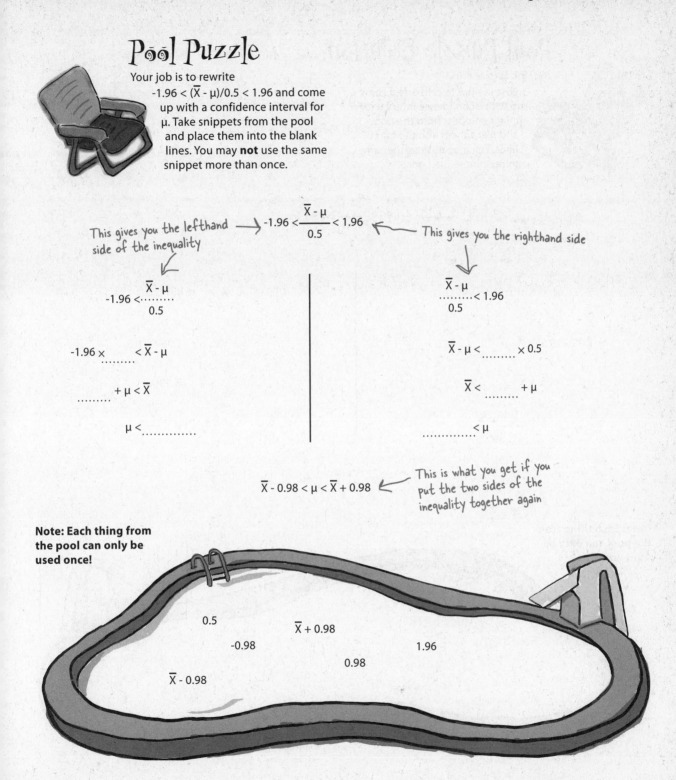

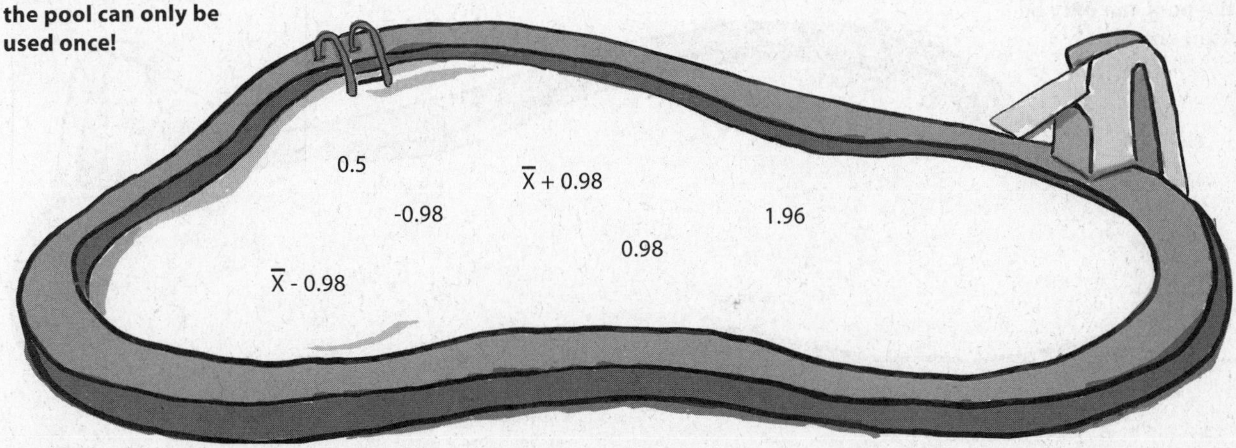

Pool Puzzle Solution

Your job is to rewrite $-1.96 < (\bar{X} - \mu)/0.5 < 1.96$ and come up with a confidence interval for μ. Take snippets from the pool and place them into the blank lines. You may **not** use the same snippet more than once.

$$-1.96 < \frac{\bar{X} - \mu}{0.5} < 1.96$$

This gives you the left hand side of the inequality. This gives you the right hand side.

Left hand side	Right hand side
$-1.96 < \frac{\bar{X} - \mu}{0.5}$	$\frac{\bar{X} - \mu}{0.5} < 1.96$
$-1.96 \times \underline{0.5} < \bar{X} - \mu$	$\bar{X} - \mu < \underline{1.96} \times 0.5$
$\underline{-0.98} + \mu < \bar{X}$	$\bar{X} < \underline{0.98} + \mu$
$\mu < \underline{\bar{X} + 0.98}$	$\underline{\bar{X} - 0.98} < \mu$

$$\bar{X} - 0.98 < \mu < \bar{X} + 0.98$$

Note: Each thing from the pool can only be used once!

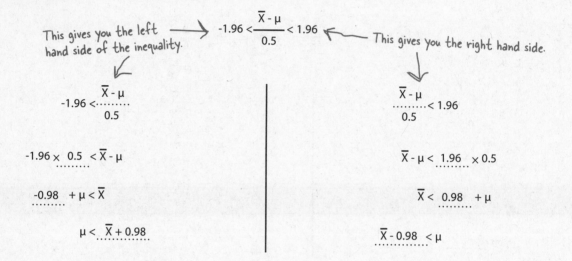

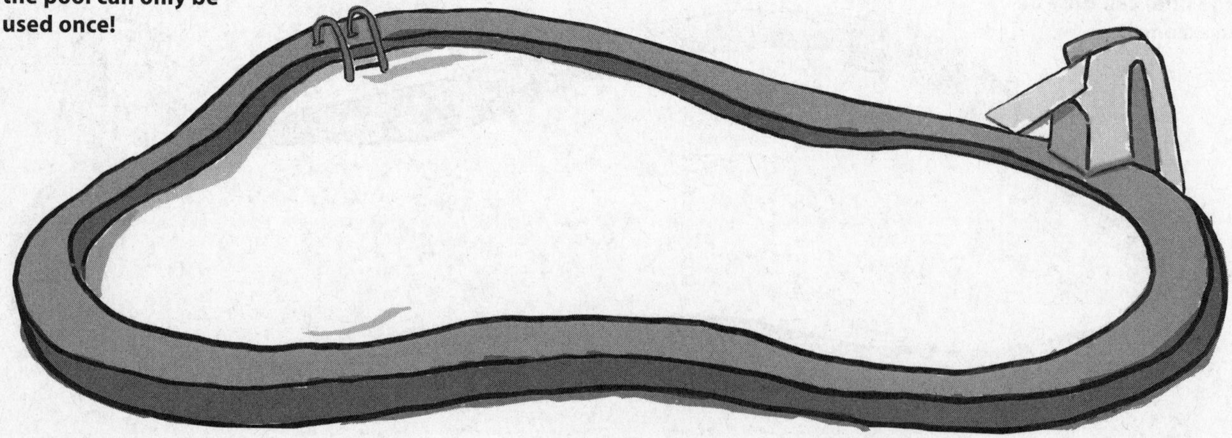

Finally, find the value of $\overline{X}$

Now that we've rewritten the inequality, we're very close to finding a confidence interval for μ that describes the amount of time gumball flavor typically lasts for. In other words, we use

$$P(\overline{X} - 0.98 < \mu < \overline{X} + 0.98) = 0.95$$

Here's a quick sketch.

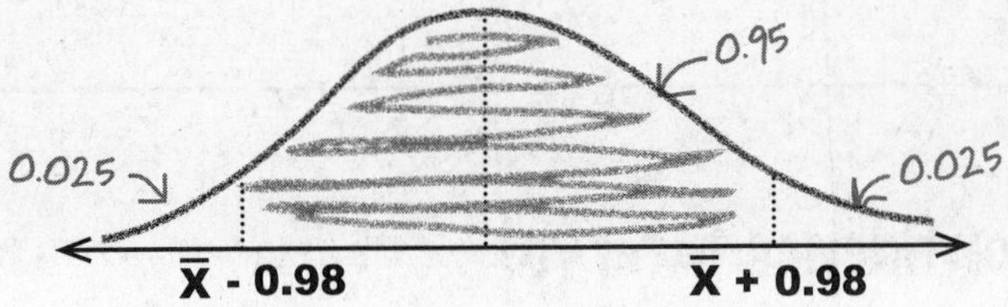
0.95
0.025
0.025
X̄ - 0.98
X̄ + 0.98

Our confidence limits are given by $\overline{X} - 0.98$ and $\overline{X} + 0.98$. If we knew what to use as a value for $\overline{X}$, we'd have values for the confidence limits.

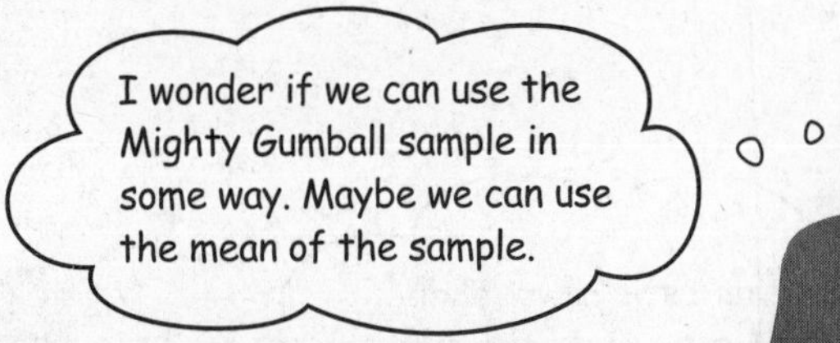
I wonder if we can use the Mighty Gumball sample in some way. Maybe we can use the mean of the sample.

$\overline{X}$ is the distribution of sample means, so we can use the value of $\overline{x}$ from the Mighty Gumball sample.

Sharpen your pencil

The confidence limits are given by $\overline{X} - 0.98$ and $\overline{X} + 0.98$. For the Mighty Gumball sample, $\overline{x}$ is given by 62.7. Use this to come up with values for the confidence limits.

Sharpen your pencil Solution

The confidence limits are given by $\overline{X}$ - 0.98 and $\overline{X}$ + 0.98. For the Mighty Gumball sample, $\overline{x}$ is given by 62.7. Use this to come up with values for the confidence limits.

The confidence limits are given by $\overline{X}$ – 0.98 and $\overline{X}$ + 0.98. If we substitute in the mean of the sample, we get confidence limits of 62.7 – 0.98 and 62.7 + 0.98. In other words, our confidence interval is (61.72, 63.68).

You've found the confidence interval

Congratulations! You've found your first confidence interval. You found that there's a 95% chance that the interval (61.72, 63.68) contains the population mean for flavor duration.

That's fantastic news! That means I can update the fine print on our advertisements. That should handle any lawsuits.

Using confidence intervals in the television advertisement rather than point estimators means that the CEO can give an accurate and precise estimate for how long flavor lasts, but without having to give a precise figure. It makes allowances for any margin of error there might be in the sample.

Let's summarize the steps

Let's look back at the steps we went through in order to construct the confidence interval.

The first thing we did was **choose the population statistic** that we needed to construct a confidence interval for. We needed to find a confidence interval for the mean duration of gumball flavor, and this meant that we needed to construct a confidence interval for μ.

Once we'd figured out which population we needed to construct a confidence interval for, we had to **find its sampling distribution**. We found the expectation and variance of the sampling distribution of means, substituting in values for every statistic except for μ. We then figured out that we could use a normal distribution for $\overline{X}$.

After that, we decided on the **level of confidence** we needed for the confidence interval. We decided to use a confidence level of 95%.

Finally, we had to **find the confidence limits** for the confidence interval. We used the level of confidence and sampling distribution to come up with a suitable confidence interval.

So does that mean I have to go through the same process every single time I want to construct a confidence interval?

We can take some shortcuts.

Constructing confidence intervals can be a repetitive process, so there are some shortcuts you can take. It all comes down to the level of confidence you want and the distribution of the test statistic.

Let's take a look at some of the shortcuts we can take.

Handy shortcuts for confidence intervals

Here are some of the shortcuts you can take when you calculate confidence intervals. All you need to do is look at the population statistic you want to find, look at the distribution of the population and the conditions, and then slot in the population statistic or its estimator. The value c depends on the level of confidence

Population statistic	Population distribution	Conditions	Confidence interval
μ	Normal	You know what σ^2 is n is large or small $\bar{x}$ is the sample mean	$\left(\bar{x} - c\frac{\sigma}{\sqrt{n}}, \bar{x} + c\frac{\sigma}{\sqrt{n}}\right)$
μ	Non-normal	You know what σ^2 is n is large (at least 30) $\bar{x}$ is the sample mean	$\left(\bar{x} - c\frac{\sigma}{\sqrt{n}}, \bar{x} + c\frac{\sigma}{\sqrt{n}}\right)$
μ	Normal or non-normal	You don't know what σ^2 is n is large (at least 30) $\bar{x}$ is the sample mean s^2 is the sample variance	$\left(\bar{x} - c\frac{s}{\sqrt{n}}, \bar{x} + c\frac{s}{\sqrt{n}}\right)$
p	Binomial	n is large p_s is the sample proportion q_s is 1 - p_s	$\left(p_s - c\sqrt{\frac{p_s q_s}{n}}, p_s + c\sqrt{\frac{p_s q_s}{n}}\right)$

The value of c depends on the level of confidence you need. These values work whenever you use the normal distribution as the basis of your test statistic.

What's the interval in general?

In general, the confidence interval is given by

statistic ± (margin of error)

Level of confidence	Value of c
90%	1.64
95%	1.96
99%	2.58

The margin of error is given by the value of c multiplied by the standard deviation of the test statistic.

margin of error = c × (standard deviation of statistic)

Exercise

Mighty Gumball took a sample of 50 gumballs and found that in the sample, the proportion of red gumballs is 0.25. Construct a 99% confidence interval for the proportion of red gumballs in the population.

Exercise Solution

Mighty Gumball took a sample of 50 gumballs and found that in the sample, the proportion of red gumballs is 0.25. Construct a 99% confidence interval for the proportion of red gumballs in the population.

The confidence interval for the population proportion is given by

$$\left(p_s - c\sqrt{\frac{p_s q_s}{n}},\ p_s + c\sqrt{\frac{p_s q_s}{n}}\right)$$

We need to find the 99% confidence interval so $c = 2.58$. The proportion of red gumballs is 0.25, so $p_s = 0.25$ and $q_s = 0.75$. $n = 50$. This gives us

$$\left(p_s - c\sqrt{\frac{p_s q_s}{n}},\ p_s + c\sqrt{\frac{p_s q_s}{n}}\right) = \left(0.25 - 2.58\sqrt{\frac{0.25 \times 0.75}{50}},\ 0.25 + 2.58\sqrt{\frac{0.25 \times 0.75}{50}}\right)$$

$$= (0.25 - 2.58 \times 0.0612,\ 0.25 + 2.58 \times 0.0612)$$

$$= (0.25 - 0.158,\ 0.25 + 0.158)$$

$$= (0.092,\ 0.408)$$

there are no Dumb Questions

Q: When we found the expectation and variance for $\bar{X}$ earlier, why did we substitute in the point estimator for σ^2 and not μ?

A: We didn't substitute $\bar{x}$ for μ because we needed to find the confidence interval for μ. We needed to find some sort of expression involving μ that we could use to find the confidence interval.

Q: Why did we use $\bar{x}$ as the value of $\bar{X}$?

A: The distribution of $\bar{X}$ is the sampling distribution of means. You form it by taking every possible sample of size n from the population, and then forming a distribution out of all the sample means.

$\bar{x}$ is the particular value of the mean taken from our sample, so we use it to help us find the confidence interval.

Q: What's the difference between the confidence interval and the confidence level?

A: The confidence interval is the probability that your statistic is contained within the confidence interval. It's normally given as a percentage, for example, 95%. The confidence interval gives the lower and upper limit of the interval itself, the actual range of numbers.

Q: We've found that the 95% confidence interval for μ is (61.72, 63.68). What does that really mean?

A: What it means is that if you were to take many samples of the same size and construct confidence intervals for all of them, then 95% of your confidence intervals would contain the true population mean. You know that 95% of the time, a confidence interval constructed in this way will contain the population mean.

Q: In the shortcuts, do the values of c apply to every confidence interval?

A: They apply to all of the shortcuts we've shown you so far because all of these shortcuts are based on the normal distribution. This is because the sampling distribution in all of these cases follows the normal distribution.

Q: I've sometimes seen "a" instead of "c" in the shortcuts for the confidence intervals. Is that wrong?

A: Not at all. The key thing is that whether you refer to it as "a" or "c", it represents a value that you can substitute into your confidence interval to give you the right confidence level. The values stay the same no matter what you call it.

Q: So are all confidence intervals based on the normal distribution?

A: No, they're not. We'll look at intervals based on other distributions later on.

Q: Why did we go through all those steps when all we have to do is slot values into the shortcuts?

A: We went through the steps so that you could see what was going on underneath and understand how confidence intervals are constructed. Most of the time, you'll just have to substitute in values.

Q: Do I need continuity corrections when I'm working with confidence intervals?

A: Theoretically, you do, but in practice, they're generally omitted. This means that you can just substitute values into the shortcuts to come up with confidence intervals.

I have one more problem I need your help with. Think you can help me out?

Just one more problem...

Mighty Gumball has one last problem for you to sort out. One of the candy stores selling gumballs wants to determine how much gumballs typically weigh, as they find that their customers often buy gumballs based on weight rather than quantity. If the store can figure out the typical weight of a gumball, they can use this information to boost sales.

That means I need **you** to come up with a confidence interval for gumball weight, but as it's just for one store, I don't want to sample a large number of gumballs.

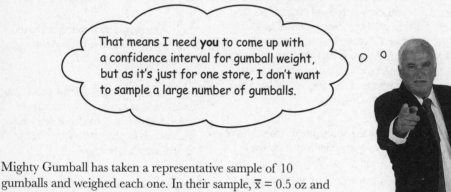

Mighty Gumball has taken a representative sample of 10 gumballs and weighed each one. In their sample, $\overline{x} = 0.5$ oz and $s^2 = 0.09$.

How do we find the confidence interval?

Step 1: Choose your population statistic

The first step is to pick the statistic we want to construct a confidence interval for. We want to construct a confidence interval for the mean weight of gumballs, so we need to construct a confidence interval for the population mean, μ.

As we need to find the confidence interval for μ, this means that the next step is to find its sampling distribution, the distribution of $\overline{X}$.

BRAIN POWER

Assuming the weight of each gumball in the population follows a normal distribution, how would you go about creating a 95% confidence interval for this data? Hint: look at the table of confidence interval shortcuts and see which situation we have here.

Step 2: Find its sampling distribution

So what's the distribution of $\overline{X}$?

That's easy. X is normal, so

The normal distribution isn't a good approximation for every situation.

All of the sampling distributions we've seen so far either follow a normal distribution or can be approximated by it. The trouble is that we can't use the normal distribution for every single confidence interval. Unfortunately, this situation is one of them.

So why can't we use the normal distribution here?

When sample sizes are large, the normal distribution is ideal for finding confidence intervals. It gives accurate results, irrespective of how the population itself is distributed.

Here we have a different situation. Even though X itself is distributed normally, $\overline{X}$ isn't.

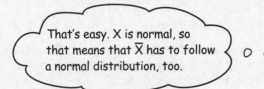

But why not?
That doesn't
make sense to me.

There are two key reasons.

The first is that we don't know what the true variance is of the population, so this means we have to estimate σ^2 using the sample data. We can easily do this using point estimators, but there's a problem: the size of the sample is so small that there are likely to be significant errors in our estimate, much larger errors than if we used a larger sample of gumballs. The potential errors we're dealing with mean that the normal distribution won't give us accurate enough probabilities for $\overline{X}$, which means it won't give us an accurate confidence interval.

So what sort of distribution does $\overline{X}$ follow? It actually follows a ***t-distribution***. Let's find out more.

$\overline{X}$ follows the t-distribution when the sample is small

The t-distribution is a probability distribution that specializes in exactly the sort of situation we have here. It's the distribution that $\overline{X}$ follows where the population is normal, σ^2 is unknown, and you only have a small sample at your disposal.

The t-distribution looks like a smooth, symmetrical curve, and it's exact shape depends on the size of the sample. When the sample size is large, it looks like the normal distribution, but when the sample size is small, the curve is flatter and has slightly fatter tails. It takes one parameter, ν, where ν is equal to $n - 1$. n is the size of the sample, and ν is called the ***number of degrees of freedom***.

We'll look at degrees of freedom in more depth in Chapter 14.

Let's take a look at this. Here's a sketch of the t-distribution for different values of ν. Can you see how the value of ν affects the shape of the distribution?

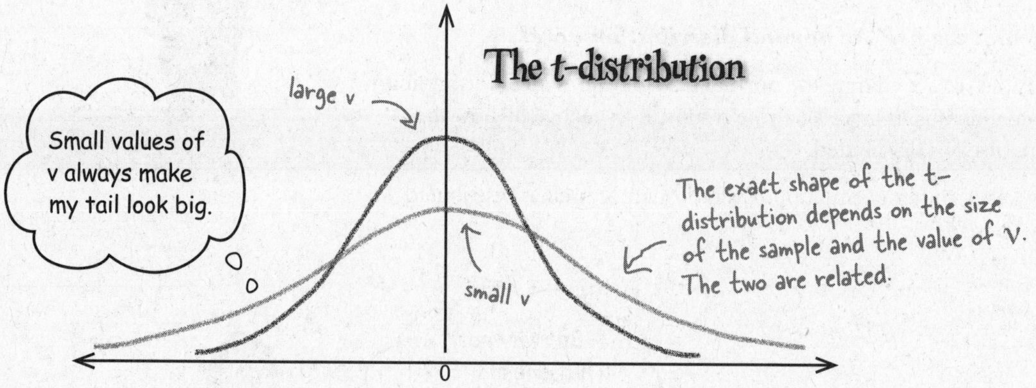
The t-distribution
large v
small v
0
Small values of v always make my tail look big.
The exact shape of the t-distribution depends on the size of the sample and the value of v. The two are related.

A shorthand way of saying that T follows the t-distribution with ν degrees of freedom is

$$T \sim t(\nu)$$

T is the test statistic. You'll see how to calculate it on the next page

$t(\nu)$ means we're using the t-distribution with ν degrees of freedom. $\nu = n - 1$.

The t-distribution works in a similar way to the normal distribution. We start off by converting the limit of the probability area into a standard score, and then we use probability tables to get the result we want.

Let's start with the standard score.

Find the standard score for the t-distribution

We calculate the standard score for the t-distribution in the same way we did for the normal distribution. As with the the normal distribution, we standardize by subtracting the expectation of the sampling distribution and then dividing by its standard deviation. The only difference is that we represent the result with T instead of Z, as we're going to use it with the t-distribution.

We need to find the distribution of $\overline{X}$, so this means we need to use the expectation and standard deviation of $\overline{X}$. The expectation of $\overline{X}$ is μ, and the standard deviation is σ/n. As we need to estimate the value of σ with s, this means that the standard score for the t-distribution is given by

$$T = \frac{\overline{X} - \mu}{s/\sqrt{n}}$$

This is the same formula as for Z—subtract the mean and divide by the standard deviation.

This is the population mean we're finding a confidence interval for.

This is the standard deviation of $\overline{X}$.

All we need to do is substitute in the values for $\overline{X}$, $\hat{\sigma}$, and n.

Sharpen your pencil

Let's see if you can apply this to the Mighty Gumball sample. There are 10 gumballs in the sample, where $\overline{x} = 0.5$oz and $s^2 = 0.09$. What's the value of v and what's T?

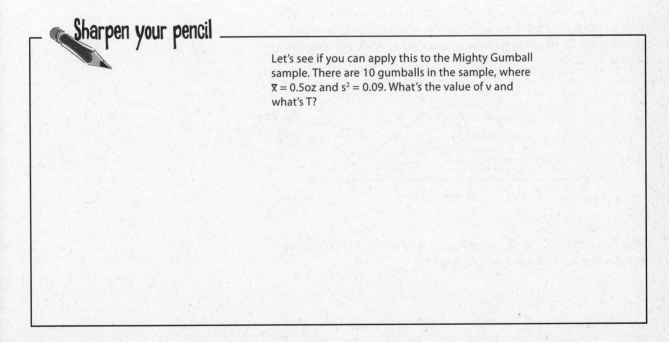

Sharpen your pencil Solution

Let's see if you can apply this to the Mighty Gumball sample. There are 10 gumballs in the sample, where $\overline{x} = 0.5$oz and $s^2 = 0.09$. What's the value of v, and what's T?

There are 10 gumballs in the sample, and $v = n - 1$. This means that the value of v is 9.

T is given by

$$T = \frac{\overline{X} - \mu}{s/\sqrt{n}}$$

$$= \frac{\overline{X} - \mu}{\sqrt{0.09/10}}$$

$$= \frac{\overline{X} - \mu}{0.0949}$$

Step 3: Decide on the level of confidence

So what level of confidence should we use for Mighty Gumball? Remember, the level of confidence says how sure you want to be that the confidence interval contains the population statistic, and it helps us figure out how wide the confidence interval needs to be. As before, let's have a confidence level of 95% for the population mean. This means that the probability of the population mean being inside the confidence interval is 0.95.

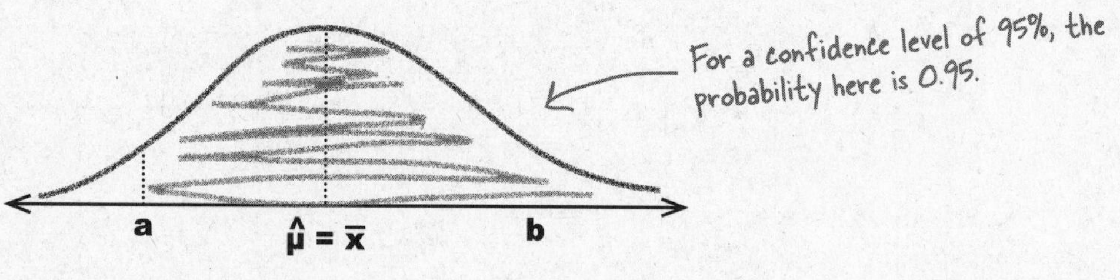
For a confidence level of 95%, the probability here is 0.95.
a
$\hat{\mu} = \overline{x}$
b

Now that we have the level of confidence, we can move onto the final step, finding the confidence interval for μ.

Step 4: Find the confidence limits

You find confidence limits with the t-distribution in a similar way to how you find them with the normal distribution. Your confidence interval is given by

$$\left(\bar{x} - t\frac{s}{\sqrt{n}},\ \bar{x} + t\frac{s}{\sqrt{n}}\right)$$

This is the same as we had before, just replace c with t.

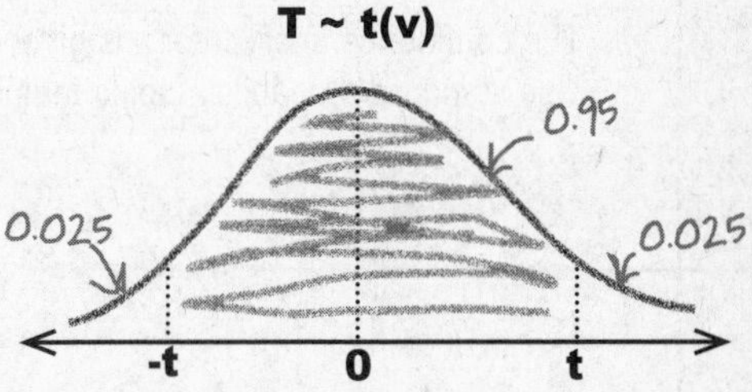

where

$$P(-t \le T \le t) = 0.95$$

This is 0.95, as we want to find the 95% confidence interval.

We can find the value of *t* using t-distribution probability tables.

Using t-distribution probability tables

t-distribution probability tables give you the value of t where P(T > t) = p. In our case, p = 0.025.

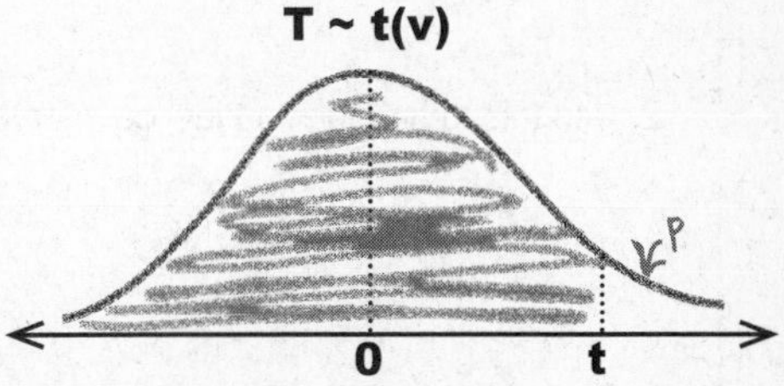

To find t, use the first column to look up ν, and the top row to look up p. The place where they intersect gives the value of t. As an example, if we look up ν = 7 and p = 0.05, we get t = 1.895.

Once you've found the value of *t*, you can use it to find your confidence interval.

p = 0.05

	Tail probability p											
ν	.25	.20	.15	.10	.05	.025	.02	.01	.005	.0025	.001	.0005
1	1.000	1.376	1.963	3.078	[illegible]	12.71	15.89	31.82	63.66	127.3	318.3	636.6
2	.816	1.061	1.386	1.886	[illegible]	4.303	4.849	6.965	9.925	14.09	22.33	31.60
3	.765	.978	1.250	1.638	[illegible]	3.182	3.482	4.541	5.841	7.453	10.21	12.92
4	.741	.941	1.190	1.533	[illegible]	2.776	2.999	3.747	4.604	5.598	7.173	8.610
5	.727	.920	1.156	1.476	[illegible]	2.571	2.757	3.365	4.032	4.773	5.893	6.869
6	.718	.906	1.134	1.440	[illegible]	2.447	2.612	3.143	3.707	4.317	5.208	5.959
7	[illegible]	[illegible]	[illegible]	[illegible]	1.895	2.365	2.517	2.998	3.499	4.029	4.785	5.408
8	.706	.889	1.108	1.397	[illegible]	2.306	2.449	2.896	3.355	3.833	4.501	5.041
9	.703	.883	1.100	1.383	[illegible]	2.262	2.398	2.821	3.250	3.690	4.297	4.781
10	.700	.879	1.093	1.372	[illegible]	2.228	2.359	2.764	3.169	3.581	4.144	4.587

ν = 7

This is where 7 and .05 meet.

See if you can find the 95% confidence interval for the average weight of gumballs. There are 10 gumballs in the sample where $\overline{x} = 0.5oz$ and $s^2 = 0.09$.

1. The confidence interval for μ is given by $(\overline{x} - t\, s/\sqrt{n}, \overline{x} + t\, s/\sqrt{n})$.
   Use standard probability tables to find the value of t.

2. Use this to find the confidence interval for μ.

The t-distribution vs. the normal distribution

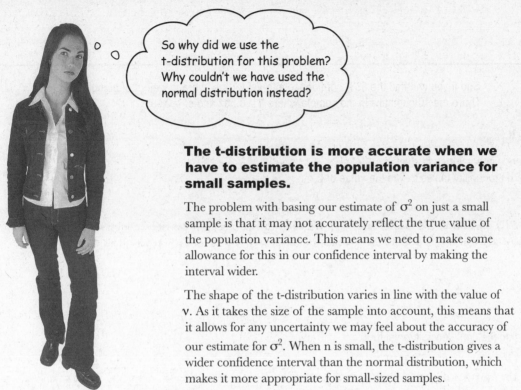

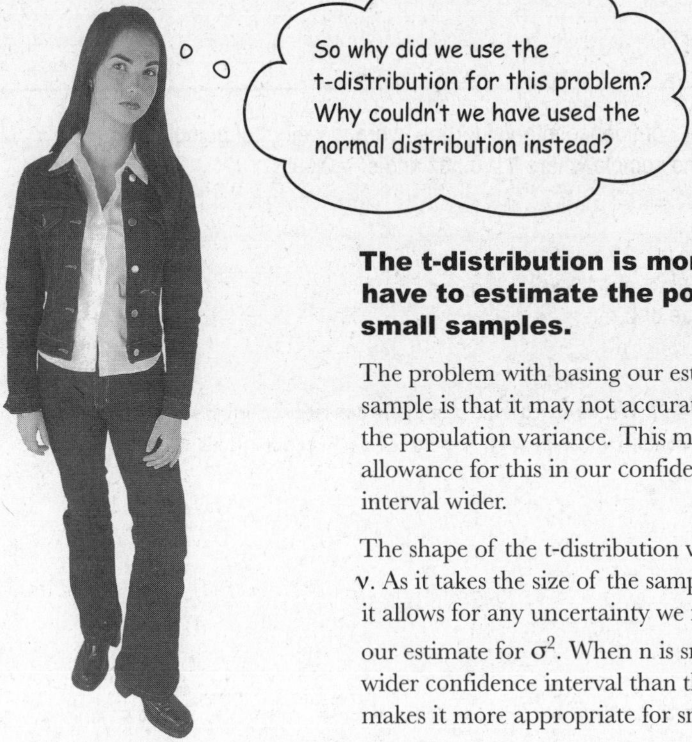

So why did we use the t-distribution for this problem? Why couldn't we have used the normal distribution instead?

The t-distribution is more accurate when we have to estimate the population variance for small samples.

The problem with basing our estimate of σ^2 on just a small sample is that it may not accurately reflect the true value of the population variance. This means we need to make some allowance for this in our confidence interval by making the interval wider.

The shape of the t-distribution varies in line with the value of ν. As it takes the size of the sample into account, this means that it allows for any uncertainty we may feel about the accuracy of our estimate for σ^2. When n is small, the t-distribution gives a wider confidence interval than the normal distribution, which makes it more appropriate for small-sized samples.

Handy shortcuts for confidence intervals - the t-distribution

Here's a quick reminder of when you need to use the t-distribution, and what the confidence interval is for μ. Just substitute in your values.

Population statistic	Population distribution	Conditions	Confidence interval
μ	Normal or non-normal	You don't know what σ^2 is n is small (less than 30) $\bar{x}$ is the sample mean s^2 is the sample variance	$\left(\bar{x} - t(\nu)\frac{s}{\sqrt{n}}, \bar{x} + t(\nu)\frac{s}{\sqrt{n}}\right)$

To find $t(\nu)$, you need to look it up in t-distribution probability tables. To do this, use $\nu = n - 1$ and your level of confidence to find the critical region.

Exercise Solution

See if you can find the 95% confidence interval for the average weight of gumballs. There are 10 gumballs in the sample where $\bar{x}$ = 0.5oz and s^2 = 0.09.

1. The confidence interval for μ is given by $(\bar{x} - t\, s/\sqrt{n},\ \bar{x} + t\, s/\sqrt{n})$.
   Use standard probability tables to find the value of t.

There are 10 gumballs in the sample, so $\nu = 9$. We want to find the 95% confidence interval, so this means we look up 0.025 in the t-distribution probability table, with 9 degrees of freedom. This gives us t = 2.262.

2. Use this to find the confidence interval for μ.

We find the confidence interval by substituting values for $\bar{x}$, t, s, and n into $(\bar{x} - t\, s/\sqrt{n},\ \bar{x} + t\, s/\sqrt{n})$. This gives us

$$
\begin{aligned}
(\bar{x} - t\, s/\sqrt{n},\ \bar{x} + t\, s/\sqrt{n}) &= (0.5 - 2.262 \times \sqrt{(0.09/10)},\ 0.5 + 2.262 \times \sqrt{(0.09/10)}) \\
&= (0.5 - 2.262 \times 0.0949,\ 0.5 + 2.262 \times 0.0949) \\
&= (0.5 - 0.215,\ 0.5 + 0.215) \\
&= (0.285,\ 0.715)
\end{aligned}
$$

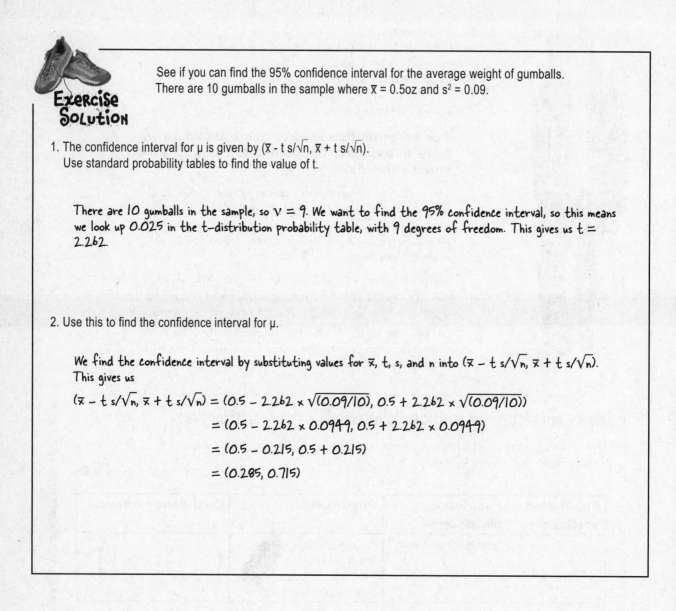

Exercise

Mighty Gumball has noticed a problem with their gumball dispensers. They have taken a sample of 30 machines, and found that the mean number of malfunctions is 15. Construct a 99% confidence interval for the number of malfunctions per month.

Exercise Solution

Mighty Gumball has noticed a problem with their gumball dispensers. They have taken a sample of 30 machines, and found that the mean number of malfunctions is 15. Construct a 99% confidence interval for the number of malfunctions per month.

The number of breakdowns per month is modelled by a Poisson distribution. As there are 30 machines, we can find the confidence interval using $(\bar{x} - cs/\sqrt{n}, \bar{x} + cs/\sqrt{n})$.

We need to find the 99% confidence interval, which means that $c = 2.58$. For the poisson distribution, the expectation and variance are both equal to λ, so $\bar{x} = 15$ and $s^2 = 15$.

The confidence interval is given by

$$
\begin{aligned}
(\bar{x} - cs/\sqrt{n}, \bar{x} + cs/\sqrt{n}) &= (15 - 2.58 \times \sqrt{(15/30)}, 15 + 2.58 \times \sqrt{(15/30)}) \\
&= (15 - 2.58 \times \sqrt{(15/30)}, 15 + 2.58 \times \sqrt{(15/30)}) \\
&= (15 - 2.58 \times 0.707, 15 + 2.58 \times 0.707) \\
&= (15 - 1.824, 15 + 1.824) \\
&= (13.176, 16.824)
\end{aligned}
$$

there are no Dumb Questions

Q: Does $\bar{X}$ follow a t-distribution?

A: $\bar{X}$ follows a t-distribution when the population is normal, the sample size is small, and you need to estimate the population variance using the sample data.

Q: In general, what happens to my confidence interval if the confidence level changes?

A: If your confidence level goes down, then your confidence interval gets narrower. If your confidence level goes up, then your confidence interval gets wider. As an example, a 95% confidence interval will be narrower than a 99% confidence interval for the same set of data.

Q: What happens to the confidence interval if the size of the sample, n, changes?

A: If n decreases, then your confidence interval gets wider, and if n increases, your confidence interval gets narrower.

Confidence intervals take the form

statistic ± margin of error

where the margin of error is equal to c times the standard deviation of the statistic.

The standard deviation of the statistic depends on the size of the sample, and it gets smaller as n gets larger. In other words, the margin of error gets smaller as n gets larger, and larger as n gets smaller.

In general, a smaller sample leads to a wider confidence interval, and a larger sample to a narrower one.

You've found the confidence intervals!

You've made a lot of progress in this chapter, and the result of it is that you now know two ways of estimating population statistics.

The first way of estimating population statistics is to use **point estimators**. Point estimators give you a way of estimating the precise value for the population statistics. It's the best guess you can possibly make based on the sample data.

You also know how to come up with **confidence intervals** for the population statistics. Rather than come up with a very precise estimate for the population statistics, you now know how to find a range of values for the population statistic that you can feel truly confident about.

You're great! I'll tell the candy shop what the confidence interval is for the mean weight of gumballs, as that's just what they wanted to know. They'll be able to sell more gumballs to their customers, and that means increased profits!

13 using hypothesis tests

Look At The Evidence

I thought you said these were **real** horses.

Not everything you're told is absolutely certain.

The trouble is, how do you know when what you're being told isn't right? ***Hypothesis tests*** give you a way of using samples to test whether or not statistical claims are likely to be true. They give you a way of ***weighing the evidence*** and testing whether extreme results can be explained by ***mere coincidence***, or whether there are darker forces at work. Come with us on a ride through this chapter, and we'll show you how you can use hypothesis tests to confirm or allay your deepest suspicions.

Statsville's new miracle drug

Statsville's leading drug company has produced a new remedy for curing snoring. Frustrated snorers are flocking to their doctors in hopes of finding nightly relief.

The drug company claims that their miracle drug cures 90% of people within two weeks, which is great news for the people with snoring difficulties. The trouble is, not everyone's convinced.

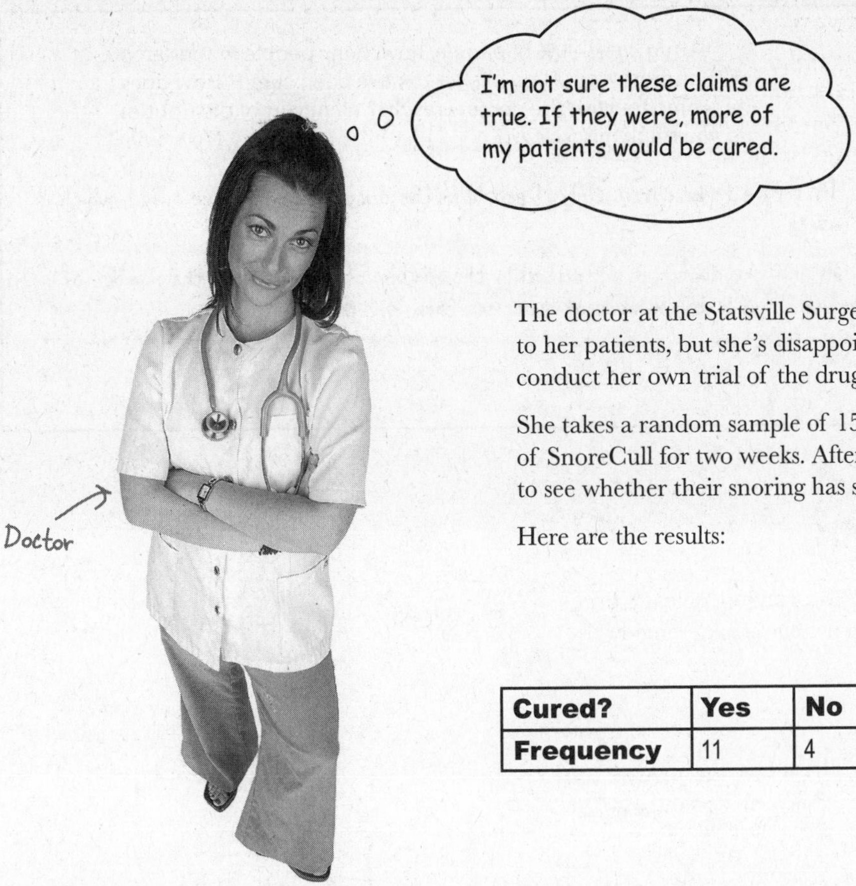

I'm not sure these claims are true. If they were, more of my patients would be cured.
Doctor

The doctor at the Statsville Surgery has been prescribing SnoreCull to her patients, but she's disappointed by the results. She decides to conduct her own trial of the drug.

She takes a random sample of 15 snorers and puts them on a course of SnoreCull for two weeks. After two weeks, she calls them back in to see whether their snoring has stopped.

Here are the results:

Cured?	Yes	No
Frequency	11	4

All the doctor records is whether or not the patients snoring has been cured.

Sharpen your pencil

If the drug cures 90% of people, how many people in the sample of 15 snorers would you expect to have been cured? What sort of distribution do you think this follows?

Sharpen your pencil Solution

If the drug cures 90% of people, how many people in the sample of 15 snorers would you expect to have been cured? How does this compare with the doctors results? What sort of distribution do you think this follows?

90% of 15 is 13.5, so you'd expect 14 people to be cured. Only 11 people in the doctors sample were cured, which is much lower than the result you'd expect

There are a specific number of trials and the doctor is interested in the number of successes, so the number of successes follows a binomial distribution. If X is the number of successes then $X \sim B(15, 0.9)$.

So what's the problem?

Here's the probability distribution for how many people the drug company says should have been cured by the snoring remedy.

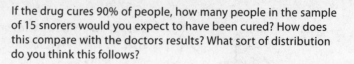

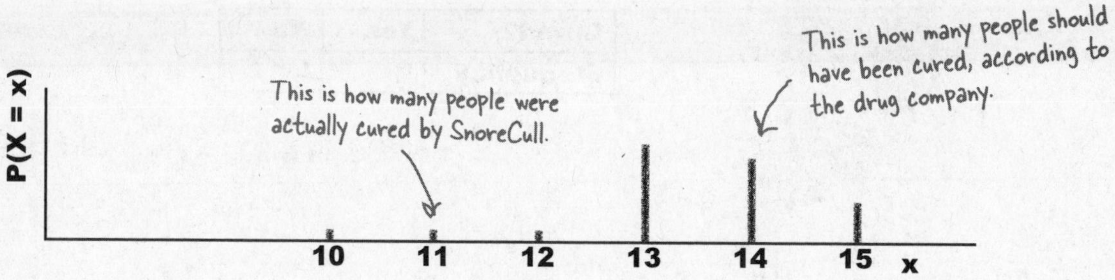
This is how many people should have been cured, according to the drug company.
This is how many people were actually cured by SnoreCull.
P(X = x)
10
11
12
13
14
15
x

The number of people cured by SnoreCull in the doctor's sample is actually much lower than you'd expect it to be. Given the claims made by the drug company, you'd expect 14 people to be cured, but instead, only 11 people have been.

So why the discrepancy?

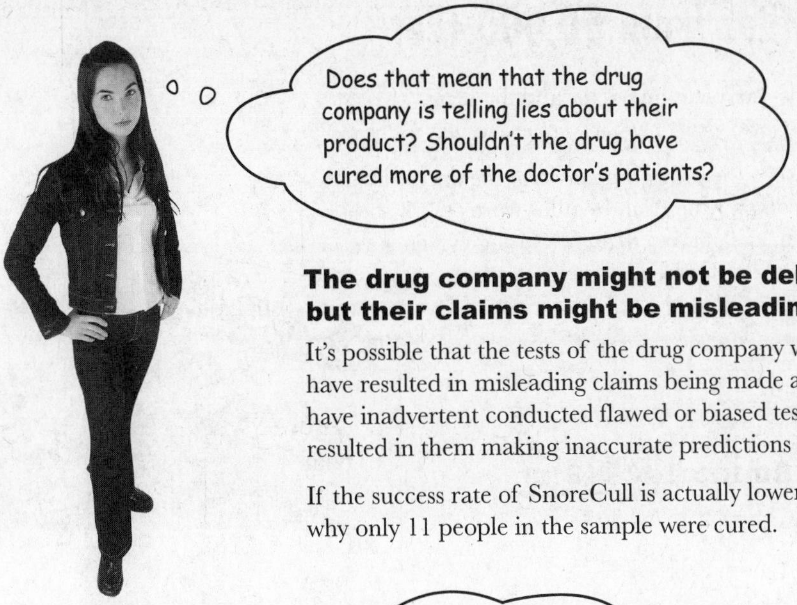

Does that mean that the drug company is telling lies about their product? Shouldn't the drug have cured more of the doctor's patients?

The drug company might not be deliberately telling lies, but their claims might be misleading.

It's possible that the tests of the drug company were flawed, and this might have resulted in misleading claims being made about SnoreCull. They may have inadvertent conducted flawed or biased tests on SnoreCull, which resulted in them making inaccurate predictions about the population.

If the success rate of SnoreCull is actually lower than 90%, this would explain why only 11 people in the sample were cured.

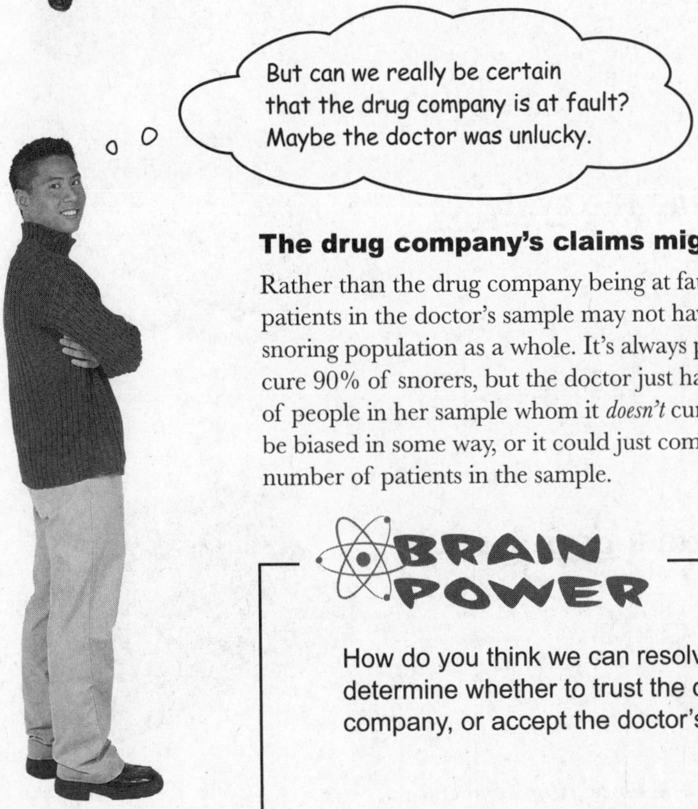

But can we really be certain that the drug company is at fault? Maybe the doctor was unlucky.

The drug company's claims might actually be accurate.

Rather than the drug company being at fault, it's always possible that the patients in the doctor's sample may not have been representative of the snoring population as a whole. It's always possible that the snoring remedy *does* cure 90% of snorers, but the doctor just happens to have a higher proportion of people in her sample whom it *doesn't* cure. In other words, her sample might be biased in some way, or it could just come down to there being a small number of patients in the sample.

BRAIN POWER

How do you think we can resolve this? How can we determine whether to trust the claims of the drug company, or accept the doctor's doubts instead?

Resolving the conflict from 50,000 feet

So how do we resolve the conflict between the doctor and the drug company? Let's take a very high level view of what we need to do.

We can resolve the conflict between the drug company and the doctor by putting the claims of the drug company on trial. In other words, we'll accept the word of the drug company by default, but if there's strong evidence against it, we'll side with the doctor instead.

Here's what we'll do:

SNORECULL
SNORECULL TABLETS: THE ULTIMATE REMEDY FOR SNORING
SNORECULL TABLETS: THE ULTIMATE REMEDY FOR SNORING
90% SUCCESS RATE!

Examine the claim

Take the claim of the drug company.

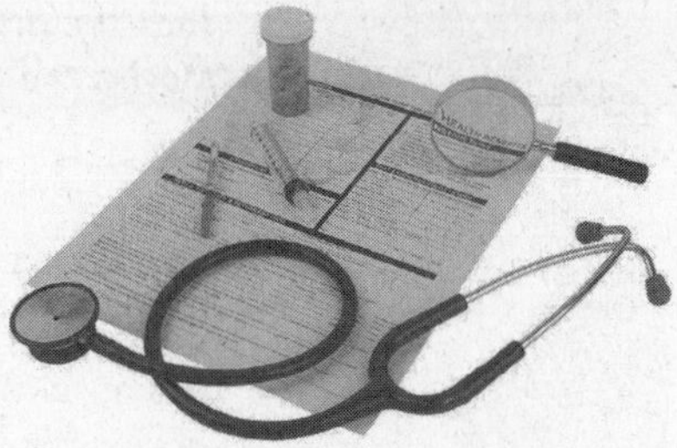

Examine the evidence

See how much evidence we need to reject the drug company's claim, and check this against the evidence we have. We do this by looking at how rare the doctors results would be if the drug company is correct.

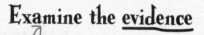

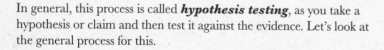

Make a decision

Depending on the evidence, accept or reject the claims of the drug company.

In general, this process is called ***hypothesis testing***, as you take a hypothesis or claim and then test it against the evidence. Let's look at the general process for this.

The six steps for hypothesis testing

Here are the broad steps that are involved in hypothesis testing. We'll go through each one in detail in the following pages.

This is the claim that we're putting on trial → **1 Decide on the hypothesis you're going to test**

2 Choose your test statistic ← We need to pick the statistic that best tests the claim.

We need a certain level of certainty. → **3 Determine the critical region for your decision**

4 Find the p-value of the test statistic ← We need to see how rare our results are, assuming the claims are true.

5 See whether the sample result is within the critical region ← We then see if it's within our bounds of certainty.

6 Make your decision

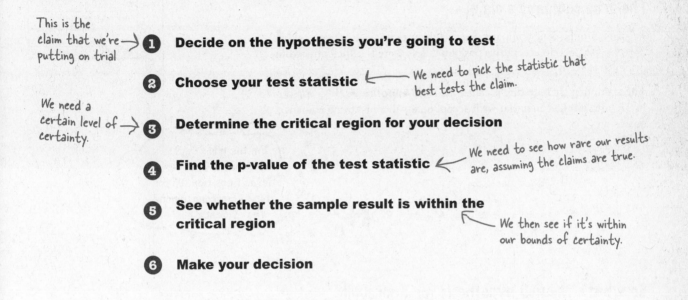

Why all the formality? It's obvious there's something going on.

We need to make sure we properly test the drug claim before we reject it.

That way we'll know we're making an impartial decision either way, and we'll be giving the claim a fair trial. What we *don't* want to to do is reject the claim if there's insufficient evidence against it, and this means that we need some way of deciding what constitutes sufficient evidence.

Step 1: Decide on the hypothesis

You are Here →

Decide on the hypothesis you're going to test
Choose your test statistic
Determine the critical region for your decision
Find the p-value of the test statistic
See whether the sample result is within the critical region
Make your decision

Let's start with step one of the hypothesis test, and look at the key claim we want to test. This claim is called a ***hypothesis***.

The drug company's claim

According to the drug company, SnoreCull cures 90% of patients within 2 weeks. We need to accept this position unless there is sufficiently strong evidence to the contrary.

The claim that we're testing is called the ***null hypothesis***. It's represented by H_0, and it's the claim that we'll accept unless there is strong evidence against it.

The null hypothesis is the claim you're going to test. It's the claim you'll accept unless there's strong evidence against it. → H_0

I'm the null hypothesis. I'm the default position. If you think I'm wrong, gimme the evidence.

So what's the null hypothesis for SnoreCull?

The null hypothesis for SnoreCull is the claim of the drug company: that it cures 90% of patients. This is the claim that we're going to go along with, unless we find strong evidence against it.

We need to test whether at least 90% of patients are cured by the drug, so this means that the null hypothesis is that p = 90%.

This is the null hypothesis for the SnoreCull trial. → H_0: p = 0.9

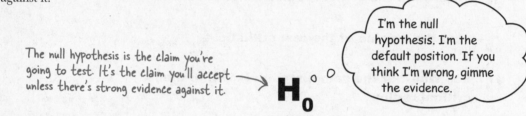

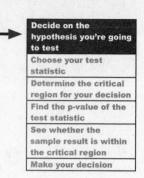

So what's the alternative?

We've looked at what the claim is we're going to test, the null hypothesis, but what if it's not true? What's the alternative?

The doctor's perspective

The doctor's view is that the claims of the drug company are too good to be true. She doesn't think that as many as 90% of patients are cured. She thinks it's far more likely that the cure rate is actually less than 90%.

The counterclaim to the null hypothesis is called the ***alternate hypothesis***. It's represented by H_1, and it's the claim that we'll accept if there's strong enough evidence to reject H_0.

The alternate hypothesis is the claim you'll accept if you reject H_0 → **H_1**

I'm the alternate hypothesis. If H_0 let's you down, then you'll have to accept that you're better off with

The alternate hypothesis for SnoreCull

The alternate hypothesis for SnoreCull is the claim you'll accept if the drug company's claim turns out to be false. If there's sufficiently strong evidence against the drug company, then it's likely that the doctor is right.

The doctor believes that SnoreCull cures less than 90% of people, so this means that the alternate hypothesis is that $p < 90\%$.

This is the alternate hypothesis for the SnoreCull trial → **H_1: $p < 0.9$**

Now that we have the null and alternate hypotheses for the SnoreCull hypothesis test, we can move onto step 2.

there are no Dumb Questions

Q: Why are we assuming the null hypothesis is true and then looking for evidence that it's false?

A: When you conduct a hypothesis test, you, in effect, put the claims of the null hypothesis on trial. You give the null hypothesis the benefit of the doubt, but then you reject it if there is sufficient evidence against it. It's a bit like putting a prisoner on trial in front of a jury. You only sentence the prisoner if there is strong enough evidence against him.

Q: Do the null hypothesis and alternate hypothesis have to be exhaustive? Should they cover all possible outcomes?

A: No, they don't. As an example, our null hypothesis is that $p = 0.9$, and our alternate hypothesis is that $p < 0.9$. Neither hypothesis allows for p being greater than 0.9.

Q: Isn't the sample size too small to do this hypothesis test?

A: Even though the sample size is small, we can still perform hypothesis tests. It all comes down to what test statistic you use — and we'll come to that on the next page.

Q: So are hypothesis tests used to prove whether or not claims are true?

A: Hypothesis tests don't give absolute proof. They allow you to see how rare your observed results actually are, under the assumption that your null hypothesis is true. If your results are extremely unlikely to have happened, then that counts as evidence that the null hypothesis is false.

When hypothesis testing, you assume the null hypothesis is <u>true</u>. If there's sufficient evidence against it, you <u>reject it</u> and accept the alternate hypothesis.

Step 2: Choose your test statistic

You are Here →

- Decide on the hypothesis you're going to test
- **Choose your test statistic**
- Determine the critical region for your decision
- Find the p-value of the test statistic
- See whether the sample result is within the critical region
- Make your decision

Now that you've determined exactly what it is you're going to test, you need some means of testing it. You can do this with a ***test statistic***.

The test statistic is the statistic that you use to test your hypothesis. It's the statistic that's most relevant to the test.

What's the test statistic for SnoreCull?

In our hypothesis test, we want to test whether SnoreCull cures 90% of people or more. To test this, we can look at the probability distribution according to the drug company, and see whether the number of successes in the sample is significant.

If we use X to represent the number of people cured in the sample, this means that we can use X as our test statistic. There are 15 people in the sample, and the probability of success according to the drug company is 0.9. As X follows a binomial distribution, this means that the test statistic is actually:

$$X \sim B(15, 0.9)$$

This is the test statistic for our hypothesis test.

We came up with this test statistic back on page 524.

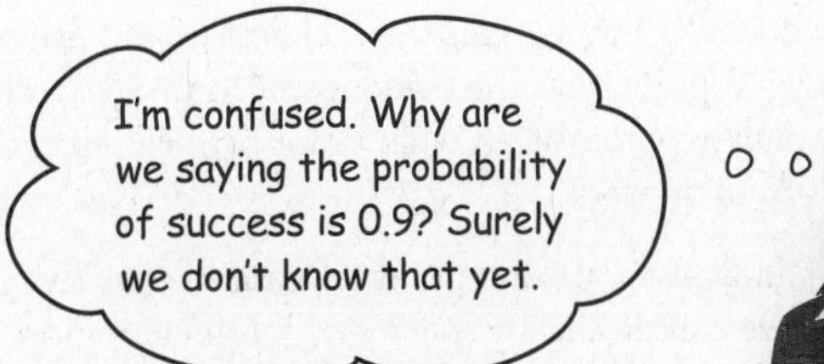
I'm confused. Why are we saying the probability of success is 0.9? Surely we don't know that yet.

We choose the test statistic according to H_0, the null hypothesis.

We need to test whether there is sufficient evidence against the null hypothesis, and we do this by first assuming that H_0 is true. We then look for evidence that contradicts H_0. For the SnoreCull hypothesis test, we assume that the probability of success is 0.9 unless there is strong evidence against this being true.

To do this, we look at how likely it is for us to get the results we did, assuming the probability of success is 0.9. In other words, we take the results of the sample and examine the probability of getting that result. We do this by finding a **critical region**.

Step 3: Determine the critical region

The *critical region* of a hypothesis test is the set of values that present the most extreme evidence against the null hypothesis.

Let's see how this works by taking another look at the doctor's sample. If 90% or more people had been cured, this would have been in line with the claims made by the drug company. As the number of people cured decreases, the more unlikely it becomes that the claims of the drug company are true.

Here's the probability distribution:

You
are
Here

| Decide on the hypothesis you're going to test |
| Choose your test statistic |
| **Determine the critical region for your decision** |
| Find the p-value of the test statistic |
| See whether the sample result is within the critical region |
| Make your decision |

If 90% of people in the sample had been cured, we could easily have assumed that the drug company's claims are correct.

The fewer people who are cured, the more likely it is that the drug company claims are wrong.

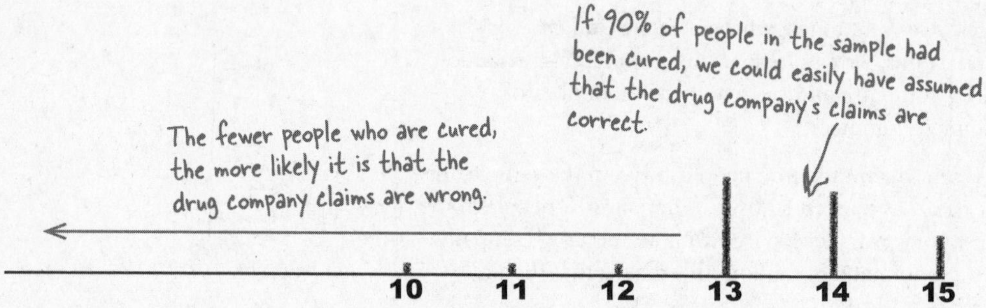

At what point can we reject the drug company claims?

The fewer people there are in the sample who are successfully cured by SnoreCull, the stronger the evidence there is against the claims of the drug company. The question is, at what point does the evidence become so strong that we confidently reject the null hypothesis? At what point can we reject the claim that SnoreCull cures 90% of snorers?

What we need is some way of indicating at what point we can reasonably reject the null hypothesis, and we can do this by specifying a critical region. If the number of snorers cured falls within the critical region, then we'll say there is sufficient evidence to reject the null hypothesis. If the number of snorers cured falls outside the critical region, then we'll accept that there isn't sufficient evidence to reject the null hypothesis, and we'll accept the claims of the drug company. We'll call the cut off point for the critical region c, the *critical value*.

So how do we choose the critical region?

If the number of people cured by SnoreCull falls in this critical region, then we can safely reject the claims of the drug company, H_o. Our alternate hypothesis is that $p < 0.9$, so we look to this region for evidence.

Cutoff point, or critical value

If the number of people cured by SnoreCull falls in this area, then we don't have sufficient evidence to reject H_o. We're only checking whether there's evidence that $p < 0.9$.

c

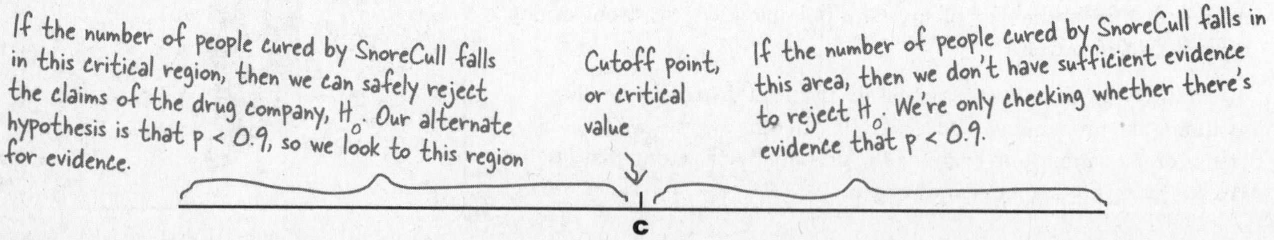

To find the critical region, first decide on the significance level

Before we can find the critical region of the hypothesis test, we first need to decide on the significance level. The ***significance level*** of a test is a measure of how unlikely you want the results of the sample to be before you reject the null hypothesis H_0. Just like the confidence level for a confidence interval, the significance level is given as a percentage.

As an example, suppose we want to test the claims of the drug company at a 5% level of significance. This means that we choose the critical region so that the probability of fewer than *c* snorers being cured is less than 0.05. It's the lowest 5% of the probability distribution.

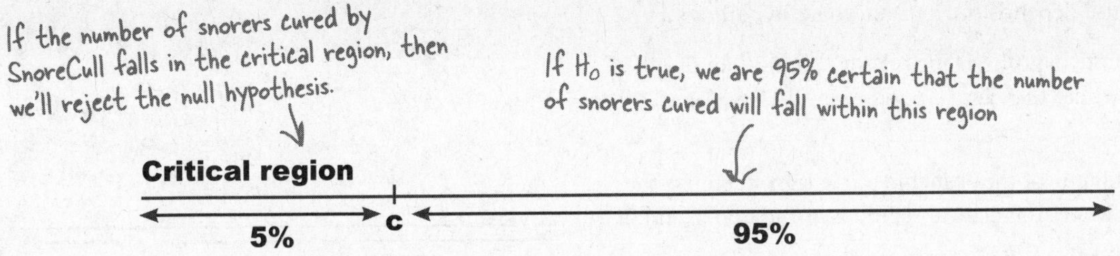
If the number of snorers cured by SnoreCull falls in the critical region, then we'll reject the null hypothesis.
If H0 is true, we are 95% certain that the number of snorers cured will fall within this region
Critical region
c
5%
95%

The significance level is normally represented by the Greek letter α. The lower α is, the more unlikely the results in your sample need to be before we reject H_0.

So what significance level should we use?

Let's use a significance level of 5% in our hypothesis test. This means that if the number of snorers cured in the sample is in the lowest 5% of the probability distribution, then we will reject the claims of the drug company. If the number of snorers cured lies in the top 95% of the probability distribution, then we'll decide there isn't enough evidence to reject the null hypothesis, and accept the claims of the drug company.

If we use X to represent the number of snorers cured, then we define the critical region as being values such that

$$P(X < c) < \alpha$$

where

$$\alpha = 5\%$$

Vital Statistics

Significance level

The significance level is represented by α. It's a way of saying how unlikely you want your results to be before you'll reject H_0.

Critical Regions Up Close

When you're constructing a critical region for your test, another thing you need to be aware of is whether you're conducting a **one-tailed** or **two-tailed** test. Let's look at the difference between the two, and what impact this has on the critical region?

One-tailed tests

A ***one-tailed test*** is where the critical region falls at one end of the possible set of values in your test. You choose the level of the test—represented by α—and then make sure that the critical region reflects this as a corresponding probability.

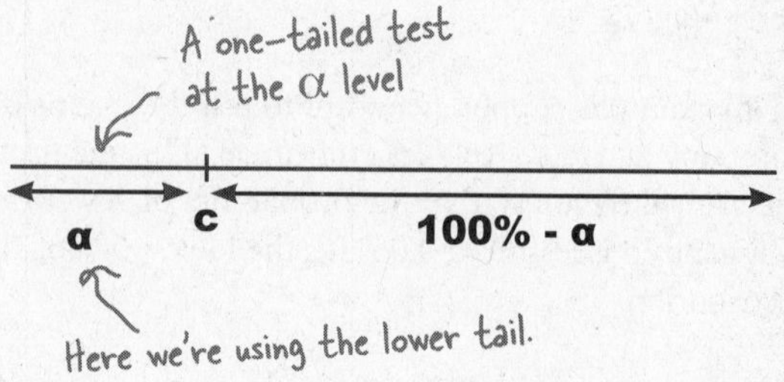

The tail can be at either end of the set of possible values, and the end you use depends on your alternate hypothesis H_1.

If your alternate hypothesis includes a < sign, then use the **lower tail**, where the critical region is at the lower end of the data.

If your alternate hypothesis includes a > sign, then use the **upper tail**, where the critical region is at the upper end of the data.

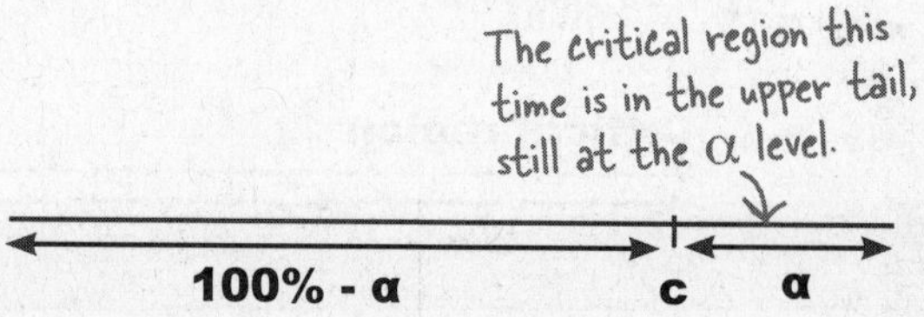

The critical region this time is in the upper tail, still at the α level.
100% - α
c
α

We're using a one-tailed test for the SnoreCull hypothesis test with the critical region in the lower tail, as our alternate hypothesis is that $p < 0.9$.

Two-tailed tests

A ***two-tailed test*** is where the critical region is split over both ends of the set of values. You choose the level of the test α, and then make sure that the overall critical region reflects this as a corresponding probability by splitting it into two. Both ends contain $\alpha/2$, so that the total is α.

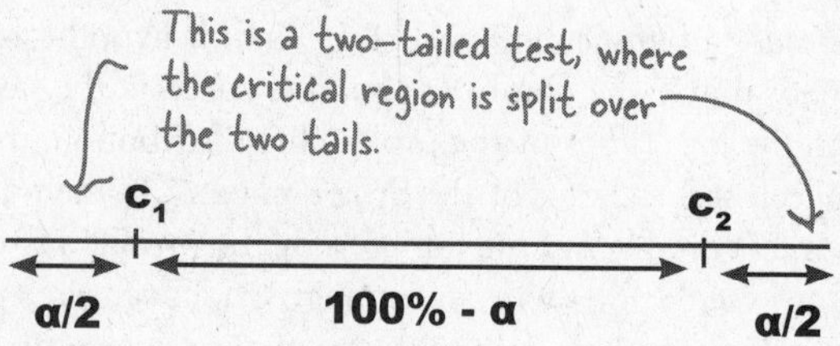

You can tell if you need to use a two-tailed test by looking at the alternate hypothesis H_1. If H_1 contains a ≠ sign, then you need to use a two-tailed test as you are looking for some change in the parameter, rather than an increase or decrease.

We would have used a two-tailed test for our SnoreCull if our alternate hypothesis had been $p \neq 0.9$. We would have had to check whether significantly more or significantly fewer than 90% of patients had been cured

Step 4: Find the p-value

- Decide on the hypothesis you're going to test
- Choose your test statistic
- Determine the critical region for your decision
- **Find the p-value of the test statistic** ← You are Here
- See whether the sample result is within the critical region
- Make your decision

Now that we've looked at critical regions, we can move on to step 4, finding the p-value.

A ***p-value*** is the probability of getting a value up to and including the one in your sample in the direction of your critical region. It's a way of taking your sample and working out whether the result falls within the critical region for your hypothesis test. In other words, we use the p-value to say whether or not we can reject the null hypothesis.

How do we find the p-value?

How we find the p-value depends on our critical region and our test statistic. For the SnoreCull test, 11 people were cured, and our critical region is the lower tail of the distribution. This means that our p-value is $P(X \leq 11)$, where X is the distribution for the number of people cured in the sample.

As the significance level of our test is 5%, this means that if $P(X \leq 11)$ is less than 0.05, then the value 11 falls within the critical region, and we can reject the null hypothesis.

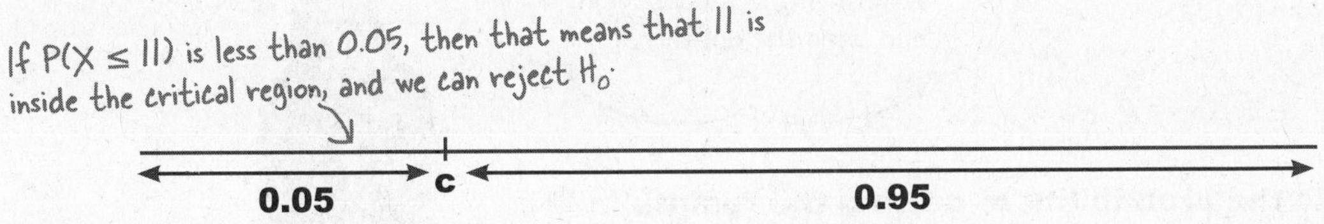
If P(X ≤ 11) is less than 0.05, then that means that 11 is inside the critical region, and we can reject H0.
c
0.05
0.95

Sharpen your pencil

We know from step 2 that X ~ B(15, 0.9). What's $P(X \leq 11)$?

Sharpen your pencil Solution

We know from step 2 that X ~ B(15, 0.9). What's $P(X \leq 11)$? Is 11 inside or outside the critical region?

$$P(X \leq 11) = 1 - P(X \geq 12)$$

$$= 1 - ({}^{15}C_{12} \times 0.1^3 \times 0.9^{12} + {}^{15}C_{13} \times 0.1^2 \times 0.9^{13} + {}^{15}C_{14} \times 0.1 \times 0.9^{14} + 0.9^{15})$$

${}^{15}C_{15} = 1$, and so does 0.1^0, so we're just left with 0.9^{15}.

$$= 1 - (0.1285 + 0.2669 + 0.3432 + 0.2059)$$

$$= 1 - 0.9445$$

$$= 0.0555$$

We've found the p-value

To find the p-value of our hypothesis test, we had to find $P(X \leq 11)$. This means that the p-value is 0.0555.

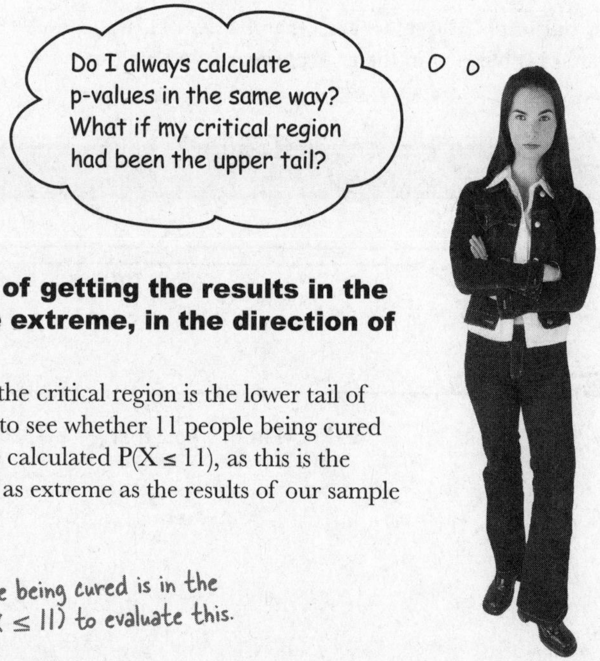
Do I always calculate p-values in the same way? What if my critical region had been the upper tail?

A p-value is the probability of getting the results in the sample, or something more extreme, in the direction of the critical region.

In our hypothesis test for SnoreCull, the critical region is the lower tail of the probability distribution. In order to see whether 11 people being cured of snoring is in the critical region, we calculated $P(X \leq 11)$, as this is the probability of getting a result at least as extreme as the results of our sample in the direction of the lower tail.

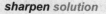

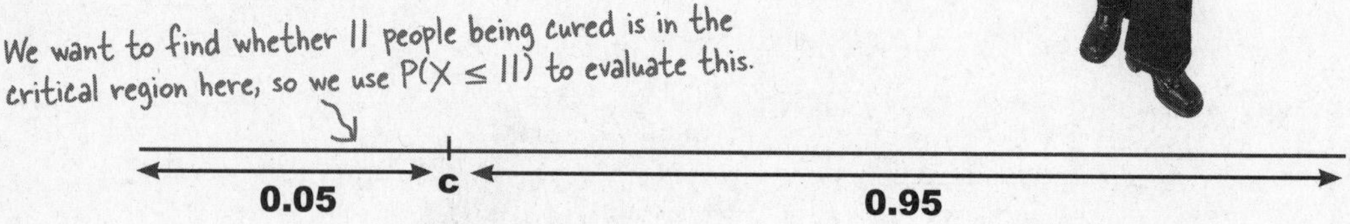
We want to find whether 11 people being cured is in the critical region here, so we use P(X ≤ 11) to evaluate this.
0.05
c
0.95

Had our critical region been the upper tail of the probability distribution instead, we would have needed to find $P(X \geq 11)$. We would have counted more extreme results as being greater than 11, as these would have been closer to the critical region.

Step 5: Is the sample result in the critical region?

Now that we've found the p-value, we can use it to see whether the result from our sample falls within the critical region. If it does, then we'll have sufficient evidence to reject the claims of the drug company.

Our critical region is the lower tail of the probability distribution, and we're using a significance level of 5%. This means that we can reject the null hypothesis if our p-value is less that 0.05. As our p-value is 0.0555, this means that the number of people cured by SnoreCull in the sample doesn't fall within the critical region.

- Decide on the hypothesis you're going to test
- Choose your test statistic
- Determine the critical region for your decision
- Find the p-value of the test statistic
- **See whether the sample result is within the critical region** ← You are Here
- Make your decision

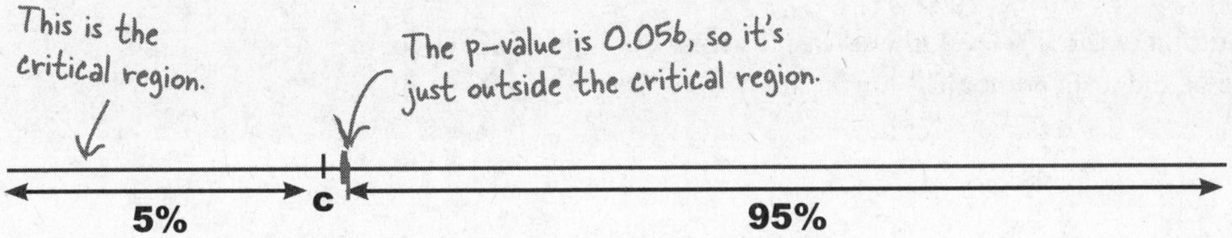
This is the critical region.
The p-value is 0.056, so it's just outside the critical region.
c
5%
95%

Step 6: Make your decision

We've now reached the final step of the hypothesis test. We can decide whether to accept the null hypothesis, or reject it in favor of the alternative.

The p-value of the hypothesis test falls just outside the critical region of the test. This means that there isn't sufficient evidence to reject the null hypothesis. In other words:

- Decide on the hypothesis you're going to test
- Choose your test statistic
- Determine the critical region for your decision
- Find the p-value of the test statistic
- See whether the sample result is within the critical region
- **Make your decision** ← You are Here

We accept the claims of the drug company

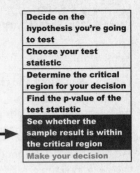

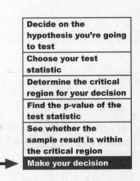

I win!
SnoreCull
SnoreCull Tablets: The Ultimate Remedy for Snoring
90% Success Rate!

So what did we just do?

Let's summarize what we just did.

First of all, we took the claims of the drug company, which the doctor had misgivings about. We used these claims as the basis of a hypothesis test. We formed a null hypothesis that the probability of curing a patient is 0.9, and then we applied this to the number of people in the doctors sample.

We then decided to conduct a test at the 5% level, using the success rate in the doctor's sample. We looked at the probability of 11 people or fewer being cured, and checked to see whether the probability of this was less than 5%, or 0.05. In other words, we looked at the probability of getting a result this extreme, or even more so.

Finally, we found that at the 5% level, there wasn't strong enough evidence to reject the claims of the drug company.

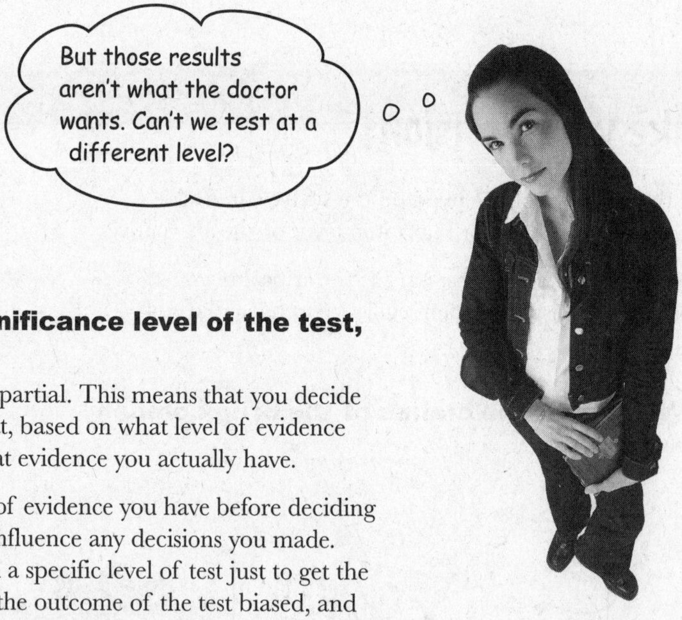

But those results aren't what the doctor wants. Can't we test at a different level?

Once you've fixed the significance level of the test, you can't change it.

The test needs to be completely impartial. This means that you decide what level you need the test to be at, based on what level of evidence you require, before you look at what evidence you actually have.

If you were to look at the amount of evidence you have before deciding on the level of the test, this could influence any decisions you made. You might be tempted to decide on a specific level of test just to get the result you want. This would make the outcome of the test biased, and you might make the wrong decision.

BULLET POINTS

- In a hypothesis test, you take a claim and test it against statistical evidence.
- The claim that you're testing is called the null hypothesis test. It's represented as H_0, and it's the claim that's accepted unless there's strong statistical evidence against it.
- The alternate hypothesis is the claim we'll accept if there's strong enough evidence against H_0. It's represented by H_1.
- The test statistic is the statistic you use to test your hypothesis. It's the statistic that's most relevant to the test. You choose the test statistic by assuming that H_0 is true.
- The significance level is represented by α. It's a way of saying how unlikely you want your results to be before you'll reject H_0.
- The critical region is the set of values that presents the most extreme evidence against the null hypothesis test. You choose your critical region by considering the significance level and how many tails you need to use.
- A one-tailed test is when your critical region lies in either the upper or the lower tail of the data. A two-tailed test is when it's split over both ends. You choose your tail by looking at your alternate hypothesis.
- A p-value is the probability of getting the result of your sample, or a result more extreme in the direction of your critical region.
- If the p-value lies in the critical region, you have sufficient reason to reject your null hypothesis. If your p-value lies outside your critical region, you have insufficient evidence.

there are no Dumb Questions

Q: What significance level should I normally test at?

A: It all depends how strong you want the evidence to be before you reject the null hypothesis. The stronger you want the evidence to be, the lower your significance level needs to be.

The most common significance level is 5%, although you sometimes see tests at the 1% level. Testing at the 1% means that you require stronger evidence than if you test at the 5% level.

Q: Does the significance level have anything in common with the level of confidence for confidence intervals?

A: Yes, they have0 a lot in common. When you construct a confidence interval for a population parameter, you want to have a certain degree of confidence that the population parameter lies between two limits. As an example, if you have a 95% level of confidence, this means that the probability that the population parameter lies between the two limits is 0.95.

The level of significance reflects the probability that values will lie outside a certain limit. As an example, a significance level of 5% means that your critical region must have a probability of 0.05.

I still have doubts. I wonder what would happen if I took a larger sample...

What if the sample size is larger?

So far the doctor has conducted her trial using a sample of just 15 people, and on the basis of this, there was insufficient evidence to reject the claims of the drug company.

It's possible that the size of the sample wasn't large enough to get an accurate result. The doctor might get more reliable results by using a larger sample.

Here are the results from the doctor's new trial:

Cured?	**Yes**	**No**
Frequency	80	20

I want to conduct a new hypothesis test using these new results.

We want to determine whether the new data will make a difference in the outcome of the test.

Let's run through another hypothesis test, this time with the larger sample.

BRAIN POWER

What's the null hypothesis of this new problem? What's the alternate hypothesis?

Hypothesis Magnets

It's time to do another hypothesis test. There are a number of steps you need to run through to perform the hypothesis test, but can you remember what the order is? Put the magnets into the right order.

Make your decision

Choose your test statistic

Decide on the hypothesis you're going to test

Determine the critical region for your decision

Find the p-value of the test statistic

See whether the test statistic is within the critical region

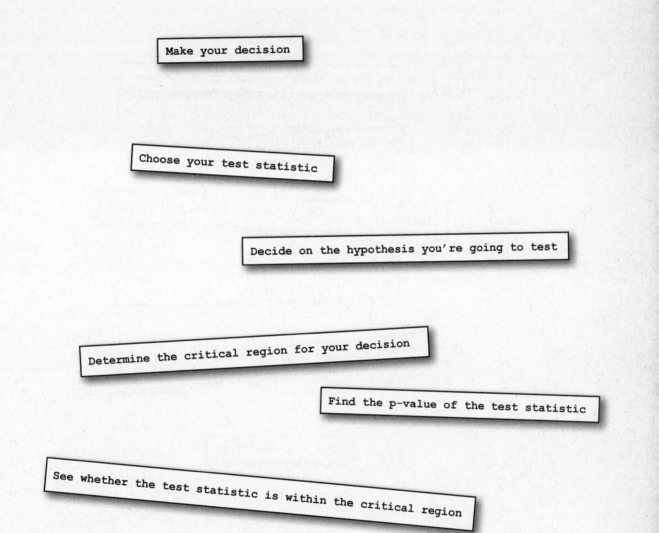

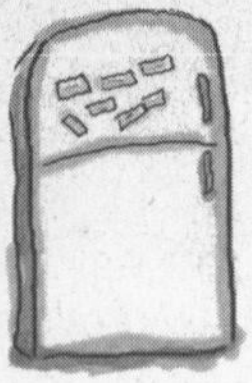

Hypothesis Magnets Solution

It's time to do another hypothesis test. There are a number of steps you need to run through to perform the hypothesis test, but can you remember what the order is? Put the magnets into the right order.

```
Decide on the hypothesis you're going to test
```

```
Choose your test statistic
```

```
Determine the critical region for your decision
```

```
Find the p-value of the test statistic
```

```
See whether the test statistic is within the critical region
```

```
Make your decision
```

- **Decide on the hypothesis you're going to test**
- Choose your test statistic
- Determine the critical region for your decision
- Find the p-value of the test statistic
- See whether the sample result is within the critical region
- Make your decision

Let's conduct another hypothesis test

The doctor still has misgivings about the claims made by the drug company. Let's conduct a hypothesis test based on the new data.

Step 1: Decide on the hypotheses

We need to start off by finding the **null hypothesis** and **alternate hypothesis** of the SnoreCull trial. As a reminder, the null hypothesis is the claim that we're testing, and the alternate hypothesis is what we'll accept if there's sufficient evidence against the null hypothesis.

So what are the null and alternate hypotheses?

It's still the same problem

For the last test, we took the claims made by the drug company and used these as the basis for the null hypothesis. We're testing the same claims, so the null hypothesis is still the same. We have

$$H_0: p = 0.9$$

The alternate hypothesis is the same too. If there is strong evidence against the claims made by the drug company, then we'll accept that the drug cures fewer than 90% of the patients. This gives us an alternate hypothesis of:

$$H_1: p < 0.9$$

So you still don't believe me? Think you can have another shot at me? Bring it on!
SnoreCull
SnoreCull Tablets: The Ultimate Remedy for Snoring
90% Success Rate!

Step 2: Choose the test statistic

As before, the next step is to choose the test statistic. In other words, we need some statistic that we can use to test the hypothesis.

For the previous hypothesis test, we conducted the test by looking at the number of successes in the sample and seeing how significant the result was. We used the binomial distribution to find the probability of getting a result *at least* as extreme as the value we got in the sample. In other words, we used a test statistic of $X \sim B(15, 0.9)$ to test whether $P(X \leq 11)$ was less than 0.05, the level of significance.

This time the number of people in the sample is 100, and we're testing the same claim, that probability of successfully curing someone is 0.9. This means that our new test statistic is $X \sim B(100, 0.9)$.

Decide on the hypothesis you're going to test
Choose your test statistic
Determine the critical region for your decision
Find the p-value of the test statistic
See whether the sample result is within the critical region
Make your decision

> Are you kidding me? If we have to calculate probabilities using the binomial distribution, we'll be here forever.

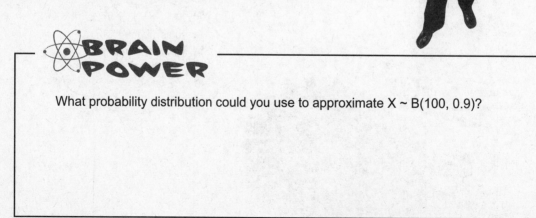

We can use another probability distribution instead of the binomial.

Using the binomial distribution for this sort of problem would be time consuming, as we'd have to calculate lots of probabilities.

Fortunately, there's another way. Rather than use the binomial distribution, we can use some other distribution instead.

BRAIN POWER

What probability distribution could you use to approximate $X \sim B(100, 0.9)$?

Exercise

To get the most out of hypothesis tests, you need to know how different variables and parameters are distributed. What distributions would you use to find probabilities for the following situations?

Hint: We covered all of these earlier in the book. If you get stuck, look back through the chapters

1. $X \sim B(n, p)$. What probability distribution could you use to approximate this if n is large, $np > 5$ and $nq > 5$?

2. $X \sim N(\mu, \sigma^2)$. You know the value of μ and σ^2. What's the distribution of $\overline{X}$?

3. $X \sim N(\mu, \sigma^2)$, and you know what μ is, but you don't know what the value of σ^2 is. The sample size is large. What's the distribution of $\overline{X}$ given the data you have?

4. $X \sim N(\mu, \sigma^2)$, you know what μ is, but you don't know what the value of σ^2 is. The sample size is small. What's the distribution of $\overline{X}$ given the data you have?

To get the most out of hypothesis tests, you need to know how different variables and parameters are distributed. What distributions would you use to find probabilities for the following situations?

Hint: We covered all of these earlier in the book. If you get stuck, look back through the chapters.

1. $X \sim B(n, p)$. What probability distribution could you use to approximate this if n is large, $np > 5$ and $nq > 5$?

If n is large, then we can approximate $X \sim B(n, p)$ using the normal distribution. As $E(X) = np$ and $Var(X) = npq$, this means we can use $X \sim N(np, npq)$. This assumes that $np > 5$ and $nq > 5$.

2. $X \sim N(\mu, \sigma^2)$. You know the value of μ and σ^2. What's the distribution of $\overline{X}$?

If we know what the value is of σ^2, $\overline{X} \sim N(\mu, \sigma^2/n)$.

3. $X \sim N(\mu, \sigma^2)$, and you know what μ is, but you don't know what the value of σ^2 is. The sample size is large. What's the distribution of $\overline{X}$ given the data you have?

If we don't know what the value is of σ^2, we estimate it using s^2. So, $\overline{X} \sim N(\mu, s^2/n)$.

4. $X \sim N(\mu, \sigma^2)$, you know what μ is, but you don't know what the value of σ^2 is. The sample size is small. What's the distribution of $\overline{X}$ given the data you have?

If we don't know what the value is of σ^2, we estimate it using s^2. If the sample size is small, we need to use the t-distribution $T \sim t(n - 1)$ where $T = \dfrac{\overline{X} - \mu}{s/\sqrt{n}}$

Use the normal to approximate the binomial in our test statistic

We still need to find a test statistic we can use in our hypothesis test, and as the number in the sample is large, this means that using the binomial distribution will be time consuming and complicated.

There are 100 people in the sample, and the proportion of successes according to the drug company is 0.9. In other words, the number of successes follows a binomial distribution, where n = 100 and p = 0.9.

As n is large, and both np and nq are greater than 5, we can use X ~ N(np, npq) as our test statistic, where X is the number of patients successfully cured. In other words, we can use

$$X \sim N(90, 9)$$

We can use this because n is large, np > 5 and nq is large.

to approximate any probabilities that we may need.

If we standardize this, we get

$$Z = \frac{X - 90}{\sqrt{9}}$$

Here we're standardizing X ~ N(90, 9).

$$= \frac{X - 90}{3}$$

This means that for our test statistic we can use

$$Z = \frac{X - 90}{3} \qquad Z \sim N(0, 1)$$

X is the number of patients cured, in our case 80.

I get it. So our test statistic is the variable we use for our test.

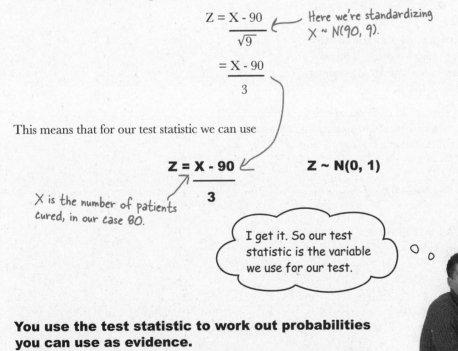

You use the test statistic to work out probabilities you can use as evidence.

This means that we use Z as our test statistic, as we can easily use it to look up probabilities and see how unlikely the results of our sample are given the claims of the drug company. We substitute our value of 80 in place of X, so we can use it to find the probability of 80 or fewer being cured.

Step 3: Find the critical region

- Decide on the hypothesis you're going to test
- Choose your test statistic
- **Determine the critical region for your decision**
- Find the p-value of the test statistic
- See whether the sample result is within the critical region
- Make your decision

Now that we have a test statistic for our test, we need to come up with a critical region. As our alternate hypothesis is $p < 0.9$, this means that our critical region lies in the lower tail just as before.

The critical region also depends on the significance level of the test. Let's choose the same significance level as before, so let's test at the 5% level.

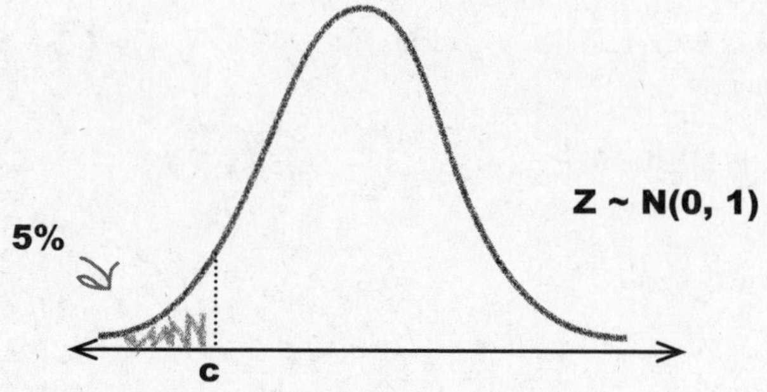

As our test statistic follows a standard normal distribution, we can use probability tables to find the critical value, c. The critical value is the boundary between whether we have strong enough evidence to reject the null hypothesis or not.

As our significance level is 5%, this means that our critical value c is the value where $P(Z < c) = 0.05$. If we look up the probability 0.05 in the probability tables, this gives us a value for c of -1.64. In other words,

$$P(Z < -1.64) = 0.05$$

This means that if our test statistic is less than -1.64, we have strong enough evidence to reject the null hypothesis.

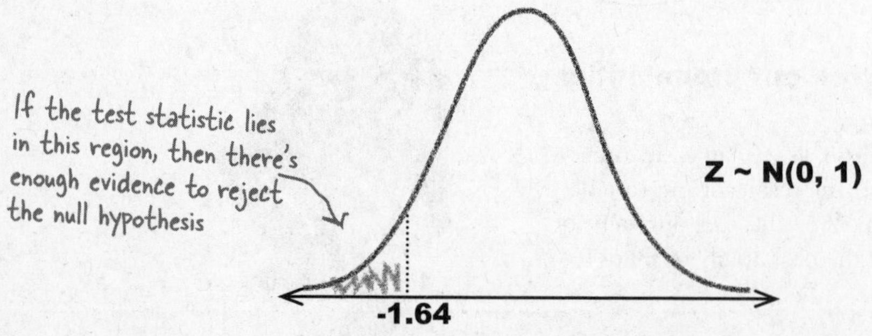

Exercise

Think you can go through the remaining steps of the hypothesis test? See if you can find the following:

Step 4: Find the p-value

The critical region is in the lower tail of the distribution. 80 people were cured, and Z = (X - 90)/3. Use this to find the p-value.

Step 5: See whether the test statistic is within the critical region

Remember that the significance level for the hypothesis test is 5%.

Step 6: Make your decision

Do you accept or reject the null hypothesis based on the evidence?

Think you can go through the remaining steps of the hypothesis test? See if you can find the following:

Step 4: Find the p-value

The critical region is in the lower tail of the distribution. 80 people were cured, and Z = (X - 90)/3. Use this to find the p-value.

Let's start by finding the standard score of 80.

$$z = (80 - 90)/3$$
$$= -10/3$$
$$= -3.33$$

The p-value is given by $P(Z < z) = P(Z < -3.33)$. Looking this up in probability tables gives us

p-value = 0.0004

Step 5: See whether the test statistic is within the critical region

Remember that the significance level for the hypothesis test is 5%.

The test statistic is on the critical region if the p-value is less than 0.05. As the p-value is equal to 0.0004, this means that the test statistic is within the critical region.

Step 6: Make your decision

Do you accept or reject the null hypothesis based on the evidence?

As the test statistic is within the critical region for the hypothesis test, this means that we have sufficient evidence to reject the null hypothesis at the 5% significance level.

SnoreCull failed the test

This time when we performed a hypothesis test on SnoreCull, there was sufficient evidence to reject the null hypothesis. In other words, we can reject the claims made by the drug company.

Busted...
SnoreCull
SnoreCull Tablets: The Ultimate Remedy for Snoring
90% Success Rate!

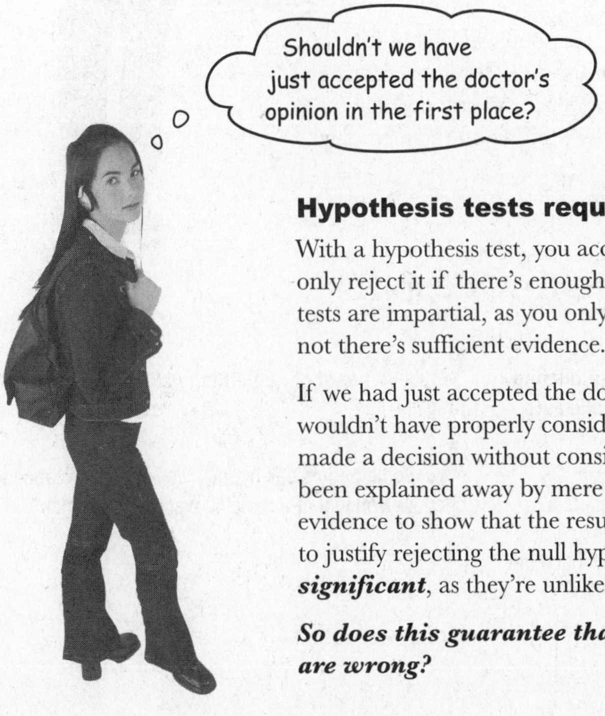
Shouldn't we have just accepted the doctor's opinion in the first place?

Hypothesis tests require evidence.

With a hypothesis test, you accept a claim and then put it on trial. You only reject it if there's enough evidence against it. This means that the tests are impartial, as you only make a decision based on whether or not there's sufficient evidence.

If we had just accepted the doctor's opinion in the first place, we wouldn't have properly considered the evidence. We would have made a decision without considering whether the results could have been explained away by mere coincidence. As it is, we have enough evidence to show that the results of the sample are extreme enough to justify rejecting the null hypothesis. The results are ***statistically significant***, as they're unlikely to have happened by chance.

So does this guarantee that the claims of the drug company are wrong?

Mistakes can happen

So far we've looked at how we can use the results of a sample as evidence in a hypothesis test. If the evidence is sufficiently strong, then we can use it to justify rejecting the null hypothesis.

We've found that there is strong evidence that the claims of the drug company are wrong, but is this guaranteed?

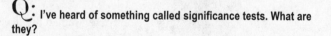

Of course it is. We've done a hypothesis test, and we've used it to prove that the drug company is lying.

Even though the evidence is strong, we can't absolutely guarantee that the drug company claims are wrong.

Even though it's unlikely, we could still have made the wrong decision. We can examine evidence with a hypothesis, and we can specify how certain we want to be before rejecting the null hypothesis, but it doesn't prove with absolute certainty that our decision is right.

The question is, how do we know?

Conducting a hypothesis test is a bit like putting a prisoner on trial in front of a jury. The jury assumes that the prisoner is innocent unless there is strong evidence against him, but even considering the evidence, it's still possible for the jury to make wrong decisions. Have a go at the exercise on the next page, and you'll see how.

there are no Dumb Questions

Q: How can we make the wrong decision if we're conducting a hypothesis test? Don't we do a hypothesis test to make sure we don't?

A: When you conduct a hypothesis test, you can only make a decision based on the evidence that you have. Your evidence is based on sample data, so if the sample is biased, you may make the wrong decision based on biased data.

Q: I've heard of something called significance tests. What are they?

A: Some people call hypothesis tests significance tests. This is because you test at a certain level of significance.

Sharpen your pencil

A prisoner is on trial for a crime, and you're on the jury. The jury's task is to assume the prisoner is innocent, but if there's enough evidence against him, they need to convict him.

1. In the trial, what's the null hypothesis?

2. What's the alternate hypothesis?

3. In what ways can the jury make a verdict that's correct?

4. In what ways can the jury make a verdict that's incorrect?

Sharpen your pencil Solution

A prisoner is on trial for a crime, and you're on the jury. The jury's task is to assume the prisoner is innocent, but if there's enough evidence against him, they need to convict him.

1. In the trial, what's the null hypothesis?

The null hypothesis is that the prisoner is innocent, as that is what we have to assume until there's proof otherwise.

2. What's the alternate hypothesis?

The alternate hypothesis is that the prisoner is guilty. In other words, if there's sufficient proof that the prisoner is not innocent, then we'll accept that he's guilty and convict him.

3. In what ways can the jury make a verdict that's correct?

We can make a correct verdict if:

The prisoner is innocent, and we find him innocent.

The prisoner is guilty, and we find him guilty.

4. In what ways can the jury make a verdict that's incorrect?

We can make an incorrect verdict if

The prisoner is innocent, and we find him guilty.

The prisoner is guilty, and we find him innocent.

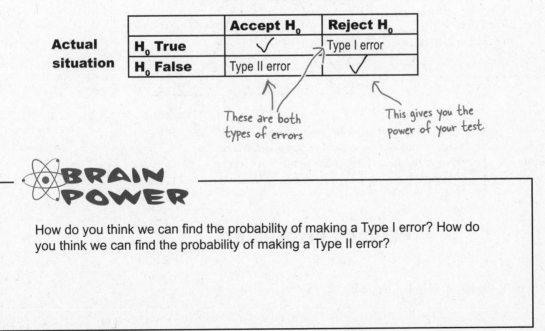

So what does putting prisoners on trial have to do with hypothesis testing?

The errors we can make when conducting a hypothesis test are the same sort of errors we could make when putting a prisoner on trial.

Hypothesis tests are basically tests where you take a claim and put it on trial by assessing the evidence against it. If there's sufficient evidence against it, you reject it, but if there's insufficient evidence against it, you accept it.

You may correctly accept or reject the null hypothesis, but even considering the evidence, it's also possible to make an error. You may reject a valid null hypothesis, or you might accept it when it's actually false.

Statisticians have special names for these types of errors. A ***Type I error*** is when you wrongly reject a true null hypothesis, and a ***Type II error*** is when you wrongly accept a false null hypothesis.

The ***power*** of a hypothesis test is the probability that that you will correctly reject a false null hypothesis.

Decision from hypothesis test

Actual situation		Accept H_0	Reject H_0
	H_0 True	✓	Type I error
	H_0 False	Type II error	✓

These are both types of errors

This gives you the power of your test.

BRAIN POWER

How do you think we can find the probability of making a Type I error? How do you think we can find the probability of making a Type II error?

Let's start with Type I errors

A Type I error is what you get when you reject the null hypothesis when the null hypothesis is actually correct. It's like putting a prisoner on trial and finding him guilty when he's actually innocent.

A Type I error is when you reject H_0 when actually it's correct.
But I'm innocent

So what's the probability of getting a Type I error?

If you get a Type I error, then this means that the null hypothesis must have been rejected. In order for the null hypothesis to have been rejected, the results of your sample must be in the critical region.

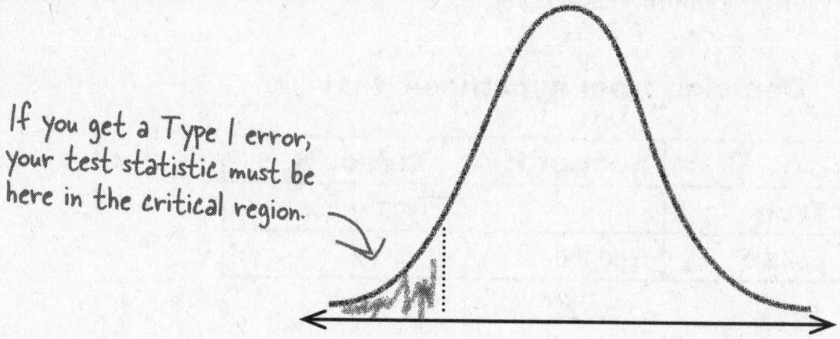
If you get a Type I error, your test statistic must be here in the critical region.

The probability of getting a Type I error is the probability of your results being in the critical region. As the critical region is defined by the significance level of the test, this means that if the significance level of your test is α, the probability of getting a Type I error must be also be α.

In other words,

$$\textbf{P(Type I error)} = \alpha$$

where α is the significance level of the test.

What about Type II errors?

A Type II error is what you get when you accept the null hypothesis, and the null hypothesis is actually wrong. It's like putting a prisoner on trial and finding him innocent when he's actually guilty.

Got away with it
A Type II error is when you accept H_0 when actually it's wrong.

The probability of getting a Type II error is normally represented by the Greek letter β.

$$P(\text{Type II error}) = \beta$$

So how do we find β?

Finding the probability of a Type II error is more difficult than finding the probability of getting a Type I error. Here are the steps that are involved, and we'll show you how to go through them on the next page.

1. **Check that you have a specific value for H_1.**
   Without this, you can't calculate the probability of getting a Type II error.

2. **Find the range of values outside the critical region of your test.**
   If your test statistic has been standardized, the range of values must be de-standardized.

3. **Find the probability of getting this range of values, assuming H_1 is true.**
   In other words, we find the probability of getting the range of values outside the critical region, but this time, using the test statistic described by H_1 rather than H_0.

Finding errors for SnoreCull

Let's see if we can find the probability of getting Type I and Type II errors for the SnoreCull hypothesis test. As a reminder, our standardized test statistic is

$$Z = \frac{X - 90}{3}$$

where X is the number of people cured in the sample. The significance level of the test is 5%.

Let's start with the Type I error

A Type I error is what you get when you reject the null hypothesis when actually it's true. The probability of getting this sort of error is the same as the significance level of the test, so this means that

P(Type I error) = 0.05

This gives you the probability of rejecting the null hypothesis that 90% of people are cured when it's true.

So what about the Type II error?

A Type II error is what you get when you accept the null hypothesis when the alternate hypothesis is true. We can only calculate this if H_1 specifies a single specific value, so let's use an alternate hypothesis of $p = 0.8$, as this is the proportion of successes in the doctor's sample. This means that our hypotheses become

$$H_0: p = 0.9$$
$$H_1: p = 0.8$$

This time we'll use H_1: $p = 0.8$ instead of H_1: $p < 0.8$. We can only calculate the probability of getting a Type II error if we have a single specific value for the alternate hypothesis.

The reason why H_1 must specify an exact value for p is so that we can calculate probabilities using it. If we used an alternate hypothesis of $p < 0.9$, we wouldn't be able to use it to calculate the probability of getting a Type II error.

To look up probabilities using the alternate hypothesis probability distribution, we need an exact value for p.

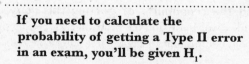

If you need to calculate the probability of getting a Type II error in an exam, you'll be given H_1.

This means that you won't have to decide on the alternate hypothesis yourself. If you need to calculate this sort of error, it will be given to you.

We need to find the range of values

Now that the alternate hypothesis H_1 gives a specific value for p, we can move on to the next step. We need to find the values of X that lie outside the critical region of the hypothesis test.

We saw back on page 548 that the critical region for the test is given by $Z < -1.64$—in other words, $P(Z < -1.64) = 0.05$. This means that values that fall outside the critical region are given by $Z \geq -1.64$.

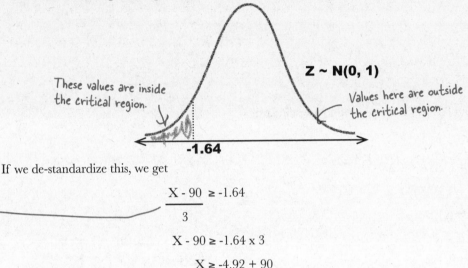

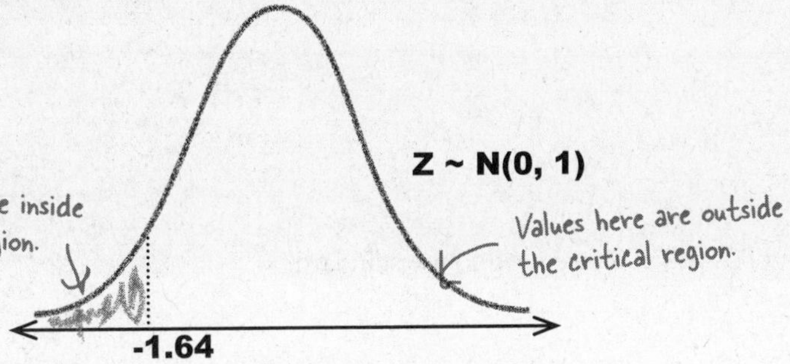
Z ~ N(0, 1)
These values are inside the critical region.
Values here are outside the critical region.
-1.64

If we de-standardize this, we get

$$\frac{X - 90}{3} \geq -1.64$$

$$X - 90 \geq -1.64 \times 3$$

$$X \geq -4.92 + 90$$

$$X \geq 85.08$$

In other words, we would have accepted the null hypothesis if 85.08 people or more had been cured by SnoreCull.

The final thing we need to do is work out $P(X \geq 85.08)$, assuming that H_1 is true. That way, we'll be able to work out the probability of accepting the null hypothesis when actually H_1 is true instead. As we're using the normal distribution to approximate X, we need to use a probability distribution $X \sim N(np, npq)$, where $n = 100$ and $p = 0.8$. This gives us

$$X \sim N(80, 16)$$

This means that if we can calculate $P(X \geq 85.08)$ where $X \sim N(80, 16)$, we'll have found the probability of getting a Type II error.

We calculate this in the same way we calculate other normal distribution probabilities, by finding the standard score and then looking up the value in standard normal probability tables.

Find P(Type II error)

We can find the probability of getting a Type II error by calculating $P(X \geq 85.08)$ where $X \sim N(80, 16)$. Let's start off by finding the standard score of 85.08.

$$z = \frac{85.08 - 80}{\sqrt{16}} = \frac{5.08}{4} = 1.27$$

This is the usual way of calculating the standard score; just subtract the expectation, and divide by the standard deviation.

This means that in order to find $P(X \geq 85.08)$, we need to use standard probability tables to find $P(Z \geq 1.27)$.

$$P(Z \geq 1.27) = 1 - P(Z < 1.27) = 1 - 0.8980 = 0.102$$

In other words,

P(Type II error) = 0.102

This gives you the probability of accepting the null hypothesis that 90% of people are cured when actually 80% of people are.

there are no Dumb Questions

Q: Why is it so much harder to find P(Type II error) than P(Type I error)?

A: It's because of the way they're defined. A Type I error is what you get when you wrongly reject the null hypothesis. The probability of getting this sort of error is the same as α, the significance level of the test.

A Type II error is the error you get when you accept the null hypothesis when actually the alternate hypothesis is true. To find the probability of getting this sort of error, you need to start by finding the range of values in your sample that would mean you accept the null hypothesis. Once you've found these values, you then have to calculate the probability of getting them assuming that H_1 is true.

Q: Do I need to use the normal distribution every time I want to find the probability of getting a Type II error?

A: The probability distribution you use all depends on your test statistic. In this case, our test statistic followed a normal distribution, so that's the distribution we used to find P(Type II error). If our test statistic had followed, say, a Poisson distribution, we would have used a Poisson distribution instead.

Introducing power

So far we've looked at the probability of getting different types of error in our hypothesis test. One thing that we haven't looked at is power.

The ***power*** of a hypothesis test is the probability that we will reject H_0 when H_0 is false. In other words, it's the probability that we will make the correct decision to reject H_0.

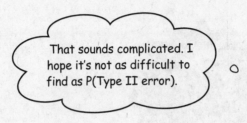

That sounds complicated. I hope it's not as difficult to find as P(Type II error).

Once you've found P(Type II error), calculating the power of a hypothesis test is easy.

Rejecting H_0 when H_0 is false is actually the opposite of making a Type II error. This means that

$$\textbf{Power} = 1 - \beta$$

where β is the probability of making a Type II error.

So what's the power of SnoreCull?

We've found the probability of getting a Type II error is 0.102. This means that we can find the power of the SnoreCull hypothesis test by calculating

$$\begin{aligned} \text{Power} &= 1 - P(\text{Type II error}) \\ &= 1 - 0.102 \\ &= 0.898 \end{aligned}$$

In other words, the power of the SnoreCull hypothesis test is 0.898. This means that the probability that we will make the correct decision to reject the null hypothesis is 0.898.

The doctor's happy

In this chapter, you've run through two hypothesis tests, and you've proved that there's sufficient evidence to reject the claims made by the drug company. You've been able to show that based on the doctor's sample, there's sufficient evidence that SnoreCull doesn't cure 90% of snorers, as the drug company claims.

I thought that the claims sounded too good to be true, and you've proved that there are strong statistical grounds for showing I'm right. I'll sleep quieter at night knowing that.

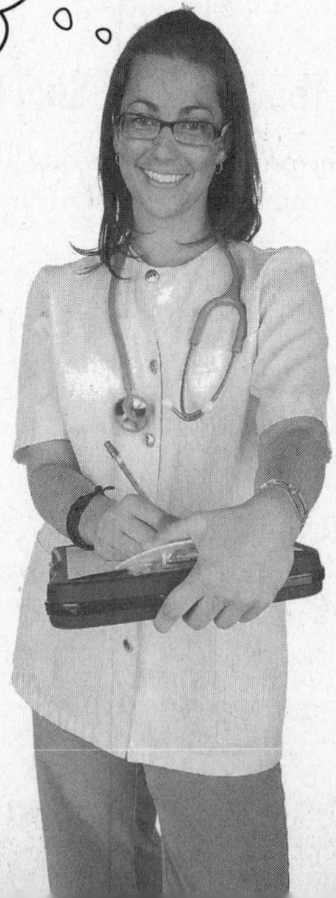

But it doesn't stop there

Keep reading, and we'll show you what other sorts of hypothesis tests you can use. We'll see you over at Fat Dan's Casino...

Exercise (part 1)

The drug company and their cough syrup manufacturer are having a dispute. The factory says that the amount of syrup that gets poured into their bottles follows a distribution $X \sim N(355, 25)$, where X is the amount of syrup in the bottle measured in mL. The drug company conducted tests on a large sample and found that the mean amount of syrup in 100 bottles is 356.5 mL. Test the hypothesis that the factory mean is correct at a 1% level of significance against the alternative that the mean amount of syrup in a bottle is greater than 355 mL.

We're going to guide you through this exercise in two parts. Here are the first three steps.

Step 1: Decide on the hypothesis you're going to test. What's the null hypothesis? What's the alternate hypothesis?

Step 2: Choose your test statistic.

Hint: Your hypothesis concerns the mean, so what's the distribution of $\overline{X}$? How do you standardize this?

Step 3: Determine the critical region for your decision. Does the critical region lie in the lower or upper tail of the distribution? What's the significance level? What's the critical value?

Exercise Solution (part 1)

The drug company and their cough syrup manufacturer are having a dispute. The factory says that the amount of syrup that gets poured into their bottles follows a distribution $X \sim N(355, 25)$, where X is the amount of syrup in the bottle measured in mL. The drug company conducted tests on a large sample and found that the mean amount of syrup in 100 bottles is 356.5 mL. Test the hypothesis that the factory mean is correct at a 1% level of significance against the alternative that the mean amount of syrup in a bottle is greater than 355 mL.

We're going to guide you through this exercise in two parts. Here are the first three steps.

Step 1: Decide on the hypothesis you're going to test. What's the null hypothesis? What's the alternate hypothesis?

We want to test whether the mean amount of syrup in the bottles is 355 mL like the factory says. This gives us

$H_0: \mu = 355$

$H_1: \mu > 355$

Step 2: Choose your test statistic.

$\overline{X} \sim N(\mu, \sigma^2/n)$, so this means that under the null hypothesis, $\overline{X} \sim N(355, 25/100)$ or $\overline{X} \sim N(355, 0.25)$.

If we standardize this, we get

$$Z = \frac{\overline{X} - 355}{\sqrt{0.25}}$$

$$= \frac{\overline{X} - 355}{0.5}$$

Step 3: Determine the critical region for your decision. Does the critical region lie in the lower or upper tail of the distribution? What's the significance level? What's the critical value?

The alternate hypothesis is $\mu > 355$, which means the critical region lies in the upper tail. We want to test at the 1% significance level, so the critical region is defined by $P(Z > c) = 0.01$. Using probability tables, this gives us $c = 2.32$. In other words, the critical region is given by $Z > 2.32$.

This exercise continues where the last left off. Here are the final three steps of the hypothesis test. What do you conclude?

Step 4: Find the p-value of the test statistic. Use the distribution $Z = (\overline{X} - 355)/0.5$, the mean amount of syrup in the sample, and remember that this time you're seeing if your test statistic lies in the upper tail of the distribution, as this is where the critical region is.

Step 5: See whether the sample result is within the critical region. Remember that you're testing at the 1% significance level.

Step 6: Make your decision. Is there enough evidence to reject the null hypothesis at the 1% level of significance?

Exercise Solution (part 2)

This exercise follows on from the last. Here are the final three steps of the hypothesis test. What do you conclude?

Step 4: Find the p-value of the test statistic. Use the distribution $Z = (\overline{X} - 355)/0.5$, the mean amount of syrup in the sample, and remember that this time you're seeing if your test statistic lies in the upper tail of the distribution as this is where the critical region is.

$$Z = (\overline{X} - 355)/0.5$$
$$= (356.5 - 355)/0.5$$
$$= 1.5/0.5$$
$$= 3$$

The p-value for this is given by $P(Z > 3)$, as the critical region is in the upper tail. Looking this up in probability tables gives us

p-value = 0.0013

Step 5: See whether the sample result is within the critical region. Remember that you're testing at the 1% significance level.

The p-value 0.0013 is less than 0.01, the significance level, so that means that the sample result is within the critical region

Step 6: Make your decision. Is there enough evidence to reject the null hypothesis at the 1% level of significance?

As the sample result lies in the critical region, there's sufficient evidence to reject the null hypothesis. We can accept the alternate hypothesis that $\mu > 355$ ml.

BULLET POINTS

- A Type I error is when you reject the null hypothesis when it's actually correct. The probability of getting a Type I error is α, the significance level of the test.
- A Type II error is when you accept the null hypothesis when it's wrong. The probability of getting a Type II error is represented by β.
- To find β, your alternate hypothesis must have a specific value. You then find the range of values outside the critical region of your test, and then find the probability of getting this range of values under H_1.

14 the χ^2 distribution

There's Something Going On...

I thought his success with girls would follow a binomial distribution with p = 0.8. I was so wrong...

Sometimes things don't turn out quite the way you expect.

When you model a situation using a particular probability distribution, you have a good idea of how things are likely to turn out long-term. But what happens if there are differences between **what you expect and what you get?** How can you tell whether your discrepancies come down to normal fluctuations, or whether they're a sign of an underlying problem with your probability model instead? In this chapter, we'll show you how you can use the χ^2 distribution to **analyze your results** and sniff out **suspicious results**.

There may be trouble ahead at Fat Dan's Casino

Fat Dan's is used to making a tidy profit from its casino-goers, but this week there's a problem. The slot machines keep hitting the jackpot, the roulette wheel keeps landing on 12, the dice are loaded, and too many people are winning off one of the blackjack tables.

The casino can't support the loss for much longer, and Fat Dan suspects foul play. He needs **your** help to get to the bottom of what's going on.

Let's start with the slot machines

As you've seen before, Fat Dan's Casino has a full row of bright, shiny slot machines, just waiting to be played. The trouble is that people keep on playing them—and winning.

Here's the expected probability distribution for one of the slot machines, where X represents the net gain from each game played:

It's $2 per game, so if you don't win anything, you lose your $2.

If you hit the jackpot, your net gain is $98.

x	-2	23	48	73	98
P(X = x)	0.977	0.008	0.008	0.006	0.001

The casino has collected statistics showing the number of times people get each outcome. Here are the frequencies for the observed net gains per game:

The frequency shows you how many games had which net gain.

x	-2	23	48	73	98
Frequency	965	10	9	9	7

Sharpen your pencil

The observed frequency is what we actually get.

We need to compare the actual frequency of each value of x with what you'd expect the frequency to be based on the probability distribution. Fill in the table below. What do you notice?

x	Observed frequency	Expected frequency
-2	965	977
23	10	
48	9	
73	9	
98	7	

Hint: The total observed frequency is 1000, as this is what you get if you add all the observed frequencies. Use the probability distribution to work out what you'd expect the frequencies to be.

Sharpen your pencil Solution

We need to compare the actual frequency of each value of x with what you'd expect the frequency to be based on the probability distribution. Fill in the table below. What do you notice?

x	Observed frequency	Expected frequency
-2	965	977
23	10	8
48	9	8
73	9	6
98	7	1

We found the expected frequencies by multiplying the probability of each outcome by 1000, the total frequency.

There's a difference between the number of people you'd expect to win the jackpot, based on the probability distribution, and the number of people actually winning it. What we don't know is how significant these differences are.

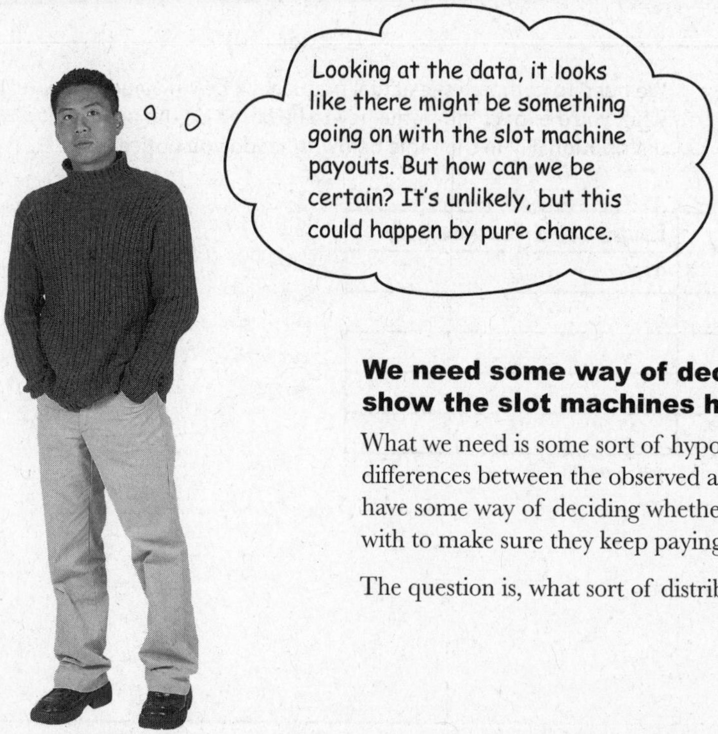
Looking at the data, it looks like there might be something going on with the slot machine payouts. But how can we be certain? It's unlikely, but this could happen by pure chance.

We need some way of deciding whether these results show the slot machines have been rigged.

What we need is some sort of hypothesis test that we can use to test the differences between the observed and expected frequencies. That way, we'll have some way of deciding whether the slot machines have been tampered with to make sure they keep paying out lots of money.

The question is, what sort of distribution can we use for this hypothesis test?

The χ^2 test assesses difference

There's a new sort of probability distribution that does exactly what we want; it's called the χ^2 distribution. χ is pronounced "kye", and it's the uppercase Greek letter chi. It uses a test statistic to look at the difference between what we *expect* to get and what we *actually* get, and then returns the probability of getting observed frequencies as extreme.

Let's start with the test statistic. To find the test statistic, first make a table featuring the observed and expected frequencies for your problem. When you've done that, use your observed and expected frequencies to compute the following statistic, where O stands for the observed frequency, and E for the expected frequency:

$$X^2 = \sum \frac{(O - E)^2}{E}$$

O refers to the observed frequency, while E refers to the expected frequency.

In other words, for each probability in the probability distribution, you take the difference between the frequency you expect and the frequency you actually get. You square the result, divide by the expected frequency, and then add all of these results up together.

So what's the test statistic for the slot machine problem?

Sharpen your pencil

Use the table of observed and expected frequencies you just worked out on the previous page for Fat Dan's slot machines to compute the test statistic. What result do you get?

What do you think a low value tells you? What about a high value?

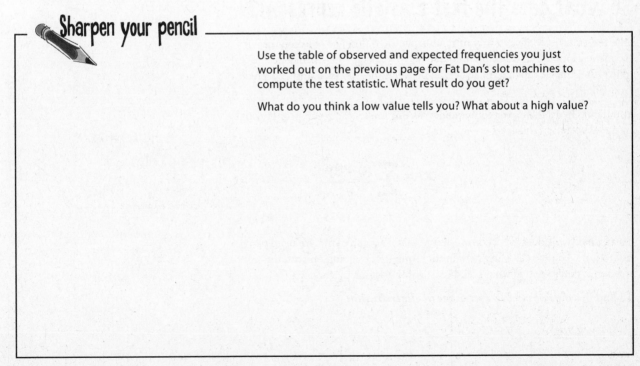

Sharpen your pencil Solution

Use the table of observed and expected frequencies you just worked out on the previous page for Fat Dan's slot machines to compute the test statistic. What result do you get?

What do you think a low value tells you? What about a high value?

$$X^2 = (965 - 977)^2/977 + (10 - 8)^2/8 + (9 - 8)^2/8 + (9 - 6)^2/6 + (7 - 1)^2/1$$
$$= (-12)^2/977 + 2^2/8 + 1^2/8 + 3^2/6 + 6^2$$
$$= 144/977 + 4/8 + 1/8 + 9/6 + 36$$
$$= 0.147 + 0.5 + 0.125 + 1.5 + 36$$
$$= 38.272$$

If the value of X^2 is low, then this means there's a less significant difference between the observed and expected frequencies. The higher X^2 is, the more significant the differences become.

So what does the test statistic represent?

The test statistic X^2 gives a way of measuring the difference between the frequencies we observe and the frequencies we expect. The smaller the value of X^2, the smaller the difference overall between the observed and expected frequencies.

You divide by E, the expected frequency, as this makes the result proportional to the expected frequency.

$$X^2 = \sum \frac{(O - E)^2}{E}$$

The smaller the differences between O and E, the smaller X^2 is.

Dividing by E makes the difference proportional to the expected frequency.

So at what point does X^2 become so large that it's significant? We need to figure out when we can fairly certain that something's going on with the slot machines that's beyond what could reasonably happen by chance.

To find this out, we need to look at the χ^2 ***distribution***.

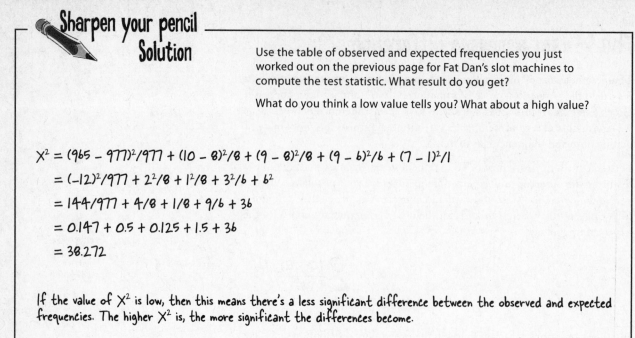

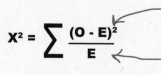

Two main uses of the χ^2 distribution

The χ^2 probability distribution specializes in detecting when the results you get are significantly different from the results you expect. The probability distribution does this using the X^2 test statistic you saw earlier.

The χ^2 distribution has two key purposes.

First of all, it's used to test ***goodness of fit***. This means that you can use it to test how well a given set of data fits a specified distribution. As an example, we can use it to test how well the observed frequencies for the slot machine winnings fits the distribution we expect.

Another use of the χ^2 distribution is to test the ***independence*** of two variables. It's a way of checking whether there's some sort of association.

The χ^2 distribution takes one parameter, the Greek letter ν, pronounced "new." Let's take a look at the effect that ν has on the shape of the probability distribution.

When ν is 1 or 2

When ν has a value of 1 or 2, the shape of the χ^2 distribution follows a smooth curve, starting off high and getting lower. It's shape is like a reverse J. The probability of getting low values of the test statistic X^2 is much higher than getting high values. In other words, observed frequencies are likely to be close to the frequency you expect.

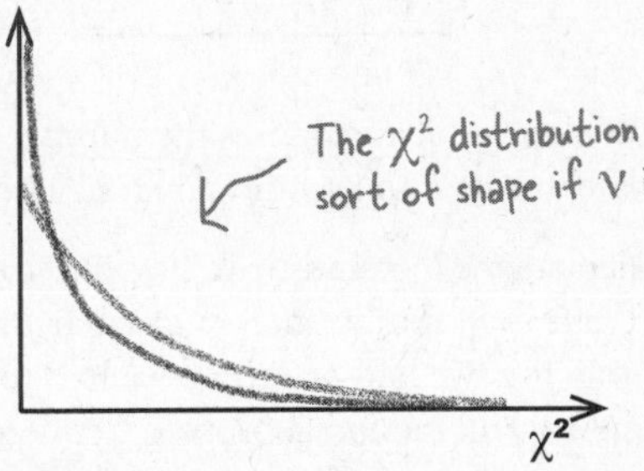

When ν is greater than 2

When ν has a value that's greater than 2, the shape of the χ^2 distribution changes. It starts off low, gets larger, and then decreases again as X^2 increases. The shape is positively skewed, but when ν is large, it's approximately normal.

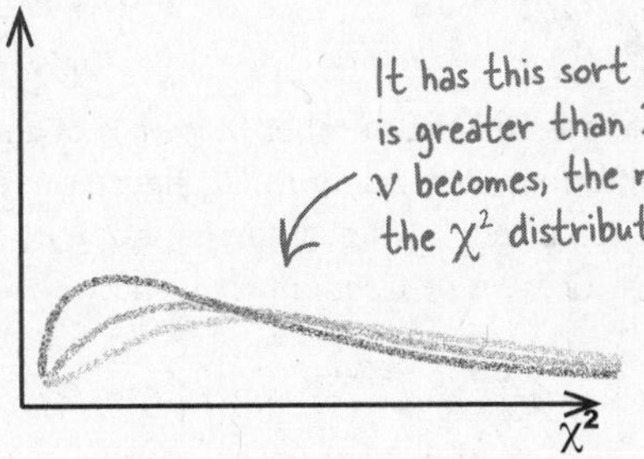

A shorthand way of saying that you're using the test statistic X^2 with the χ^2 distribution that has a particular value of ν is

$$X^2 \sim \chi^2(\nu)$$

X^2 follows a χ^2 distribution with a given value of ν.

It's like an X, but curvier.

ν represents degrees of freedom

You've seen how the shape of the χ^2 distribution depends on the value of ν, but how do we find what ν is?

ν is the number of ***degrees of freedom***. It's the number of independent variables used to calculate the test statistic X^2, or the number of independent pieces of information. Let's see what this means in practice.

Here's another look at the table of observed and expected frequencies for the slot machines:

x	Observed frequency	Expected frequency
-2	965	977
23	10	8
48	9	8
73	9	6
98	7	1

The number of degrees of freedom is the number of expected frequencies we have to calculate, taking into account any restrictions we have upon us.

In order to calculate the test statistic X^2, we had to calculate all of the expected frequencies. This meant that we had to calculate five expected frequencies. While calculating this, we had one thing we had to bear in mind: the total expected frequency and the total observed frequency had to add up to the same amount. In other words, we had one restriction on us in our calculations.

So what's ν?

To calculate ν, we take the number of pieces of information we calculated, and subtract the number of restrictions. To figure out the test statistic X^2, we had to calculate five separate pieces of information, with 1 restriction. This means that the number of degrees of freedom is given by

$$\nu = 5 - 1$$
$$= 4$$

Another way of looking at this is that we had to calculate four of the expected frequencies using the probability distribution. We could work out the final frequency by looking at what the total expected frequency should be.

In general,

ν = (number of classes) - (number of restrictions)

What's the significance?

So how can we use the χ^2 distribution to say how significant the discrepancy is between the observed and expected frequencies? As with other hypothesis tests, it all depends on the level of significance.

When you conduct a test using the χ^2 distribution, you conduct a one-tailed test using the upper tail of the distribution as your critical region. This way, you can specify the likelihood of your results coming from the distribution you expect by checking whether the test statistic lies in the critical region of the upper tail.

If you conduct a test at significance level α, then you write this as

$$\chi^2{}_\alpha(\nu)$$

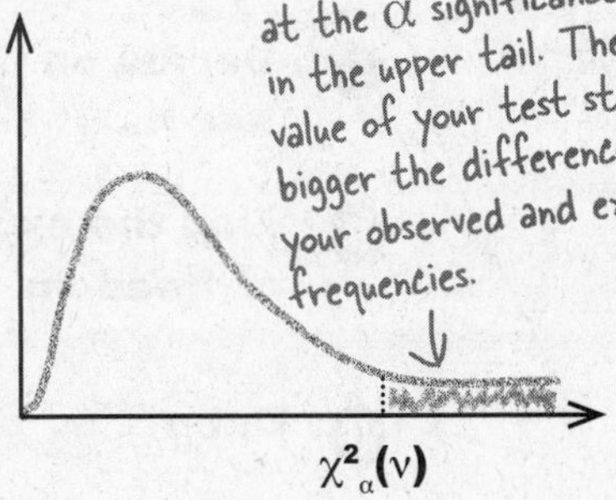

So how do we find the critical region for the χ^2 distribution? We can use χ^2 probability tables.

The critical region for testing at the α significance level lies in the upper tail. The higher the value of your test statistic, the bigger the differences between your observed and expected frequencies.

$\chi^2{}_\alpha(\nu)$

How to use χ^2 probability tables

To find the critical value, start off with the degrees of freedom, ν, and the significance level, α. Use the first column to look up ν, and the top row to look up α. The place where they intersect gives the value x, where $P(\chi^2{}_\alpha(\nu) \geq x) = \alpha$. In other words, it gives you the critical value.

As an example, if you wanted to find the critical value for testing at the 5% level with 8 degrees of freedom, you'd find 8 in the first column, 0.05 in the top row, and read off a value of 15.51. In other words, if our test statistic X^2 was greater than 15.51, it would be in the critical region at the 5% level with 8 degrees of freedom.

Here's the column for 0.05.

ν	**Tail probability** α										
	.25	**.20**	**.15**	**.10**	**.05**	**.025**	**.02**	**.01**	**.005**	**.0025**	**.001**
1	1.32	1.64	2.07	2.71	3.84	5.02	5.41	6.63	7.88	9.14	10.83
2	2.77	3.22	3.79	4.61	5.99	7.38	7.82	9.21	10.60	11.98	13.82
3	4.11	4.64	5.32	6.25	7.81	9.35	9.84	11.34	12.84	14.32	16.27
4	5.39	5.99	6.74	7.78	9.49	11.14	11.67	13.28	14.86	16.42	18.47
5	6.63	7.29	8.12	9.24	11.07	12.83	13.39	15.09	16.75	18.39	20.51
6	7.84	8.56	9.45	10.64	12.59	14.45	15.03	16.81	18.55	20.25	22.46
7	9.04	9.80	10.75	12.02	14.07	16.01	16.62	18.48	20.28	22.04	24.32
8	10.22	11.03	12.03	13.36	15.51	17.53	18.17	20.09	21.95	23.77	26.12
9	11.39	12.24	13.29	14.68	16.92	19.02	19.68	21.67	23.59	25.46	27.88

Here's the row for $\nu = 8$.

This is where 8 and 0.05 meet.

Hypothesis testing with χ^2

Here are the broad steps that are involved in hypothesis testing with the χ^2 distribution.

1. **Decide on the hypothesis you're going to test, and its alternative**
2. **Find the expected frequencies and the degrees of freedom**
3. **Determine the critical region for your decision**
4. **Calculate the test statistic X^2**
5. **See whether the test statistic is within the critical region**
6. **Make your decision**

These steps are just like the ones we had before

These steps are different from the ones you saw before

Look familiar? Most of these steps are exactly the same as for other hypothesis tests. In other words, it's exactly the same process as before.

there are no Dumb Questions

Q: So are χ^2 tests really just a special kind of hypothesis test?

A: Yes, they are. You go through pretty much all the steps you had to go through before.

Q: Do I always use the upper tail for my test?

A: Yes, if you're conducting a hypothesis test, you always use the upper tail. This is because the higher the value of your χ^2 test statistic, the more your observed frequencies differ from the expected frequencies.

Q: I think I've heard the term degrees of freedom before. Have I?

A: Yes, you have. Remember when we looked at how we can use the t-distribution to create confidence intervals? Well, the t-distribution uses degrees of freedom, too.

Q: I think I've seen degrees of freedom referred to as df rather than v. Is that wrong?

A: Not at all. Different text books use different conventions, and we're using v. At the end of the day, they have the same meaning.

Q: I want to look for information about the χ^2 distribution on the Internet. How do I find it? Do I need to type in Greek?

A: You should be able to find any information you need by searching for the term "chi square." The χ^2 distribution is also written "chi-squared."

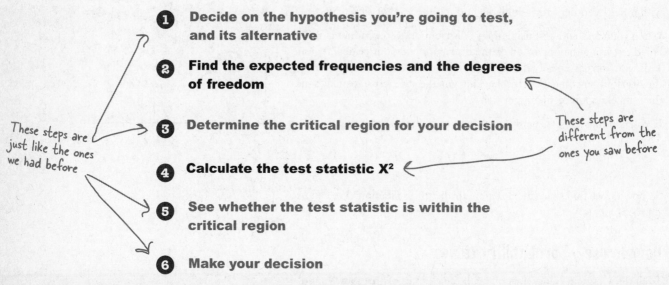

Exercise

It's your job to see whether there's sufficient evidence at the 5% level to say that the slot machines have been rigged. We'll guide you through the steps.

1. What's the null hypothesis you're going to test? What's the alternate hypothesis?

2. There are 4 degrees of freedom. What's the region for the 5% level?

3. What's the test statistic? ← Hint: You calculated this earlier.

4. Is your test statistic inside or outside the critical region?

5. Will you accept or reject the null hypothesis?

It's your job to see whether there's sufficient evidence at the 5% level to say that the slot machines have been rigged. We'll guide you through the steps.

1. What's the null hypothesis you're going to test? What's the alternate hypothesis?

H_0: The slot machine winnings per game follow the described probability distribution.

x	-2	23	48	73	98
P(X = x)	0.977	0.008	0.008	0.006	0.001

H_1: The slot machine winnings per game do not follow this probability distribution.

2. There are 4 degrees of freedom. What's the region for the 5% level?

From probability tables, $\chi^2_{5\%}(4) = 9.49$. This means that the critical region is where $X^2 > 9.49$.

3. What's the test statistic?

The test statistic is X^2. You found this earlier; its value is 38.272.

4. Is your test statistic inside or outside the critical region?

The value of X^2 is 38.27, and as the critical region is $X^2 > 9.49$, this means that X^2 is inside the critical region.

5. Will you accept or reject the null hypothesis?

The value of X^2 is inside the critical region, so this means that we reject the null hypothesis. In other words, there is sufficient evidence to reject the hypothesis that the slot machine winnings follow the described probability distribution.

You've solved the slot machine mystery

Thanks to your careful use of the χ^2 probability distribution, you've found out that there's sufficient evidence that the slot machine isn't following the probability distribution that the casino expects it to. Fat Dan is very grateful to you, as this means you've come up with evidence that the slot machine has been rigged in some way. He's shut them down, so he doesn't lose any more money.

FAT DAN'S CASINO
WIN BIG!
Out of Order

Let's summarize the steps you went through to discover this.

First of all, you took a set of observed frequencies for the slot machine and calculated what you expected the frequencies to be, assuming they followed a particular probability distribution. You then calculated the degrees of freedom and calculated the test statistic X^2, which gave you an indication of the total discrepancy between the observed frequencies and those you expected.

After this, you used the χ^2 probability tables to find the critical region of the distribution at the 5% level of significance. You checked this against your test statistic and found that there was sufficient evidence to say that the slot machine has been rigged to pay out more money.

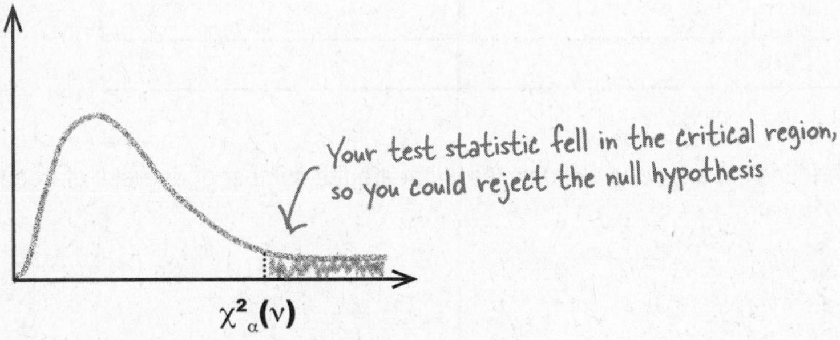
Your test statistic fell in the critical region, so you could reject the null hypothesis
$\chi^2_\alpha(\nu)$

This sort of hypothesis test is called a ***goodness of fit*** test. It tests whether observed frequencies actually fit in with an assumed probability distribution. You use this sort of test whenever you have a set of values that should fit a distribution, and you want to test whether the data actually does.

Long Exercise

Fat Dan thinks that the dice in the dice games are loaded. Take a look at the following observed frequencies for one six-sided die, and test whether there's enough evidence to support the claim that the die isn't fair at the 1% significance level. We'll guide you through the steps.

Here are the observed frequencies:

Value	**1**	**2**	**3**	**4**	**5**	**6**
Frequency	107	198	192	125	132	248

Step 1: Decide on the hypothesis you're going to test, and its alternative.

Step 2: Find the expected frequencies and the degrees of freedom.

Start off by completing the expected frequencies for the die. You'll need to take into account how many times the die is thrown in total, and the probability of getting each value. X represents the value of one toss of the die.

x	**Observed frequency**	**Expected frequency**
1	107	
2	198	
3	192	
4	125	
5	132	
6	248	

Once you've found the expected frequencies, what are the number of degrees of freedom?

You find this the same way you found the degrees of freedom for the slot machines.

Step 3: Determine the critical region for your decision.

You'll need to use the significance level and number of degrees of freedom

Step 4: Calculate the test statistic X^2.

You can calculate this using your observed and expected frequencies from step 2.

Step 5: See whether the test statistic is within the critical region.

Step 6: Make your decision.

Long Exercise Solution

Fat Dan thinks that the dice in the dice games are loaded. Take a look at the following observed frequencies for one six-sided die, and test whether there's enough evidence to support the claim that the die isn't fair at the 1% significance level. We'll guide you through the steps.

Here are the observed frequencies:

Value	**1**	**2**	**3**	**4**	**5**	**6**
Frequency	107	198	192	125	132	248

Step 1: Decide on the hypothesis you're going to test, and its alternative.

To test whether the die is fair, we have to determine whether there's sufficient evidence that it isn't. This gives you

H_0: The die is fair, and every value has an equal chance of being thrown. This means the probability of getting each value is 1/6.

H_1: The die isn't fair.

Step 2: Find the expected frequencies and the degrees of freedom.

Start off by completing the expected frequencies for the die. You'll need to take into account how many times the die is thrown in total, and the probability of getting each value. X represents the value of one toss of the die.

x	**Observed frequency**	**Expected frequency**
1	107	167
2	198	167
3	192	167
4	125	167
5	132	167
6	248	167

The total expected frequency needs to match the total observed frequency. If you add the observed frequencies together, you get 1002.

The probability of getting each value is 1/6. This means the expected frequency of each value is 1002/6 = 167.

Once you've found the expected frequencies, what are the number of degrees of freedom?

We had to find 6 expected frequencies, and their total had to equal 1002. In other words, we had to find 6 pieces of information with 1 restriction. This gives us

$$\nu = 6 - 1$$
$$= 5$$

Step 3: Determine the critical region for your decision.

You'll need to use the significance level and number of degrees of freedom

From probability tables, $\chi^2_{1\%}(5) = 15.09$. This means that the critical region is where $X^2 > 15.09$.

Step 4: Calculate the test statistic X^2.

You can calculate this using your observed and expected frequencies from step 2.

$$X^2 = \sum \frac{(O - E)^2}{E}$$

$$= (107-167)^2/167 + (198-167)^2/167 + (192-167)^2/167 + (125-167)^2/167 + (132-167)^2/167 + (248-167)^2/167$$

$$= (-60)^2/167 + (31)^2/167 + (25)^2/167 + (-42)^2/167 + (-35)^2/167 + (81)^2/167$$

$$= (3600 + 961 + 625 + 1764 + 1225 + 6561)/167$$

$$= 14736/167$$

$$= 88.24$$

Step 5: See whether the test statistic is within the critical region.

The critical region is given by $X^2 > 15.09$. As $X^2 = 88.24$, the test statistic is within the critical region.

Step 6: Make your decision.

As your test statistic lies within the critical region, this means that there is sufficient evidence at the 1% level to reject the null hypothesis. In other words, you accept the alternate hypothesis that the die isn't fair.

So can you use the χ^2 goodness of fit test with any underlying probability distribution?

o O

The χ^2 goodness of fit test works for pretty much any probability distribution.

You can use the χ^2 distribution to test the goodness of fit of *any* probability distribution, just as long as you have a set of observed frequencies, and you can work out what you expect the frequencies to be.

The hardest thing is working out what the degrees of freedom for ν should be. Here are the degrees of freedom for some of the most common probability distributions you'll want to use with the χ^2 goodness of fit.

p is the probability of success, or the proportion of successes in the population.

Distribution	Condition	ν
Binomial	You know what p is	$\nu = n - 1$
	You don't know what p is, and you have to estimate it from the observed frequencies	$\nu = n - 2$
Poisson	You know what λ is	$\nu = n - 1$
	You don't know what λ is, and you have to estimate it from the observed frequencies	$\nu = n - 2$
Normal	You know what μ and σ^2 are	$\nu = n - 1$
	You don't know what μ and σ^2 are, and you have to estimate them from the observed frequencies	$\nu = n - 3$

n is the total number of observed frequencies.

λ is the rate of occurrences in an interval.

Fat Dan has another problem

So far you've investigated whether the slot machines seem to be rigged in some way, by using a goodness of fit test to see whether the observed frequencies you have correspond to the expected probability distribution. Fat Dan has other problems, though, and this time it's his staff.

Fat Dan thinks he's losing more money than he should from one of the croupiers on the blackjack tables. Can you determine whether there's significant evidence to show whether or not Fat Dan's right?

Here are the three croupiers who man the tables:

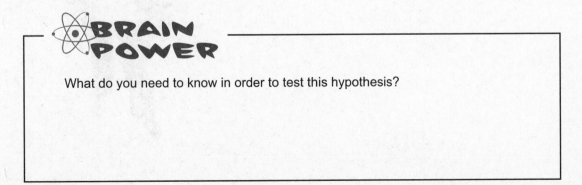

These are the outcomes you get for each of the games.

	Croupier A	Croupier B	Croupier C
Win	43	49	22
Draw	8	2	5
Lose	47	44	30

These are the observed results for each of the croupiers.

What we need is some way of testing whether the outcome of the game is dependent on which croupier is leading the game.

BRAIN POWER

What do you need to know in order to test this hypothesis?

the χ^2 distribution can test for independence

So far we've looked at the χ^2 distribution in terms of performing goodness of fit tests. This isn't the only use of the χ^2 distribution. The χ^2 distribution can also be used to perform tests of ***independence***.

A χ^2 test for independence is a test to see whether two factors are independent, or whether there seems to be some sort of association between them. This is just the situation we have with the croupiers. We want to test whether the croupier leading a game of blackjack has any impact on the outcome. In other words, we assume that the choice of croupier is independent of the outcome, unless there's sufficient evidence against it.

You conduct a test for independence in the same way you conduct a goodness of fit test. You set up a hypothesis, use the observed and expected frequencies to calculate the X^2 test statistic, and then see if it falls within the critical region.

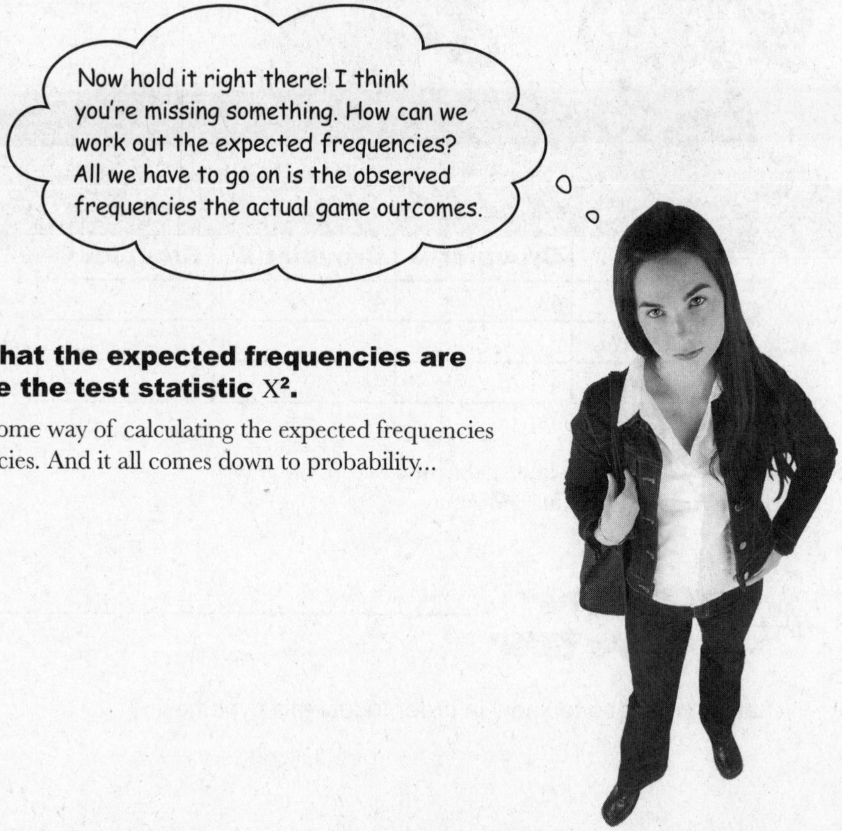

Now hold it right there! I think you're missing something. How can we work out the expected frequencies? All we have to go on is the observed frequencies the actual game outcomes.

We need to know what the expected frequencies are in order to calculate the test statistic X^2.

This means that we need some way of calculating the expected frequencies from the observed frequencies. And it all comes down to probability...

You can find the expected frequencies using probability

There are a few steps you need to go through to find the expected frequencies.

To start off, calculate the total frequencies for the outcomes and the croupiers, and also the grand total. You can show the results in a table like this, called a ***contingency table***.

	Croupier A	Croupier B	Croupier C	Total
Win	43	49	22	114
Draw	8	2	5	15
Lose	47	44	30	121
Total	98	95	57	250

Total number of wins

Total for croupier A

Grand total

Now we can use this information to find the expected number of wins for each croupier.

Let's start by finding the expected frequency for the number of wins with croupier A.

First off, we can use these grand totals to find the probability of getting a particular outcome, or a particular croupier. As an example, to find the probability of winning, you divide the total number of wins by the grand total:

$$P(\text{Win}) = \frac{\text{Total Wins}}{\text{Grand Total}}$$

Similarly, you can find the probability of playing against croupier A by dividing the total for croupier A by the grand total.

$$P(A) = \frac{\text{Total A}}{\text{Grand Total}}$$

Now if the croupier and the outcome of the game are independent, as we assume they are, this means that you can find the probability of getting a win with croupier A by multiplying together these two probabilities. In other words:

$$P(\text{Win and A}) = \frac{\text{Total Wins}}{\text{Grand Total}} \times \frac{\text{Total A}}{\text{Grand Total}}$$

Back in chapter 4, we saw that for independent events, $P(A \cap B) = P(A) \times P(B)$.

BRAIN POWER

How can we use this to find the expected number of wins for croupier A?

So what are the frequencies?

So far, we've found that the probability of winning with croupier A, and we want to use this to find the expected frequency of wins. To do this, all we just need to multiply the probability of winning with croupier A by the grand total. This gives us

$$\text{Expected frequency} = \cancel{\text{Grand Total}} \times \frac{\text{Total Wins}}{\cancel{\text{Grand Total}}} \times \frac{\text{Total A}}{\text{Grand Total}}$$

$$= \frac{\text{Total Wins} \times \text{Total A}}{\text{Grand Total}}$$

In other words, to find the expected frequency of wins with croupier A, multiply the total number of wins by the total number of games with croupier A, and divide by the grand total.

How do we find the frequencies in general?

You can generalize this so that you have a nice, easy result you can apply to every frequency you need to find. To find the expected frequency for a particular row and column combination, multiply the total for the row by the total for the column, and divide by the grand total.

$$\textbf{Expected frequency} = \frac{\textbf{Row Total} \times \textbf{Column Total}}{\textbf{Grand Total}}$$

Once you've figured out what all the expected frequencies are, you can use this to calculate the test statistic X^2. It's the same test statistic as before, so you need to calculate

$$X^2 = \sum \frac{(O - E)^2}{E}$$

For every observed frequency, subtract the expected frequency, square the result, and divide by the expected frequency. Then add your results together.

The key is to ensure you include every observed frequency and every corresponding expected frequency.

Exercise

Here's the table showing the observed frequencies for the croupiers. Your task is to figure out all the expected frequencies.

These are the observed frequencies. →

	Croupier A	Croupier B	Croupier C	Total
Win	43	49	22	114
Draw	8	2	5	15
Lose	47	44	30	121
Total	98	95	57	250

(Row total × column total)/grand total

Work out each of the expected frequencies here. →

	Croupier A	Croupier B	Croupier C
Win	(114×98)/250=44.688		
Draw	(15×98)/250=5.88		
Lose	(121×98)/250=47.432		

Once you've found all the expected frequencies, calculate the test statistic X^2. Use the table below to help you. The first column gives all the observed frequencies, the second column is for the corresponding expected frequencies, and if you add together all the numbers in the third column, it gives you your test statistic.

Use the values in the first two columns to help you calculate this.

	Observed	Expected	$\frac{(O - E)^2}{E}$
A	43	44.688	$(43-44.688)^2/44.688 = 2.85/44.688 = 0.064$
	8	5.88	$(8-5.88)^2/5.88 = 4.4944/5.88 = 0.764$
	47	47.432	$(47-47.432)^2/47.432 = 0.187/47.432 = 0.004$
B	49		
	2		
	44		
C	22		
	5		
	30		
	$\Sigma O = 250$	$\Sigma E =$	$\sum \frac{(O - E)^2}{E} =$

Exercise Solution

Here's the table showing the observed frequencies for the croupiers. Your task is to figure out all the expected frequencies.

Observed frequencies

	Croupier A	Croupier B	Croupier C	Total
Win	43	49	22	114
Draw	8	2	5	15
Lose	47	44	30	121
Total	98	95	57	250

Expected frequencies

	Croupier A	Croupier B	Croupier C
Win	(114×98)/250=44.688	(114×95)/250=43.32	(114×57)/250=25.992
Draw	(15×98)/250=5.88	(15×95)/250=5.7	(15×57)/250=3.42
Lose	(121×98)/250=47.432	(121×95)/250=45.98	(121×57)/250=27.588

Once you've found all the expected frequencies, calculate the test statistic X^2. Use the table below to help you. The first column gives all the observed frequencies, the second column is for the corresponding expected frequencies, and if you add together all the numbers in the third column, it gives you your test statistic.

	Observed	Expected	$\frac{(O - E)^2}{E}$
A	43	44.688	$(43-44.688)^2/44.688 = 2.85/44.688 = 0.064$
	8	5.88	$(8-5.88)^2/5.88 = 4.4944/5.88 = 0.764$
	47	47.432	$(47-47.432)^2/47.432 = 0.187/47.432 = 0.004$
B	49	43.32	$(49-43.32)^2/43.32 = 5.68/43.32 = 0.131$
	2	5.7	$(2-5.7)^2/5.7 = 13.69/5.7 = 2.402$
	44	45.98	$(44-45.98)^2/45.98 = 3.9204/45.98 = 0.085$
C	22	25.992	$(22-25.992)^2/25.992 = 15.936/25.992 = 0.613$
	5	3.42	$(5-3.42)^2/3.42 = 2.4964/3.42 = 0.730$
	30	27.588	$(30-27.588)^2/27.588 = 5.817/27.588 = 0.211$
	ΣO = 250	**ΣE =** 250	$\sum \frac{(O - E)^2}{E} = 5.004$

This is your test statistic

We still need to calculate degrees of freedom

Before we can use the χ^2 distribution to find the significance of the observed frequencies, there's just one more thing we need to find. We need to find ν, the number of degrees of freedom.

You saw earlier that the number of degrees of freedom is the number of pieces of independent information we are free to choose, taking into account any restrictions. This means that we look at how many expected frequencies we have to calculate independently, and subtract the number of restrictions.

First of all, let's look at the total number of expected frequencies we had to calculate. We had to figure out the expected frequencies for the three croupiers and the three possible outcomes. This means that we worked out $3 \times 3 = 9$ expected frequencies.

We had to figure out 3x3 = 9 expected frequencies.

	Croupier A	Croupier B	Croupier C
Win			
Draw			
Lose			

Now for each row and for each column, we only actually had to calculate two of the expected frequencies. We knew what the total frequency should be, so we could choose the third to make sure that the frequencies added up to the right result. In other words, we only actually had to calculate 4 of the expected frequencies; the other 5 had to fit in with the total frequencies we already knew about.

We only had to calculate these expected frequencies – we could figure out the others using the total frequency of each row and column.

	Croupier A	Croupier B	Croupier C
Win			
Draw			
Lose			

We could figure the last row and column out using the totals.

Since we had to calculate 4 expected frequencies, this makes the number of degrees of freedom. There were 4 pieces of independent information we had to calculate; once we'd done that, the rest were known automatically. In other words, $\nu = 4$.

Another way of looking at this is that we needed to find 9 values overall, and there were 5 values that we didn't have to calculate independently. Using our formula from before, this gives us $\nu = 9 - 5 = 4$.

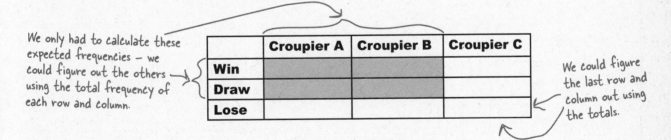

Long Exercise

Conduct a hypothesis test with a 1% significance level to see whether the outcome of the game is independent of the croupier manning the table. Here's a reminder of the steps, but remember you've worked out some of these already.

1. Decide on the hypothesis you're going to test, and its alternative.
2. Find the expected frequencies and the degrees of freedom.
3. Determine the critical region for your decision.
4. Calculate the test statistic X^2.
5. See whether the test statistic is within the critical region.
6. Make your decision.

We've left you lots of space for your calculations.

Long Exercise Solution

Conduct a hypothesis test with a 1% significance level to see whether the outcome of the game is independent of the croupier manning the table. Here's a reminder of the steps, but remember you've worked out some of these already.

1. Decide on the hypothesis you're going to test, and its alternative.
2. Find the expected frequencies and the degrees of freedom.
3. Determine the critical region for your decision.
4. Calculate the test statistic X^2.
5. See whether the test statistic is within the critical region.
6. Make your decision.

Step 1:

We want to test whether the outcome of the game is independent of the croupier manning the table. This means we can use:

H_0: There is no relationship between the outcome of the game and the croupier manning the table.

H_1: There is a relationship between the outcome of the game and the croupier manning the table

Step 2:

We found the expected frequencies in the exercise back on page 590, and we've just seen that the number of degrees of freedom is 4.

Step 3:

From probability tables, $\chi^2_{1\%}(4) = 13.28$. This means that the critical region is given by $X^2 > 13.28$.

Step 4:

We also calculated the test statistic X^2 using the expected frequencies back on page 590. We found that $X^2 = 5.004$.

Step 5:

The critical region is given by $X^2 > 13.28$, so this means that X^2 is outside the critical region.

Step 6:

As X^2 is outside the critical region, we accept the null hypothesis. There is insufficient evidence that there's a relationship between game outcome and croupier.

there are no Dumb Questions

Q: I'm still not sure I understand how you found the degrees of freedom for the croupiers. Why are there four degrees of freedom?

A: We found the degrees of freedom by looking at how many expected frequencies we had to calculate, and working out how many of these we could have calculated by just looking at the total observed frequencies for each row and column.

There are three croupiers and three outcomes. If you use a contingency table to calculate these, the row and column totals for the expected frequencies must match those of the observed frequencies. This means that once you've calculated the first 2 expected frequencies for any row or column, the final one is determined by the overall total. Therefore, you only need to calculate 2×2 expected frequencies from scratch. This gives you your four degrees of freedom.

Q: Are there any other uses of the χ^2 distribution besides testing goodness of fit and independence?

A: These are the two main uses of the χ^2 distribution. The thing to remember is that you can use it to test the goodness of fit of virtually any probability distribution. As an example, you can use it to test whether observed frequencies fit a particular binomial distribution.

Q: Should I test at any particular level?

A: It depends on your situation. Just as with other hypothesis tests, the smaller the level of significance, the stronger you need your evidence to be before you reject your null hypothesis.

Testing at the 5% and 1% level of significance is common.

I wonder what happens if you have a different size contingency table? How do you find the number of degrees of freedom then?

BRAIN POWER

Take a look at how we calculated the degrees of freedom for a 3x3 table. How do you think we could generalize this? See if you can work this out, then turn the page.

Generalizing the degrees of freedom

So far we've looked at the degrees of freedom for a 3×3 contingency table, but how do we generalize the result?

Imagine you're comparing two variables, and you have *h* rows of one variable and *k* columns of another. You know what the row and column totals should be. Now imagine you want to find the number of degrees of freedom.

	Column 1	...	Column k-1	Column k
Row 1				
...				
Row h-1				
Row h				

For each row, there are *k* columns. You know what the total of each row should be, so you only actually need to calculate the expected frequency of $(k - 1)$ of the columns. You automatically know what the *k*th column is because you know the total frequency of the row.

	Column 1	...	Column k-1	Column k
Row 1				

You need to calculate these.

You can figure out column k using the row total.

It's a similar process for the columns. Each column has *h* rows, and you know what the total of each column should be. This means that you have to calculate $(h - 1)$ of the rows for each column. You automatically know what the value of the *h*th row is because you know the total frequency of the column.

	Column 1
Row 1	
...	
Row h – 1	
Row h	

You need to calculate the frequencies of these h–1 rows.

You can figure out row h using the column total.

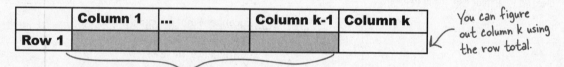

And the formula is...

If we put this together, the total number of expected frequencies you have to calculate is $(k - 1) \times (h - 1)$. In other words, if you have a table with dimensions h by k, you can find the degrees of freedom by calculating

$$\nu = (h - 1) \times (k - 1)$$

	Column 1	...	Column k-1	Column k
Row 1				
...				
Row h-1				
Row h				

You have to calculate (h–1) × (k–1) expected frequencies, so there are (h–1) × (k–1) degrees of freedom.

Sharpen your pencil

Fat Dan has hired two more croupiers. What are the degrees of freedom now? The outcomes of the game remain the same.

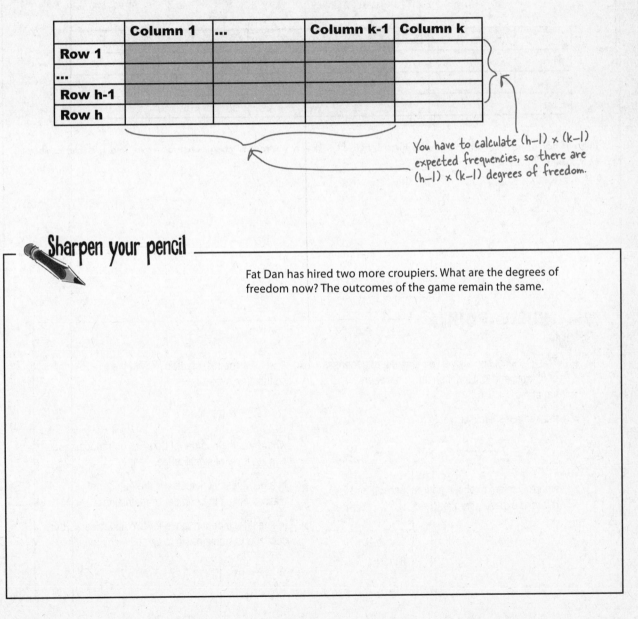

Sharpen your pencil Solution

Fat Dan has hired two more croupiers. What are the degrees of freedom now? The outcomes of the game remain the same.

As Fat Dan has hired 2 more croupiers, this means that we now have a 3x5 contingency table.

A, B, and C are the original croupiers, and Fat Dan has hired two more.

	Croupier A	Croupier B	Croupier C	Croupier D	Croupier E
Win					
Draw					
Lose					

The number of degrees of freedom is given by (h-1) x (k-1), where h is the number of rows, and k is the number of columns. This gives us

$$v = 2 \times 4$$
$$= 8$$

BULLET POINTS

- The χ^2 distribution allows you to conduct goodness of fit tests and test independence between variables.
- It takes a test statistic

$$X^2 = \sum \frac{(O - E)^2}{E}$$

where O refers to observed frequencies, and E refers to expected frequencies.

- If we're using test statistic X^2 with the χ^2 distribution, we write

$$X^2 \sim \chi^2_\alpha(v)$$

where v is the number of degrees of freedom, and α is the level of significance.

- In a goodness of fit test, v is the number of classes minus the number of restrictions.
- In a test for independence for two variables, if your contingency table has h rows and k columns,

$$v = (h - 1) \times (k - 1)$$

You've saved the casino

Thanks to your mastery of the χ^2 distribution, you've managed to unearth which of the casino games look like they've been rigged. You discerned explainable discrepancies between what you got and what you expected, and you also detected suspicious activity at certain levels of significance.

Fat Dan is delighted with your efforts. Thanks to you, he knows which of his casino games need to be investigated, and the blackjack croupiers get to keep their jobs. Next time you're in town, tell Fat Dan—he'll supply you with extra chips, all on the house.

Well done!

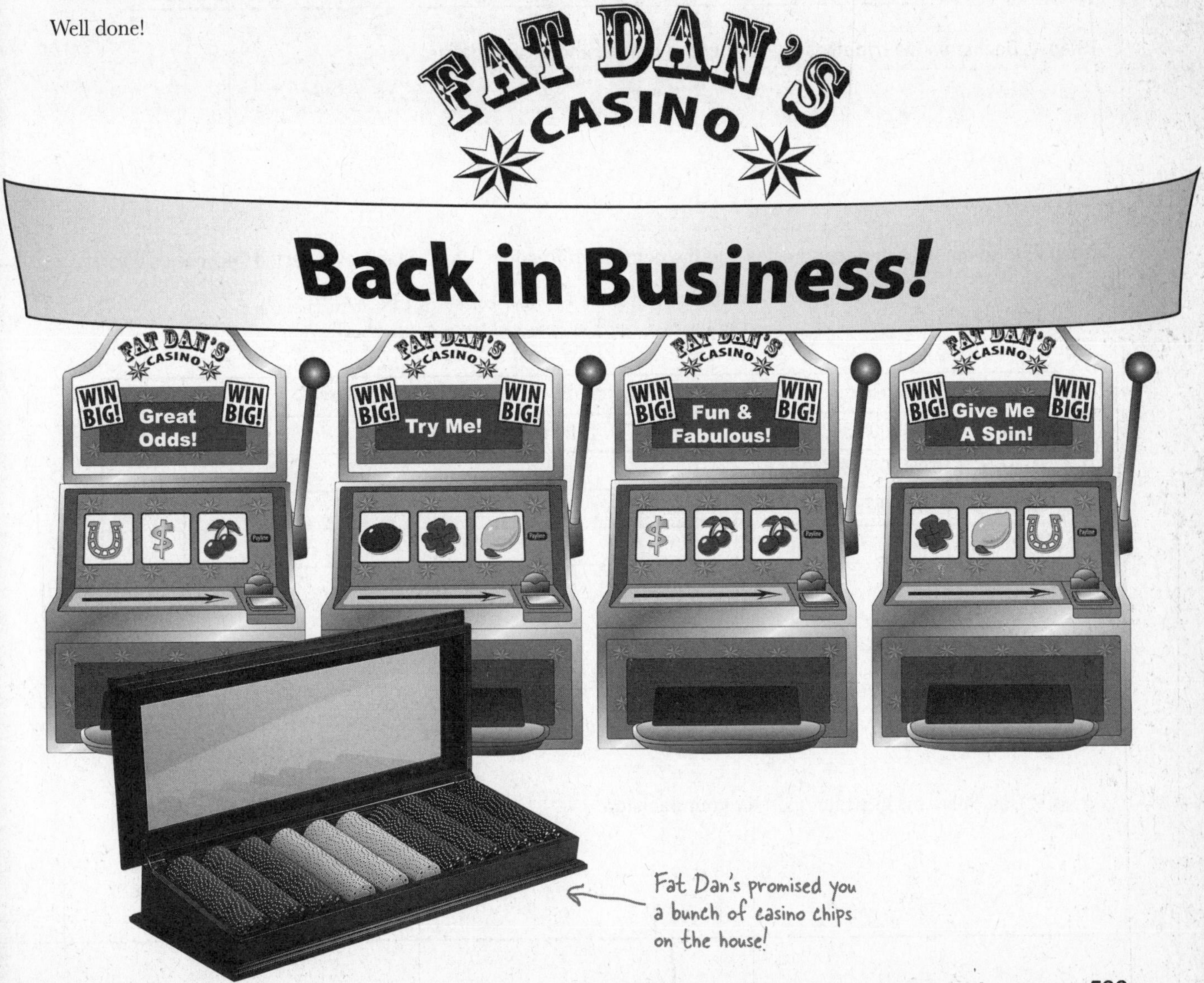

Fat Dan's promised you a bunch of casino chips on the house!

Long Exercise

Fat Dan thinks that one or more of his croupiers are somehow influencing the results of the roulette wheel. Here's data showing the observed frequency with which the ball lands in each color pocket for each of the croupiers. Conduct a test at the 5% level to see whether pocket color and croupier are independent, or whether there is sufficient evidence to show there might be something going on.

	Croupier A	Croupier B	Croupier C
Red	375	367	357
Black	379	336	362
Green	46	37	41

Step 1: Decide on the hypothesis you're going to test, and its alternative.

Step 2: Find the expected frequencies and the degrees of freedom. Use the table of expected frequencies below.

Hint: Complete the row and column totals first these are the same as for the observed frequencies above.

	Croupier A	Croupier B	Croupier C	Total
Red	1099×800/2300=382.3	1099×740/2300=353.6		
Black	1077×800/2300=374.6			
Green	124×800/2300=43.1			
Total	800			

Step 3: Determine the critical region for your decision.

Step 4: Calculate the test statistic X^2. Use the table below to help you.

	Observed	**Expected**	$\frac{(O - E)^2}{E}$
	375	382.3	$(375-382.3)^2/382.3 = 53.29/382.3 = 0.139$
A	379	374.6	$(379-374.6)^2/374.6 = 19.36/374.6 = 0.005$
	46	43.1	$(46-43.1)^2/43.1 = 8.41/43.1 = 0.195$
	367	353.6	$(367-353.6)^2/353.6 = 179.56/353.6 = 0.508$
B	336		
	37		
	357		
C	362		
	41		
	ΣO =	**ΣE =**	$\sum \frac{(O - E)^2}{E} =$

Step 5: See whether the test statistic is within the critical region.

Step 6: Make your decision.

Fat Dan thinks that one or more of his croupiers are somehow influencing the results of the roulette wheel. Here's data showing the observed frequency with which the ball lands in each color pocket for each of the croupiers. Conduct a test at the 5% level to see whether pocket color and croupier are independent, or whether there is sufficient evidence to show there might be something going on.

	Croupier A	Croupier B	Croupier C
Red	375	367	357
Black	379	336	362
Green	46	37	41

Step 1: Decide on the hypothesis you're going to test, and its alternative.

You want to test whether or not pocket color is independent of croupier. This gives

H_0: Roulette wheel pocket color and croupier are independent.

H_1: Pocket color and croupier are not independent.

Step 2: Find the expected frequencies and the degrees of freedom. Use the table of expected frequencies below.

You find the expected frequencies by multiplying each row and column total, and dividing by the grand total.

	Croupier A	Croupier B	Croupier C	Total
Red	1099×800/2300=382.3	1099×740/2300=353.6	1099×760/2300=363.1	1099
Black	1077×800/2300=374.6	1077×740/2300=346.5	1077×760/2300=355.9	1077
Green	124×800/2300=43.1	124×740/2300=39.9	124×760/2300=41.0	124
Total	800	740	760	2300

There are 3 columns and 3 rows, and we find the number of degrees of freedom by multiplying together (number of rows – 1) and (number of columns – 1). This gives us

$$\nu = 2 \times 2$$
$$= 4$$

Step 3: Determine the critical region for your decision.

From probability tables, $\chi^2_{5\%}(4) = 9.49$. This means that the critical region is given by $X^2 > 9.49$.

Step 4: Calculate the test statistic X^2. Use the table below to help you.

	Observed	Expected	$\frac{(O - E)^2}{E}$
	375	382.3	$(375-382.3)^2/382.3 = 53.29/382.3 = 0.139$
A	379	374.6	$(379-374.6)^2/374.6 = 19.36/374.6 = 0.005$
	46	43.1	$(46-43.1)^2/43.1 = 8.41/43.1 = 0.195$
	367	353.6	$(367-353.6)^2/353.6 = 179.56/353.6 = 0.508$
B	336	346.5	$(336-346.5)^2/346.5 = 110.25/346.5 = 0.318$
	37	39.9	$(37-39.9)^2/39.9 = 8.41/39.9 = 0.211$
	357	363.1	$(357-363.1)^2/363.1 = 37.21/363.1 = 0.102$
C	362	355.9	$(362-355.9)^2/355.9 = 37.21/355.9 = 0.105$
	41	41.0	$(41-41)^2/41 = 0/41 = 0$
	$\Sigma O = 2300$	$\Sigma E = 2300$	$\sum \frac{(O - E)^2}{E} = 1.583$

This means that the test statistic is given by $X^2 = 1.583$.

Step 5: See whether the test statistic is within the critical region.

The critical region is given by $X^2 > 9.48$. As $X^2 = 1.583$, the test statistic is outside the critical region.

Step 6: Make your decision.

As your test statistic lies outside the critical region, this means that there is insufficient evidence at the 5% level to reject the null hypothesis. In other words, you accept the null hypothesis that pocket color and croupier are independent.

15 correlation and regression

What's My Line?

The more I use this sandpaper, the less chance there is of him noticing my stubble.

Have you ever wondered how two things are connected?

So far we've looked at statistics that tell you about just one variable—like men's height, points scored by basketball players, or how long gumball flavor lasts—but there are other statistics that tell you about the **connection between variables**. Seeing how things are connected can give you a lot of information about the real world, information that you can use to your advantage. Stay with us while we show you the **key to spotting connections**: correlation and regression.

Never trust the weather

Concerts are best when they're in the open air—at least that's what these groovy guys think. They have a thriving business organizing open-air concerts, and ticket sales for the summer look promising.

Today's concert looks like it will be one of their best ones ever. The band has just started rehearsing, but there's a cloud on the horizon...

Feel that funky rhythm, baby.
Sweet! But is that a rain cloud I see up there?

Before too long the sky's overcast, temperatures are dipping, and it looks like rain. Even worse, ticket sales are hit. The guys are in trouble, and they can't afford for this to happen again.

What the guys want is to be able to predict what concert attendance will be given predicted hours of sunshine. That way, they'll be able to gauge the impact an overcast day is likely to have on attendance. If it looks like attendance will fall below 3,500 people, the point where ticket sales won't cover expenses, then they'll cancel the concert

They need **your** help.

Let's analyze sunshine and attendance

Here's sample data showing the predicted hours of sunshine and concert attendance for different events. How can we use this to estimate ticket sales based on the predicted hours of sunshine for the day?

Sunshine (hours)	1.9	2.5	3.2	3.8	4.7	5.5	5.9	7.2
Concert attendance (100's)	22	33	30	42	38	49	42	55

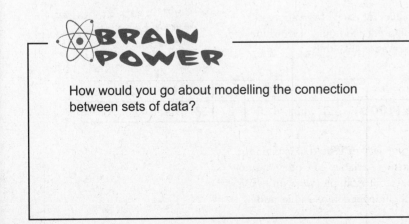

That's easy. We can find the mean and standard deviation and look at the distribution. That will tell us everything.

Most of the time, that's exactly the sort of thing we'd need to do to predict likely outcomes.

The problem this time is, what would we find the mean and standard deviation of? Would we use the concert attendance as the basis for our calculations, or would we use the hours of sunshine? Neither one of them gives us all the information that we need. Instead of considering just *one* set of data, we need to look at *both*.

So far we've looked at independent random variables, but not ones that are dependent. We can assume that if the weather is poor, the probability of high attendance at an open air concert will be lower than if the weather is sunny. But how do we model this connection, and how do we use this to predict attendance based on hours of sunshine?

It all comes down to the type of data.

BRAIN POWER

How would you go about modelling the connection between sets of data?

Exploring types of data

Up until now, the sort of data we've been dealing with has been univariate.

Univariate data concerns the frequency or probability of a single variable. As an example, univariate data could describe the winnings at a casino or the weights of brides in Statsville. In each case, just one thing is being described.

What univariate data can't do is show you connections between sets of data. For example, if you had univariate data describing the attendance figures at an open air concert, it wouldn't tell you anything about the predicted hours of sunshine on that day. It would just give you figures for concert attendance.

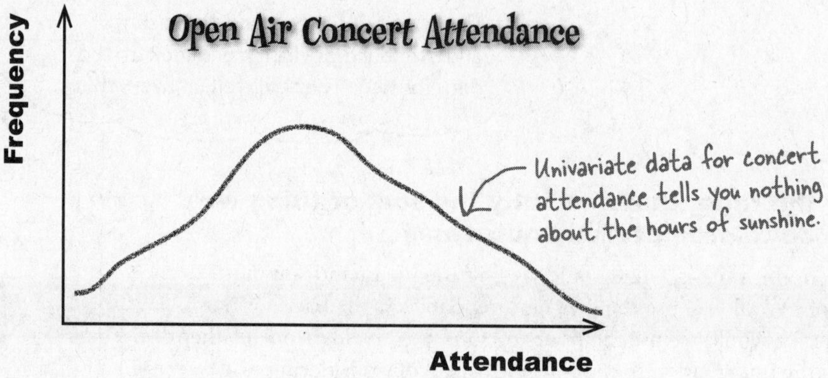

So what if we do need to know what the connection is between variables? While univariate data can't give us this information, there's another type of data that can—**bivariate data**.

All about bivariate data

Bivariate data gives you the value of *two* variables for each observation, not just one. As an example, it can give you both the predicted hours of sunshine and the concert attendance for a single event or observation, like this.

Bivariate data gives you the value of two variables for each observation.

Sunshine (hours)	1.9	2.5	3.2	3.8	4.7	5.5	5.9	7.2
Concert attendance (100's)	22	33	30	42	38	49	42	55

If one of the variables has been controlled in some way or is used to explain the other, it is called the ***independent*** or ***explanatory*** variable. The other variable is called the ***dependent*** or ***response*** variable. In our example, we want to use sunshine to predict attendance, so sunshine is the independent variable, and attendance is the dependent.

Visualizing bivariate data

Just as with univariate data, you can draw charts for bivariate data to help you see patterns. Instead of plotting a value against its frequency or probability, you plot one variable on the x-axis and the other variable against it on the y-axis. This helps you to visualize the connection between the two variables.

This sort of chart is called a ***scatter diagram*** or ***scatter plot***, and drawing one of these is a lot like drawing any other sort of chart.

Start off by drawing two axes, one vertical and one horizontal. Use the x-axis for one variable and the y-axis for the other. The independent variable normally goes along the x-axis, leaving the dependent variable to go on the y-axis. Once you've drawn your axes, you then take the values for each observation and plot them on the scatter plot.

Here's a scatter plot showing the number of hours of sunshine and concert attendance figures for particular events or observations. As the predicted number of hours sunshine is the independent variable, we've plotted it on the x-axis. The concert attendance is the dependent variable, so that's on the y-axis.

Hours sunshine goes on the x-axis, attendance on the y-axis.

x (sunshine)	1.9	2.5	3.2	3.8	4.7	5.5	5.9	7.2
y (attendance)	22	33	30	42	38	49	42	55

Here's the data.

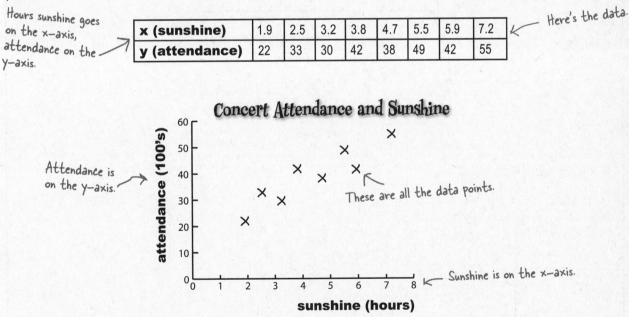

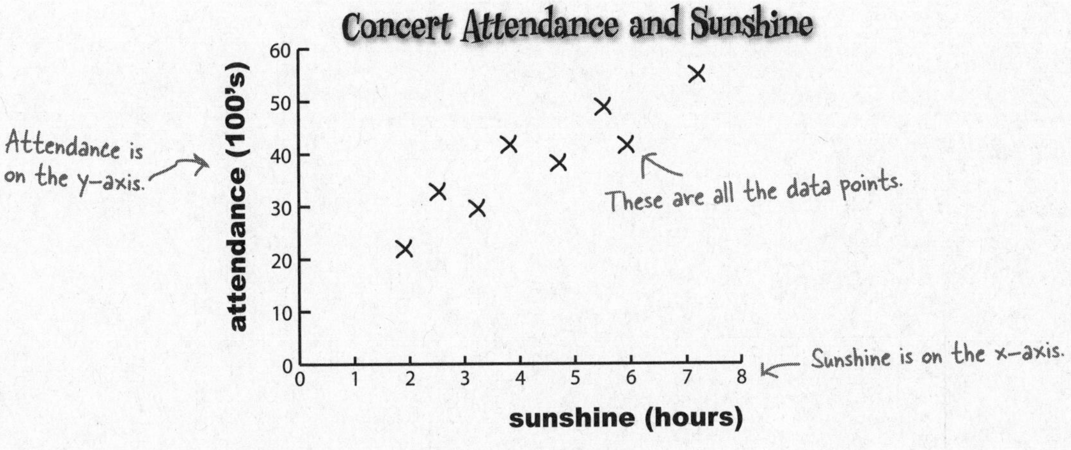

Can you see how the scatter diagram helps you visualize patterns in the data? Can you see how this might help us to define the connection between open air concert attendance and predicted number of hours sunshine for the day?

Sharpen your pencil

We know we haven't shown you how to analyze bivariate data yet, but see how far you get in analyzing the scatter diagram for the concert organizers.

What sort of patterns do you see in the chart? How can you relate this to the underlying data? What do you expect open air concert attendance to be like if it's sunny? What about if it's overcast?

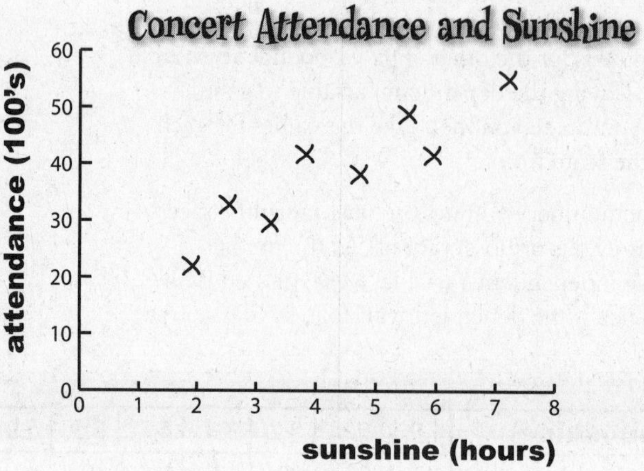

Five Minute Mystery

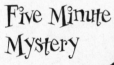

The Case of the High Sunscreen Sales

An intern at a sunscreen manufacturer has been given the task of looking at sunscreen sales in order to see how they can best market their particular brand.

He's been given a pile of generated scatter diagrams that model sunscreen sales against various other factors. He's been asked to pull out ones where there seems to be some relationship between the two factors on the diagram, as this will help the sales team.

The first diagram that the intern finds plots sunscreen sales for the day against pollen count. He's surprised to see that when there's a high pollen count, sales of sunscreen are significantly higher, and he decides to tell the sales team that they need to think about using pollen count in their advertising.

When the sales team hears his suggestion, they look at him blankly. What do you think the sales team should do?

Does a high pollen count make people buy sunscreen?

Sharpen your pencil Solution

We know we haven't shown you how to analyze bivariate data yet, but see how you get on with analyzing the scatter diagram for the concert organizers.

What sort of patterns do you see in the chart? How can you relate this to the underlying data? What do you expect open air concert attendance to be like if it's sunny? What about if it's overcast?

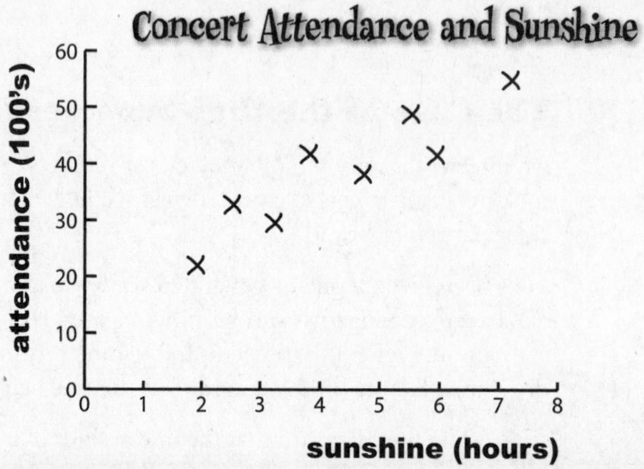

First of all, the chart shows that the data points are clustered around a straight line on the chart, and this line slopes upwards. It looks like, if the predicted number of hours of sunshine in a day is relatively low, then the concert attendance is low too. If the number of hours sunshine is high, then we can expect concert attendance to be high too. This basically means that the sunnier the weather, the more people you can expect to go to the open air concert.

One thing that's important to note is that we can only be confident about saying this within the range of the data. We have no data to say what the pattern is like if the number of hours of sunshine is below 2 hours or above 7.5 hours.

Scatter diagrams show you patterns

As you can see, scatter diagrams are useful because they show the actual pattern of the data. They enable you to more clearly visualize what connection there is between two variables, if indeed there's any connection at all.

The scatter diagram for the concert data shows a distinct pattern—the data points are clustered along a straight line. We call this a ***correlation***.

Linear Correlations Up Close

Scatter diagrams show the ***correlation*** between pairs of values.

Correlations are mathematical relationships between variables. You can identify correlations on a scatter diagram by the distinct patterns they form. The correlation is said to be ***linear*** if the scatter diagram shows the points lying in an approximately straight line.

Let's take a look at a few common types of correlation between two variables:

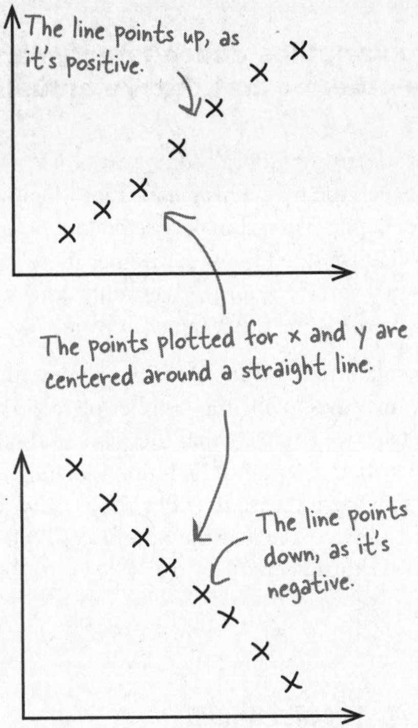

The line points up, as it's positive.
The points plotted for x and y are centered around a straight line.
The line points down, as it's negative.

Positive linear correlation

Positive linear correlation is when low values on the x-axis correspond to low values on the y-axis, and higher values of x correspond to higher values of y. In other words, y tends to increase as x increases.

Negative linear correlation

Negative linear correlation is when low values on the x-axis correspond to high values on the y-axis, and higher values of x correspond to lower values of y. In other words, y tends to decrease as x increases.

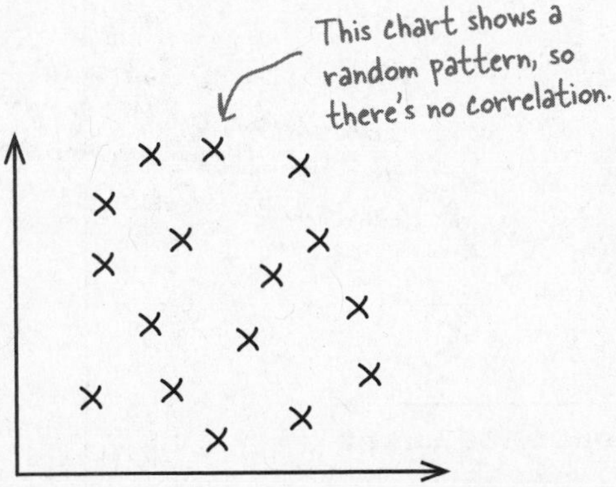

No correlation

If the values of x and y form a random pattern, then we say there's no correlation.

Correlation vs. causation

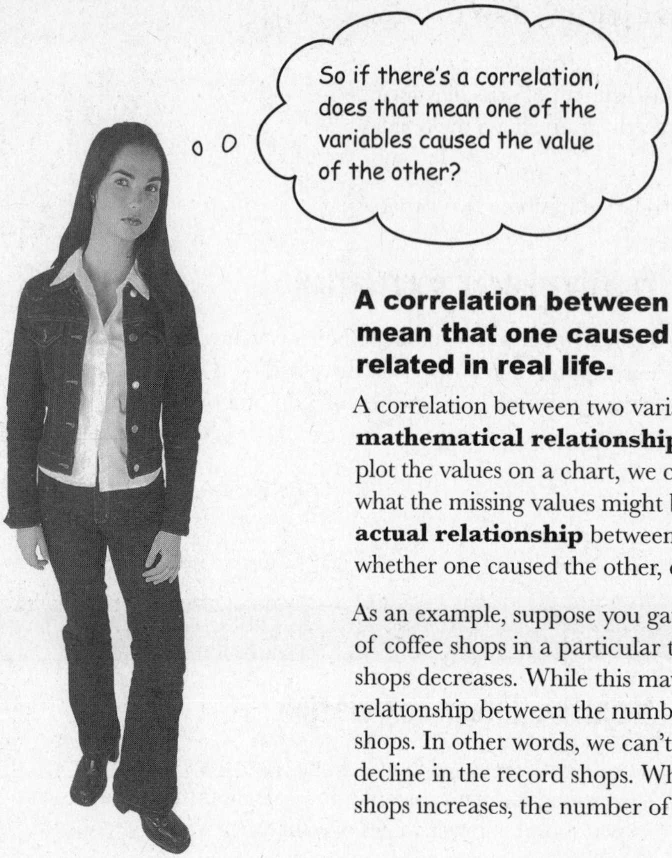

So if there's a correlation, does that mean one of the variables caused the value of the other?

A correlation between two variables doesn't necessarily mean that one caused the other or that they're actually related in real life.

A correlation between two variables means that there's some sort of **mathematical relationship** between the two. This means that when we plot the values on a chart, we can see a pattern and make predictions about what the missing values might be. What we don't know is whether there's an **actual relationship** between the two variables, and we certainly don't know whether one caused the other, or if there's some other factor at work.

As an example, suppose you gather data and find that over time, the number of coffee shops in a particular town increases, while the number of record shops decreases. While this may be true, we can't say that there is a real-life relationship between the number of coffee shops and the number of record shops. In other words, we can't say that the increase in coffee shops caused the decline in the record shops. What we *can* say is that as the number of coffee shops increases, the number of record shops decreases.

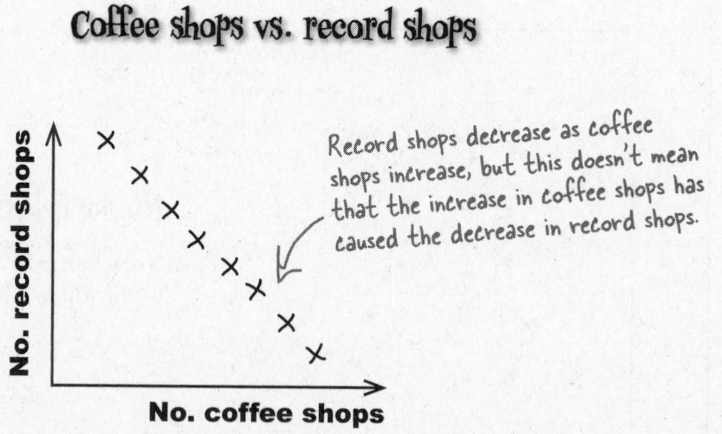

Solved: The Case of the High Sunscreen Sales

Does a high pollen count make people buy sunscreen?

One of the sales team members walks over to the intern.

"Thanks for the idea," she says, "but we're not going to use it in our advertising. You see, the high pollen count doesn't make people buy more sunscreen."

The intern looks at her, confused. "But it's all here on this scatter diagram. As pollen count increases, so do sunscreen sales."

"That's true," says the salesperson, "but that doesn't mean that the high pollen count has caused the high sales. The days when the pollen count is high are generally days when the weather is sunny, so people are going outside more. They're buying more sunscreen because they're spending the day outside."

Far out, dude, I'm liking the way the sunshine and attendance connect.

there are no Dumb Questions

Q: So are we saying that the predicted sunshine causes low ticket sales?

A: The bivariate data shows that there is a mathematical relationship between the two variables, but we can't use it to demonstrate cause and effect. It's intuitively possible that more people will go to open air concerts when it's sunny, but we can't say for certain that sunshine causes this. We'd need to do more research, as there may be other factors.

Q: Other factors? Like what?

A: One example would be the popularity of the artist performing. If a well-known artist is holding a concert, then fans may want to go to the concert no matter what the weather. Similarly, an unpopular artist is unlikely to have the same dedication from fans.

Q: Do scatter diagrams use populations or samples of data?

A: They can use either. A lot of the time, you'll actually be using samples, but the process of plotting a scatter diagram is the same irrespective of whether you have a sample or a population.

Q: If there's a correlation between two variables, does it have to be linear?

A: Correlation measures linear relationships, but not all relationships are linear. As an example, a strong relationship between two variables could be a distinctive curve, such as $y = x^2$. In this chapter, we're only going to be dealing with linear relationships, though.

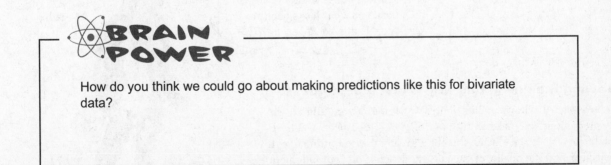

But hold on, man! How can we predict concert attendance based on predicted sunshine? If the concert attendance drops below 3,500, we'll have to bail out, and that'd be a burn.

We need to predict the concert attendance

So far we've looked at what bivariate data is, and how scatter diagrams can show whether there's a mathematical relationship between the two variables. What we *haven't* looked at yet is how we can use this to make predictions.

What we need to do next is see how we can use the data to make predictions for concert attendance, based on predicted hours of sunshine.

BRAIN POWER

How do you think we could go about making predictions like this for bivariate data?

Predict values with a line of best fit

So far you've seen how scatter diagrams can help you see whether there's a correlation between values, by showing you if there's some sort of pattern. But how can you use this to predict concert attendance, based on the predicted amount of sunshine? How would you use your existing scatter diagram to predict the concert attendance if you know how many hours of sunshine are expected for the day?

One way of doing this is to draw a straight line through the points on the scatter diagram, making it fit the points as closely as possible. You won't be able to get the straight line to go through every point, but if there's a linear correlation, you should be able to make sure every point is reasonably close to the line you draw. Doing this means that you can read off an estimate for the concert attendance based on the predicted amount of sunshine.

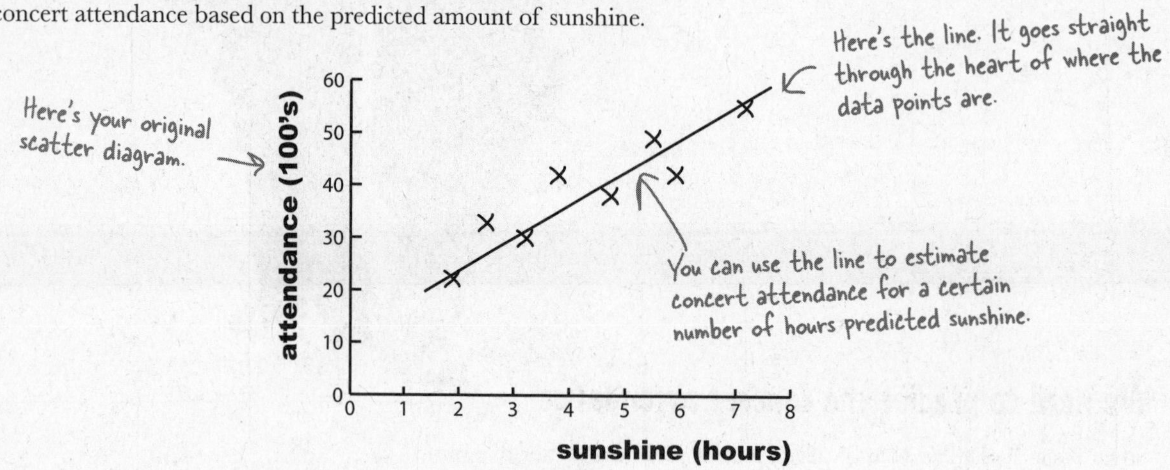

Here's the line. It goes straight through the heart of where the data points are.
Here's your original scatter diagram.
attendance (100's)
60
50
40
30
20
10
0
0
1
2
3
4
5
6
7
8
sunshine (hours)
You can use the line to estimate concert attendance for a certain number of hours predicted sunshine.

The line that best fits the data points is called the ***line of best fit***.

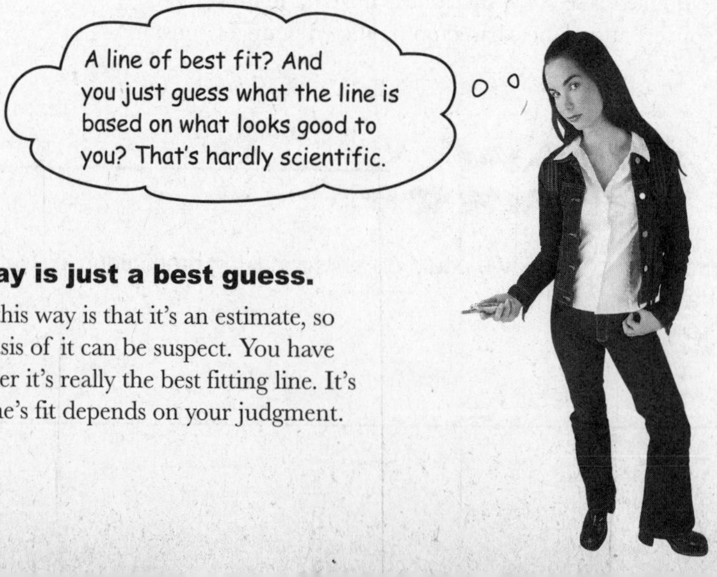

A line of best fit? And you just guess what the line is based on what looks good to you? That's hardly scientific.

Drawing the line in this way is just a best guess.

The trouble with drawing a line in this way is that it's an estimate, so any predictions you make on the basis of it can be suspect. You have no precise way of measuring whether it's really the best fitting line. It's subjective, and the quality of the line's fit depends on your judgment.

Your best guess is still a guess

Imagine if you asked three different people to draw what each of them think is the line of best fit for the open air concert data. It's quite likely that each person would come up with a slightly different line of best fit, like this:

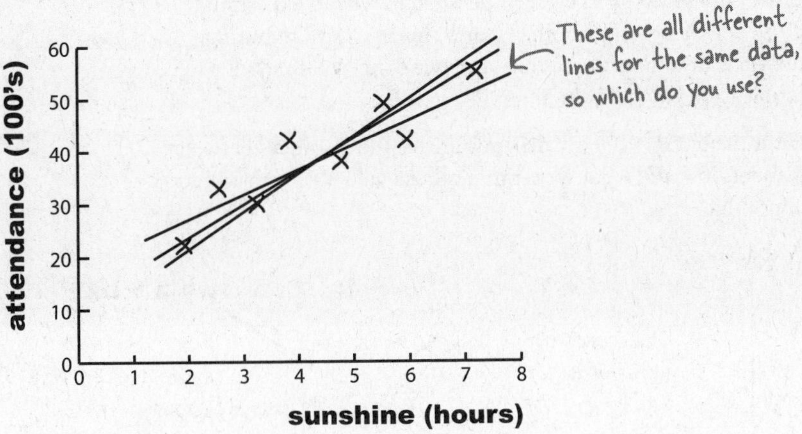
60
50
40
30
20
10
0
attendance (100's)
0
1
2
3
4
5
6
7
8
sunshine (hours)
These are all different lines for the same data, so which do you use?

All three lines could conceivably be a line of best fit for the data, but what we can't tell is which one's really best.

What we really need is some alternative to drawing the line of best fit by eye. Instead of guessing what the line should be, it will be more reliable if we had a mathematical or statistical way of using the data we have available to find the line that fits best.

We need to find the equation of the line

The equation for a straight line takes the form y = a + bx, where a is the point where the line crosses the y-axis, and b is the slope of the line. This means that we can write the line of best fit in the form y = a + bx.

In our case, we're using x to represent the predicted number of hours of sunshine, and y to represent the corresponding open air concert figures. If we can use the concert attendance data to somehow find the most suitable values of a and b, we'll have a reliable way to find the equation of the line, and a more reliable way of predicting concert attendance based on predicted hour of sunshine.

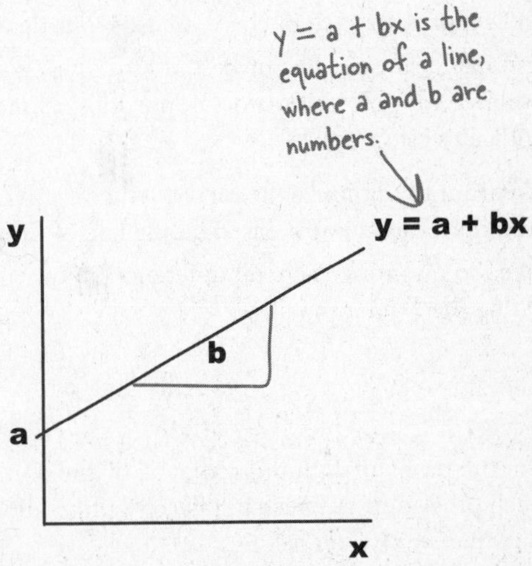
y = a + bx is the equation of a line, where a and b are numbers.
y
y = a + bx
b
a
x

We need to minimize the errors

Let's take a look at what we need from the line of best fit, $y = a + bx$.

The best fitting line is the one that most accurately predicts the true values of all the points. This means that for each known value of x, we need each of the y variables in the data set to be as close as possible to what we'd estimate them to be using the line of best fit. In other words, given a certain number of hours sunshine, we want our estimates for open air concert attendance to be as close as possible to the actual values.

The line of best fit is the line $y = a + bx$ that minimizes the distances between the actual observations of y and what we estimate those values of y to be for each corresponding value of x.

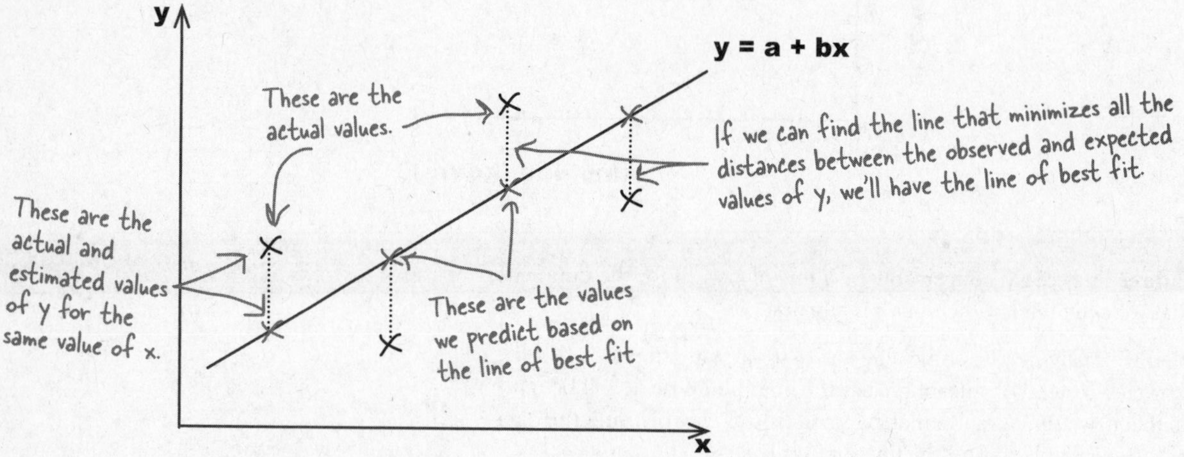
y
x
y = a + bx
These are the actual values.
If we can find the line that minimizes all the distances between the observed and expected values of y, we'll have the line of best fit.
These are the actual and estimated values of y for the same value of x.
These are the values we predict based on the line of best fit

Let's represent each of the y values in our data set using y_i, and its estimate using the line of best fit as $\hat{y}_i$. This is the same notation that we used for point estimators in previous chapters, as the ^ symbol indicates estimates.

We want to minimize the total distance between each actual value of y and our estimate of it based on the line of best fit. In other words, we need to minimize the total differences between y_i and $\hat{y}_i$. We could try doing this by minimizing

$$\Sigma(y_i - \hat{y}_i)$$

but the problem with this is that all of the distances will actually cancel each other out. We need to take a slightly different approach, and it's one that we've seen before.

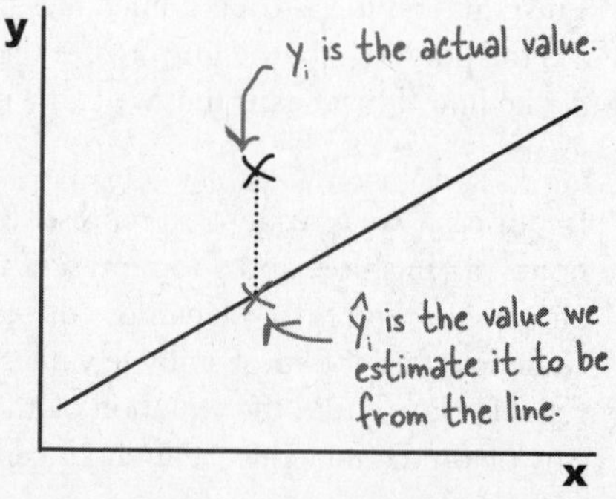
y
x
y_i is the actual value.
$\hat{y}_i$ is the value we estimate it to be from the line.

Introducing the sum of squared errors

Can you remember when we first derived the variance? We wanted to look at the total distance between sets of values and the mean, but the total distances cancelled each other out. To get around this, we added together all the distances squared instead to ensure that all values were positive.

We have a similar situation here. Instead of looking at the total distance between the actual and expected points, we need to add together the distances *squared*. That way, we make sure that all the values are positive.

The total sum of the distances squared is called the **sum of squared errors**, or **SSE**. It's given by:

The sum of squared errors →

$$SSE = \Sigma(y - \hat{y})^2$$

← The difference between the real values of y, and what we predict from the line of best fit

In other words, we take each value of y, subtract the predicted value of y from the line of best fit, square it, and then add all the results together.

The SSE reminds me of the variance. The variance uses squared distances from the mean, and the SSE uses squared distances from the line.

The variance and SSE are calculated in similar ways.

The SSE isn't the variance, but it *does* deal with the distance squared between two particular points. It gives the total of the distances squared between the actual value of y and what we predict the value of y to be, based on the line of best fit.

What we need to do now is use the data to find the values of a and b that minimize the SSE, based on the line y = a + bx.

Find the equation for the line of best fit

We've said that we want to minimize the sum of squared errors, $\Sigma(y - \hat{y})^2$, where $y = a + bx$. By doing this, we'll be able to find optimal values for a and b, and that will give us the equation for the line of best fit.

Let's start with b

The value of b for the line $y = a + bx$ gives us the ***slope***, or steepness, of the line. In other words, b is the slope for the line of best fit.

We're not going to show you the proof for this, but the value of b that minimizes the SSE $\Sigma(y - \hat{y})^2$ is given by

Each value of x, minus the mean of the x values, multiplied by the corresponding value of y, minus the mean of the y values

$$b = \frac{\Sigma((x - \bar{x})(y - \bar{y}))}{\Sigma(x - \bar{x})^2}$$

This bit's similar to how you find the variance of x. For each value of x, subtract the mean of the x values and square the result.

Are you sure? That looks complicated.

The calculation looks tricky at first, but it's not that difficult with practice.

First of all, find $\bar{x}$ and $\bar{y}$, the means of the x and y values for the data that you have. Once you've done that, calculate $(x - \bar{x})$ multiplied by $(y - \bar{y})$ for every observation in your data set, and add the results together. Finally, divide the whole lot by $\Sigma(x - \bar{x})^2$. This last part of the equation is very similar to how you calculate the variance of a sample. The only difference is that you don't divide by $(n - 1)$. You can also get software packages that work all of this out for you.

Let's take a look at how you use this in practice.

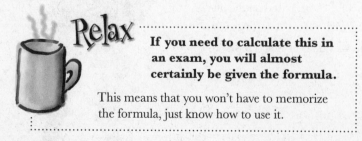

Relax

If you need to calculate this in an exam, you will almost certainly be given the formula.

This means that you won't have to memorize the formula, just know how to use it.

Finding the slope for the line of best fit

Let's see if we can use this to find the slope of the line y = a + bx for the concert data. First of all, here's a reminder of the data:

x (sunshine)	1.9	2.5	3.2	3.8	4.7	5.5	5.9	7.2
y (attendance)	22	33	30	42	38	49	42	55

Let's start by finding the values of $\bar{x}$ and $\bar{y}$, the sample means of the x and y values. We calculate these in exactly the same way as before, so

$\bar{x} = (1.9 + 2.5 + 3.2 + 3.8 + 4.7 + 5.5 + 5.9 + 7.2)/8$

$= 34.7/8$

$= 4.3375$

Use the values of x to find $\bar{x}$, and the values of y to find $\bar{y}$.

$\bar{y} = (22 + 33 + 30 + 42 + 38 + 49 + 42 + 55)/8$

$= 311/8$

$= 38.875$

Now that we've found $\bar{x}$ and $\bar{y}$, we can use them to help us find the value of b using the formula on the opposite page.

We use $\bar{x}$ and $\bar{y}$ to help us find b

The first part of the formula is $\Sigma(x - \bar{x})(y - \bar{y})$. To find this, we take the x and y values for each observation, subtract $\bar{x}$ from the x value, subtract $\bar{y}$ from the y value, and then multiply the two together. Once we've done this for every observation, we then add the whole lot up together.

$x - \bar{x}$ $y - \bar{y}$

$\Sigma(x - \bar{x})(y - \bar{y}) = (1.9 - 4.3375)(22 - 38.75) + (2.5 - 4.3375)(33 - 38.75) + (3.2 - 4.3375)(30 - 38.75) + (3.8 - 4.3375)(42 - 38.75) + (4.7 - 4.3375)(38 - 38.75) + (5.5 - 4.3375)(49 - 38.75) + (5.9 - 4.3375)(42 - 38.75) + (7.2 - 4.3375)(55 - 38.75)$

$(x - \bar{x})(y - \bar{y})$

Add these together for every set of values.

$= (-2.4375)(-16.75) + (-1.8375)(-5.875) + (-1.1375)(-8.875) + (-0.5375)(3.125) + (0.3625)(-0.875) + (1.1625)(10.125) + (1.5625)(3.125) + (2.8625)(16.125)$

$= 40.828125 + 10.7953125 + 10.0953125 - 1.6796875 - 0.3171875 + 11.7703125 + 4.8828125 + 46.1578125$

$= 122.53$ (to 2 decimal places)

Finding the slope for the line of best fit, part ii

Here's a reminder of the formula.

$$b = \frac{\Sigma(x - \bar{x})(y - \bar{y})}{\Sigma(x - \bar{x})^2}$$

Here's a reminder of the data for concert attendance and predicted hours of sunshine:

x (sunshine)	1.9	2.5	3.2	3.8	4.7	5.5	5.9	7.2
y (attendance)	22	33	30	42	38	49	42	55

We're part of the way through calculating the value of b, where y = a + bx. We've found that $\bar{x} = 4.3375$, $\bar{y} = 38.875$, and $\Sigma(x - \bar{x})(y - \bar{y}) = 122.53$. The final thing we have left to find is $\Sigma(x - \bar{x})^2$. Let's give it a go

We find $\Sigma(x - \bar{x})^2$ using the x values. It's a bit like finding the variance of a sample, but without dividing by n–1.

$$\Sigma(x - \bar{x})^2 = (1.9 - 4.3375)^2 + (2.5 - 4.3375)^2 + (3.2 - 4.3375)^2 + (3.8 - 4.3375)^2 + (4.7 - 4.3375)^2 + (5.5 - 4.3375)^2 + (5.9 - 4.3375)^2 + (7.2 - 4.3375)^2$$

$(x - \bar{x})^2$

$$= (-2.4375)^2 + (-1.8375)^2 + (-1.1375)^2 + (-0.5375)^2 + (0.3625)^2 + (1.1625)^2 + (1.5625)^2 + (2.8625)^2$$

$$= 23.02 \text{ (to 2 decimal places)}$$

Note, we don't use y or $\bar{y}$ in this part of the equation.

We find the value of b by dividing $\Sigma(x - \bar{x})(y - \bar{y})$ by $\Sigma(x - \bar{x})^2$. This gives us

$$b = 122.53/23.02$$
$$= 5.32$$

We've found b. This gives the slope for the line of best fit.

In other words, the line of best fit for the data is y = a + 5.32x. But what's a?

there are no Dumb Questions

Q: It looks like the formulas you've given are for samples rather than populations. Is that right?

A: That's right. We've used samples rather than populations because the data we've been given is a sample. There's nothing to stop you using a population if you have the data, just use μ instead of $\bar{x}$.

Q: Is the value of b always positive?

A: No, it isn't. Whether b is positive or negative actually depends on the type of linear correlation. For positive linear correlation, b is positive. For negative linear correlation, b is negative.

Q: I've heard of the term gradient. What's that?

A: Gradient is another term for the slope of the line, b.

Q: What about if there's no correlation? Can I still work out b?

A: If there's no correlation, you can still technically find a line of best fit, but it won't be an effective model of the data, and you won't be able to make accurate predictions using it.

Q: Is there an easy way of calculating b?

A: Calculating b is tricky if you have lots of observations, but you can get software packages to calculate this for you.

We've found b, but what about a?

So far we've found what the optimal value of b is for the line of best fit $y = a + bx$. What we don't know yet is the value of a.

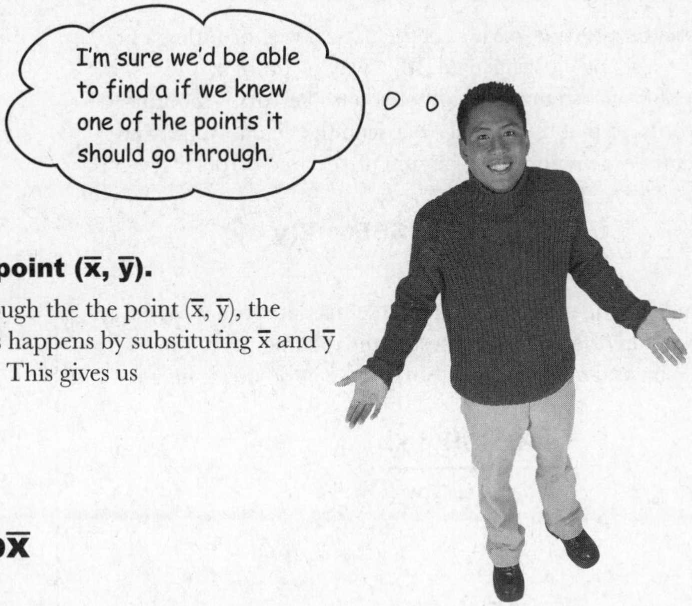

I'm sure we'd be able to find a if we knew one of the points it should go through.

The line needs to go through point $(\bar{x}, \bar{y})$.

It's good for the line of best fit to go through the the point $(\bar{x}, \bar{y})$, the means of x and y. We can make sure this happens by substituting $\bar{x}$ and $\bar{y}$ into the equation for the line $y = a + bx$. This gives us

$$\bar{y} = a + b\bar{x}$$

or

$$\mathbf{a = \bar{y} - b\bar{x}}$$

We've already found values for $\bar{x}$, $\bar{y}$, and b. Substituting in these values gives us

$$a = 38.875 - 5.32(4.3375)$$
$$= 38.875 - 23.0755$$
$$= 15.80 \text{ (to 2 decimal places)}$$

This means that the line of best fit is given by

$$y = 15.80 + 5.32x$$

If you're taking a statistics exam, it's likely you'll be given this formula.

This means that you're unlikely to have to memorize it, you just need to know how to use it.

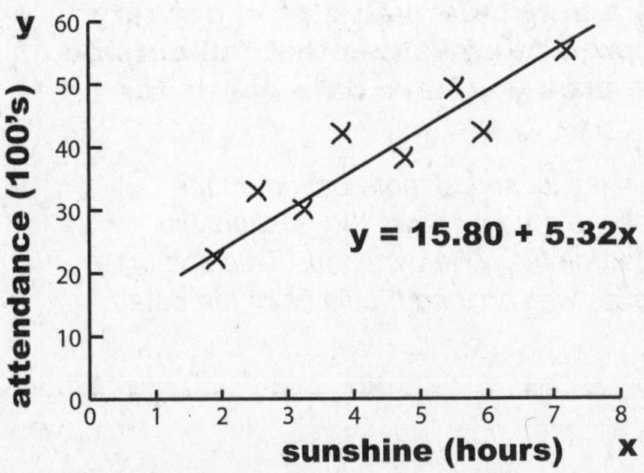

Least Squares Regression Up Close

The mathematical method we've been using to find the line of best fit is called **least squares regression**.

Least squares regression is a mathematical way of fitting a line of best fit to a set of bivariate data. It's a way of fitting a line y = a + bx to a set of values so that the sum of squared errors is minimized—in other words, so that the distance between the actual values and their estimates are minimized. The sum of squared errors is given by

$$\mathbf{SSE = \Sigma(y - \hat{y})^2}$$

To perform least squares regression on a set of data, you need to find the values of a and b that best fit the data points to the line y = a + bx and minimizes the SSE. You can do this using:

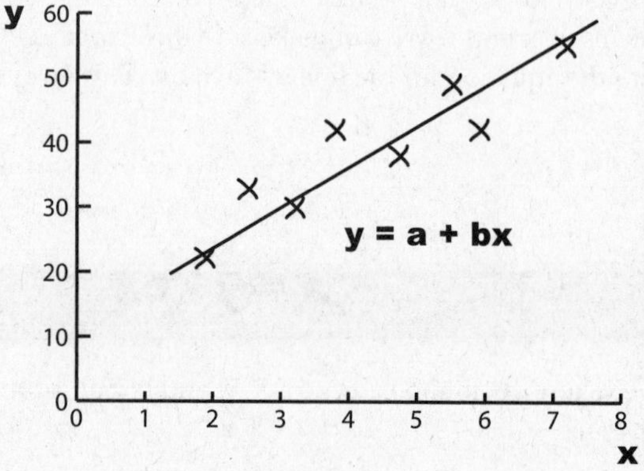

$$\mathbf{b = \frac{\Sigma(x - \bar{x})(y - \bar{y})}{\Sigma(x - \bar{x})^2}}$$

and

$$\mathbf{a = \bar{y} - b\bar{x}}$$

Once you've found the line of best fit, y = a + bx, you can use it to predict the value of y, given a value b. To do this, just substitute your x value into the equation y = a + bx.

The line y = a + bx is called the ***regression line***.

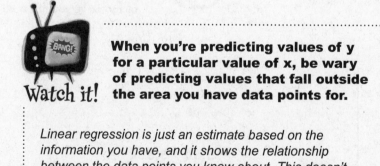

Watch it!

When you're predicting values of y for a particular value of x, be wary of predicting values that fall outside the area you have data points for.

Linear regression is just an estimate based on the information you have, and it shows the relationship between the data points you know about. This doesn't mean that it applies well beyond the limits of the data

Sharpen your pencil

We've found an equation for the regression line, so now the concert organizers have a couple of questions for you. As a reminder, the regression line is given by

$$y = 15.80 + 5.32x$$

where x is the predicted hours of sunshine, and y is the concert attendance in 100's.

The predicted amount of sunshine on the day of the next concert is 6 hours. What do you expect concert attendance to be?

If concert attendance looks like it's dropping below 3,500, the concert organizers won't make a profit and will have to cancel the concert. What's the corresponding number of hours of predicted sunshine?

Sharpen your pencil Solution

We've found an equation for the regression line, so now the concert organizers have a couple of questions for you. As a reminder, the regression line is given by

$$y = 15.80 + 5.32x$$

where x is the predicted hours of sunshine, and y is the concert attendance in 100's.

The predicted amount of sunshine on the day of the next concert is 6 hours. What do you expect concert attendance to be?

As x is the predicted number of hours of sunshine, this means that $x = 6$. We need to find the corresponding prediction for concert attendance, so this means we need to find y for this value of x.

$$\begin{aligned} y &= 15.80 + 5.32x \\ &= 15.80 + 5.32 \times 6 \\ &= 15.80 + 31.92 \\ &= 47.72 \end{aligned}$$

As y is in 100s, this means that the expected concert attendance is $47.72 \times 100 = 4772$.

If concert attendance looks like it's dropping below 3,500, the concert organizers won't make a profit and will have to cancel the concert. What's the corresponding number of hours of predicted sunshine?

This time, we want to find the value of x for a particular value of y.

The concert attendance is 3,500, which means that $y = 35$. This gives us

$$\begin{aligned} y &= 15.80 + 5.32x \\ 35 &= 15.80 + 5.32x \\ 35 - 15.80 &= 5.32x \\ 19.2 &= 5.32x \\ x &= 19.2/5.32 \\ &= 3.61 \text{ (to 2 decimal places)} \end{aligned}$$

In other words, we'd predict concert attendance to be below 3,500 if the predicted hours of sunshine is below 3.61 hours.

You've made the connection

So far you've used linear regression to model the connection between predicted hours of sunshine and concert attendance. Once you know what the predicted amount of sunshine is, you can predict concert attendance using y = a + bx.

Being able to predict attendance means you'll be able to really help the concert organizers know what they can expect ticket sales to be, and also what sort of profit they can reasonably expect to make from each event.

That's awesome, dude! But just one question. How accurate is this exactly?

It's the line of best fit, but we don't know how accurate it is.

The line y = a + bx is the best line we could have come up with, but how accurately does it model the connection between the amount of sunshine and the concert attendance? There's one thing left to consider, the strength of correlation of the regression line.

What would be really useful is if we could come up with some way of indicating how far the points are dispersed away from the line, as that will give an indication of how accurate we can expect our predictions to be based on what we already know.

Let's look at a few examples.

BRAIN POWER

Why do you think it's important to know the strength of the correlation? What difference do you think this would make to the concert organizers?

Let's look at some correlations

The line of best fit of a set of data is the best line we can come up with to model the mathematical relationship between two variables.

Even though it's the line that fits the data best, it's unlikely that the line will fit precisely through every single point. Let's look at some different sets of data to see how closely the line fits the data.

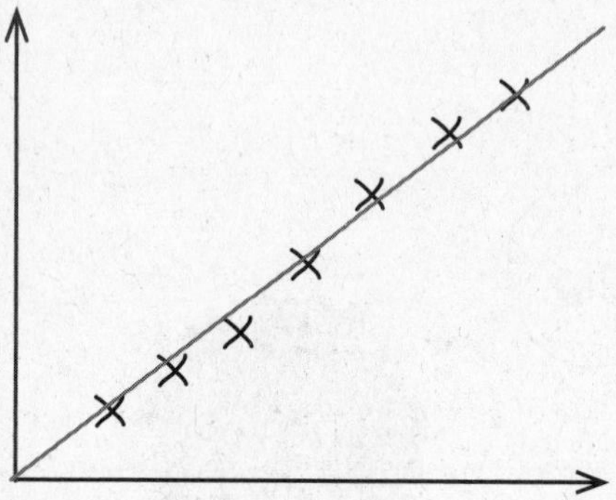

Accurate linear correlation

For this set of data, the linear correlation is an accurate fit of the data. The regression line isn't 100% perfect, but it's very close. It's likely that any predictions made on the basis of it will be accurate.

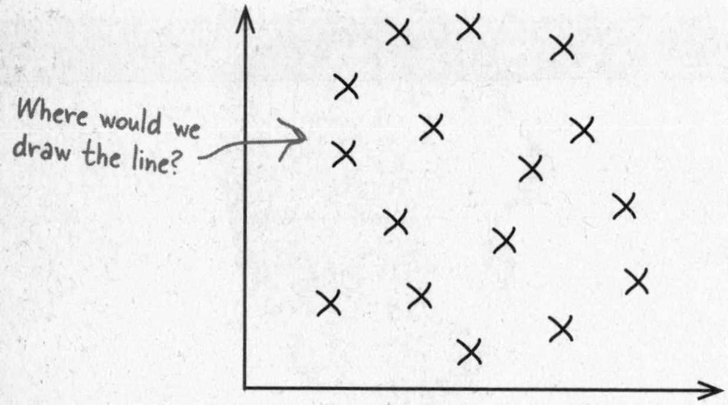
Where would we draw the line?

No linear correlation

For this set of data, there is no linear correlation. It's possible to calculate a regression line using least squares regression, but any predictions made are unlikely to be accurate.

Can you see what the problem is?

Both sets of data have a regression line, but the actual fit of the data varies quite a lot. For the first set of data, the correlation is very tight, but for the second, the points are scattered too widely for the regression line to be useful.

Least squares estimates can be used to predict values, which means they would be helpful if there was some way of indicating how tightly the data points fit the line, and how accurate we can expect any predictions to be as a result.

There's a way of calculating the fit of the line, called the **correlation coefficient**.

The correlation coefficient measures how well the line fits the data

The ***correlation coefficient*** is a number between -1 and 1 that describes the scatter of data points away from the line of best fit. It's a way of gauging how well the regression line fits the data. It's normally represented by the letter r.

If r is -1, the data is a **perfect negative linear correlation**, with all of the data point in a straight line. If r is 1, the data is a **perfect positive linear correlation**. If r is 0, then there is **no correlation**.

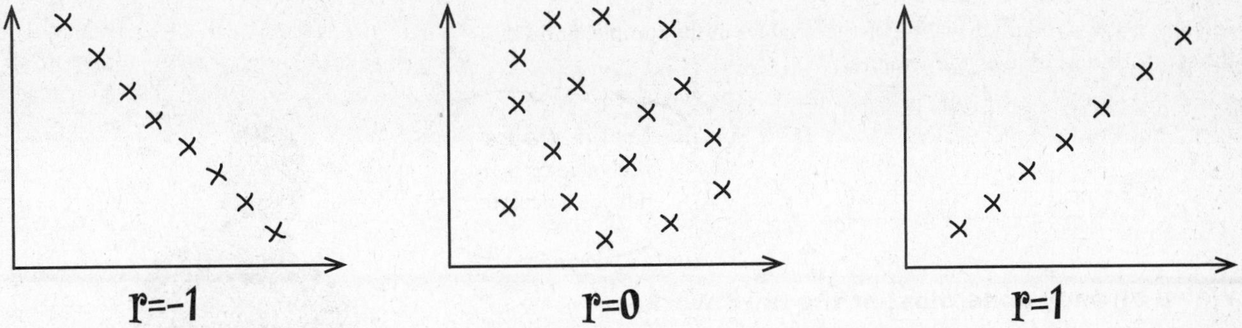
r=-1
r=0
r=1

Usually r is somewhere between these values, as -1, 0, and 1 are all extreme.

If r is negative, then there's a **negative linear correlation** between the two variables. The closer r gets to -1, the stronger the correlation, and the closer the points are to the line.

If r is positive, then there's a **positive linear correlation** between the variables. The closer r gets to 1, the stronger the correlation.

In general, as r gets closer to 0, the **linear correlation gets weaker**. This means that the regression line won't be able to predict y values as accurately as when r is close to 1 or -1. The pattern might be random, or the relationship between the variables might not be linear.

If we can calculate r for the concert data, we'll have an idea of how accurately we can predict concert attendance based on the predicted hours of sunshine. So how do we calculate r? Turn the page and we'll show you how.

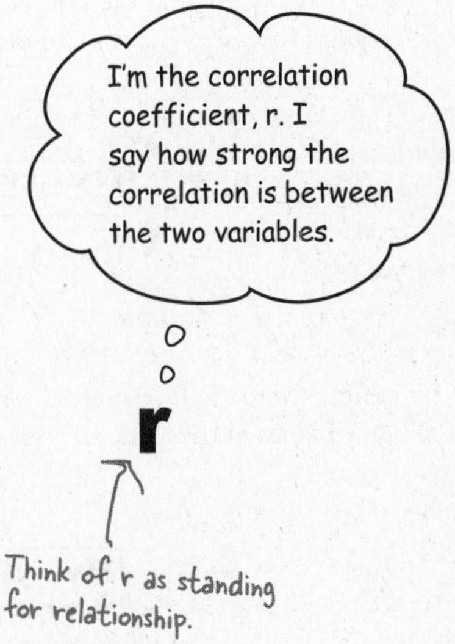
I'm the correlation coefficient, r. I say how strong the correlation is between the two variables.
r
Think of r as standing for relationship.

There's a formula for calculating the correlation coefficient, r

So how do we calculate the correlation coefficient, r?

We're not going to show you the proof for this, but the correlation coefficient r is given by

$$r = b\frac{s_x}{s_y}$$

b is the slope of the line of best fit that you've already found.

s_x is the standard deviation of the x values in the sample. s_y is the standard deviation of the y values.

where s_x is the standard deviation of the x values in the sample, and s_y is the standard deviation of the y values.

I get it. We use the value of b to help us calculate r.

We've already done most of the hard work.

Since we've already calculated b, all we have left to find is s_x and s_y. What's more, we're already most of the way towards finding s_x.

When we calculated b, we needed to find the value of $\Sigma(x - \bar{x})^2$. If we divide this by n - 1, this actually gives us the sample variance of the x values. If we then take the square root, we'll have s_x. In other words,

This is the standard deviation of the x values in the sample, it's the same formula you've seen before

$$s_x = \sqrt{\frac{\Sigma(x - \bar{x})^2}{n - 1}}$$

You calculated this bit earlier, so there's no need to calculate it again.

The only remaining piece of the equation we have to find is s_y, the standard deviation of the y values in the sample. We calculate this in a similar way to finding s_x.

$$s_y = \sqrt{\frac{\Sigma(y - \bar{y})^2}{n - 1}}$$

This is the standard deviation of the x values in the sample, and you've done these sorts of calculations before.

Let's try finding what r is for the concert attendance data.

Find r for the concert data

Let's use the formula to find the value of r for the concert data. First of all, here's a reminder of the data:

x (sunshine)	1.9	2.5	3.2	3.8	4.7	5.5	5.9	7.2
y (attendance)	22	33	30	42	38	49	42	55

To find r, we need to know the values of b, s_x, and s_y so that we can use them in the formula on the opposite page. So far we've found that

$b = 5.32$ ← This is the slope of the line we found earlier.

but what about s_x and s_y?

Let's start with s_x. We found earlier that $\Sigma(x - \bar{x})^2 = 23.02$, and we know that the sample size is 8. This means that if we divide 23.02 by 7, we'll have the sample variance of x. To find s_x, we take the square root.

$$s_x = \sqrt{(23.02/7)}$$
$$= \sqrt{3.28857}$$
$$= 1.81 \text{ (to 2 decimal places)}$$

← This is the standard deviation of the x values. It's a sample, so we divide by n – 1.

The only piece of the formula we have left to find is s_y. We already know that $\bar{y} = 38.875$, as we found it earlier on, so this means that

$$\Sigma(y - \bar{y})^2 = (22 - 38.875)^2 + (33 - 38.875)^2 + (30 - 38.875)^2 + (42 - 38.875)^2 + (38 - 38.875)^2 + (49 - 38.875)^2 + (42 - 38.875)^2 + (55 - 38.875)^2$$
$$= (-16.875)^2 + (-5.875)^2 + (-8.875)^2 + (3.125)^2 + (-0.875)^2 + (10.125)^2 + (3.125)^2 + (16.125)^2$$
$$= 780.875 \text{ (to 2 decimal places)}$$

We can now use this to find s_y, by dividing by n - 1 and taking the square root.

$$s_y = \sqrt{(780.875/7)}$$
$$= \sqrt{111.55357}$$
$$= 10.56 \text{ (to 2 decimal places)}$$

← Finally, we use the y values in the sample to find s_y, the standard deviation of y.

All we need to do now is use b, s_x, and s_y to find the value of the correlation coefficient r.

Find r for the concert data, continued

Now that we've found that b = 5.32, s_x = 1.81, and s_y = 10.56, we can put them together to find r.

$$r = bs_x/s_y$$
$$= 5.32 \times 1.81/10.56$$
$$= 0.91 \text{ (to 2 decimal places)}$$

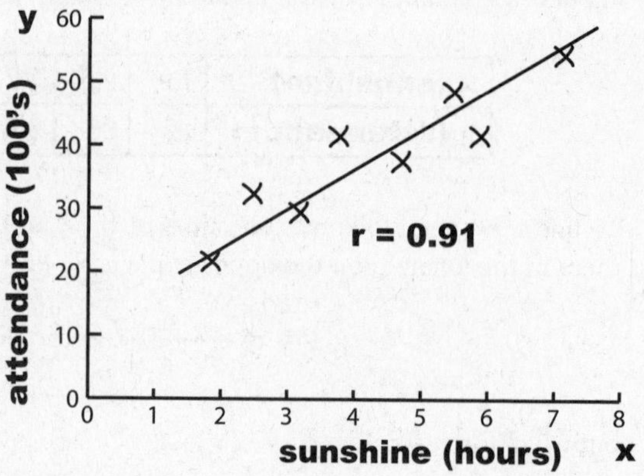

As r is very close to 1, this means that there's strong positive correlation between open air concert attendance and hours of predicted sunshine. In other words, based on the data that we have, we can expect the line of best fit, y = 15.80 + 5.32x, to give a reasonably good estimate of the expected concert attendance based on the predicted hours of sunshine.

there are no Dumb Questions

Q: I've seen other ways of calculating r. Are they wrong?

A: There are several different forms of the equation for finding r, but underneath, they're basically the same. We've used the simplest form of the equation so that it's easier to see what you've already calculated through finding b.

Q: Are the results accurate with such a small sample?

A: A larger sample would definitely be better, but we used a small sample just to make the calculations easier to follow.

Q: You haven't proved or derived why you calculate the values of b and r in this way. Why not?

A: Deriving the formula for b and r is quite complex and involved, so we've decided not to go through this in the book. The key thing is that you understand when and how to use them.

Q: What's the expected concert attendance if the predicted hours of sunshine is 0?

A: We can't say for certain because this is quite a way outside the range of data we have. The line of best fit is a pretty good estimate for the range of data that we have, but we can't say with any certainty what the concert attendance will be like outside this range. The data might follow a different pattern outside this range, so any estimate we gave would be unreliable.

Q: When we were looking at averages, we saw that univariate data can have outliers. What about bivariate data?

A: Yes, bivariate data can have outliers too. Outliers are points that lie a long way from your regression line. If you have outliers, then this can mean that you have anomalies in your data set, or alternatively, that your regression line isn't a good fit of the data.

Q: I've heard of influential observations. What are they?

A: Influential observations are points that lie a long way horizontally from the rest of the data. Because of this, they have the effect of pulling the regression line towards them.

Q: So is an influential observation the same as an outlier?

A: No. Outliers lie a long way from the **line**. Influential observations lie a long way horizontally from the **data**.

You've saved the day!

The concert organizers are amazed at the work you've done with their concert data. They now have a way of predicting what attendance will be like at their concerts based on the weather reports, which means they have a way of maximizing their profits.

Long Exercise

The evil Swindler has been collecting data on the effect radiation exposure has on Captain Amazing's super powers. Here is the number of minutes of exposure to radiation, paired with the number of tons Captain Amazing is able to lift:

Radiation exposure (minutes)	3	3.5	4	4.5	5	5.5	6	6.5	7
Weight (tons)	14	14	12	10	8	9.5	8	9	6

Your job is to use least squares regression to find the line of best fit, and then find the correlation coefficient to describe the strength of the relationship between your line and the data. Sketch the scatter diagram too.

If Swindler exposes Captain Amazing to radiation for 5 minutes, what weight do you expect Captain Amazing to be able to lift?

We've left you plenty of space for your calculations

Long Exercise Solution

The evil Swindler has been collecting data on the effect radiation exposure has on Captain Amazing's super powers. Here is the number of minutes of exposure to radiation, paired with the number of tons Captain Amazing is able to lift:

Radiation exposure (minutes)	4	4.5	5	5.5	6	6.5	7
Weight (tons)	12	10	8	9.5	8	9	6

Your job is to use least squares regression to find the line of best fit, and then find the correlation coefficient to describe the strength of the relationship between your line and the data. Sketch the scatter diagram too.

If Swindler exposes Captain Amazing to radiation for 5 minutes, what weight do you expect Captain Amazing to be able to lift?

Let's use x to represent minutes of radiation exposure and y to represent weight in tons. We need to find the regression line $y = a + bx$, so let's start by calculating $\bar{x}$ and $\bar{y}$.

$$\bar{x} = (4 + 4.5 + 5 + 5.5 + 6 + 6.5 + 7)/7$$
$$= 38.5/7$$
$$= 5.5$$

$$\bar{y} = (12 + 10 + 8 + 9.5 + 8 + 9 + 6)/7$$
$$= 62.5/7$$
$$= 8.9 \text{ (to 2 decimal places)}$$

Next, let's calculate $\Sigma(x - \bar{x})(y - \bar{y})$ and $\Sigma(x - \bar{x})^2$, and then b.

$$\Sigma(x - \bar{x})(y - \bar{y}) = (4-5.5)(12-8.9) + (4.5-5.5)(10-8.9) + (5-5.5)(8-8.9) + (5.5-5.5)(9.5-8.9) + (6-5.5)(8-8.9) + (6.5-5.5)(9-8.9) + (7-5.5)(6-8.9)$$
$$= (-1.5)(3.1) + (-1)(1.1) + (-0.5)(-0.9) + (0)(0.6) + (0.5)(-0.9) + (1)(0.1) + (1.5)(-2.9)$$
$$= -4.65 - 1.1 + 0.45 + 0 - 0.45 + 0.1 - 4.35$$
$$= -10$$

$$\Sigma(x - \bar{x})^2 = (4-5.5)^2 + (4.5-5.5)^2 + (5-5.5)^2 + (5.5-5.5)^2 + (6-5.5)^2 + (6.5-5.5)^2 + (7-5.5)^2$$
$$= (-1.5)^2 + (-1)^2 + (-0.5)^2 + 0^2 + 0.5^2 + 1^2 + 1.5^2$$
$$= 2.25 + 1 + 0.25 + 0 + 0.25 + 1 + 2.25$$
$$= 7$$

$$b = \frac{\Sigma(x - \bar{x})(y - \bar{y})}{\Sigma(x - \bar{x})^2}$$
$$= -10/7$$
$$= -1.43 \text{ (to 2 decimal places)}$$

Now that we've found b, let's use it to find a.

$$a = \bar{y} - b\bar{x}$$
$$= 8.9 + 1.43 \times 5.5$$
$$= 8.9 + 7.86$$
$$= 16.76$$

This means that the line of best fit is given by $y = 16.76 - 1.43x$

The correlation coefficient, r, is given by $r = bs_x/s_y$ where s_x and s_y are the standard deviations of the x and y variables. We've found b, so we need to find s_x and s_y.

$$s_x = \sqrt{\frac{\Sigma(x - \bar{x})^2}{n - 1}}$$
$$= \sqrt{7/6}$$
$$= 1.08$$

$$\Sigma(y - \bar{y})^2 = (12-8.9)^2 + (10-8.9)^2 + (8-8.9)^2 + (9.5-8.9)^2 + (8-8.9)^2 + (9-8.9)^2 + (6-8.9)^2$$
$$= 3.1^2 + 1.1^2 + (-0.9)^2 + 0.6^2 + (-0.9)^2 + 0.1^2 + (-2.9)^2$$
$$= 9.61 + 1.21 + 0.81 + 0.36 + 0.81 + 0.01 + 8.41$$
$$= 21.22$$

$$s_y = \sqrt{\frac{\Sigma(y - \bar{y})^2}{n - 1}}$$
$$= \sqrt{21.77/6}$$
$$= 1.90$$

Putting this together gives us

$$r = bs_x/s_y$$
$$= -1.43 \times 1.08/1.9$$
$$= -0.81 \text{ (to 2 decimal places)}$$

If $x = 5$, then we find y by calculating

$$y = 16.76 - 1.43x$$
$$= 16.76 - 1.43 \times 5$$
$$= 9.61$$

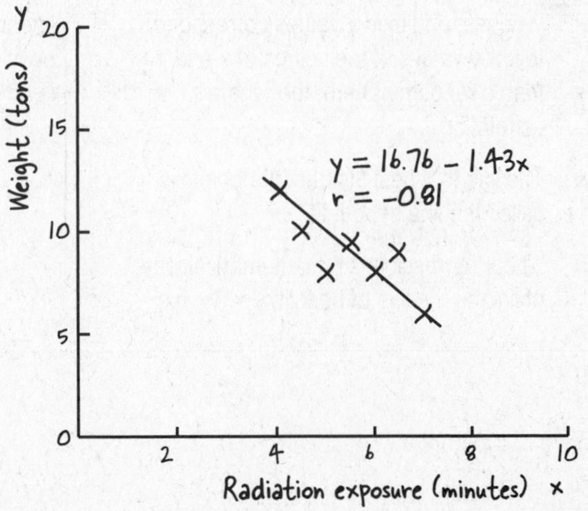

In other words, after 5 minutes of exposure to radiation, we'd expect Captain Amazing to be able lift 9.61 tons.

BULLET POINTS

- Univariate data deals with just one variable. Bivariate data deals with two variables.
- A scatter diagram shows you patterns in bivariate data.
- Correlations are mathematical relationships between variables. It does not mean that one variable causes the other. A linear correlation is one that follows a straight line.
- Positive linear correlation is when low x values correspond to low y values, and high x values correspond to high y values. Negative linear correlation is when low x values correspond to high y values, and high x values correspond to low y values. If the values of x and y form a random pattern, then there's no correlation.
- The line that best fits the data points is called the line of best fit.
- Linear regression is a mathematical way of finding the line of best fit, $y = a + bx$.
- The sum of squared errors, or SSE, is given by $\Sigma(y - \hat{y})^2$.
- The slope of the line $y = a + bx$ is

$$b = \frac{\Sigma(x - \bar{x})(y - \bar{y})}{\Sigma(x - \bar{x})^2}$$

- The value of a is given by

$$a = \bar{y} - b\bar{x}$$

- The correlation coefficient, r, is a number between -1 and 1 that describes the scatter of data away from the line of best fit. If r = -1, there is perfect negative linear correlation. If r = 1, there is perfect positive linear correlation. If r = 0, there is no correlation. You find r by calculating

$$r = \frac{b\, s_x}{s_y}$$

Leaving town...

It's been great having you here in Statsville!

We're sad to see you leave, but there's nothing like taking what you've learned and putting it to use. There are still a few more gems for you in the back of the book, some handy probability tables, and an index to read though, and then it's time to take all these new ideas and put them into practice. We're dying to hear how things go, so ***drop us a line*** at the Head First Labs web site, **www.headfirstlabs.com**, and let us know how Statistics is paying off for **YOU**!

appendix i: leftovers

The Top Ten Things (we didn't cover)

Even after all that, there's still a bit more.

There are just a few more things we think you need to know. We wouldn't feel right about ignoring them, even though they **only need a brief mention**, and we really wanted to give you a book you'd be able to lift without extensive training at the local gym. So before you put the book down, **take a read through these tidbits**.

#1. Other ways of presenting data

We showed you a number of charts in the first chapter, but here are a couple more that might come in useful.

Dotplots

A ***dotplot*** shows your data on a chart by representing each value as a dot. You put each dot in a stacked column above the corresponding value on the horizontal axis like this:

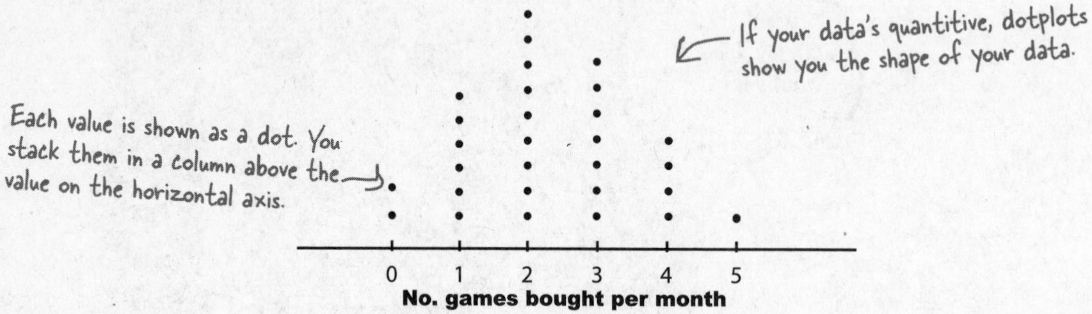
If your data's quantitive, dotplots show you the shape of your data.
Each value is shown as a dot. You stack them in a column above the value on the horizontal axis.
0
1
2
3
4
5
No. games bought per month

Stemplots

A ***stemplot*** is used for quantitive data, usually when your data set is fairly small. Stemplots show each exact value in your data set in such a way that you can easily see the shape of your data. Here's an example:

16	17	22	23	23	24	25	26	26	27	28
29	29	30	31	31	32	32	33	34	34	35
36	37	37	38	39	40	41	42	42	43	43
44	45	45	49	50	50	50	51	55	58	60

Here's your raw data.

Here's a stemplot based on the data.

Key: 10 | 6 = 16

60	0
50	0 0 0 1 5 8
40	0 1 2 2 3 3 4 5 5 9
30	0 1 1 2 2 3 4 4 5 6 7 7 8 9
20	2 3 3 4 5 6 6 7 8 9 9
10	6 7

A stemplot has a shape that is similar to a histogram's, but flipped onto its side.

The entries on the left are called ***stems***, and the entries on the right are called ***leaves***. In this stemplot, the stem shows tens, and the leaves show units. To find each value in the raw data, you take each leaf and add it to its stem. As an example, take the line

10 | 6 7

This represents two numbers, 16 and 17. You get 16 by adding the leaf 6 to its stem 10. Similarly, you get value 17 by adding the leaf 7 to the stem 10.

There's usually a key to help you interpret the stemplot correctly. In this case, the key is 10 | 6 = 16.

#2. Distribution anatomy

There are two rules that tell you where most of your data values lie in a probability distribution.

The empirical rule for normal distributions

The ***empirical rule*** applies to any set of data that follows a normal distribution. It states that almost all of the values lie within three standard deviations of the mean. In particular,

- **About 68% of your values lie within 1 standard deviation of the mean.**
- **About 95% of your values lie within 2 standard deviations of the mean.**
- **About 99.7% of your values lie within 3 standard deviations of the mean.**

The empirical rule tells you what percentage of your values you can expect to lie in which area of a normal distribution.

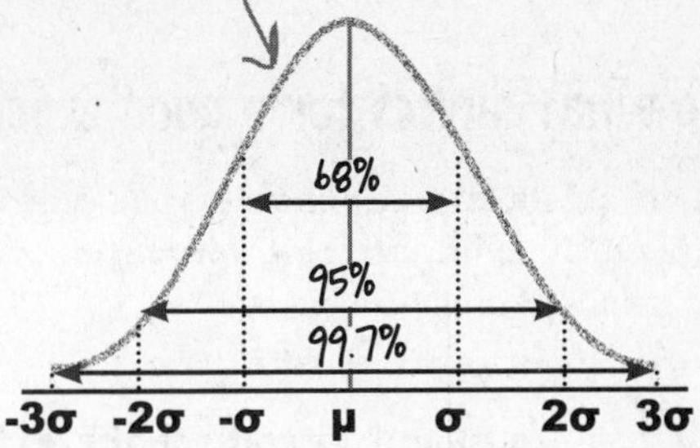

Just knowing the number of standard deviations from the mean can give you a rough idea about the probability.

Chebyshev's rule for any distribution

A similar rule applies to any set of data called ***Chebyshev's rule***, or ***Chebyshev's inequality***. It states that for any distribution

- **At least 75% of your values lie within 2 standard deviations of the mean.**
- **At least 89% of your values lie within 3 standard deviations of the mean.**
- **At least 94% of your values lie within 4 standard deviations of the mean.**

Chebychev's rule isn't as precise as the empirical rule, as it only gives you the minimum percentages, but it still gives you a rough idea of where values fall in the probability distribution. The advantage of Chebyshev's rule is that it **applies to any distribution**, while the empirical rule just applies to the normal distribution.

#3. Experiments

Experiments are used to test cause and effect relationships between variables. As an example, an experiment could test the effect of different doses of SnoreCull on snorers.

In an experiment, ***indpendent variables*** or factors are manipulated so that we can see the effect on ***dependent variables***. As an example, we might want to examine the effect that different doses of SnoreCull have on the number of hours spent snoring in a night. The doses of SnoreCull would be the independent variable, and the number of hours spent snoring would be the dependent variable.

The subjects that you use for your experiment are called ***experimental units***—in this case, snorers.

So what makes for a good experiment?

There are three basic principles you need to bear in mind when you design an experiment: ***controls***, ***randomization***, and ***replication***. Just as with sampling, a key aim is to minimize bias.

- **You need to control the effects of external influences or natural variability.**
  When you conduct an experiment, you need to minimize effects that are not part of the experiment. To do this, the first thing is to have a ***control group***, a neutral group that receives no treatments, or only neutral treatments. You can assess the effectiveness of the treatment by comparing the results of your treated groups with the results of your control group.

  A ***placebo*** is a neutral treatment, one that has no effect on the dependent variable. Sometimes the subjects of your experiment can respond differently to having a neutral treatment as opposed to having no treatment at all, so giving a placebo to a group is a way of controlling this effect. If the group taking a placebo doesn't know that it's a placebo, then this is called ***blinding***, and it's called ***double blinding*** if even those administering the treatments don't know.

- **You need to assign subjects to treatments at random.**
  You'll see more about this on the next page.

- **You need to replicate treatments.**
  Each treatment should be given to many subjects. You need to use many snorers per treatment to gauge the effects, not just one snorer.

Another factor to be aware of is confounding. ***Confounding*** occurs when the controls in an experiment don't eliminate other possible causes for the effect on the dependent variable. As an example, imagine if you gave doses to SnoreCull to men, but placebos to women. If you compared the results of the two groups, you wouldn't be able to tell whether the effect on the men was because of the drug, or because one gender naturally snores more than another.

Designing your experiment

We said earlier that you need to randomly assign subjects to experiments. But what's the best way of doing this?

Completely randomized design

One option is to use a ***completely randomized design***. For this, you literally assign treatments to subjects at random. If we were to conduct an experiment testing the effect of doses of SnoreCull on snorers, we would randomly assign snorers to particular treatment groups. As an example, we could give half of the snorers a placebo and the other half a single dose of SnoreCull.

Completely randomized design is similar to simple random sampling. Instead of choosing a sample at random, you assign treatments at random.

Placebo	SnoreCull
500	500

If there were 1,000 subjects, we could give half a placebo and the other half a dose of SnoreCull

Randomized block design

Another option is to use ***randomized block design***. For this, you divide the subjects into similar groups, or blocks. As an example, you could split the snorers into males and females. Within each block, you assign treatments at random, so for each gender, you could give half the snorers a dose of SnoreCull and give the other half a placebo. The aim of this is to minimize confounding, as it reduces the effect of gender.

Randomized block design is similar to stratified random sampling. Instead of splitting your population into strata, you split your subjects into blocks.

	Placebo	SnoreCull
Male	250	250
Female	250	250

If there were 500 men and 500 women, we could give half of each gender a placebo and the other half a dose of SnoreCull

Matched pairs design

Matched pairs design is a special case of randomized block design. You can use it when there are only two treatment conditions and subjects can be grouped into like pairs. As an example, the SnoreCull experiment could have two treatment conditions, to give a placebo or to give a single dose, and snorers could be grouped into similar pairs according to gender and age. You then give one of each pair a placebo, and the other a dose of SnoreCull. If one pair consisted of two men aged 30, for instance, you would give one of the men a placebo and the other man a dose of SnoreCull.

	Placebo	SnoreCull
Male 30	1	1
Male 30	1	1
Female 30	1	1
Female 30	1	1
...	...	...

You could also form matched pairs using gender and age to negate confounding due to these variables.

#4. Least square regression alternate notation

In Chapter 15 you saw how a least squares regression line takes the form y = a + bx, where

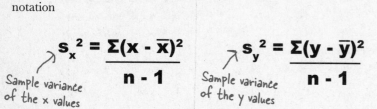

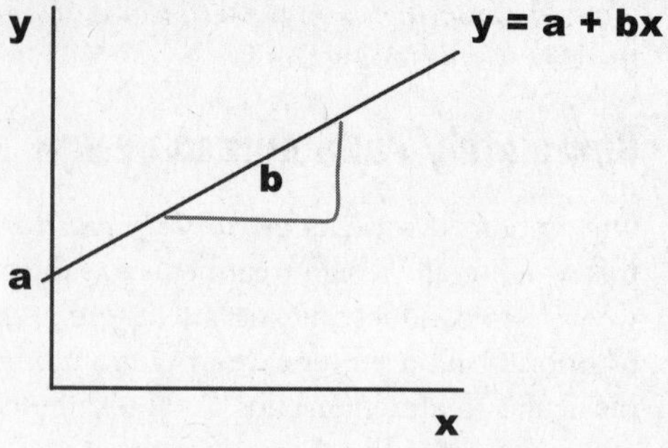

$$b = \frac{\Sigma(x - \bar{x})(y - \bar{y})}{\Sigma(x - \bar{x})^2}$$

This is the formula for the slope of the line.

There's another form of writing this that a lot of people find easier to remember, and that's to rewrite it in terms of variances. If we use the notation

$$s_x^2 = \frac{\Sigma(x - \bar{x})^2}{n - 1}$$

Sample variance of the x values

$$s_y^2 = \frac{\Sigma(y - \bar{y})^2}{n - 1}$$

Sample variance of the y values

$$s_{xy} = \frac{\Sigma(x - \bar{x})(y - \bar{y})}{n - 1}$$

then you can rewrite the formula for the slope of a line as

$$b = \frac{s_{xy}}{s_x^2}$$

This is the same calculation written in a different way.

You can do something similar with the correlation coefficient. Instead of writing

$$r = b\frac{s_x}{s_y}$$

you can write the equation for the correlation coefficient as

$$r = \frac{s_{xy}}{s_x s_y}$$

This is the formula for the correlation coefficient.

s_{xy} is called the ***covariance***. Just as the variance of x describes how the x values vary, and the variance of y describes how the y values vary, the covariance of x and y is a measure of how x and y vary together.

#5. The coefficient of determination

The ***coefficient of determination*** is given by $\mathbf{r^2}$ or $\mathbf{R^2}$. It's the percentage of variation in the y variable that's explainable by the x variable. As an example, you can use it to say what percentage of the variation in open-air concert attendance is explainable by the number of hours of predicted sunshine.

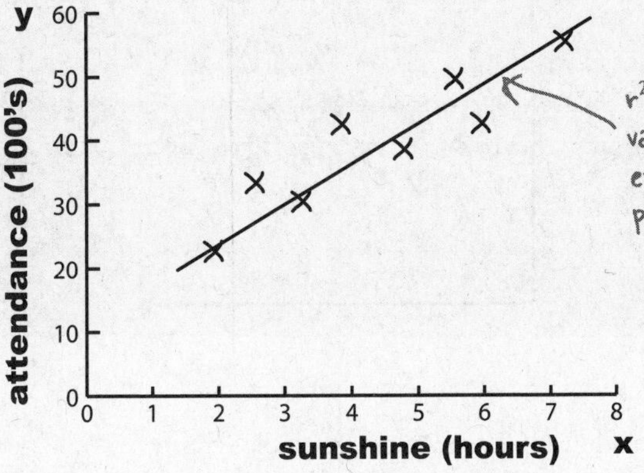

r^2 tells you what percentage of the variation in concert attendance is explainable by the number of hours predicted sunshine.

If $r^2 = 0$, then this means that you can't predict the y value from the x value.

If $r^2 = 1$, then you can predict the y value from the x value without any errors.

Usually r^2 is between these two extremes. The closer the value of r^2 is to 1, the more predictable the value of y is from x, and the closer to r^2 it is, the less predictable the value of y is.

Calculating r²

There are two ways of calculating r^2. The first way is to just square the correlation coefficient r.

This is just the correlation coefficient squared.

$$r^2 = \left(\frac{s_{xy}}{s_x s_y}\right)^2$$

Another way of calculating it is to add together the squared distances of the y values to their estimates, and then divide by the result of adding together squared distances of the y values to $\bar{y}$.

$$r^2 = \frac{\Sigma(y - \hat{y})^2}{\Sigma(y - \bar{y})^2}$$

This gives you the same value as above; it's just a different way of calculating it.

#6. Non-linear relationships

If two variables are related, their relationship isn't necessarily linear. Here are some examples of scatter plots where there's a clear mathematical relationship between x and y, but it's non-linear:

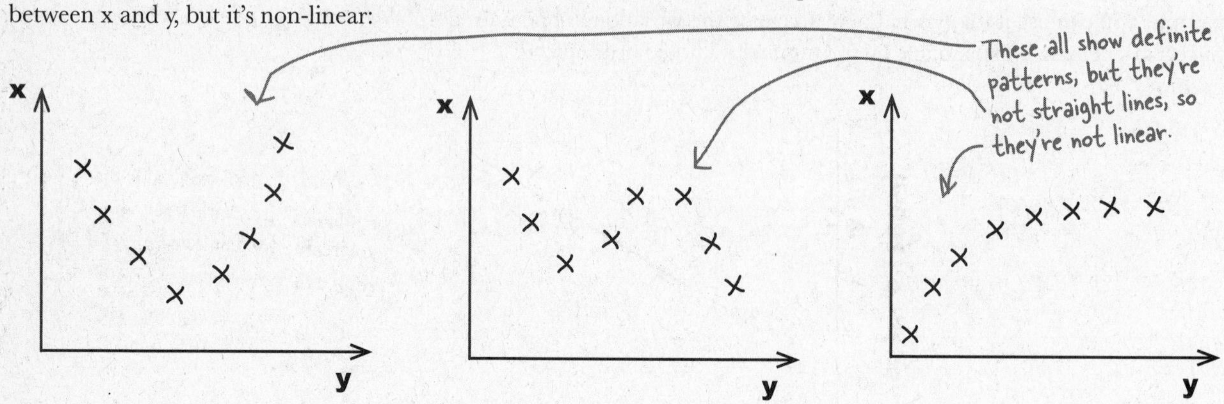

These all show definite patterns, but they're not straight lines, so they're not linear.
x
y
x
y
x
y

Linear regression assumes that the relationship between two variables can be described by a straight line, so performing least squares regression on raw data like this won't give you a good estimate for the equation of the line.

There is a way around this, however. You can sometimes transform x and y in such a way that the transformation is close to being linear. You can then perform linear regression on the transformation to find the values of a and b. The big trick is to try and transform your non-linear equation of the line so that it takes the form

$$y' = a + bx'$$

where y' and x' are functions of x.

If your line of best fit isn't linear, you can sometimes transform it to a linear form.

As an example, you might find that your line of best fit takes the form

$$y = 1/(a + bx)$$

This can be rewritten as

It's now in the form y' = a + bx, so we can use linear regression.

$$1/y = a + bx$$

so that $y' = 1/y$. In other words, you can perform least squares regression using the line $y' = a + bx$, where $y' = 1/y$. Once you've transformed your y values, you can use least squares regression to find the values of a and b, then substitute these back into your original equation.

This is just a quick overview, so you know what's possible.

#7. The confidence interval for the slope of a regression line

You've seen how you can find confidence intervals for μ and σ^2. Well, you can also find one for the slope of the regression line $y = a + bx$.

The confidence interval for b takes the form

$$\hat{b} \pm \text{(margin of error)}$$

But what's the margin of error?

The margin of error for b

The margin of error is given by

margin of error = t(v) x (standard deviation of b)

where $v = n - 2$, and n is the number of observations in your sample. To find the value of $t(v)$, use t-distribution probability tables to look up v and your confidence level.

The standard deviation of the sampling distribution of b is given by

This is the standard deviation of the sampling distribution of b →

$$s_b = \frac{\sqrt{\dfrac{\Sigma(y - \hat{y})^2}{n - 2}}}{\sqrt{\Sigma(x - \bar{x})^2}}$$

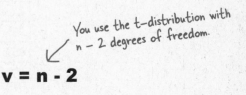

If you're taking a statistics exam where you have to use s_b, the formula will be given to you.

This means that you don't have to memorize it; you just need to know how to apply it.

To calculate this, add together the differences squared between each actual y observation and what you estimate it to be from the regression line. Then divide by $n - 2$, and take the square root. Once you've done this, divide the whole lot by the square root of the total differences squared between each x observation and $\bar{x}$.

This gives us a confidence interval of

$$(\hat{b} - t(v)\, s_b,\ \hat{b} + t(v)\, s_b) \qquad v = n - 2$$

You use the t-distribution with n – 2 degrees of freedom.

Knowing the standard deviation of b has other uses too. As an example, you can also use it in hypothesis tests to test whether the slope of a regression line takes a particular value.

#8. Sampling distributions - the difference between two means

Sometimes it's useful to know what the sampling distribution is like for the difference between the means of two normally distributed populations. You may want to use this to construct a confidence interval or conduct a hypothesis test. As an example, you may want to conduct a hypothesis test based on the means of two normally distributed populations being equal.

If $X \sim N(\mu_x, \sigma_x^2)$ and $Y \sim N(\mu_y, \sigma_y^2)$ where X and Y are independent, then the expectation and variance of the distribution $\overline{X} - \overline{Y}$ are given by

$$E(\overline{X} - \overline{Y}) = \mu_x - \mu_y$$

This is because $E(\overline{X} - \overline{Y}) = E(\overline{X}) - E(\overline{Y})$

$$Var(\overline{X} - \overline{Y}) = \frac{\sigma_x^2}{n_x} + \frac{\sigma_y^2}{n_y}$$

Similarly, $Var(\overline{X} - \overline{Y}) = Var(\overline{X}) + Var(\overline{Y})$

If the population variances σ_x^2 and σ_y^2 are known, then $\overline{X} - \overline{Y}$ is distributed normally. In other words

$$\overline{X} - \overline{Y} \sim N\left(\mu_x - \mu_y, \frac{\sigma_x^2}{n_x} + \frac{\sigma_y^2}{n_y}\right)$$

You can use this to find a confidence interval for $\overline{X} - \overline{Y}$. Confidence intervals take the form (statistic) ± (margin of error), so in this case, the confidence interval is given by

$$\bar{x} - \bar{y} \pm c\sqrt{Var(\overline{X} - \overline{Y})}$$

This is your confidence interval for $\overline{X} - \overline{Y}$.

The value of c depends on the level of confidence you need for your confidence interval:

Level of confidence	Value of c
90%	1.64
95%	1.96
99%	2.58

Your level of confidence gives you your value of c

If σ_x^2 and σ_y^2 are unknown, then you will need to approximate them with s_x^2 and s_y^2. If the samples sizes are large, then you can still use the normal distribution. If the sample sizes are small, then you will need to use the t-distribution instead.

#9. Sampling distributions - the difference between two proportions

There's also a sampling distribution for the difference between the proportions of two binomial populations. You can use this to construct a confidence interval or conduct a hypothesis test. As an example, you may want to conduct a hypothesis test based on the proportions of two populations being equal.

If $X \sim B(n_x, p_x)$ and $Y \sim B(n_y, p_y)$ where X and Y are independent, then the expectation and variance of the distribution $P_x - P_y$ are given by

$$E(P_x - P_y) = p_x - p_y$$

As before, $E(P_x - P_y) = E(P_x) - E(P_y)$

$$Var(P_x - P_y) = \frac{p_x q_x}{n_x} + \frac{p_y q_y}{n_y}$$

$Var(P_x - P_y) = Var(P_x) + Var(P_y)$

If np and nq are both greater than 5 for each population, then $P_x - P_y$ can be approximated with a normal distribution. In other words

$$P_x - P_y \sim N\left(p_x - p_y, \frac{p_x q_x}{n_x} + \frac{p_y q_y}{n_y}\right)$$

You can use this to find a confidence interval for $P_x - P_y$. Confidence intervals take the form (statistic) ± (margin of error), so in this case the confidence interval is given by

$$p_x - p_y \pm c\sqrt{Var(P_x - P_y)}$$

This is your confidence interval for $P_x - P_y$

The value of c depends on the level of confidence you need for your confidence interval. They're the same values of c as on the opposite page.

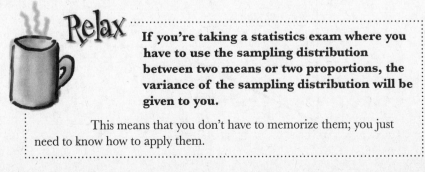

If you're taking a statistics exam where you have to use the sampling distribution between two means or two proportions, the variance of the sampling distribution will be given to you.

This means that you don't have to memorize them; you just need to know how to apply them.

#10. E(X) and Var(X) for continuous probability distributions

When we found the expectation and variance of ***discrete*** probability distributions, we used the equations

$$E(X) = \Sigma x P(X = x)$$

$$Var(X) = \Sigma x^2 P(X = x) - E^2(X)$$

When your probability distribution is ***continuous***, you find the expectation and variance using area.

As an example, suppose you have a continuous probability distribution where the probability density function is given by

$$f(x) = 0.05 \qquad 0 \leq x \leq 20$$

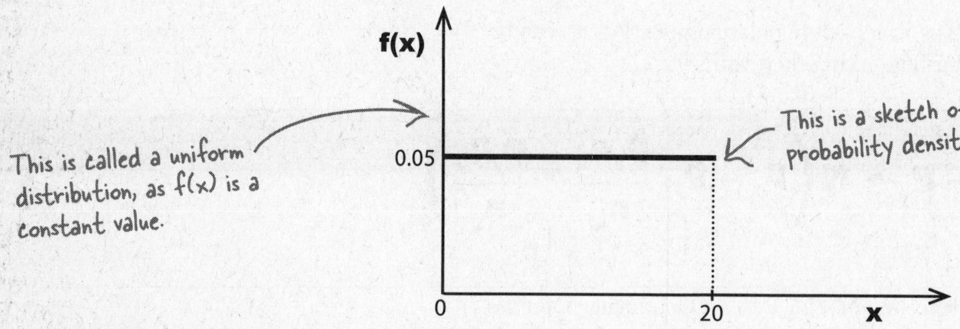

f(x)
0.05
0
20
x
This is called a uniform distribution, as f(x) is a constant value.
This is a sketch of f(x), the probability density function.

Finding E(X)

Multiply x by the original function to get xf(x).

To find the expectation, we'd need to find the area under the curve xf(x) for the range of the probability distribution. Here we need to find the area under the line 0.05x where x is between 0 and 20

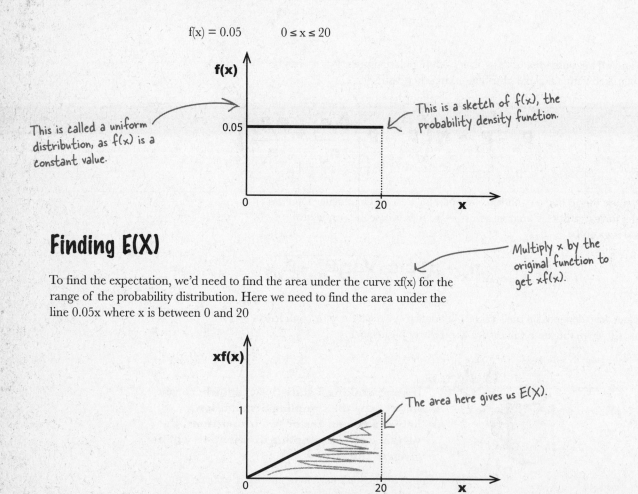

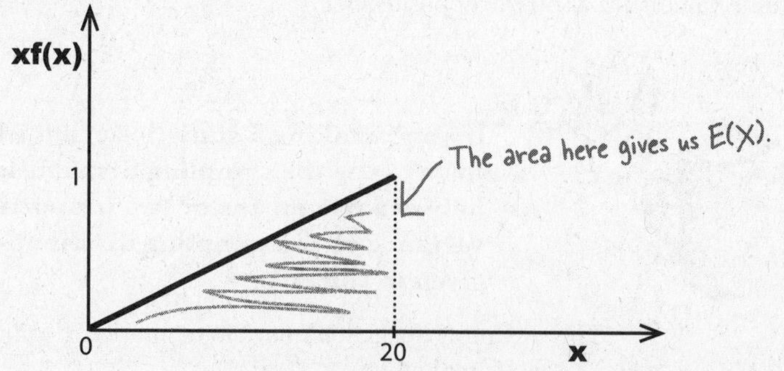

xf(x)
1
0
20
x
The area here gives us E(X).

Finding Var(X)

Relax

You don't often need to find the expectation and variance of a continuous random variable.

A lot of the time you'll be working with distributions like the normal, and in this, case the expectation and variance are given to you.

To find the variance, you need to find the area under the curve $x^2f(x)$ and subtract $E^2(X)$. In other words, we need to find the area under the curve $0.05x^2$ between 0 and 20 and subtract the square of E(X).

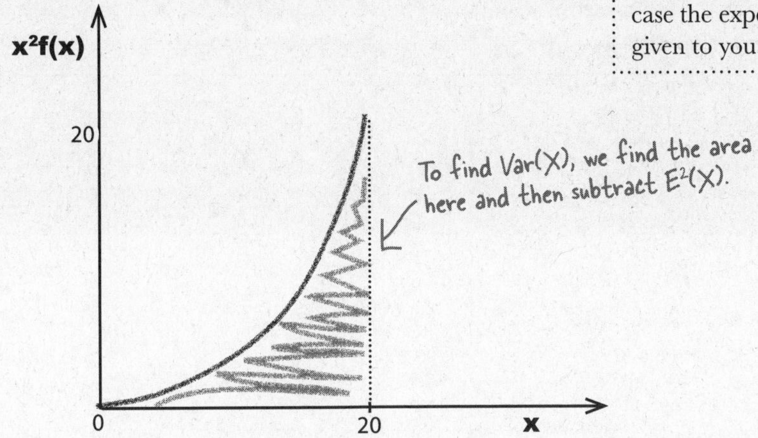

x²f(x)

To find Var(X), we find the area here and then subtract E²(X).

In general, you can find the expectation and variance of a continuous random variable using

$$E(X) = \int xf(x)dx$$

$$Var(X) = \int x^2f(x)dx - E^2(X)$$

Finding the expectation and variance of a continuous random variable often involves using calculus.

[Note from Marketing: Can we put in a plug for Head First Calculus—coming soon]

over the entire range of x.

Vital Statistics

Uniform Distribution

If X follows a uniform distribution then

$f(x) = 1/(b - a)$ where $a \le x \le b$

$E(X) = (a + b)/2$

$Var(X) = (b - a)^2/12$

appendix i: statistics tables

Looking Things Up

Where would you be without your trusty probability tables?

Understanding your probability distributions isn't quite enough. For some of them, you need to be able to **look up your probabilities** in standard **probability tables**. In this appendix, you'll find tables for the **normal, t, and X^2 distributions,** so you can look up probabilities to your heart's content.

#1. Standard normal probabilities

This table gives you the probability of finding P(Z < z) where Z ~ N(0, 1). To find the P(Z < z), look up your value of z to 2 decimal places, then read off the probability.

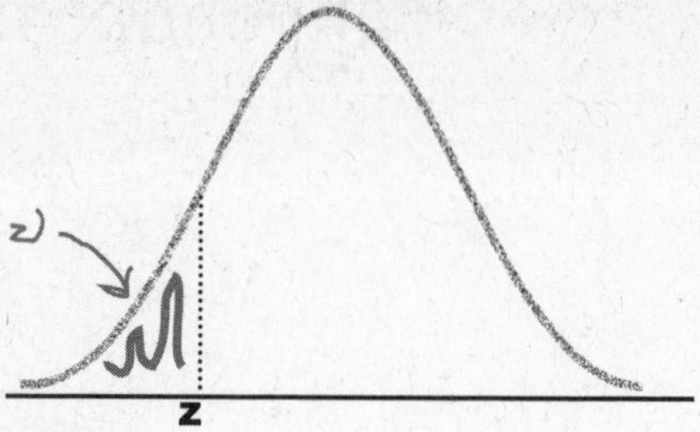

Look up the value of z using the first column and first row...

...then read off the probability from the table.

These are the probabilities for P(Z < z) where z is negative.

z	.00	.01	.02	.03	.04	.05	.06	.07	.08	.09
-3.4	.0003	.0003	.0003	.0003	.0003	.0003	.0003	.0003	.0003	.0002
-3.3	.0005	.0005	.0005	.0004	.0004	.0004	.0004	.0004	.0004	.0003
-3.2	.0007	.0007	.0006	.0006	.0006	.0006	.0006	.0005	.0005	.0005
-3.1	.0010	.0009	.0009	.0009	.0008	.0008	.0008	.0008	.0007	.0007
-3.0	.0013	.0013	.0013	.0012	.0012	.0011	.0011	.0011	.0010	.0010
-2.9	.0019	.0018	.0018	.0017	.0016	.0016	.0015	.0015	.0014	.0014
-2.8	.0026	.0025	.0024	.0023	.0023	.0022	.0021	.0021	.0020	.0019
-2.7	.0035	.0034	.0033	.0032	.0031	.0030	.0029	.0028	.0027	.0026
-2.6	.0047	.0045	.0044	.0043	.0041	.0040	.0039	.0038	.0037	.0036
-2.5	.0062	.0060	.0059	.0057	.0055	.0054	.0052	.0051	.0049	.0048
-2.4	.0082	.0080	.0078	.0075	.0073	.0071	.0069	.0068	.0066	.0064
-2.3	.0107	.0104	.0102	.0099	.0096	.0094	.0091	.0089	.0087	.0084
-2.2	.0139	.0136	.0132	.0129	.0125	.0122	.0119	.0116	.0113	.0110
-2.1	.0179	.0174	.0170	.0166	.0162	.0158	.0154	.0150	.0146	.0143
-2.0	.0228	.0222	.0217	.0212	.0207	.0202	.0197	.0192	.0188	.0183
-1.9	.0287	.0281	.0274	.0268	.0262	.0256	.0250	.0244	.0239	.0233
-1.8	.0359	.0351	.0344	.0336	.0329	.0322	.0314	.0307	.0301	.0294
-1.7	.0446	.0436	.0427	.0418	.0409	.0401	.0392	.0384	.0375	.0367
-1.6	.0548	.0537	.0526	.0516	.0505	.0495	.0485	.0475	.0465	.0455
-1.5	.0668	.0655	.0643	.0630	.0618	.0606	.0594	.0582	.0571	.0559
-1.4	.0808	.0793	.0778	.0764	.0749	.0735	.0721	.0708	.0694	.0681
-1.3	.0968	.0951	.0934	.0918	.0901	.0885	.0869	.0853	.0838	.0823
-1.2	.1151	.1131	.1112	.1093	.1075	.1056	.1038	.1020	.1003	.0985
-1.1	.1357	.1335	.1314	.1292	.1271	.1251	.1230	.1210	.1190	.1170
-1.0	.1587	.1562	.1539	.1515	.1492	.1469	.1446	.1423	.1401	.1379
-0.9	.1841	.1814	.1788	.1762	.1736	.1711	.1685	.1660	.1635	.1611
-0.8	.2119	.2090	.2061	.2033	.2005	.1977	.1949	.1922	.1894	.1867
-0.7	.2420	.2389	.2358	.2327	.2296	.2266	.2236	.2206	.2177	.2148
-0.6	.2743	.2709	.2676	.2643	.2611	.2578	.2546	.2514	.2483	.2451
-0.5	.3085	.3050	.3015	.2981	.2946	.2912	.2877	.2843	.2810	.2776
-0.4	.3446	.3409	.3372	.3336	.3300	.3264	.3228	.3192	.3156	.3121
-0.3	.3821	.3783	.3745	.3707	.3669	.3632	.3594	.3557	.3520	.3483
-0.2	.4207	.4168	.4129	.4090	.4052	.4013	.3974	.3936	.3897	.3859
-0.1	.4602	.4562	.4522	.4483	.4443	.4404	.4364	.4325	.4286	.4247
-0.0	.5000	.4960	.4920	.4880	.4840	.4801	.4761	.4721	.4681	.4641

#1. Standard normal probabilities (cont.)

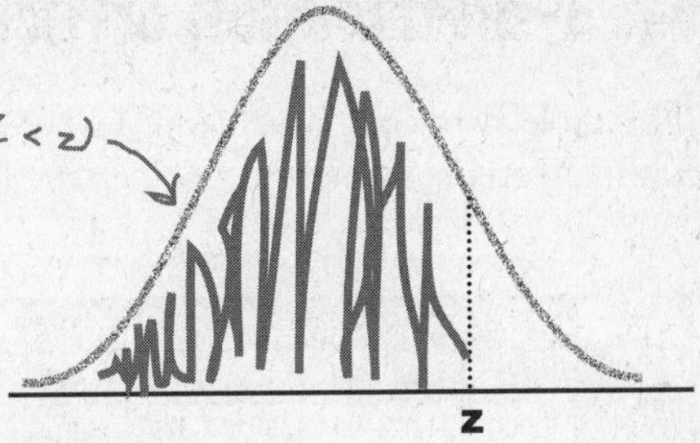

These are the probabilities for P(Z < z) where z is positive.

z	.00	.01	.02	.03	.04	.05	.06	.07	.08	.09
0.0	.5000	.5040	.5080	.5120	.5160	.5199	.5239	.5279	.5319	.5359
0.1	.5398	.5438	.5478	.5517	.5557	.5596	.5636	.5675	.5714	.5753
0.2	.5793	.5832	.5871	.5910	.5948	.5987	.6026	.6064	.6103	.6141
0.3	.6179	.6217	.6255	.6293	.6331	.6368	.6406	.6443	.6480	.6517
0.4	.6554	.6591	.6628	.6664	.6700	.6736	.6772	.6808	.6844	.6879
0.5	.6915	.6950	.6985	.7019	.7054	.7088	.7123	.7157	.7190	.7224
0.6	.7257	.7291	.7324	.7357	.7389	.7422	.7454	.7486	.7517	.7549
0.7	.7580	.7611	.7642	.7673	.7704	.7734	.7764	.7794	.7823	.7852
0.8	.7881	.7910	.7939	.7967	.7995	.8023	.8051	.8078	.8106	.8133
0.9	.8159	.8186	.8212	.8238	.8264	.8289	.8315	.8340	.8365	.8389
1.0	.8413	.8438	.8461	.8485	.8508	.8531	.8554	.8577	.8599	.8621
1.1	.8643	.8665	.8686	.8708	.8729	.8749	.8770	.8790	.8810	.8830
1.2	.8849	.8869	.8888	.8907	.8925	.8944	.8962	.8980	.8997	.9015
1.3	.9032	.9049	.9066	.9082	.9099	.9115	.9131	.9147	.9162	.9177
1.4	.9192	.9207	.9222	.9236	.9251	.9265	.9279	.9292	.9306	.9319
1.5	.9332	.9345	.9357	.9370	.9382	.9394	.9406	.9418	.9429	.9441
1.6	.9452	.9463	.9474	.9484	.9495	.9505	.9515	.9525	.9535	.9545
1.7	.9554	.9564	.9573	.9582	.9591	.9599	.9608	.9616	.9625	.9633
1.8	.9641	.9649	.9656	.9664	.9671	.9678	.9686	.9693	.9699	.9706
1.9	.9713	.9719	.9726	.9732	.9738	.9744	.9750	.9756	.9761	.9767
2.0	.9772	.9778	.9783	.9788	.9793	.9798	.9803	.9808	.9812	.9817
2.1	.9821	.9826	.9830	.9834	.9838	.9842	.9846	.9850	.9854	.9857
2.2	.9861	.9864	.9868	.9871	.9875	.9878	.9881	.9884	.9887	.9890
2.3	.9893	.9896	.9898	.9901	.9904	.9906	.9909	.9911	.9913	.9916
2.4	.9918	.9920	.9922	.9925	.9927	.9929	.9931	.9932	.9934	.9936
2.5	.9938	.9940	.9941	.9943	.9945	.9946	.9948	.9949	.9951	.9952
2.6	.9953	.9955	.9956	.9957	.9959	.9960	.9961	.9962	.9963	.9964
2.7	.9965	.9966	.9967	.9968	.9969	.9970	.9971	.9972	.9973	.9974
2.8	.9974	.9975	.9976	.9977	.9977	.9978	.9979	.9979	.9980	.9981
2.9	.9981	.9982	.9982	.9983	.9984	.9984	.9985	.9985	.9986	.9986
3.0	.9987	.9987	.9987	.9988	.9988	.9989	.9989	.9989	.9990	.9990
3.1	.9990	.9991	.9991	.9991	.9992	.9992	.9992	.9992	.9993	.9993
3.2	.9993	.9993	.9994	.9994	.9994	.9994	.9994	.9995	.9995	.9995
3.3	.9995	.9995	.9995	.9996	.9996	.9996	.9996	.9996	.9996	.9997
3.4	.9997	.9997	.9997	.9997	.9997	.9997	.9997	.9997	.9997	.9998

#2. t-distribution critical values

This table gives you the values of t where P(T > t) = p. T follows a t-distribution with ν degrees of freedom. Look up the values of ν and p and look up t.

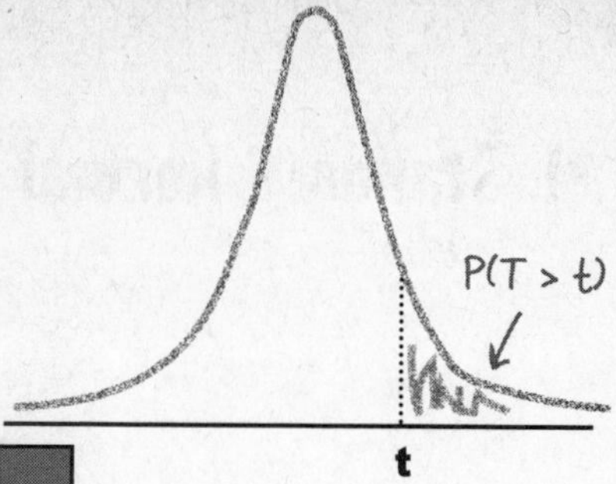

Look up ν in the first column... ...look up p in the first row...

ν	.25	.20	.15	.10	.05	.025	.02	.01	.005	.0025	.001	.0005
	Tail probability p											
1	1.000	1.376	1.963	3.078	6.314	12.71	15.89	31.82	63.66	127.3	318.3	636.6
2	.816	1.061	1.386	1.886	2.920	4.303	4.849	6.965	9.925	14.09	22.33	31.60
3	.765	.978	1.250	1.638	2.353	3.182	3.482	4.541	5.841	7.453	10.21	12.92
4	.741	.941	1.190	1.533	2.132	2.776	2.999	3.747	4.604	5.598	7.173	8.610
5	.727	.920	1.156	1.476	2.015	2.571	2.757	3.365	4.032	4.773	5.893	6.869
6	.718	.906	1.134	1.440	1.943	2.447	2.612	3.143	3.707	4.317	5.208	5.959
7	.711	.896	1.119	1.415	1.895	2.365	2.517	2.998	3.499	4.029	4.785	5.408
8	.706	.889	1.108	1.397	1.860	2.306	2.449	2.896	3.355	3.833	4.501	5.041
9	.703	.883	1.100	1.383	1.833	2.262	2.398	2.821	3.250	3.690	4.297	4.781
10	.700	.879	1.093	1.372	1.812	2.228	2.359	2.764	3.169	3.581	4.144	4.587
11	.697	.876	1.088	1.363	1.796	2.201	2.328	2.718	3.106	3.497	4.025	4.437
12	.695	.873	1.083	1.356	1.782	2.179	2.303	2.681	3.055	3.428	3.930	4.318
13	.694	.870	1.079	1.350	1.771	2.160	2.282	2.650	3.012	3.372	3.852	4.221
14	.692	.868	1.076	1.345	1.761	2.145	2.264	2.624	2.977	3.326	3.787	4.140
15	.691	.866	1.074	1.341	1.753	2.131	2.249	2.602	2.947	3.286	3.733	4.073
16	.690	.865	1.071	1.337	1.746	2.120	2.235	2.583	2.921	3.252	3.686	4.015
17	.689	.863	1.069	1.333	1.740	2.110	2.224	2.567	2.898	3.222	3.646	3.965
18	.688	.862	1.067	1.330	1.734	2.101	2.214	2.552	2.878	3.197	3.611	3.922
19	.688	.861	1.066	1.328	1.729	2.093	2.205	2.539	2.861	3.174	3.579	3.883
20	.687	.860	1.064	1.325	1.725	2.086	2.197	2.528	2.845	3.153	3.552	3.850
21	.686	.859	1.063	1.323	1.721	2.080	2.189	2.518	2.831	3.135	3.527	3.819
22	.686	.858	1.061	1.321	1.717	2.074	2.183	2.508	2.819	3.119	3.505	3.792
23	.685	.858	1.060	1.319	1.714	2.069	2.177	2.500	2.807	3.104	3.485	3.768
24	.685	.857	1.059	1.318	1.711	2.064	2.172	2.492	2.797	3.091	3.467	3.745
25	.684	.856	1.058	1.316	1.708	2.060	2.167	2.485	2.787	3.078	3.450	3.725
26	.684	.856	1.058	1.315	1.706	2.056	2.162	2.479	2.779	3.067	3.435	3.707
27	.684	.855	1.057	1.314	1.703	2.052	2.158	2.473	2.771	3.057	3.421	3.690
28	.683	.855	1.056	1.313	1.701	2.048	2.154	2.467	2.763	3.047	3.408	3.674
29	.683	.854	1.055	1.311	1.699	2.045	2.150	2.462	2.756	3.038	3.396	3.659
30	.683	.854	1.055	1.310	1.697	2.042	2.147	2.457	2.750	3.030	3.385	3.646
40	.681	.851	1.050	1.303	1.684	2.021	2.123	2.423	2.704	2.971	3.307	3.551
50	.679	.849	1.047	1.299	1.676	2.009	2.109	2.403	2.678	2.937	3.261	3.496
60	.679	.848	1.045	1.296	1.671	2.000	2.099	2.390	2.660	2.915	3.232	3.460
80	.678	.846	1.043	1.292	1.664	1.990	2.088	2.374	2.639	2.887	3.195	3.416
100	.677	.845	1.042	1.290	1.660	1.984	2.081	2.364	2.626	2.871	3.174	3.390
1000	.675	.842	1.037	1.282	1.646	1.962	2.056	2.330	2.581	2.813	3.098	3.300
∞	.674	.841	1.036	1.282	1.645	1.960	2.054	2.326	2.576	2.807	3.091	3.291
	50%	60%	70%	80%	90%	95%	96%	98%	99%	99.5%	99.8%	99.9%
	Confidence level C											

...then read off the value of t from the table.

#3. X^2 critical values

This table gives you the value of x where $P(X \geq x) = \alpha$. X has a χ^2 distribution with ν degrees of freedom. Look up the values of ν and α, and read off x.

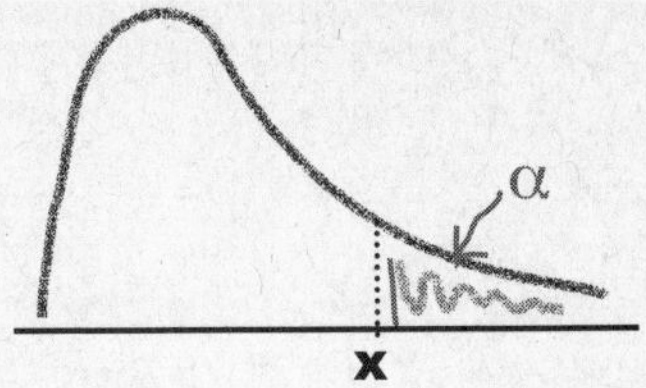

Look up the value of ν in the first column...

...look up the value of α in the first row...

...then read off the value of x from the table.

	Tail probability α										
ν	.25	.20	.15	.10	.05	.025	.02	.01	.005	.0025	.001
1	1.32	1.64	2.07	2.71	3.84	5.02	5.41	6.63	7.88	9.14	10.83
2	2.77	3.22	3.79	4.61	5.99	7.38	7.82	9.21	10.60	11.98	13.82
3	4.11	4.64	5.32	6.25	7.81	9.35	9.84	11.34	12.84	14.32	16.27
4	5.39	5.99	6.74	7.78	9.49	11.14	11.67	13.28	14.86	16.42	18.47
5	6.63	7.29	8.12	9.24	11.07	12.83	13.39	15.09	16.75	18.39	20.51
6	7.84	8.56	9.45	10.64	12.59	14.45	15.03	16.81	18.55	20.25	22.46
7	9.04	9.80	10.75	12.02	14.07	16.01	16.62	18.48	20.28	22.04	24.32
8	10.22	11.03	12.03	13.36	15.51	17.53	18.17	20.09	21.95	23.77	26.12
9	11.39	12.24	13.29	14.68	16.92	19.02	19.68	21.67	23.59	25.46	27.88
10	12.55	13.44	14.53	15.99	18.31	20.48	21.16	23.21	25.19	27.11	29.59
11	13.70	14.63	15.77	17.28	19.68	21.92	22.62	24.72	26.76	28.73	31.26
12	14.85	15.81	16.99	18.55	21.03	23.34	24.05	26.22	28.30	30.32	32.91
13	15.98	16.98	18.20	19.81	22.36	24.74	25.47	27.69	29.82	31.88	34.53
14	17.12	18.15	19.41	21.06	23.68	26.12	26.87	29.14	31.32	33.43	36.12
15	18.25	19.31	20.60	22.31	25.00	27.49	28.26	30.58	32.80	34.95	37.70
16	19.37	20.47	21.79	23.54	26.30	28.85	29.63	32.00	34.27	36.46	39.25
17	20.49	21.61	22.98	24.77	27.59	30.19	31.00	33.41	35.72	37.95	40.79
18	21.60	22.76	24.16	25.99	28.87	31.53	32.35	34.81	37.16	39.42	42.31
19	22.72	23.90	25.33	27.20	30.14	32.85	33.69	36.19	38.58	40.88	43.82
20	23.83	25.04	26.50	28.41	31.41	34.17	35.02	37.57	40.00	42.34	45.31
21	24.93	26.17	27.66	29.62	32.67	35.48	36.34	38.93	41.40	43.78	46.80
22	26.04	27.30	28.82	30.81	33.92	36.78	37.66	40.29	42.80	45.20	48.27
23	27.14	28.43	29.98	32.01	35.17	38.08	38.97	41.64	44.18	46.62	49.73
24	28.24	29.55	31.13	33.20	36.42	39.36	40.27	42.98	45.56	48.03	51.18
25	29.34	30.68	32.28	34.38	37.65	40.65	41.57	44.31	46.93	49.44	52.62
26	30.43	31.79	33.43	35.56	38.89	41.92	42.86	45.64	48.29	50.83	54.05
27	31.53	32.91	34.57	36.74	40.11	43.19	44.14	46.96	49.64	52.22	55.48
28	32.62	34.03	35.71	37.92	41.34	44.46	45.42	48.28	50.99	53.59	56.89
29	33.71	35.14	36.85	39.09	42.56	45.72	46.69	49.59	52.34	54.97	58.30
30	34.80	36.25	37.99	40.26	43.77	46.98	47.96	50.89	53.67	56.33	59.70
40	45.62	47.27	49.24	51.81	55.76	59.34	60.44	63.69	66.77	69.70	73.40
50	56.33	58.16	60.35	63.17	67.50	71.42	72.61	76.15	79.49	82.66	86.66
60	66.98	68.97	71.34	74.40	79.08	83.30	84.58	88.38	91.95	95.34	99.61
80	88.13	90.41	93.11	96.58	101.9	106.6	108.1	112.3	116.3	120.1	124.8
100	109.1	111.7	114.7	118.5	124.3	129.6	131.1	135.8	140.2	144.3	149.4

Index

Symbols

A

B

C

D

I

K

L

M

N

O

P

Q

R

S

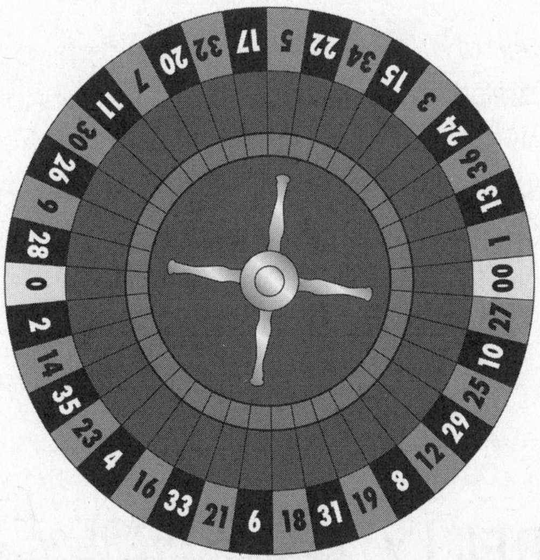

FAT DAN'S CASINO
WIN BIG!
WIN BIG!
Jackpot!